D0082863

THE IOWA STATE UNIVERSITY PRESS SERIES
IN THE HISTORY OF TECHNOLOGY AND SCIENCE

ROBERT E. SCHOFIELD / *Editor*

BOOKS IN THE

IOWA STATE UNIVERSITY PRESS SERIES IN THE HISTORY OF TECHNOLOGY AND SCIENCE

Christiaan Huygens' *The Pendulum Clock or Geometrical Demonstrations Concerning the Motion of Pendula as Applied to Clocks*

Translated with Notes by Richard J. Blackwell
Introduction by H. J. M. Bos

The History of Modern Science:
A Guide to the Second Scientific Revolution, 1800–1950

by Stephen G. Brush

THE HISTORY OF MODERN SCIENCE

THE HISTORY OF MODERN SCIENCE

A Guide to the Second Scientific Revolution, 1800–1950

STEPHEN G. BRUSH

IOWA STATE UNIVERSITY PRESS, AMES

TO PHYLLIS

Stephen G. Brush is Professor, Department of History and Institute for Physical Science and Technology, University of Maryland at College Park

Composed by Iowa State University Press
Printed in the United States of America

First edition, 1988

Brush, Stephen G.
The history of modern science.

(The Iowa State University Press series in the history of technology and science)
Bibliography: p.
Includes index.
1. Science—History. 2. Anthropology—History.
3. Psychology—History. I. Title. II. Series: Iowa State University Press series in the history of technology and science.
Q125.B88 1988 509′.034 87-26177
ISBN 0–8138–0883–9

CONTENTS

IT IS WITH PLEASURE that I introduce the second volume to be published in the Iowa State University Press Series in the History of Technology and Science. In establishing guidelines for this series the editor's considered intention was to promote publications that would include translations, as did the first volume, Richard J. Blackwell's English translation of Christiaan Huygens's *The Pendulum Clock*; republications of rare and standard works (for which there is now a need) in the [illegible] History of Science and Technology [illegible]; memoirs and biographies of significant figures in engineering, technological, or science [illegible]; and finally, [illegible] scholarly [illegible].

Stephen G. Brush's *The History of Modern Science: A Guide to the Second Scientific Revolution, 1800–1950* does not fall [illegible] [illegible] analysis of the [illegible] [illegible] the relationship to one another and to the [illegible] context in which they arose. [illegible] teachers in the specialized disciplines, and [illegible] and extensive citations to the [illegible] history of science make [illegible] essential reference.

[illegible] [illegible] With this broad approach the author [illegible] [illegible] space permit. His [illegible]

EDITOR'S INTRODUCTION

IT IS WITH PLEASURE that I introduce the second volume to be published in the Iowa State University Press Series in the History of Technology and Science. In establishing guidelines for this series the editor's considered intention was to promote publications that would include translations (as did the first volume: Richard J. Blackwell's English translation of Christiaan Huygens' *The Pendulum Clock*); republications of texts and standard works (for which there is now a Press Reprint Series in the History of Science and Technology edited by Professor Brush); memoirs and biographies of significant figures in engineering, technological, or science's history; and finally, the scholarly monograph.

Stephen G. Brush's *The History of Modern Science: A Guide to the Second Scientific Revolution, 1800–1950* does not fall precisely into any of these specific categories; instead, it adds another dimension in the series' exploration and analysis of the history of technology and science. Brush's *Guide* truly conforms to the series' avowed intention of "increasing public understanding of the nature of technological and scientific creativity and their relationship to one another and to the social, cultural, political, and economic context in which they arise." In one sense it will serve as a textbook for teachers in this specialized discipline, and its bibliographic information and extensive citations to the book and journal literature in the history of science make the *Guide* an essential reference.

Dr. Brush provides an informed and analytical guide through his extensive coverage of subjects, interpretations, and sources that make up the fields of the sciences: physical, earth, biological, social, and behavioral, since 1800. With this broad approach, the subjects covered are treated as fully as time and space permit. His work is immensely useful, even for persons who have studied and taught the subject for years, and his interpretations are always interesting, and to some, controversial. And most important, the availability of the *Guide* should open the field of the history of science to the interest and enthusiasm of many people who have lacked the time or the courage to investigate it.

Robert E. Schofield
Professor, History of Technology and Science, Iowa State University

PREFACE

THIS BOOK, which was planned as a result of discussions among members of the Committee on Undergraduate Education of the History of Science Society, is intended to assist teachers who wish to discuss the history of modern science in their courses; it also provides an introduction to the literature for advanced students, scientists, historians, and others who want to learn about the subject on their own.

Following an overview of the subject and a survey of textbooks and research publications, the book is organized as a series of chapters covering selected major developments in biology, anthropology, psychology, physics, mathematics, and astronomy between 1800 and 1950, along with philosophical and sociological approaches to these develoments. It is not a textbook or a treatise on the history of science, but an outline of topics with brief expository essays, detailed suggestions about materials that might be suitable for student reading, and references to major sources and recent scholarly analyses. Each chapter is relatively self-contained so that the instructor may rearrange them and use as many as appropriate for a particular course; taken all together they would cover a full year course.

Since a section on the sixteenth- to eigtheenth-century background of modern science could be used by instructors who follow different routes through the other topics, I have placed this section in this introductory chapter (**1.4**).

At the beginning of each section I have listed several readings that might be suitable for students with little or no background in science and mathematics. These include three kinds of publications: general textbooks that may cover this topic only very briefly but are useful because they can be used for several other topics that one might want to include in the course; chapters in more specialized (but fairly elementary) books, for example on evolution or atomic physics; and articles in magazines. In addition, I have occasionally suggested advanced readings appropriate for students who do have some scientific background, since a number of courses are designed for majors in a particular science.

In the bibliographies at the end of the sections I have, in general, included only works published within the last ten years, along with standard editions of primary sources, autobiographies, bibliographies, and articles in the *Dictionary of Scientific Biography*. These reference works enable the reader to locate most of the important earlier publications. In a few cases I have included older monographs and articles that were recommended by my colleagues or are the only publications I could find that deal adequately with certain topics.

Since most of the books recommended for student reading are mentioned several times, I have followed the practice of citing these books by author and title, using a short title after first citation within each chapter. Complete citations of these books, including 1985–1986 prices for books still in print then, may be found in the Book List at the end. This list also indicates in boldface type the sections for which each book is suggested as possible student reading, so that one may judge which would be most useful as textbooks for a course covering a certain selection of topics. Books cited at the end of each section are more specialized works for the instructor, and are not included in the Book List. Since articles in a wide variety of periodicals in different fields have been cited, full titles of these periodicals have been used rather than abbreviations. Thus the bibliography for each section is self-contained.

The following colleagues served as consultants during the preparation of an earlier version of this book and provided much valuable advice and criticism: Alice Baxter, Robert Fox, Lucille Garmon, William Haskett, Lois Magner, Robert Nadeau, and William Woodward. I am also grateful for comments and suggestions from Michele Aldrich, James W. Atkinson, Carl Jay Bajema, W. H. Brock, Joe Burchfield, Hamilton Cravens, Lindley Darden, Richard Duschl, Richard C. Frey, Brian Gee, C. Stewart Gillmor, I. Grattan-Guinness, Joel Hagen, Bert Hansen, Norriss S. Hetherington, Gerald Holton, John Lankford, Stephen Mason, Philip D. Jordan, Kathryn Olesko, Emanuel D. Rudolph, Virendra Singh, David Topper, and Louis Weinberg. I regret that I could not follow all their recommendations (especially the request to add sections on other topics), and of course none of them is responsible for the remaining errors of fact and interpretation.

Preparation of the first draft of this book was supported by a grant from the Division of Education Programs of the National Endowment for the Humanities. Since the book includes many evaluations and opinions, I emphasize that they are entirely my own and should not be attributed to the History of Science Society or the National Endowment for the Humanities; no official of either organization has seen the final version of this book before publication.

THE HISTORY OF MODERN SCIENCE

1 INTRODUCTION

1.1 THE SECOND SCIENTIFIC REVOLUTION 1800–1950

Now that it's over we can try to figure out what it means. The transformations in modern science brought about by John Dalton, Charles Darwin, Albert Einstein, Niels Bohr, Sigmund Freud, and many others are now complete even if not universally understood or accepted; and we have lived long enough since their heroic deeds to recognize that their "discoveries" are not necessarily immutable facts about the world but provisional knowledge, extremely useful and illuminating yet still subject to revision. In this section I present an overview—a personal interpretation—of the historical phenomenon to which the rest of the book provides a detailed guide for study and possible alternative interpretations.

RESULTS OF THE FIRST SCIENTIFIC REVOLUTION

The mechanistic worldview that dominated science at the end of the eighteenth century is often called the "clockwork universe" or the "Newtonian world-machine." According to this paradigm, God was supposed to be such a clever craftsman that He created the universe as a perfect machine whose parts never wear out (conservation of matter) and which never runs down (conservation of motion). Although this view could accommodate a certain amount of purpose or design in the original creation, it rejected any need for divine intervention in the affairs of the world after its creation. Its counterpart in theology is the "deism" of the French Enlightenment.

Another feature of the mechanistic worldview was *determinism,* articulated most clearly by Pierre-Simon Laplace. The motion of every particle of matter is completely determined by the state of the universe at any instant of time, so that if God knows the positions and velocities of all the particles at any time, together with the forces they exert on each other, He would know in complete detail the entire past and future. It was not explained how such an intelligent deity would be able to survive this condition of perfect boredom for all eternity, or why He would bother to create such a universe in the first place; those were not scientific questions.

Isaac Newton himself rejected this view of God's role in the world on the very reasonable grounds that it would lead scientists to atheism. It was Robert Boyle and Gottfried Wilhelm Leibniz who defended it as a more worthy and dignified conception of the deity than one that expected Him to be continually on call to make minor repairs on an imperfectly designed machine.

What I call not the Newtonian but the mechanistic worldview had two other important features that need to be mentioned because they were rejected in the nineteenth and twentieth centuries. First, in accordance with René Descartes's "dualistic" philosophy, there was a complete separation between the world of the spirit and the world of matter and motion. Human consciousness is not part of the physical world and can interact with that world only indirectly through the brain. The physical world exists independently of our awareness of it. And it is our possession of a spirit that sets us apart from animals. Second, the phenomena of nature are to be explained in terms of a fixed set of kinds of substance. In biology this means a collection of immutable species; in chemistry it means a list of elements; in physics it means weightless fluids postulated to explain heat, electricity, magnetism, etc., in addition to ordinary matter.

The phrase "scientific revolution" currently has several meanings, as explained in I. Bernard Cohen's recent book, *Revolution in Science* (1985). I will use it, capitalized, to mean a fundamental change in all the sciences, such as occurred in the period 1500–1800 as a result of the work of Nicolaus Copernicus, Galileo Galilei, Descartes, Newton, Antoine Lavoisier, and others. This is generally known as "The" Scientific Revolution but historians of science probably would not insist that it was the only one. According to Cohen, Thomas S. Kuhn was the first in 1961 to use the term "Second Scientific Revolution," defining it as the "successful quantification of the Baconian sciences" that took place primarily in the nineteenth century [see Kuhn's *The Essential Tension* (1977), 147, 218–20]. I extend this to include what other historians of science have considered equally revolutionary developments in physics in the period 1895–1925, in biology after 1859, in psychology in the period 1895–1920, and in the explosive outcome of nuclear physics in 1945.

Western European civilization has seen only two complete Scientific Revolutions of this magnitude, and it is therefore risky to generalize about their characteristics. Yet anyone writing about or teaching a course on the history of science can hardly avoid taking some position on the thesis of the book on this subject that is best known to those outside the profession: Kuhn's *Structure of Scientific Revolutions* (1962). Unlike many of my colleagues, I am quite willing to accept Kuhn's model as a plausible approximate description of a typical revolution within a single science; it is less adequate for Revolutions that involve interactions among several sciences. Here the concept of "themata" introduced by Gerald Holton seems useful. Themata such as "atoms"—the idea of a discrete particulate structure as opposed to a continuous structure—may link theories in physics, chemistry, and biology, though their detailed interpretation is different in each science.

According to Kuhn, a scientific revolution involves not only a radical change of specific theories and techniques but also a change in the kinds of questions that theories are expected to answer and the criteria for judging those answers. More generally it may entail a basic transformation in the worldview of those people who look to science for intellectual guidance. In short, a scientific revolution involves a change from one "paradigm" to another, to use a term Kuhn himself has abandoned but which many other observers of science now find indispensable. Since a paradigm depends partly on philosophical preference and cultural environment, it cannot be proved right or wrong by scientific research alone. Hence we should not be surprised to find that adherents of an old paradigm refuse to abandon it even when the scientific community has adopted a new paradigm, and that they can justify their decision.

For example, according to the Aristotelian-Christian paradigm that was generally accepted in Western Europe at the time of Copernicus, the world was created by God, everything has its place in a harmonious whole, all events occur to fulfill some ultimate purpose, and humans have access to a spiritual world. In addition, it was widely believed that people's lives are governed by the positions of planets. Those who continue to accept those beliefs can and do reject the paradigm that was victorious in the (first) Scientific Revolution.

Similarly, the mechanistic "clockwork-universe" paradigm established by the First Scientific Revolution included the assumptions or themata mentioned above: that the world consists of independent pieces of matter whose motion in an absolute space is precisely determined by mathematical laws; that events occurring in different parts of the universe can be said to happen at the same time; that humans are qualitatively different from all other biological species and can, if they wish, liberate themselves from the influence of animal passions; and that the physical world has an objective existence independent of our observation of it. (These assumptions go back to Descartes, with some modifications by Newton.) People who accept any or all of these premises can and do reject the paradigm that was victorious in the Second Scientific Revolution. Or at least they reject the more radical views of Darwin, Freud, Einstein, and Bohr associated with that paradigm.

The difference between these two cases is that many people in the second group—those who cling to a mechanistic worldview—still think their views are consistent with "modern science." This confusing fact illustrates another characteristic of Scientific Revolutions: it takes much longer to understand and explicitly accept the metaphysical worldview associated with a new paradigm than to accept its scientific theories and procedures. It is easy enough for scientists to dismiss those who believe in creationism, astrology, and spiritualism as irrational proponents of unscientific dogma. After all, their paradigm was rejected 300 years ago and was replaced by another one which has itself become obsolete! It is much more difficult to recognize and elucidate some of the disturbing features of the new paradigm scientists now supposedly accept.

This "metaphysical lag" is illustrated by the curious fact that the scientist who played the most important role in assuring the ultimate success of each of the two Scientific Revolutions recognized and explicitly *rejected* its most radical philosophical premise. That is why in listing five leaders of each Revolution, I included the name of a man whose purely scientific contributions were somewhat less important but who articulated the radical change in philosophical worldview more boldly. Descartes was overshadowed by Newton in physics and astronomy but proposed a mechanistic "clockwork-universe" philosophy that was accepted in spite of Newton's objections. Niels Bohr's contributions to physics, though of major importance, were less significant than those of Einstein; yet he won the debate with Einstein on the reality of atomic properties. Einstein, having himself undermined a major part of the mechanistic paradigm with his theory of relativity, could not give it up entirely. Similarly, other scientists like Copernicus, Darwin, Freud, and Max Planck refused to accept some of the revolutionary implications of their own discoveries.

GOALS AND THEMATA OF THE SECOND SCIENTIFIC REVOLUTION

Scientists at the beginning of the nineteenth century were impressed by the success of the mechanistic paradigm in accounting for the motions of planets, satellites, and comets by means of Newton's law of gravity. They hoped to apply similar methods to explain other domains of natural phenomena: heat, electricity, magnetism, light, chemical reactions, physiology, even psychology and sociology. The route to scientific understanding was thought to include the collection and quantification of empirical data and the formulation and testing of general laws. Moreover, scientists were fascinated by the possibility of finding not only the laws governing individual phenomena but also the interrelationships between phenomena. Thus Charles-Augustin de Coulomb's success in showing that electric and magnetic forces followed the same mathematical law as gravity was an important vindication of the mechanistic view, but Hans Christian Oersted's discovery of the interaction between electricity and magnetism opened up a new realm that ultimately undermined that view.

Another goal was to go beyond the laws governing natural phenomena and understand the *origin* of the present state of the solar system, the Earth's surface, and the particular living organisms that now inhabit it. Not only scientists but thinkers of all kinds began to proclaim that one must study the past in order to comprehend the present.

Pursuit of these goals led by the middle of the nineteenth century to the formulation and widespread popularity of three concepts: atom, energy, and evolution. None was completely new but each was given a more specific meaning than before. Atoms and energy were conceptual tools for explaining and correlating the general laws of physical phenomena and extending those laws into the realm of biological phenomena. Evolution provided a framework for explaining how specific phenomena and structures had come to be through the action of natural laws.

I see intellectual curiosity as the major driving force in scientific research, but not the only one. Practical applications did play a role in the development of the energy concept: the need to measure and maximize the mechanical work done by the steam engine demanded some understanding of how heat is used to produce work. In fact, energy itself is the one physical concept that can be directly measured in units of money—the amount you pay for electricity is proportional to the number of kilowatt-hours (a measure of energy) you use.

Toward the end of the nineteenth century, faith in the mechanistic worldview began to wane. The extent to which this situation resulted from objective "anomalies" (in Kuhn's sense), as distinct from shifts in metaphysical attitudes linked to broader currents in the culture, is a matter of controversy among historians. Some scientists thought that a new worldview could be based on one or more of the concepts atom, energy, and evolution, even if mechanistic explanations, determinism, and the Newtonian concepts of space, time, mass, and force had to be abandoned.

Eventually, in the first quarter of the twentieth century, the new paradigm emerged. In one sense it was assembled from components already present in the nineteenth century though not seen as revolutionary. Thus energy itself was atomized in the new quantum theory of Planck and Einstein, while Darwinian evolution was combined with Gregor Mendel's atomistic genetics to form a new synthesis in biology. On a more metaphysical level, mechanism was replaced by positivism. Space, time, velocity, mass, and energy were not abolished but relativized: they no longer seemed to have an objective existence independent of the scientist who observes them. Psychology took from physics the doctrine that science should attribute reality only to observable behavior, not to hypothetical constructs—though behaviorists pushed this doctrine much farther than physicists did.

The following exposition attempts to describe the course of the Second Scientific Revolution by following one strand at a time, then coming back to pick up others. Only at the end can one hope to see how they all fit together.

Cosmic Evolution

According to T. H. Huxley, a Victorian advocate of Darwinism, the principle of evolution was first formulated by Immanuel Kant in his 1755 theory of the origin of the solar system. The theory was much better known in the version formulated by the French astronomer Laplace and the British astronomer William Herschel around 1800, called the "nebular hypothesis." In his presidential address of 1869, Huxley told the Geological Society of London that the doctrine of evolution "embraces in one stupendous analogy the growth of a solar system from molecular chaos, the shaping of the earth from the nebulous cubhood of its youth, through innumerable changes and immeasurable ages, to its present form, and the development of a living being from the shapeless mass of protoplasm we term a germ." Moreover, as Huxley pointed out in an article on the genealogy of animals published in the same year, the principle of evolution is both teleological and deterministic; it implies, as the etymology of the word itself does, an *unfolding* of an original plan. "It is not less certain that the existing world lay potentially in the cosmic vapour, and that a sufficient intelligence could, from a knowledge of the properties of the molecules of that vapour, have predicted, say the state of the fauna of Britain in 1869, with as much certainty as one can say what will happen to the vapour of the breath on a cold winter's day."

For most of the nineteenth century the nebular hypothesis, rather than natural selection, was the paradigm of evolution. It carried the prestige associated with the names of Laplace and William Herschel, the two leading astronomers at the beginning of the nineteenth century; and of course astronomy itself was one of the most prestigious of the sciences. Geology was strongly influenced by the assumption that the Earth was formed as a hot fluid sphere in accordance with Laplace's mechanism; the process of cooling and contraction was widely thought to have determined its surface features; volcanoes and earthquakes were considered vivid demonstrations of the internal heat still remaining from the Earth's fiery beginnings. It seemed reasonable to suppose that new and more complex forms of life appeared as it reached a more moderate temperature; thus a "progressivist" or "directionalist" synthesis of astronomy, geology, and biology began to form in the first half of the nineteenth century.

This progressivist synthesis did not lead to Darwinian evolution, although it may have encouraged scientists to accept Darwin's theory in a general way. Darwin based his own ideas about Earth history on the "uniformitarian" geology of Charles Lyell. Lyell rejected the progressivist synthesis, urging geologists to forget speculative cosmogonic schemes and explain the present surface of the Earth by invoking only those physical processes that could now be seen in operation. Moreover, he asserted that the rate or intensity of processes such as erosion and uplifting was not much different in the past; one therefore had to assume that these processes had been going on for millions of years or longer to produce the observed results.

Lyell's eventual acceptance of transmutation of species and the attacks on his theory by the physicist Lord Kelvin neutralized the strongest *scientific* opposition to the evolutionary worldview. Religious opposition had focussed for a time on the nebular hypothesis; the same charges of atheism later levelled against Darwin were hurled at Laplace. But the astronomers were fighting on secure ground and managed to repel these attacks. The net result of this encounter, as shown by Ronald Numbers in his *Creation by Natural Law* (1977), was that the intellectual community was able to come to terms with an evolutionary perspective in the physical world, thus facilitating the later acceptance of evolution in the life sciences.

Darwin and Darwinism

Evolution is not necessarily incompatible with either a religious or a mechanistic view of the world. The clockwork-universe concept does imply that all change is cyclic, so there is no progress or decay in the long run. But the postulate that biological species are immutable, and the corresponding postulate that the diversity of physical phenomena is to be explained by postulating a plethora of kinds of matter, seem to be rather arbitrary components of mechanism. All three conflict with the more fundamental principle that one should seek the most simple explanation of the facts. Evolutionary theory, as formulated in the nineteenth century, promised to do this by showing how present complexity could have arisen from past simplicity.

The particular version of evolution introduced by Charles Darwin did conflict with religious beliefs. His theory of natural selection was a direct challenge to the earlier theory that explained biological phenomena in terms of providential design—"God made it that way to accomplish a purpose." Thus one result of Darwin's work was to convert many biologists from the teleological viewpoint, already abandoned by physicists, to a naturalistic or mechanistic viewpoint. But biologists scarcely had time to adapt their thinking to a mechanistic paradigm before they had to face the fact that Darwinism really went beyond mechanism.

The challenge to mechanism came from Darwin's proposal that the *variation* of offspring from the characters of their parents is, in effect, *random*. Whereas Newton had expanded the boundaries of mechanism by introducing gravity as a new force or cause of motion, despite its unintelligibility, Darwin burst through those boundaries by assuming an effect without postulating any cause at all. (Darwin and his followers did not admit that variation is random or uncaused, but they were unable to specify any plausible cause.) This aspect of the evolutionary worldview did not become acceptable until indeterminism proved indispensable in physics in the early twentieth century.

Like Boyle and Leibniz, Darwin argued that his theory was theologically superior to one that postulated a more active divine intervention: "It is derogatory," he wrote in 1842, "that the Creator of countless systems of worlds should have created each of the myriads of creeping parasites and slimy worms . . . on this one globe." How much more elegant to say, as Loren Eiseley puts it with a nostalgic touch of mechanism, that species arise "through the working out of the natural forces implanted in that highly complicated chemical compound known as protoplasm and the response of this same protoplasm to the environmental world about it." [*Darwin's Century* (1961), 193]

Darwin's innovation was not the idea that species change—that was a commonplace though generally rejected notion in the midnineteenth century—but a mechanism by which particular changes could be explained. First, each generation of offspring contains variations from the inherited characters of the parents. Second, there is a fierce struggle for survival among these organisms and the other species competing in an environment with limited resources. Third, those variations that help the organism to survive and reproduce in a particular environment will be more likely to persist in future generations. This process, somewhat misleadingly called *natural selection,* automatically causes the species to evolve into another species better adapted to the environment. Fourth, the process of sexual reproduction favors the development of certain characters, such as bright colors, distinctive shapes and odors, that are attractive to potential mates even though they may seem to have zero or even negative survival value in a jungle full of predators; this concept of *sexual selection* can be used to explain aspects of evolution that seem inexplicable by natural selection.

The most sensational implication of Darwin's theory, and of course the one that still arouses the most intense resistance, is that we ourselves are the product of evolution and thus in some sense are just another species of animals. This conclusion could be supported by several kinds of evidence already available in the nineteenth century, especially the similarity of human embryonic development and bone structure to sequences of existing and fossil mammals. While Darwin was willing to state explicitly that the human race has evolved from "lower" animals, he did not play a direct role in bringing about the further consequences of his theory.

During the late nineteenth century, what we now call "social Darwinism" became a component of social theory in England, the United States, and some European countries. It interacted with Marxism and other doctrines that postulated evolution toward a desired (or feared) future state of society, and with the tendency to see people as members of races or other groups having distinct characteristics. Some of the phrases associated with social Darwinism, such as "struggle for survival" and "survival of the fittest," came not so much from Darwin as from Herbert Spencer, who advanced a grand scheme of cosmic and social evolution. In the following discussion I will concentrate on ideas more directly related to biological theory, because they seem to be less familiar to historians who have written on cultural and social theories.

Three major consequences of evolutionary theory were:

1. *Cultural evolution, racism, and sexism.* Primitive man represented a lower stage of evolution than modern man; the colored races and white women have not yet achieved the level of the white European male.

Perhaps there is a one-dimensional scale along which homo sapiens is evolving; the less advanced forms of humanity may eventually catch up but in the meantime should submit to the dominance and tutelage of their evolutionary betters.

2. *Hereditarianism, degeneration, and eugenics.* Each of us is born with a fixed inherited capacity to learn and achieve success. Individual variations are necessary for the advancement of the race, but natural selection must not be thwarted by softhearted social policies that prevent the unfit from being weeded out in the struggle for survival. Certain environmental influences, especially alcohol and immoral behavior, may damage the hereditary endowment of the offspring and cause the stock to degenerate. Thus the Progressive movement, especially in the United States early in the twentieth century, used Darwinism to justify the prohibition of alcoholic beverages, restriction of immigration from those countries thought to have inferior populations, and eugenic sterilization of persons showing the signs of degeneracy such as congenital idiocy and criminality.

The intelligence-test movement, originating in a concern for the proper classification of idiots and imbeciles, soon developed into an ambitious program to put *everyone* on a one-dimensional scale. A single number, the IQ, was thought to be mostly determined by heredity and in any case fixed early in life; it measures one's capacity to succeed in any intellectual endeavor. To improve the race, persons with high IQs should be encouraged to mate with each other. The recent proposal to establish a "sperm bank" funded by Nobel Laureates does not seem to have been taken seriously by very many intelligent people, yet most "selective" colleges still require applicants to attain a minimum score on an IQ test (known as the "Scholastic Aptitude Test") in order to qualify for admission.

The interaction of 1 and 2 is the race/IQ controversy, another product of the Second Scientific Revolution that is still very much with us.

3. *The human mind is governed by animal passions*—lust, fear, anger—and even the supposedly nobler emotions are a result of repressed primitive urges. This theme of course brings us directly to Sigmund Freud, whose debt to nineteenth-century biology is stressed in Frank Sulloway's recent biography.

Whereas the first two consequences of Darwinism had placed the white European male at the most human end of a continuum, looking down with pity at less advanced cultures and hereditary degenerates within his own society, Freud reduced *all* humanity to the lowest common denominator. According to his theories, the sexual impulses that lead to psychoses or perversions in extreme cases are present to some degree in all of us. The *id* fills our dreams with sexual symbolism and motivates the "Freudian slip."

For Freud there are only two kinds of people—male and female. All other differences are due to early experiences in a particular environment. The slogan "anatomy is destiny" means that males and females will of necessity have different experiences and hence different psychosexual development. The boy's first sexual object is his mother; the resulting conflict with his father leads him to internalize a conscience (superego) much more strongly than the girl. The girl suffers more strongly from confusion and repression, and her complaints of sexual abuse by male relatives are to be ascribed to fantasy or wishful thinking. (We are just beginning to realize the terrible human cost of that ascription.) Freud's own admitted confusion about the female mind led him to ask in exasperation—"What does a woman want?"

One thing women want is the opportunity to make scientific discoveries. At the beginning of the twentieth century, while Freud was reconfirming the cultural stereotype of woman's intellectual inferiority, a woman was making a major discovery for which she has never received general recognition. I refer to the American biologist Nettie M. Stevens who established in 1905 the genetic difference between male and female. Females have two of the special particles known as "X" chromosomes while males have one "X" and one of another kind called "Y." This fundamental fact was first uncovered in insects and subsequently found to be true of humans. Stevens's discovery provided the first simple *observable* case of the particulate mechanism of heredity postulated by Gregor Mendel forty years earlier and was part of the revival of mendelism that produced modern genetics.

It would be misleading to leave this topic without recognizing that the effects of Darwinism have been countered by a number of twentieth-century scientists. In anthropology, Franz Boas rejected the doctrine of cultural evolution and argued that each culture must be studied on its own terms without judging it advanced or primitive by our standards. He rejected the crude racism of nineteenth-century anthropologists and showed that supposedly immutable racial characteristics such as head shape were rapidly modified in a new environment when immigrants entered American society. His student Margaret Mead studied males and females in several primitive societies (the validity of her research has recently been challenged) and found that the Western stereotypes of sex-linked personalities did not always apply; males could display "feminine" behavior and females could display "masculine" behavior in some cases, hence anatomy is not destiny. In the 1960s William Masters and Virginia Johnson refuted some of Freud's claims about female sexuality by clinical observations, to the applause of feminists.

In current evolutionary theory it is no longer taken for granted that all changes are slow, with natural selection continuously working on small variations. Niles Eldredge and Stephen Jay Gould proposed in the 1970s that long periods of stasis are broken by short periods of rapid change, a theory they call "punctuated equilibrium" but which might also be called a "quantum theory of evolution." A convincing genetic mechanism for sudden variation is still lacking, but it may be possible to appeal to catastrophic changes in the environment triggered by the impact of large asteroids (the hypothesis suggested by Luis and Walter Alvarez).

Despite vigorous attacks Darwinism survives as a major component of the modern worldview. Fundamentalists who have tried to replace evolution by creationism in the schools have occasionally succeeded but at the cost of making their own beliefs appear antiscientific and ridiculous to rational people. Other theologians, unable to defeat evolution, have decided to co-opt it. A recent example of this trend is the writings of Ralph Wendell Burhoe, who has attempted to show that "evolutionary theory, once considered the arch enemy of religion, now can be interpreted properly to show the inevitable necessity of religion" (Templeton Foundation news release, March 1980).

Evolution in Chemistry and Astronomy

The counterpart of evolution in chemistry is the theory that the "elements" are not elementary but composite structures formed by combining hydrogen atoms. That theory was based on Dalton's revival of the atomic theory at the beginning of the nineteenth century. Dalton proposed to estimate the weights of atoms (relative to the lightest one, hydrogen) from the weights of substances entering into chemical reactions. In 1815 William Prout pointed out that most atomic weights are close to whole numbers, and suggested that if determined more accurately they would turn to be exactly integers. Then one could assume that every atom simply consists of a certain number of hydrogen atoms bound together.

Prout's hypothesis floundered for a century because the atomic weights of the elements are *not* quite integer multiples of that of hydrogen; more accurate measurements simply confirmed the discrepancies. Then in 1903 Ernest Rutherford reopened the question by showing that transmutation of elements occurs in radioactive decay; thus the materialistic doctrine that the elements are different kinds of substances was finally overthrown. In 1938 Hans Bethe showed how the process of forming heavier elements from light ones could account for energy generation in stars; thus the evolution of elements is part of the evolution of stars.

By this time the evolutionary approach had been applied to the universe as a whole, as a result of the expanding-universe concept based on E. P. Hubble's correlation of distances and velocities of distant galaxies. Here, in contrast to the basic process of condensation of a diffuse nebula to a compact star described by Herschel, we have an explosion of matter and energy initially concentrated in a very small region: the "Big Bang." And, contrary to those skeptics who claim that theories about origins are inherently untestable ("no one was there to observe it, and it won't happen again"), the Big Bang theory yielded a quantitative prediction that was unexpectedly confirmed a few years later: space is now filled with electromagnetic radiation at a temperature a few degrees above absolute zero.

To understand cosmic evolution and much else that seems amazing or mysterious about the world, we must go back to developments in physics at the beginning of the nineteenth century, leading to the two great revolutionary theories of the early twentieth century, relativity and quantum mechanics. These theories completed the demolition of mechanism begun by Darwin.

Light and Electromagnetism

Two major advances around 1820 began the process of undermining the mechanistic paradigm of Descartes and Newton. The first was the revival and development of the wave theory of light by Thomas Young in England and Augustin Fresnel in France. The second was the discovery of the action of electric currents on magnets by Hans Christian Oersted in Denmark, followed by the discovery of electromagnetic induction by Michael Faraday in England. Both suggested that the materialistic approach—postulating a different kind of substance to explain each kind of phenomenon—should be replaced by an attempt to describe all phenomena as different manifestations of an underlying force or as different forms of motion of a single universal substance. Heat, for example, could now be attributed to vibrations of the same medium that propagates light waves—this explanation works best for the phenomenon known as "radiant heat." The attempt culminated in James Clerk Maxwell's brilliant synthesis, which described light and radiant heat in terms of oscillating electric and magnetic fields. At the same time the elucidation of electromagnetic phenomena by physicists created a new technology: the dynamo, electric motor, telegraph, telephone, radio, and television.

These two advances violated Newtonian ideas about the nature of light (Newton advocated a particle theory) and about the nature of forces. Magnets and electric charges interact only when they are in motion relative to each other; the forces between them are neither attractive nor repulsive but directed at right angles to the line between them. Yet Maxwell's theory suggested that light and electromagnetism could be forced into the mechanistic mold by defining electric and magnetic fields in terms of motions in a rather peculiar type of ether substance filling all space.

Energy and the Kinetic Worldview

The discovery that electricity, magnetism, light, and other forms of radiation are physically related to each other proved ultimately more significant than the difficulty in understanding exactly how these relations worked. The 1840s saw the establishment of a general principle of conservation and interconvertibility of force, asserting that all the forces of nature—electricity, magnetism, heat, gravity, mechanical work, etc.—are nothing but different forms of one underlying force in nature. The word "force" as used in this context seemed somewhat ambiguous and was eventually replaced by "energy." (The ambiguity may itself have played a role in the establishment of the principle, according to Yehuda Elkana.) The principle states that energy may be transformed from one form to another, provided that the total amount of energy remains constant. The first law of thermodynamics is a special case of this principle that applies to transformations of heat and mechanical work.

Since all forms of energy can be *converted* into the kinetic energy of motion of pieces of matter, some scientists jumped to the conclusion that all forms of energy *are* just kinetic energy. This "reductionist" conclusion, which would have revived the mechanistic paradigm in its original form, was already anticipated in the 1820s and 1830s by A. M. Ampère, who proposed that magnetism can be reduced to the circular motion of electric charges, and heat can be reduced to the vibrational motion of the ether. As a general proposition, the conclusion might be called the "kinetic worldview."

The kinetic theory of gases, first proposed by Daniel Bernoulli in 1738 and revived after 1844 by John Herapath, J. J. Waterston, J. P. Joule, A. Kronig, and R. Clausius, is a particular version of the general idea that heat is simply the energy of molecular motion. It includes the further assumption that, most of the time, molecules in gases move in straight lines at constant speed until they strike another molecule or the sides of the container; at ordinary pressures the space occupied by the molecules themselves is only a very small part of the entire gas volume. The gas exerts pressure through the enormously frequent impacts of molecules; thus, in Cartesian fashion, the "force" of gas pressure is reduced to contact interactions.

A major triumph of the kinetic theory of gases was the first reliable estimate of the size of a molecule, made by Josef Loschmidt in 1865 and soon confirmed by G. J. Stoney and William Thomson (Lord Kelvin). As Thomson pointed out, despite the long history of the atomic concept, scientists found it difficult to take atoms seriously when they were impossible to see or measure; that they could now be assigned quantitative properties made them seem real.

We have here another of the major themes of the Second Scientific Revolution: the quantification of everything. It was of course Lord Kelvin himself who declared in 1883:

> I often say that when you can measure what you are speaking about and express it in numbers you know something about it; but when you cannot measure it, when you cannot express it in numbers, your knowledge is of a meagre and unsatisfactory kind: it may be the beginning of knowledge, but you have scarcely, in your thoughts, advanced to the stage of Science, whatever the matter may be. [*Popular Lectures and Addresses,* 2d ed. (1883) 1:80]

But another of Kelvin's famous sayings, expressing the primacy of the mechanistic paradigm, now strikes us as a limitation of his own powers of imagination—a limitation shared, to be sure, by many nineteenth-century scientists:

> I never satisfy myself until I can make a mechanical model of a thing. If I can make a mechanical model I can understand it. [*Notes of Lectures on Molecular Dynamics and the Wave Theory of Light* (1884), 270]

Einstein

One feature of Kuhn's theory of scientific revolutions seems to be accepted even by those who reject its other features: a revolution is preceded by a breakdown of confidence in the established paradigm because of the persistent *anomalies*—objective failures of the theory, despite repeated patching up, to account for important facts. Thus scientists are "forced" to consider radical new theories and ultimately a new paradigm.

The transition from geocentric to heliocentric astronomy is often seen as an example of a scientific revolution caused by an anomaly: astronomers had to keep adding more and more epicycles as new data showed up the inadequacies of the Ptolemaic system, until finally the geocentric theory had become so complicated and implausible that a simpler, more accurate theory had to be found. Kuhn seemed to endorse this scenario in *The Structure of Scientific Revolutions* (1962, 1970, 67–68).

Other historians of science have argued that the reason why Copernicus proposed his heliocentric theory was *not* to achieve a more accurate fit to the data or to replace a theory that was too complicated. Copernicus himself stated that the geocentrists had deviated from the axiomatic principle that the motions of celestial bodies must be explained by combinations of uniform circular motions. To construct a theory "pleasing to the mind" respecting that principle, Copernicus made the Sun, not the Earth, the center of the circular orbits. But in all other respects he followed the Aristotelian paradigm and certainly did not think of himself as a revolutionary. Thus, as Owen Gingerich has pointed out, the Copernican Revolution began not with a crisis resulting from discrepancies between theory and observation but with an esthetic problem resulting from discrepancies between a basic principle of a theory and its mathematical realization.

In the case of the origin of Einstein's relativity theory, it has long been believed that the crisis in ether theory precipitated by the Michelson-Morley experiment—which seemed to show that one could never detect motion relative to the ether or to an "absolute space"—was the fateful anomaly in Newtonian physics. But Gerald Holton has shown that Einstein did not invent his theory in order to account for the result of the Michelson-Morley experiment. Instead Einstein, like Copernicus, was disturbed by an esthetic defect in the orthodox theory. He noted that electromagnetic induction is described differently depending on whether the magnet or the conducting wire is moving, even though the result depends only on their relative motion. If we assume that the laws of physics are the same for different observers moving at constant relative velocity, and that the speed of light is found to be the same by any such observer, it follows that space and time are no longer absolute entities with independent existence. They depend on the motion—the "frame of reference"—of the observer.

Similarly in the case of quantum theory, it was previously believed (and is still stated in many textbooks) that Max Planck introduced the hypothesis that radiation is emitted and absorbed only in discrete amounts—"quanta"—in order to resolve the anomaly in Newtonian physics known as the "ultraviolet catastrophe." The catastrophe is a hypothetical phenomenon: all the energy put into the ether goes into the highest (ultraviolet) modes of vibration, and it would take an infinite amount of energy to raise the temperature of the ether by one degree. But there is no evidence that Planck was aware of this catastrophe or, if he was, that he would have considered it important. Moreover, as Kuhn himself has shown, Planck did not really propose a quantum hypothesis in 1900, although his mathematical formula was later interpreted in that sense. It was Einstein rather than Planck who was the primary inventor of the quantum theory; in another 1905 paper he proposed explicitly that light and other forms of electromagnetic radiation have a particulate nature in addition to a wave nature.

Just as Darwin denied that he was postulating randomness as a basic principle in evolution even though his theory of natural selection seemed to provide no alternative to the assumption that variations occur by chance, Einstein supplied the basis for indeterminism in physics yet insisted that "God does not play dice." In a third paper published in 1905, Einstein treated the irregular motion of microscopic particles in a fluid ("Brownian movement") as a random process dependent on chance impacts of molecules. Eleven years later he presented a theory of spontaneous and stimulated emission of radiation—a theory that ultimately inspired the invention of the laser—in which he assumed that atomic processes are governed by chance. While Einstein himself was unhappy with this assumption, other physicists made it the basis of the quantum theory of the atom.

Randomness or indeterminism, another major theme of the Second Scientific Revolution, did not originate in quantum theory but emerged in nineteenth-century debates about the temporal direction of irreversible processes. Kelvin had proposed in 1852 a "universal tendency in nature to the dissipation of mechanical energy"; this became associated with the second law of thermodynamics, as the statement that "entropy" always increases or remains constant. (Entropy was originally defined as heat transfer divided by temperature.) Maxwell recognized that the natural tendency of heat to flow from hot to cold—the simplest example of energy dissipation or entropy increase—is equivalent to a tendency for molecules to become more and more mixed up as time goes on. Ludwig Boltzmann quantified this insight with a mathematical relation between entropy and disorder and reformulated the kinetic theory of gases so as to imply that all natural processes involve some degree of randomness at the atomic level.

Entropy was thereby liberated from its moorings in physics and entered common language as a synonym for randomness. In the twentieth century we no longer fear randomness as a threat of chaos but welcome it as a possible haven for free will and a guarantee of "fairness" in statistical surveys and military draft selection procedures.

In "quantum mechanics" (the mature version of quantum theory), randomness reached new heights of abstraction as mechanism retreated. Erwin Schrödinger proposed a mathematical equation from which can be deduced the observable properties of atoms, molecules, and (with a powerful enough computer or accurate approximate methods) any material system under ordinary conditions. But the symbols in the Schrödinger equation have no direct physical meaning. According to Max Born one of these symbols, the "wave function," represents the probability that the system will follow a certain path or be found in a certain state. There is no longer a causal law to determine how the positions and velocities of atoms evolve; instead there is a causal law that determines how their probability distributions evolve.

Following the publication in 1926 of Schrödinger's equation and Max Born's statistical interpretation of it, there ensued a friendly but intense debate between Einstein and Niels Bohr, the philosophical leader of the new quantum mechanicians. The major issue was not so much randomness as *realism.* Einstein argued that quantum mechanics is not wrong but incomplete because it fails to account for some atomic properties that have a real existence. For example, according to Werner Heisenberg's principle, the position and momentum of a particle cannot be simultaneously determined: the more accurately one of these quantities is pinned down, the more indeterminate the other must be. To call this an "uncertainty principle" as is usually done implies that the position and momentum actually have values but we can't find out what they are because any attempt to measure them must disturb the system. But Bohr and Heisenberg, the proponents of the "Copenhagen interpretation" of quantum mechanics, claimed that these properties cannot be said to have an objective existence. It is only the observation that gives them reality. Thus the term "indeterminacy principle" is more accurate.

Einstein argued that in a certain type of thought experiment, first described by him in a joint paper with Boris Podolsky and Nathan Rosen in 1935, it would be possible to determine the position or the momentum of one particle by measurements on another particle so far away that the observation could not possibly disturb the first particle. Hence both position and momentum must really exist before we measure them, contrary to the Copenhagen interpretation.

Most people would probably sympathize with Einstein's "realist" position; after all, why do science at all if you are not going to discover something that existed "out there" before you looked at it? Nevertheless

Bohr and Heisenberg won the debate because Einstein could not provide any satisfactory alternative to their interpretation. Moreover, it is now possible to perform experiments that are similar to the one discussed by Einstein, Podolsky, and Rosen, and the results confirm the conclusion of Bohr and Heisenberg.

It appears that the outcome of the revolution in physics is complementary to that of Darwinism. Not only are humans part of nature, but nature cannot be separated from humans; we cannot, at the most fundamental level, draw a sharp distinction between the observer and the things observed. Not only are we a product of nature, but nature is in some sense a product of ourselves. This is a rather weird conclusion (advanced most forcefully in recent years by the physicist John Wheeler), and I am not quite ready to believe it. But I am confident that by the time this view of the world has been thoroughly understood and accepted, a Third Scientific Revolution will be underway to undermine it.

1.2 BOOKS FOR BEGINNERS

Historians of science are frequently asked to recommend an elementary introduction to their subject. Unfortunately there are very few such books that can be considered both reliable and readable by the average student; most of them were written two or three decades ago and do not reflect the considerable amount of new research and revision of professional standards that have recently taken place. We have become acutely aware of the inadequacy of books written from secondary sources, now that so much of what used to be taken for granted has been shown to be wrong or misleading; yet we cannot escape the need for some kind of survey.

GENERAL WORKS

Probably the best short survey by professionals published in the last twenty years is *A Brief History of Science* (1964) by A. Rupert Hall and Marie Boas Hall. Unfortunately the authors are forced to race through the modern period in order to give an adequate account of early science in the allotted space. Even worse, this book has remained out of print for several years while publishers have flooded the market with inferior works. A reprint is in preparation (1988).

Of the textbooks currently available, the most popular is *A History of the Sciences* by Stephen Mason, first published in 1953 under the title

Main Currents of Scientific Thought; the paperback reprint is reasonably priced. It has about 600 small and closely printed pages unrelieved by any diagrams or portraits, and supported by a short bibliography of respectable sources, all published before 1960. Mason covers a large field, including ancient and medieval science and the development of national scientific traditions as well as the major advances since the Renaissance. I find the quantity of detailed information overwhelming, but my students have generally reacted favorably to the book. [Stauffer's review of the first edition appeared in *Isis* 45 (1954): 201–2.]

My own preference for a course in the history of *modern* science is A. E. E. McKenzie's *The Major Achievements of Science,* first published in 1960 in 2 volumes, which have been combined into one in the most recent edition (1973). McKenzie provides in 350 pages a narrative account of astronomy, physics, biology, and chemistry from the sixteenth to the twentieth centuries, with brief sections on geology and medicine and on the social-philosophical aspects of science. This is followed by about 200 pages of extracts from sources. McKenzie is more concerned to explain the scientific ideas than is Mason and less aware of some of the questions that interest the historian. But notwithstanding Robert Schofield's criticisms [*Isis* 53(1962):394–95] I think the book can provide a good foundation for the instructor who can fill in the corrections and interpretations provided by recent scholarship.

Another text reflecting more recent historiographic views is John Marks's *Science and the Making of the Modern World* (1983). It touches on the social sciences and social aspects of modern science and technology as well as more traditional topics.

Any comprehensive course on the history of science would be well served by the three books by Stephen Toulmin and June Goodfield: *The Fabric of the Heavens* (1961), *The Architecture of Matter* (AM) (1962), and *The Discovery of Time* (DT) (1965). These were planned as parts of a four-volume series on "The Ancestry of Science" but the fourth volume never appeared. *AM* and *DT* cover most of the history of nineteenth-century science, picking up some themes in chemistry, geology, and biology from earlier periods and following them through to a conclusion in the twentieth century; *DT* also has some excellent chapters on attitudes toward and writing of history in general. The organization of topics is unusual but has its own logic; the integration of scientific concepts with ideas from philosophy and history yields a coherent narrative that is both readable and sophisticated. These books must be supplemented with readings on psychology, anthropology, and planetary science.

Another older book that still has considerable value is W. C. Dampier's *A History of Science and Its Relations with Philosophy and*

Religion (1929, 1948, reprinted with a postscript by I. B. Cohen, 1961). About 300 pages are devoted to the period 1800–1930, touching on most areas of science.

A relatively recent comprehensive book prepared by two historians of science, L. Pearce Williams and Henry Steffens, is *The History of Science in Western Civilization* (1978); volume 3 covers the period 1700–1900. It includes extensive extracts from primary and a few secondary sources.

Instructors who want to stress the social relations of science might consider the series by H. J. Fyrth and Maurice Goldsmith, *Science, History and Technology* (1965–1969). Book 1, *A.D. 800 to the 1840s,* will not be considered here. Book 2 is in three parts: I, *The Age of Confidence: The 1840s to the 1880s;* II, *The Age of Uncertainty: The 1880s to the 1940s;* III, *The Age of Choice: The 1940s to the 1960s.* Each part of Book 2 is bound separately and has 60 to 68 pages of text followed by numerous chronological charts. Book 2, Part I, is on applications of science and evolution; Part II on nuclear physics, heredity, Pavlov, Freud, modernism in art, and the atom bomb; Part III is on Big Science, space flight, politics, automation, education. While the treatment of scientific topics is superficial the authors make an unusual attempt to correlate these with technological and cultural developments.

Science and the Rise of Technology since 1800 (AST 281) (1973), is a series of short books prepared for the Open University correspondence course in England. I give a rather detailed description of this series because it seems especially suitable for an undergraduate history of science course, yet it is not easily available in the United States and many American instructors have not even had a chance to look at the books. (I understand that this course has been withdrawn from the Open University "menu" and a totally different course on a roughly comparable theme is now in preparation.) The emphasis is on electricity, chemistry, and their relations to industry, with major sections on the social history of science. Block 1, "The Historical Perspective," would provide an excellent introduction to any course on the history of modern science; it consists of three 40-page units—Arthur Marwick, "Introduction to the History of Science"; Colin Russell, "From Natural Philosopher to Scientist"; David Goodman, "Science and Technology in the Industrial Revolution." Block 2 deals with coal as a basis for nineteenth-century technology (separately bound). Block 3 again contains three 40-page units: Russell and Goodman, "Atomism"; Russell, "Early Electrochemistry"; Goodman, "Faraday and Electromagnetism." Block 4 contains two units: Russell, "The Organization of Science in England" and Goodman, "Scientific and Technical Communication." Blocks 5 and 6 deal with public health and the chemical industry. Block 7, on science and engineering, includes one unit that would go well with **8.2** and **8.4**: G. R. M. Garratt, "The Communications Explosion in the Nineteenth Century: Some Contributions of Electrical Engineering" (telegraph, telephone, origins of radio). Block 8 addresses "The Wider Issues": Goodman on "Nationalism in Science" and Steven Rose on "Social Responsibility in Science and Technology: State, Science and Society in the Twentieth Century." The latter discusses World War II, "The Physicists' War," and various science policy issues, from a British viewpoint.

Another Open University series that would also be an excellent text for this course if it were available at a reasonable price is *Science and Belief: From Copernicus to Darwin* (AMST 283) (1974). It has a useful historiographic introduction by Colin A. Russell, "Some Approaches to the History of Science," including an analysis of the works of J. W. Draper, A. D. White, etc.; this is followed, in the same volume (Block 1) by R. Hooykaas, "The Impact of the Copernican Transformation" and D. C. Goodman, "Galileo and the Church." The next block, "Towards a Mechanistic Philosophy," contains Goodman's unit on Descartes and John Hedley Brooke's on Newton. Block 3 contains R. Hooykaas, "Puritanism and Science" (with study guide by Clive Lawless); Goodman, "The Enlightenment: Deists and 'Rationalists' "; Noel Coley, Clive Lawless, and Gerrylynn Roberts, "Nonconformity and the Growth of Technology." Block 4 contains Brooke, "Natural Theology in Britain from Boyle to Paley" and R. Hooykaas, "Genesis and Geology" (study guide by Lawless). Block 5 contains Brooke, "Precursors of Darwin?" and "Darwin," and Alan Richardson, "Religious Thought and the Idea of Evolution." Block 6 gives an overview of the impact of science on the modern worldview: Hooykaas, "Nature and History" (study guide by Lawless) and Russell, "The End of an Era?" (on Darwinism).

A third Open University series, *Science and Belief: From Darwin to Einstein* (A381) (1981), also has sections on evolution, concentrating on its philosophical and scientific aspects in the period after 1859. It has major sections on thermodynamics and randomness, modern physics and the problem of knowledge, and selected topics in the social sciences. Two anthologies have been published to accompany it: Colin Chant and John Fauvel, eds., *Darwin to Einstein: Historical Studies on Science and Belief* (1980), reprinting several articles by historians of science, and Noel G. Coley and Vance M. D. Hall, eds., *Darwin to Einstein: Primary Sources on Science and Belief* (1980). This is all first-rate material whose usefulness is severely limited by its exorbitant price and inaccessibility in the United States.

The North American distributor for Open University printed

course materials (texts) is: Taylor and Francis, 242 Cherry Street, Philadelphia, PA 19106–1906. 800/821–8312.

Audiovisual materials have also been produced to accompany the Open University courses but seem to be expensive and difficult to obtain in the United States. Instructors who are interested in using Open University materials should request current information on prices and availability from the above address or Open University Publications, Walton Hall, Milton Keynes, Bucks. MK7 6AA, England, UK.

Some other general works in the history of science are the following:

Jacob Bronowski, *The Ascent of Man* (1973), written in conjunction with the well-known TV series. The book is made attractive by its illustrations and the author's style; it lacks a systematic treatment of most areas of science since 1800. By using the companion volume of "Sources and Interpretations" (1975), ed. J. F. Henahan, together with viewings of the original Bronowski presentations (available on videocassette or film), one could make up quite a respectable course. There is also a series of "Viewer's Guides" prepared by Maud H. Chaplin, Eleanor R. Webster, and Dorothea J. Widmayer at Wellesley College; for information on current prices and availability of these booklets contact Professor Chaplin at Wellesley College, Wellesley, MA 02181.

Science 84, November 1984, special issue on "20 Discoveries that Shaped Our Lives" by H. F. Judson et al., is a collection of short articles on topics in twentieth-century science and technology.

John G. Burke, ed., *Science & Culture in the Western Tradition: Sources and Interpretations* (1987), is a mixture of primary and secondary extracts, including sections on Darwinism, modern physics, and the philosophy of science.

Derek Gjertsen, *The Classics of Science* (1984), includes chapters on books by Dalton, Lyell, and Darwin.

Jerzy Neyman, ed., *The Heritage of Copernicus* (1974), a collection of articles by scientists on twentieth-century science, mostly accessible to students.

Loren R. Graham, *Between Science and Values* (1981), primarily an exposition of the significance of twentieth-century physics and biology.

J. D. Bernal, *Science in History,* 3d ed. (1965), first published in 1954, was issued as a four-volume paperback by MIT Press in 1971. Volume 2 deals with "The Scientific and Industrial Revolutions" (sixteenth through nineteenth centuries) and volume 3 with the physical and biological sciences in the twentieth century. These two could be used for a course emphasizing the relations between science and technology, but the instructor would have to be prepared to cope with Bernal's Marxist bias. Volume 4, on the "Social Sciences," omits most of the topics in psychology and anthropology that I find most essential and consists mainly of the author's long-winded presentation of his opinions on the relations between science and society. My students had a very negative reaction to volumes 2, 3, and 4 of this series when I once used them as required texts. For a sympathetic but critical assessment see J. R. Ravetz, *Isis* 72 (1981): 393–402.

René Taton, ed., *Science in the Nineteenth Century* and *Science in the Twentieth Century* (1961–1964), translated from the French editions of 1961–1964. These books tend to overwhelm the reader with information but don't explain most of it very well or fit it into a coherent story. They might be useful reference works but their bibliographies are generally inadequate.

J. G. Crowther, *A Short History of Science* (1969), stresses the relations between science and industry.

F. Sherwood Taylor, *An Illustrated History of Science* (1955), emphasizes experiments; 105–67 are on the period 1800–1930 (chemistry, physics, biology); probably suitable for high school students.

Emanuel Hackel and Dennis W. Strawbridge, eds., *The Search for Explanation: Studies in Natural Science,* 3 vols. (1967–1969), the first on astronomy, physics, and atomic theories from the Greeks to the twentieth century; the second on biology and heredity; the third on geology and evolution.

Physical Sciences

The following books deal with the history of *physical science* (including the twentieth century) and include chapters that could be assigned in a course on the history of modern science for students with little or no scientific background.

Cecil J. Schneer, *The Evolution of Physical Science* (1960), provides a readable, large qualitative account of developments in the seventeenth through nineteenth centuries with brief chapters on relativity and atomic physics.

Gerald Holton, *Introduction to Concepts and Theories in Physical Science* (1952), second edition revised and with new material by Stephen G. Brush (1973). This is intended as a textbook for a college course in physics or physical science for nonscience majors; the students are expected to be willing to follow some elementary derivations, so the exposition is not quite at the "physics for poets" level. The emphasis is on the explanation of physical concepts in their historical development, with some discussion of models for the growth of science. About half of the book deals with physics since 1800, through quantum mechanics and relativity.

Gerald Holton et al., *Project Physics Text* (1970, 1975). Intended as a textbook in a high school physics course but can also be used in a college course for nonscience majors; the approach and some of the individual sections are similar to those in the above described book by Holton and Brush. The following parts deal with the development of physical science in the nineteenth and twentieth centuries, and are available separately as paperbacks: *The Triumph of Mechanics, Light and Electromagnetism, Models of the Atom, The Nucleus,* and *Discoveries in Physics.*

Cecil J. Schneer, *Mind and Matter, Man's Changing Concepts of the Material World* (1969), is primarily a history of chemistry, including its relations with physics, and probably more interesting to nonchemists than most books on this subject.

John G. Taylor, *The New Physics* (1972), is a good semihistorical survey of twentieth-century physics and cosmology.

Eugene Hecht, *Perspectives in Physics* (1980), is a historically oriented textbook.

Heinz Pagels, *The Cosmic Code* (1982), is primarily on quantum physics and its interpretation.

Isaac Asimov, *The Universe, from Flat Earth to Quasar* (1966, 1971). The emphasis is on stellar astronomy and cosmology rather than the solar system and no account is taken of historical research, but these defects are compensated by the fact that most students actually enjoy reading the book, and learn something from it. (The same could be said of Asimov's other works on history of science, which should be given some consideration since they are easily available in stores and libraries.)

F. Durham and R. D. Purrington, *The Frame of the Universe: A History of Physical Cosmology* (1983). According to historian of astronomy Norriss Hetherington, contains several historical errors and omissions but gives a good description of the physical phenomena for students [*Annals of Science* 41(1984):501–3].

Albert Einstein and Leopold Infeld, *The Evolution of Physics* (1938), is highly recommended as a text even though its authors are not historians of science. It was written more than forty years ago, contains some inaccuracies and misleading statements, and is not even an "original source" in the usual sense. It does have the essential virtue that many students will read it and learn something about the development of the most important theories of modern physics as well as acquiring some enthusiasm for the subject. Perhaps they are turned on by the experience of actually being able to understand a book with the forbidding name of Einstein on the title page. In any case, this book has gotten favorable reactions from my students every time I have assigned it, whereas most texts written by professional historians of science are judged dull or incomprehensible.

Joan Solomon, *The Structure of Space: The Growth of Man's Ideas on the Nature of Forces, Fields, and Waves* (1973). Most of this 220-page book deals with light, electromagnetism, relativity, and cosmology since 1800.

Joan Solomon, *The Structure of Matter: The Growth of Man's Ideas on the Nature of Matter* (1973), concerns atomic theory, mostly from Dalton to the atom bomb.

Armin Hermann, *The New Physics* (1979), is mostly on physics in Germany, 1905–1955; it is richly illustrated with photos and reproductions of documents.

Morris Kline, *Mathematics in Western Culture* (1953), has chapters on various applications of mathematics to science, including vibrating strings, electromagnetic waves, relativity, statistical phenomena, as well as pure mathematics (paradoxes of the infinite) and relations of mathematics to art.

George E. Owen, *The Universe of the Mind* (1971), is a survey of topics in the history of mathematics and its applications to physics.

Mendel Sachs, *The Search for a Theory of Matter* (1971), includes chapters on relativity and quantum mechanics.

Barry M. Casper and Richard J. Noer, *Revolutions in Physics* (1972), is a textbook with study questions and problems; concentrates on development of Newton's theory of motion and Einstein's conception of space and time, with sections on the Copernican revolution and ether physics.

Gerard de Vaucouleurs, *Discovery of the Universe: An Outline of the History of Astronomy from its Origins to 1956* (1957).

Charles A. Whitney, *The Discovery of Our Galaxy* (1971), is a detailed account of research on nebulae, galaxies, etc., since the seventeenth century; the author is an astronomer who has used good historical sources and presents an interesting story.

Daniel Kevles, *The Physicists: The History of a Scientific Community in Modern America* (1977), is primarily a social history, but quite useful as supplementary reading.

W. H. Brock, M. Chapple, and M. A. Hewson, *Studies in Physics* (1972), contains chapters on "Thomas Young"; "Heat and Energy"; "A Scientist among the Spirits: William Crookes and the Radiometer"; "Buns and Billiards: I. The Philosopher's Atom and the Scientist's Atom" and "II. The Atom from Inside"; "Radioactivity: Some Early Instruments"; "Aston, Isotopes and the Mass Spectrograph"; "Vertical Transport"; "Time." These are case histories designed for students in science or history/philosophy of science courses, with discussion questions.

Bernard Jaffe, *Crucible: The Story of Chemistry from Ancient Alchemy to Nuclear Fission* (1930, 1976). The revised edition has about 100 pages on nineteenth-century chemists, and 160 on twentieth-century radiochemistry and atomic/nuclear physics.

Walter J. Lehmann, *Atomic and Molecular Structure: The Development of Our Concepts* (1972).

Stephen G. Brush, *Statistical Physics and the Atomic Theory of Matter, from Boyle and Newton to Landau and Onsager* (1983), is intended for students with a strong background in physics.

D. L. Hurd and J. J. Kipling, eds., *The Origins and Growth of Physical Science* (1964), is a good anthology of sources on electromagnetism, chemical atomic theory, physics up to 1905.

Finally, I mention two books that deal with nineteenth-century physical science:

Peter Harman, *Energy, Force, and Matter* (1982), although intended as a textbook, may be too difficult for undergraduates to read; a good example of the approach of a modern professional historian of science, based on the latest scholarship in the field.

Harold I. Sharlin, *The Convergent Century: The Unification of Science in the Nineteenth Century* (1966), arranged by topics such as heat, electricity, light, cathode rays, etc.

Biology

Since my course on history of modern science deals with only a few topics in biology—evolution, genetics, and sex differences—I find that most general texts on history of science are adequate and need to be supplemented only by one or two monographs or articles. These will be discussed in the appropriate sections of this *Guide* (Chaps. 2 and 4). I have not made a survey of books on the history of biology but can suggest the following titles for those who wish to give more time to this area:

Lois Magner, *A History of the Life Sciences* (1979).

Garland Allen, *Life Science in the Twentieth Century* (1975).

William Coleman, *Biology in the Nineteenth Century: Problems of Form, Function, and Transformation* (1971).

Charles Singer, *History of Biology to about the Year 1900* (3d ed. 1959).

Peter J. Bowler, *Evolution: The History of an Idea* (1984).

Anthropology

Similarly, books on the history of anthropology will be discussed in Chapter 3 in connection with the special topics of interest there (evolutionary theories, cultural relativism, and race differences) but a few general texts may be noted here:

Annemarie De Waal Malefijt, *Images of Man: A History of Anthropological Thought* (1974).

Abram Kardiner and Edward Preeble, *They Studied Man* (1961).

H. R. Hays, *From Ape to Angel: An Informal History of Social Anthropology* (1958).

Stephen Jay Gould, *The Mismeasure of Man* (1981), is a critique of nineteenth- and twentieth-century anthropologists and psychologists who have tried to quantify intelligence and race differences; despite its polemical tone, I find it useful as the only readable account of several topics I want to cover.

Psychology

My treatment of psychology is also very selective (Freud and the fate of his theories, behaviorism, the IQ test) so many instructors will probably want to cover other topics. Fortunately there are many excellent books on the history of psychology:

Ernest R. Hilgard, *Psychology in America: A Historical Survey* (1987), is a definitive work by a successful author of introductory textbooks.

B. R. Hergenhahn, *An Introduction to the History of Psychology* (1986), includes the clinical and experimental traditions as well as Darwinian influences and psychoanalysis.

David Hothersall, *History of Psychology* (1984), includes chapters on psychoanalysis, American psychology, intelligence tests, and behaviorism.

Duane Schultz, *A History of Modern Psychology* (3d ed. 1981), is a widely used text.

Merle Curti, *Human Nature in American Thought* (1980), contains chapters on William James and the "new" psychology (circa 1900), genetics and the IQ controversies, Freud and the unconscious, behaviorism.

J. P. Chaplin and T. S. Krawiek, *Systems and Theories of Psychology* (4th ed. 1979).

Raymond Fancher, *Pioneers of Psychology* (1979), I found convenient to use in my course because of the selection of topics and reasonable price; it received favorable responses from students.

Michael Wertheimer, *A Brief History of Psychology* (2d ed., 1979).

Thomas Leahey, *A History of Psychology* (1979), is recommended if you want to put more emphasis on behaviorism; Leahey comments extensively on whether psychology has undergone Kuhnian revolutions.

Leslie S. Hearnshaw, *A Short History of British Psychology, 1840–1940* (1964).

G. S. Brett, *Brett's History of Psychology* (abridged ed. 1953, reprinted 1965).

Robert I. Watson, *The Great Psychologists* (1962, 4th ed. 1978).

G. Murphy and J. K. Kovach, *Historical Introduction to Modern Psychology* (1949, 3d ed. 1972).

Edna Heidbreder, *Seven Psychologies* (1933, 1963).

Robert Thomson, *The Pelican History of Psychology* (1968), is primarily on the period 1880–1940, with chapters on evolution, psychology in the United States, psychoanalysis, behaviorism, mental testing, etc.

A. A. Roback, *History of Psychology and Psychiatry* (1961).

Philosophy of Science

Among the many undergraduate textbooks on philosophy of science, the one most appropriate as collateral reading for a history of science course is John Losee, *A Historical Introduction to the Philosophy of Science* (1972), since it gives a systematic account. But for this particular course I would recommend using selected chapters from Ronald Giere's *Understanding Scientific Reasoning* (1979); his examples include Newtonian physics (64–72) compared with Mendelian genetics (73–77, 103–5). Another useful work is Larry Laudan's *Progress and Its Problems* (1978), a synthesis of recent philosophical views that does justice to both rational and irrational aspects of science.

Social History of Science

There has recently been considerable interest in the social history of science. Most of the professional literature is written in sociologese, a dialect that my students find difficult to read. Steven Shapin has prepared a useful syllabus for an undergraduate course, published in *Social Studies of Science* 10(1980):231–58. The following books might be considered:

Hilary Rose and Steven Rose, *Science and Society* (1969), discusses primarily twentieth-century issues from a British perspective.

Joseph Ben-David, *The Scientist's Role in Society: A Comparative Study* (1971, 1984), contains chapters on the social organization of science in France, Germany, and the United States in the nineteenth and twentieth centuries.

Derek de Solla Price, *Science since Babylon* (1975), includes several self-contained chapters that could be assigned as collateral reading, e.g., 'mutations of science" (X and N rays) and "diseases of science" (exponential growth). The author pioneered the quantitative approach to history of science, but his own writing is not limited to that approach.

Science in a Social Context (1977, 1979), is a series of pamphlets prepared by the "SISCON" group, including: Leonard Isaacs, "Darwin to Double Helix: The Biological Theme in Science Fiction"; Kenneth Green and Clive Morphet, "Research and Technology as Economic Activities"; Keith Pavitt and Michael Worboys, "Science, Technology and the Modern Industrial State"; Iain Cameron and David Edge, "Scientific Images and Their Social Uses: An Introduction to the Concept of Scientism"; Joan Lipscombe and Bill Williams, "Are Science and Technology Neutral?"; Margaret Gowing and Lorna Arnold, "The Atomic Bomb." The SISCON pamphlets are distributed in the United States by Butterworths (Boston, Mass.).

1.3 RESOURCES FOR INSTRUCTORS AND ADVANCED STUDENTS

This section lists only those items that are relevant to several topics or to the history of science in general; more specialized works will be mentioned in the appropriate sections. In addition to comprehensive histories and reference works, I have included a few of the many anthologies of selections from primary sources and collections of essays by historians. Those marked by an asterisk are recommended for purchase by any prospective teacher of a course on history of modern science on the basis of importance, utility, and reasonable price; others should be ordered for the library. (For current information on publishers and prices of asterisked books, see the Book List following Chapter 13.) Some information about history of science as an academic discipline is also included in this section.

GENERAL

The Open University course materials mentioned in **1.2** provide an excellent introduction on a fairly sophisticated level for an instructor who does not have formal training in the history of modern science. I especially recommend *Science and the Rise of Technology since 1800* (AST 281), Units 1, 5, 7, 14; *Science and Belief: From Copernicus to*

Darwin (AMST 283), Units 1, 11–16; and *Science and Belief: From Darwin to Einstein* (A381) as being relevant to many of the topics covered by this course.

Historiographic works that have been influential and should be familiar to the instructor include, in addition to Thomas Kuhn's well-known **Structure of Scientific Revolutions* (1962, 1970), Joseph Agassi's *Towards an Historiography of Science* (The Hague: Mouton, 1963), and the essays of Imre Lakatos which appeared in the volume *Criticism and the Growth of Knowledge,* ed. I. Lakatos and A. Musgrave (New York: Cambridge Univ. Press, 1970) and in **Method and Appraisal in the Physical Sciences* (1976, ed. C. Howson). A recent addition is the important study by I. B. Cohen, *Revolution in Science* (Cambridge: Harvard Univ. Press, 1985). An excellent presentation of the growth of science from the seventeenth through nineteenth centuries according to the earlier orthodox style (but more suitable for advanced students and instructors than for beginners) is **The Edge of Objectivity* (1960) by Charles Gillispie. Derek Price initiated the "science of science" and the quantitative approach to history of science in his short classic work **Little Science, Big Science* (1963).

H. Kragh, *An Introduction to the Historiography of Science* (New York: Cambridge Univ. Press, 1987) is a very useful survey of different viewpoints.

*Paul T. Durbin et al., eds., *A Guide to the Culture of Science, Technology, and Medicine* (1980, 1984), includes: Arnold Thackray, "History of Science," 3–69; C. W. Pursell, Jr., "History of Technology," 70–120; A. C. Michalos, "Philosophy of Science: Historical, Social, and Value Aspects," 197–281. These are critical surveys of the current state of these disciplines, with selective bibliographies.

Pietro Corsi and Paul Weindling, eds., *Information Sources in the History of Science and Medicine* (Boston: Butterworth Scientific, 1983) is a collection of short essays and bibliographies.

*W. F. Bynum et al., *Dictionary of the History of Science* (1981), has short essays on concepts, theories, personalities.

*David Steele, ed., *The History of Scientific Ideas: A Teachers' Guide* (1970), has sections on "The Nature of the Universe," "Matter Theory," and "Scientific Ideas in Biology," each presenting an exposition of the topic from ancient to modern times, with suggested readings, demonstrations, discussion questions. Very useful for teachers.

Philip P. Wiener, ed., *Dictionary of the History of Ideas* (New York: Scribner, 1973), is a four-volume encyclopedia with many valuable essays.

ISIS Guide to the History of Science (revised every three or four years), is published by the History of Science Society (HSS) and distributed free to its members. It includes listings of graduate programs and their faculties, members of HSS with their addresses, phone numbers, specialties, etc.; subject and geographical indexes to members; list of scholarly journals that publish articles on history of science, etc.

BIBLIOGRAPHIES

Magda Whitrow, ed., *ISIS Cumulative Bibliography: A Bibliography of the History of Science Formed from ISIS Critical Bibliographies 1–90, 1913–1965*. Vol. 1, Part I, and vol. 2, Part I, *Personalities;* vol. 2, Part II, *Institutions;* vol. 3, *Subjects* (1971–1976); vols. 4 and 5, *Civilizations and Periods* [through the nineteenth century]; vol. 6, *Author Index* to vols. 1–5 (London: Mansell, 1971–1984). Allows quick retrieval of almost all historical books and articles on any person or topic in the history of science, including biographies, reprints, and translations of sources, etc., published in the period 1912–1964.

John Neu, ed., *ISIS Cumulative Bibliography 1966–1975: A Bibliography of the History of Science Formed from ISIS Critical Bibliographies 91–100 Indexing Literature Published from 1965 through 1974.* Vol. 1, *Personalities and Institutions;* vol. 2, *Subjects* (London: Mansell, 1980–1983), is a continuation of Whitrow's volumes.

*David Knight, *Sources for the History of Science, 1660–1914* (1975), contains chapters that discuss historiography, manuscripts, journals, books, and objects in a format more suitable for continuous reading than quick reference.

*S. A. Jayawardene, *Reference Books for the Historian of Science: A Handlist* (1982).

Diane Weisz and Carlos Kruytbosch, *Studies of Scientific Disciplines: An Annotated Bibliography* (Washington, D.C.: Office of Planning and Policy Analysis, National Science Foundation, 1982).

*Bruce Eastwood, *Directory of Audio-Visual Sources: History of Science, Medicine and Technology,* (1979).

T. H. Leith, *Bibliography for the Preparation of Research Papers in the History and Philosophy and Sociology of Science, Biography of Scientists, Science and Religion, Science and the Humanities, and Education in Science* (Downsview, Ont., Can.: Department of Natural Science, Atkinson College, York University, 1976).

Francois Russo, *Elements de Bibliographie de l'Histoire des Sciences et des Techniques,* 2d ed. (Paris: Hermann, 1969).

BIOGRAPHICAL REFERENCE WORKS

C. C. Gillispie, ed., *Dictionary of Scientific Biography,* 16 vols. (New York: Scribner, 1970–1980). A comprehensive encyclopedia of the history of science with many outstanding articles; the last volume is a detailed index to subjects. It includes only scientists who had died before the 1970s. This work is absolutely essential for any library (cited in this *Guide* as *DSB*). There is also a less expensive one-volume edition available: *Concise Dictionary of Scientific Biography* (New York: Scribner, 1981). A supplement, edited by F. L. Holmes, is currently being prepared.

Trevor I. Williams, ed., *A Biographical Dictionary of Scientists,* 3d ed. (New York: Wiley, 1981), contains 1100 short biographies and a list of anniversaries.

John Daintith et al., eds., *A Biographical Encyclopedia of Scientists* (New York: Facts on File, 1981), contains about 2000 biographies, chronology of major discoveries and publications, short list of "books and papers particularly influential in the development of science."

Clark A. Elliott, *Biographical Dictionary of American Science: The Seventeenth through the Nineteenth Centuries* (Westport, Conn.: Greenwood, 1979).

Allen G. Debus, ed., *World Who's Who in Science: A Biographical Dictionary of Notable Scientists from Antiquity to the Present* (Chicago: Marquis Who's Who, 1968), is useful for checking basic data on living scientists not covered in *DSB*.

J. C. Poggendorff et al., eds., *Biographisch-Literarisches Handwörterbuch zur Geschichte der exacten Wissenschaften* . . . (Berlin: Akademie-Verlag, 1863–), is a continuing series, based partly on questionnaires sent to all major scientists, giving biographical data and publications, from eighteenth century to present.

Isaac Asimov, *Asimov's Biographical Encyclopedia of Science and Technology,* 2d ed. (New York: Doubleday, 1972), is useful for anecdotes (if one is not concerned about their historical accuracy), pronunciation of names, and basic facts, especially for living scientists not included in the *DSB*.

COLLECTIONS OF ESSAYS BY HISTORIANS

Starting here, I adopt the practice followed in bibliographies in later sections: the most recent works are listed first, and (with a few exceptions) those published more than ten years ago, or concerned primarily with science before 1800, are omitted.

*David C. Lindberg and Ronald L. Numbers, eds. (1986), *God and*

Nature: Historical Essays on the Encounter between Christianity and Science.

Arnold Thackray, ed. (1985–1986), *Osiris,* new series, vols. 1 and 2 (Philadelphia: History of Science Society). The first volume in the revival of what was once a major history of science journal is devoted to surveys of the history of American science. The second volume contains several long essays on various topics.

Roger Cooter et al. (1985), "What Is the History of Science?" *History Today* 35(April):32–40, 35(May):46–53, contains short pieces by twelve British historians of science.

Spencer R. Weart and Melba Phillips, eds. (1985), *History of Physics* (New York: American Institute of Physics), contains forty-seven articles reprinted from *Physics Today.* Although most of the authors are not professional historians, the collection is an excellent source for twentieth-century physics.

Peter Achinstein and Owen Hannaway, eds. (1985), *Observation, Experiment, and Hypothesis in Modern Physical Science* (Cambridge: MIT Press), is a collection of essays by historians and philosophers of science.

David J. Depew and Bruce H. Weber, eds. (1985), *Evolution at a Crossroads: The New Biology and the New Philosophy of Science* (Cambridge: Bradford/MIT Press), includes the gene as a case history, and the difference between biology and physical science.

Edward Garber, ed. (1985), *Genetic Perspectives in Biology and Medicine* (Chicago: Univ. of Chicago Press), includes eight historical articles; the others are surveys by biologists, with some historical content.

Victor Weisskopf et al. (1984), "Modern Masters of Science," *Social Research* 51, no. 3:581–835, includes essays on N. Bohr, T. H. Morgan, H. J. Muller, and E. Schrödinger.

Loren Graham, Wolf Lepenies, and Peter Weingart, eds. (1983), *Functions and Uses of Disciplinary Histories* (Hingham, Mass.: Reidel), is mostly articles by historians of science discussing how scientists use the history of their own discipline.

Friedrich Herneck (1984), *Wissenschaftsgeschichte: Vorträge und Abhandlungen* (Berlin: Akademie-Verlag), includes several articles on physicists.

Rutherford Aris et al., eds. (1983), *Springs of Scientific Creativity* (Minneapolis: Univ. of Minnesota Press), includes articles on Joule, Maxwell, Einstein, Schrödinger, J. von Neumann, and others.

Joseph W. Dauben and Virginia S. Sexton, eds. (1983), *History and*

Philosophy of Science: Selected Papers (New York: New York Academy of Sciences), contains articles on medicine, mathematics, psychology, and industrial science, originally presented at the New York Academy of Sciences.

Joseph Agassi and Robert S. Cohen, eds. (1982), *Scientific Philosophy Today* (Hingham, Mass.: Reidel), includes articles on Mach, realistic history of science, and thermodynamics.

Gabriel A. Almond et al., eds. (1982), *Progress and Its Discontents* (Berkeley: Univ. of California Press), covers progress in science and historiography.

Harry Woolf, ed. (1981), *The Analytical Spirit: Essays in the History of Science* (Ithaca, N.Y.: Cornell Univ. Press), is a collection of essays in honor of Henry Guerlac.

Colin Chant and John Fauvel, eds. (1980), *Darwin to Einstein: Historical Studies on Science and Belief* (New York: Longmans), contains articles selected to be read in conjunction with the Open University course (see **1.2**).

*Nathan Reingold, ed. (1979), *The Sciences in the American Context: New Perspectives,* is on astronomy, earth sciences, genetics, physics, etc., mostly nineteenth and early twentieth century.

Barry Barnes and Steven Shapin, eds. (1979), *Natural Order* (London: Sage), is on theories of the earth, social Darwinism, non-Euclidean geometry, physics and psychics in Victorian Britain, biometry-Mendelism controversy, heredity-environment debate.

*Gerald Holton, *The Scientific Imagination: Case Studies* (1978), selects most of his examples from modern physics (Millikan, Fermi, Einstein); includes reflections on "themata," the measurement of science, and psychology of scientists.

*Thomas S. Kuhn, *The Essential Tension: Selected Studies in Scientific Tradition and Change* (1977), contains reprints of historiographic studies on "mathematical versus experimental tradition" and energy conservation, and "metahistorical studies" on revolutions and paradigms.

ANTHOLOGIES OF ORIGINAL SOURCES (MISCELLANEOUS)

Noel G. Coley and Vance D. Hall, eds. (1980), *Darwin to Einstein: Primary Sources on Science and Belief* (New York: Longmans), is on science and metaphysics in Victorian Britain; time, chance, and thermodynamics; problems of knowledge and modern physics; the mystery of life; miscellaneous problems in biological and human sciences (prepared for use with the Open University course, see **1.2**).

L. Pearce Williams and Henry John Steffens (1978), *The History of Science in Western Civilization,* vol. 3, *Modern Science, 1700–1900* (Lanham, Md.: Univ. Press of America).

Richard G. Olson, ed. (1971), *Science as Metaphor: The Historical Role of Scientific Theories in Forming Western Culture* (Belmont, Calif.: Wadsworth), includes Darwinism, thermodynamics, Freudian psychoanalysis, Zola on the experimental novel, B. F. Skinner, relativity, and quantum mechanics. This anthology comes closer to my own choice of topics than any other.

OTHER LITERATURE

Physical Science

Simon Schaffer (1983), "History of Physical Science," in *Information Sources in the History of Science and Medicine,* ed. P. Corsi and P. Weindling (Boston: Butterworth Scientific), 285–314.

*C. Howson, ed., *Method and Appraisal in the Physical Sciences* (1976), contains a concise statement by Imre Lakatos of his historiography, followed by case studies applying it to atomism and thermodynamics, Young's optics, oxygen vs. phlogiston, Einstein vs. Lorentz, Avogadro's hypotheses; concludes with a critique by Paul Feyerabend.

H. A. Boorse and L. Motz, eds. (1966), *The World of the Atom* (New York: Basic Books), is a major two-volume anthology of sources in physics and chemistry, including some original translations. Most of the selections are not too technical. There are good introductory/biographical essays.

D. L. Hurd and J. J. Kipling, eds. (1964), *The Origins and Growth of Physical Science* (Baltimore: Penguin Books), is based on an earlier anthology by G. Schwartz and P. W. Bishop, *Moments of Discovery* (1958). If this were available in a paperback edition it would be suitable for student reading in a course that stressed physical science and use of some original sources.

Physics

Stephen G. Brush (1987), "Resource Letter HP-1: History of Physics," *American Journal of Physics* 55:683–91.

W. D. Hackmann (1985), "History of Physics," in *Information Sources in Physics,* ed. Dennis F. Shaw (Boston: Butterworth Scientific), 209–32.

R. W. Home (1984), *The History of Classical Physics: A Selected,*

Annotated Bibliography (New York: Garland), is an analysis of 1210 publications covering the period from 1700 to 1900.

Bruce R. Wheaton (1984), "Inventory of Sources for History of Twentieth-Century Physics," *Isis* 75:153–57.

Stephen G. Brush and Lanfranco Belloni (1983), *History of Modern Physics: An International Bibliography* (New York: Garland), is intended to complement the Heilbron-Wheaton list (next item) by including more recent and more obscure publications.

John L. Heilbron and Bruce R. Wheaton (1981), *Literature on the History of Physics in the 20th Century* (Berkeley: Office for History of Science and Technology, Univ. of California), is a very comprehensive list of books and articles for the period 1900–1950.

Lewis Pyenson (1981), "History of Physics," in *Encyclopedia of Physics,* ed. Rita G. Lerner and G. L. Trigg (Reading, Mass.: Addison-Wesley), 404–14, surveys both technical and social aspects of the development of modern physics.

Chemistry

George B. Kauffman, ed. (1971), *Teaching the History of Chemistry: A Symposium* (Budapest: Akademiai Kiado), covers proceedings of a symposium held in San Francisco; contains useful ideas and resources.

Aaron J. Ihde (1964), *The Development of Modern Chemistry* (New York: Harper and Row).

J. R. Partington (1964), *A History of Chemistry,* 4 vols. Volume 4 (New York: St. Martin's Press) covers the nineteenth and early twentieth centuries in 967 pages of text and many footnote references.

Astronomy

Owen Gingerich, ed. (1984), *General History of Astronomy* (New York: Cambridge Univ. Press), 4 vols. Volume 4, Part A, covers selected topics in astrophysics and twentieth-century astronomy to 1950; other volumes in preparation.

Dieter B. Hermann (1984), *The History of Astronomy from Herschel to Hertzsprung* (New York: Cambridge Univ. Press).

David H. DeVorkin (1982), *The History of Modern Astronomy and Astrophysics: A Selected, Annotated Bibliography* (New York: Garland).

Kenneth R. Lang and Owen Gingerich, eds. (1979), *Source Book in Astronomy and Astrophysics, 1900–1975* (Cambridge: Harvard Univ. Press), is far superior to most "source books" since it gives complete or nearly complete reprints of original articles.

Biology

Judith Overmier (1986), *The History of Biology: An Annotated Bibliography* (New York: Garland).

Ernst Mayr (1982), *The Growth of Biological Thought* (Cambridge: Harvard Univ. Press). Every instructor should read the first three chapters of this excellent book, "How to Write History of Biology," "The Place of Biology in the Sciences and Its Conceptual Structure," and "The Changing Intellectual Milieu of Biology." Most of the rest of the book deals with evolution and heredity, but the last chapter presents reflections on the "Science of Science."

Pieter Smit (1974), *History of the Life Sciences: An Annotated Bibliography* (New York: Hafner).

Anthropology

M. J. Leaf (1979), *Man, Mind, and Science* (New York: Columbia Univ. Press).

Psychology

Mitchell G. Ash and William R. Woodward, eds. (in press), *Psychology in Twentieth-Century Thought and Society* (New York: Cambridge Univ. Press), contains chapters on the social context of investigative practice, programmatic research in the United States, the Protestant origins of Jean Piaget, women and professionalization, psychology and politics in the Third Reich, and psychology in India and China.

Sigmund Koch and David Leary, eds. (1985), *A Century of Psychology as Science* (New York: McGraw-Hill), includes retrospective statements by major contemporary psychologists, arranged by problem area; fields include the new look in cognitive psychology, theories of learning and ethology, language, and culture, etc.

Claude Buxton, ed. (1985), *Points of View in the History of Modern Psychology* (New York: Academic Press), includes chapters on Gestalt, cognitive psychology, and biological psychology.

Josef Brozek, ed. (1984), *Explorations in the History of Psychology in the United States* (Lewisburg, Pa.: Bucknell Univ. Press), is primarily on the period before 1930.

B. Lomov and V. Schustikov, eds. (1984), *Soviet Psychology* (Moscow: Nauka Publishers), includes chapters on "methodological and theoretical problems" and on history of psychology.

Leonard Zusne (1984), *Biographical Dictionary of Psychology* (Westport, Conn.: Greenwood Press).

William R. Woodward and Mitchell Ash, eds. (1982), *The Problematic Science: Psychology in Nineteenth Century Thought* (New York:

Praeger), has several chapters on psychophysics, evolution and voluntary behavior, medical psychology.

Michael M. Sokal and Patrice A. Rafail (1982), *A Guide to Manuscript Collections in the History of Psychology and Related Areas* (Millwood, N.Y.: International Publishers).

Albert R. Gilgen (1982), *American Psychology since World War II* (Westport, Conn.: Greenwood).

John D. Lawry (1981), *Guide to the History of Psychology* (Totowa, N.J.: Littlefield, Adams), is a chronological account of innovators with summary of what each did; section on "who was the first to . . . ?"; list of classic works, bibliography of secondary sources.

Josef Brozek and Ludwig J. Pongratz, eds. (1980), *Historiography of Modern Psychology: Aims, Resources, Approaches* (Toronto: Hogrefe), includes several externalist treatments of the professionalization of psychology in Germany and America at the turn of this century; the chapter by W. R. Woodward contains bibliography, sources of funding, and ideas for teaching.

R. W. Rieber and Kurt Salzinger, eds. (1980), *Psychology: Theoretical-Historical Perspectives* (New York: Academic Press).

Wayne Viney, Michael Wertheimer, and Marilyn Lou Wertheimer (1979), *History of Psychology: A Guide to Information Sources* (Detroit: Gale Research).

E. G. Boring (1929, 1950), *A History of Experimental Psychology* (New York: Appleton-Century-Crofts), is the classic comprehensive text; in spite of its title it has useful sections on theories and the relations of psychology to philosophy, physiology, evolution, and psychiatry. Though some of Boring's interpretations are considered obsolete, he is a lively writer and provides much valuable information in readable form.

Mathematics

Joseph W. Dauben (1985), *The History of Mathematics from Antiquity to the Present: A Selective Bibliography* (New York: Garland).

H. N. Jahnke and M. Otte, eds. (1981), *Epistemological and Social Problems of the Sciences in the Early Nineteenth Century* (Boston: Reidel), is mostly on mathematics; topics in physiology and science education are also treated.

I. Grattan-Guinness, ed. (1980), *From the Calculus to Set Theory, 1630–1910* (London: Duckworth), has essays on history of analysis, set theory, integration, foundational studies, etc.

Kenneth O. May (1973), *Bibliography and Research Manual of the History of Mathematics* (Buffalo, N.Y.: Univ. of Toronto Press), includes a 30-page introductory section on Information Retrieval and Personal Information Storage that should be read by people doing research in any area of the history of science.

Morris Kline (1972), *Mathematical Thought from Ancient to Modern Times* (New York: Oxford Univ. Press), is very comprehensive and reasonably priced for its length (1300 pages).

*Carl B. Boyer (1968, reprinted 1985), *A History of Mathematics,* is more concise (700 pp.) than Kline but almost as good for the nineteenth and twentieth centuries.

Technology

Stephen H. Cutcliffe, ed. (1983), *The Machine in the University* (Bethlehem, Pa.: Science, Technology and Society Program, Lehigh University), contains twenty-seven representative course syllabi in the history of technology.

Carroll W. Pursell, Jr., ed. (1981), *Technology in America: A History of Individuals and Ideas* (Cambridge: MIT Press), includes articles on Alexander Graham Bell, Thomas Alva Edison, technology and women, Henry Ford, Lindbergh, Enrico Fermi and nuclear power, Robert H. Goddard and space flight.

Eugene S. Ferguson (1968), *Bibliography of the History of Technology* (Cambridge: MIT Press).

Philosophy of Science

Guttorm Fløistad, ed. (1982), *Contemporary Philosophy: A New Survey,* vol. 2, *Philosophy of Science* (The Hague: Nijhoff).

Peter D. Asquith and Henry E. Kyburg (1978), *Current Research in Philosophy of Science* (East Lansing, Mich.: Philosophy of Science Association).

*Frederick Suppe, ed. (1974, 1977), *The Structure of Scientific Theories,* includes a long introductory survey by Suppe on the history of modern philosophy of science and major papers (presented at a 1969 symposium) by I. B. Cohen, T. S. Kuhn, D. Shapere, and others. The 1977 edition has an "Afterword" by Suppe on more recent developments.

Sociological Approaches

John Ziman (1985), *An Introduction to Science Studies: The Philosophical and Social Aspects of Science and Technology* (New York: Cambridge Univ. Press), according to Paul Durbin's review in *Science* 229(1985):458–59, is "a masterly summary of the whole range of the science studies literature, from philosophy of science to sociology of science to science and technology policy studies . . . it would also make a fine textbook for an introductory course in a science, technology, and society program."

Mary Douglas, ed. (1982), *Essays in the Sociology of Perception* (Boston: Routledge and Kegan Paul), makes various applications of her "grid-group" anthropological analysis to science.

Everett Mendelsohn and Yehuda Elkana, eds. (1981), *Sciences and Cultures: Anthropological and Historical Studies of the Sciences* (Boston: Reidel).

Karin D. Knorr et al., eds. (1980), *The Social Process of Scientific Investigation* (Boston: Reidel).

Henry Menard (1971), *Science: Growth and Change* (Cambridge: Harvard Univ. Press), is a good introduction to statistical studies of the history of science.

INSTITUTIONAL SUPPORT AND PROFESSIONAL SOCIETIES

It has long been customary for teachers of a science to give courses in the history of that science and for older professors to write books and articles surveying the development of their field. "General education" programs in the 1930s and 1940s often used the historical approach in science courses intended for nonscience majors. After World War II, especially in the United States and Great Britain, historians of science attempted to establish their field as a specialty within history rather than as an adjunct to science education. A few of the larger universities created separate departments of the history of science (e.g., Harvard, Wisconsin, Oklahoma, Johns Hopkins), or of the history and philosophy of science (Indiana, Pittsburgh), or more recently history and sociology of science (Pennsylvania). More frequently when there are only one or two historians of science at an institution, they are found in the history department. In Britain and other countries, historians of science are more likely to be associated with philosophy or a science department. (See the *ISIS Guide* mentioned above.)

Historians of science are also found, along with historians of technology, in museums, especially those of the Smithsonian Institution in Washington, D.C. Several government agencies and a few corporations employ historians to document their own activities.

Grants for research (and occasional special conferences) are provided by the History and Philosophy of Science Program of the National Science Foundation (NSF), and by several divisions of the National Endowment for the Humanities (NEH).

Degrees in the history of science are generally given only at the M.A. and Ph.D. level, with only a few exceptions. A list of graduate programs may be found in the *ISIS Guide to the History of Science.* I have given a more detailed discussion of the nature of these programs in my article "Education of Historians of Science in the U.S.A." [*Synthesis,* 3, no. 2(1975):6–19].

During the past few years opportunities have been provided for teachers without degrees in history of science to attend summer institutes, short courses, etc., sponsored by NSF, American Association for the Advancement of Science, and NEH. Under the current administration such opportunities have been sharply reduced, although the NEH Summer Seminars for College Teachers are still continuing and generally include two or three history of science seminars each year. Information can be obtained from the Fellowships Division, National Endowment for the Humanities, Washington, D.C., or from one of the departments listed in the *ISIS Guide.*

The History of Science Society was founded in the United States in 1924; it now has approximately 1600 members. Membership is open to all persons interested in the history of science; dues ($29 in 1987) include a subscription to *Isis* and to the *Newsletter.* The fifth number of *Isis* each year contains the *Critical Bibliography* of publications on the history of science in the preceding year. A *Guide* is prepared every three or four years (see above). The HSS has an annual meeting (often in conjunction with another group having related interests), and awards prizes to the authors of outstanding books and articles (including books suitable for teaching and public understanding of history of science as well as research monographs and student essays). Because of the substantial benefits provided for the rather low annual dues, membership in HSS is a "best buy" for anyone with even a part-time interest in the subject. Write to History of Science Society c/o *Isis,* University of Pennsylvania, 215 South 34th Street/D6, Philadelphia, PA 19104 (phone 215/243–5575).

The British Society for the History of Science serves similar purposes in the United Kingdom. Dues ($27 for U.S. members in 1986–1987) include a subscription to *The British Journal for the History of Science* (published four times a year) and a newsletter (three times a year). An annual list of dissertations is also available on request. The BSHS has a programme of general and topical meetings and recently started a monograph series. Application for membership should be sent to Executive Secretary, Wing-Commander G. Bennett, "Southide," 31 High St., Stanford-in-the-Vale, Faringdon, Oxfordshire SN7 8LH, England, U.K. Subscriptions are handled by Blackwell Scientific Publications Ltd., PO Box 88, Oxford, England, U.K.

An "International Congress of the History of Science" is held every four years under the auspices of the International Union of the History and Philosophy of Science (IUHPS); the next one will be in Hamburg and Munich, 1–9 August 1989. For information write to: ICHS Congress 1989, COP Hanser Service, Postfach 1221, D-2000 Hamburg-Barsbüttel, Federal Republic of Germany. A smaller "International Conference on the History and Philosophy of Science" has been held biennially since 1976, also sponsored by IUHPS.

Some of the scientific societies have historical divisions and/or centers, e.g., the Center for History of Physics at the American Institute of Physics in New York and the Division of History of Physics of the American Physical Society. The American Chemical Society, which has had a historical division for many years, recently started a Center for History of Chemistry at the University of Pennsylvania. A mathematics archive is being established at the University of Texas. Information about these can be obtained from the main offices of the societies.

To get a bird's eye view of where the discipline "history of science" has come from, at least in the past seventy-five years, and where it seems to be going, one should look at the following articles:

George Sarton (1936), *The Study of the History of Science,* a 52-page pamphlet (bound with *The Study of the History of Mathematics* in a Dover reprint), presents the more traditional view as developed by the founder of the discipline in the United States.

Arnold Thackray and Robert K. Merton (1972), "On Discipline Building: The Paradoxes of George Sarton," *Isis* 63:473–95, is a sociobiographical treatment; see also Thackray (1980), "The Pre-History of an Academic Discipline: The Study of the History of Science in the United States 1891–1941," *Minerva* 18:448–73.

Marshall Clagett, ed. (1959), *Critical Problems in the History of Science* (Madison: Univ. of Wisconsin Press), includes: D. Stimson, "The Place of the History of Science in a Liberal Arts Curriculum"; H. Guerlac, "History of Science for Engineering Students at Cornell"; major papers on nineteenth-century physics by T. S. Kuhn and I. B. Cohen; papers on nineteenth-century biology by J. W. Wilson and J. C. Greene. Then look at the "Retrospective Review" of this book by M. Rudwick, W. Coleman, E. Sylla, and L. Daston (1981) in *Isis* 72:267–83. In his introduction to the review, A. Thackray points out that the contributors to this book were the "first wave" of professional historians of science and that they have occupied the presidency of the History of Science Society for twenty of the twenty-five years from 1953 to 1978.

T. S. Kuhn (1968), "The History of Science," *International Encyclo-*

pedia of the Social Sciences 14:74–83, focuses on different directions emerging in the 1950s and 1960s.

A. R. Hall (1969), "Can the History of Science Be History?" *British Journal for the History of Science* 4:207–20.

C. Truesdell (1972), "The Scholar's Workshop and Tools," *Centaurus* 17:1–10, is a defense of the view that the historian of science must have competence in science.

S. G. Brush (1974), "Should the History of Science Be Rated X?" *Science* 183:1164–72, reflects on the relation between recent writings on the history and philosophy of science and traditional ideas about science as presented to students.

Arnold Thackray (1980), "History of Science," in **A Guide to the Culture of Science, Technology and Medicine,* ed. Paul T. Durbin, 3–69; also, "History of Science in the 1980s: Science, Technology, and Medicine," *Journal of Interdisciplinary History* 12:299–314. Thackray is a recent editor of *Isis* and represents the British influence, somewhat more oriented toward social or "externalist" historical problems, though like many externalists he has also done valuable "internalist" research (e.g., on the origin of Dalton's atomic theory).

Nathan Reingold (1980), "Through Paradigm-Land to a Normal History of Science," *Social Studies of Science* 10:475–96, explains why historians of science have largely ignored T. S. Kuhn's theory despite its enormous influence on other disciplines.

Carolyn Merchant (1982), "Isis' Consciousness Raised," *Isis* 73:398–409, gives feminist perspectives on history of science.

Frederick Gregory (1985), "The Historical Investigation of Science in North America," *Zeitschrift für allgemeine Wissenschaftstheorie* 16:151–66, describes recent developments and includes a list of graduate programs; David M. Knight (1984), "The History of Science in Britain: A Personal View," ibid. 15:343–53, describes how historians of science have turned (for inspiration and methodology) from scientists to philosophers to historians to sociologists, and wonders if they will next try psychologists or theologians.

1.4 THE SIXTEENTH- TO EIGHTEENTH-CENTURY BACKGROUND FOR MODERN SCIENCE

i. The Astronomical Revolution, Copernicus to Galileo
ii. The Seventeenth-Century Scientific Revolution: Bacon, Descartes, Harvey, Newton

iii. The "world machine" or "clockwork universe" philosophy
iv. Newtonian mechanics as a paradigm for other sciences; development and maturing of "descriptive" sciences
v. Romantic "nature philosophy" as a reaction against mechanism
vi. Rise of historicist/evolutionary viewpoint
vii. Growth of scientific institutions and national traditions

READINGS

Mason, *A History of the Sciences,* 127–64 (on *i*), 165–266 (*ii*), 269–372 (*iii–vi*), with additional material on *vii* in 256–66, 435–48. McKenzie, *Major Achievements of Science,* 16–31 (on *i*), 32–90 (*ii*), only a brief discussion of the philosophical aspects of the other topics, 122–32. Marks, *Science and the Making of the Modern World,* 1–42 (*i*), 44–76 (*ii*), 87–101 (*iii*), 103–15 (*iv*), 78–85 (*vii*). Toulmin and Goodfield, *Fabric of the Heavens,* 161–209 (on *i*), 210–49 (*ii*), 250–71 (philosophical reflections relevant to *iii*); *Architecture of Matter,* 137–98 (on *ii*), 202–28 (eighteenth-century chemistry). Their *Discovery of Time,* 83–140 (and later section on geology) concentrates on *vi.* Dampier, *History of Science,* 97–199 (*i–iv*).

The best general introduction to *i* and *ii* is the revised edition (1985) of I. B. Cohen's *Birth of a New Physics,* 1–184, with supplements based in recent historical research; this would be too much reading for a course devoted primarily to the period 1800–1950 but is useful background for the instructor. Similarly, for more extensive coverage of the eighteenth century one should consider Thomas L. Hankins, *Science and the Enlightenment* (1985).

The following might be assigned if other chapters of the book are being used later in the course: Magner, *History of the Life Sciences,* chapter on scientific societies, 137–51; Fancher, *Pioneers of Psychology,* chapter on Descartes, 3–42; J. Ben-David, *The Scientist's Role in Society,* 88–107; J. C. Greene, *Science, Ideology and World View,* 9–25; Gillispie, *Edge of Objectivity,* 151–201; Einstein and Infeld, *The Evolution of Physics,* 3–35; C. J. Schneer, *Evolution of Physical Science,* 21–130.

The Open University course *Science and Belief: From Copernicus to Darwin* provides an excellent introduction to this topic: Units 1–3 (123 large pages) on *i,* Units 4–5 (95 pages) on *ii,* and Units 6–8 (109 pages) on aspects of the interaction of science and religion in the seventeenth and eighteenth centuries.

Stephen Toulmin's article "The Emergence of Post-Modern Science" in *Great Ideas Today* (1981): 69–114, presents an interesting and readable perspective on the path from seventeenth-century to twentieth-century science.

BIBLIOGRAPHIC ESSAY

The best available concise guide to the literature on sixteenth- and seventeen-century science is "The Scientific Revolution" by Richard S. Westfall, a leading Newtonian scholar. It is a 6-page insert published in the July 1986 issue of the *History of Science Society Newsletter* (vol. 15, no. 3) and will be available as a separate pamphlet in the series "Teaching in the History of Science: Resources and Strategies" issued by the Committee on Education of the History of Science Society. For information on this series contact Prof. Kathryn Olesko, Department of History, Georgetown University, Washington, D.C. 20057.

Until recently, the eighteenth century has been regarded as a rather uneventful period in which the spectacular advances of the seventeenth century were consolidated to provide a basis for the major discoveries of the nineteenth and twentieth centuries. Classic works like Herbert Butterfield's *The Origins of Modern Science* (New York: Free Press, reprinted 1965) and A. R. Hall's *The Scientific Revolution 1500–1800* (Boston: Beacon Press, 1954, reprinted 1962) present eighteenth-century science from this viewpoint, while E. A. Burtt, in his *Metaphysical Foundations of Modern* [sic] *Physical Science* (Garden City, N.Y.: Doubleday, 2d ed. 1932, reprinted 1954), seems to think nothing significant has happened in physics since 1700. Of the older books, the most useful is A. Wolf, *A History of Science, Technology, and Philosophy in the XVIIIth Century* (New York: Macmillan, 1938, rev. ed. 1952), especially for *iv*. The stereotype "Newtonian world-machine" picture is expounded in J. H. Randall's *The Making of the Modern Mind* (New York: Columbia Univ. Press, 1926, 1940), Book III, while Book IV touches on *v* and *vi*. Randall's *Career of Philosophy* (New York: Columbia Univ. Press, 1962, 1965) gives a more sophisticated and detailed treatment of the philosophical background of eighteenth-century science.

The view that classical mechanics was completed by Newton has been vigorously attacked by C. Truesdell, who has revived the work of Leonhard Euler and other eighteenth-century mathematicians. His interpretations and much detailed technical criticism may be found in his introductions to *Euleri Opera Omnia,* ser. 2, vols. 10, 11, 12, and 13 (Zurich: Fussli, 1954–1960), in his *Essays in the History of Mechanics* (New York: Springer-Verlag, 1968), and in *Handbuch der Physik* 3, no. 1(1960):226–793. For brief introductions see Truesdell's "History of Classical Mechanics," *Naturwissenschaften* 63(1976):63–66, 119–30, and

"Leonhard Euler, Supreme Geometer (1707–1783)" in *Studies in Eighteenth-Century Culture* 2(1972):51–95. A different view of the significance of eighteenth-century physics is presented by Y. Elkana in "The Historical Roots of Modern Physics," in *History of Twentieth Century Physics,* ed. Charles Weiner (New York: Academic Press, 1977), 197–265.

An excellent overview of recent scholarship may be found in *The Ferment of Knowledge: Studies in the Historiography of Eighteenth-Century Science,* ed. G. S. Rousseau and Roy Porter (New York: Cambridge Univ. Press, 1980). Two stimulating but quite different discussions of the relation between different scientific fields are T. S. Kuhn, "Mathematical versus Experimental Tradition in the Development of Physical Science," *Journal of Interdisciplinary History* 7(1976):1–31 [reprinted in his *Essential Tension* (Chicago: Univ. of Chicago Press, 1977)] and Colm Kiernan, *Science and Enlightenment in Eighteenth-Century France* (Banbury, Eng.: Voltaire Foundation, 1968, 1973). Other important works relevant to the history of particular sciences will be mentioned later, e.g., **2.2** (biology).

2 EVOLUTION

2.1 HISTORY OF THE EARTH

i. Early theories about the origin of the solar system, Johannes Kepler to Georges-Louis de Buffon
ii. The nebular hypothesis of Kant, Laplace, and Herschel
iii. Geology in the eighteenth century: James Hutton, Abraham Werner, and the Neptunist-Plutonist controversy
iv. Early nineteenth-century ideas about origin and cooling of Earth—J. B. J. Fourier and the contraction theory (Elie de Beaumont)
v. Charles Lyell; uniformitarian vs. catastrophic philosophy; British geology before 1859

READINGS

Note: Publication data for books recommended as "Readings" may be found in the Book List at the end.

Stephen Mason, *History of the Sciences* (1962), 395–411 (*i–iii, v*), 295 (*iv*). A. E. E. McKenzie, *Major Achievements* (1973), 107–21 (*ii, iii, v*), 413–23 (*ii, v* sources). Stephen Toulmin and June Goodfield, *Discovery of Time* (1965), 141–70, 189–94 (*i–v*). Charles C. Gillispie, *The Edge of Objectivity* (1960), 291–302 (*ii, iii, v*). C. J. Schneer, *Evolution of Physical Science* (1984), 159–85. John Marks, *Science and the Making of the Modern World* (1983), 156–64.

Vincent Cronin, *The View from Planet Earth* (1981), 179–202, gives a good overview of this subject.

Books on history of biology and on evolution frequently have short sections on this topic: Lois Magner, *A History of the Life Sciences* (1979), 374–76; William Coleman, *Biology in the Nineteenth Century* (1971), 61–65; Michael Ruse, *The Darwinian Revolution* (1979), 36–48;

Peter J. Bowler, *Evolution* (1984), 23–45, 112–17, 126–31. The most comprehensive of these is Loren Eiseley's *Darwin's Century* (1961), 41–42, 57–115 (*i–iii, v*). John Greene's *Death of Adam* (1959), 15–87 covers all these topics, including cosmogony.

Of the many books on the history of astronomy, the one with the best account of the nebular hypothesis is Charles A. Whitney, *The Discovery of Our Galaxy* (1971), 69–86, 103–54.

The Open University course on *Science and Belief: Copernicus to Darwin* includes a section by R. Hooykaas, "Genesis and Geology" (Unit 11, with 20 large pages plus a section of study questions) which would be appropriate supplemental reading for this topic but probably not suitable as an introduction for beginning students. The sequel, *Darwin to Einstein* (A381), Block III, by Colin Russell, includes an excellent account of the debate on the age of the Earth (17–27), followed by a discussion of "Religion and Science in Kelvin's Thought" (28–35). The anthology prepared by N. G. Coley and V. M. D. Hall, *Darwin to Einstein: Primary Sources on Science and Belief* (1980), includes extracts from papers by W. Thomson.

SYNOPSIS

The problem of the *origin* of the solar system could not be considered as such until after the *existence* of the solar system had been accepted as a reasonably plausible theoretical description of astronomical observations. Copernicus published his heliocentric theory in 1543, but it was Johannes Kepler who first attempted to explain, in his *Mysterium Cosmographicum* (1596), *why* the planets should occupy orbits at the distances indicated by the Copernican theory. Kepler proposed that these distances are the ones that could be obtained by nesting a series of spheres alternating with the five regular solids of Euclidean geometry. Since it is mathematically impossible to construct more than these five kinds of solids, Kepler's theory appeared to suggest a reason why there are just six planets in the system. If one is willing to accept a rather arbitrary sequence of these solids, one can also get approximately the right values for the ratios of the distances of the planets from the sun.

Unlike later planetary theorists, Kepler assumed that the solar system was created in essentially its present form rather than reaching it through an evolutionary process. His mathematical justification for the sizes of the planetary orbits has not been taken seriously by modern astronomers, although since the discovery of the Titius-Bode law (**12.1**) it has generally been thought there is some kind of mathematical regularity in these orbits that a proper theory should be able to explain. Before concluding Kepler's purely mathematical theory is not "proper," it would be well to look at some of the theories recently used by physicists to predict the existence and properties of elementary particles (**10.4**).

Kepler's later suggestion that magnetic lines of force extending from a rotating sun push the planets around in their paths was rejected as a primary cause for planetary motion after the law of inertia was adopted by Galileo, Descartes, and Newton. According to that law, the planets would continue to move indefinitely in straight lines in the absence of any force; the sun's force was needed only to explain why they deviated from a straight line to an elliptical orbit. But something very similar to Kepler's idea was proposed in the twentieth century to account for the transfer of angular momentum from the sun to the planets during the early evolution of the solar system (Alfven's "magnetic braking" effect, see **12.4**).

Descartes's vortex theory (1644) anticipates in a general way the approach to be followed by later scientists in constructing a theory of the solar system based on mechanical principles; but it was not worked out in enough detail to give a plausible explanation of the way the system could have evolved. He postulated that there is a stratification of particles in the solar system, the particles nearest the Sun being the smallest and most rapidly moving. Each planet has an equilibrium position in a vortex of moving particles, and that vortex in turn moves in the large vortex dominated by the sun. Descartes encountered an anomaly that was to plague later theories: the rotation of the sun, as indicated by the motion sunspots, is much slower than one would expect from extrapolating the motion of the vortex particles around it.

Newton devoted part of his *Principia* (1687) to proving that the Cartesian theory could not account for Kepler's laws of planetary motion; but Newton himself, while presenting a very successful theory of the present operation of the solar system, made no attempt to explain its origin.

Emanuel Swedenborg proposed, in his *Principia Rerum Naturalium* (1734), that the planets had been produced from the sun and later moved outward to assume their present orbits.

The two major theories of the origin of the solar system, which have provided the basis for almost all subsequent speculation, were proposed in the eighteenth century: the first by Buffon, the second by Kant and Laplace. It is remarkable that, in spite of the enormous amount of information which has been acquired about the solar system since the eighteenth century, scientists have not been able to choose definitely between these two theories or to come up with an essentially different alternative that might be more satisfactory (though the "supernova trigger" was for a time thought to be such an alternative). Thus the origin of the solar system is one of the oldest of the scientific problems that are still unsolved today.

Buffon (1745), like other theorists, was struck by the fact that all planets in our system revolve in the same direction around the sun in orbits that lie within a few degrees of the same plane. This could hardly be the result of chance; it must, he argued, be attributed to a single event that gave the planets an impulse in the same direction. He assumed that a comet passed near to the sun and displaced from it a small amount of material, at the same time setting this material into motion around the sun. By this time the masses of the planets were known approximately, and Buffon noted that they add up to only about 1/650th of the sun's mass; not having much definite information about the masses of comets, he thought it reasonable to suppose that a comet could remove that much mass from the sun in a close encounter.

Scientists who later adopted similar "encounter" theories criticized the assumption that a body as light as a comet could produce such an effect and assumed instead that another star was involved. But Buffon's theory did have one important point in its favor: comets had been observed to pass fairly close to the sun, whereas all other theories have had to postulate events very much different from any that we have directly observed.

The other major theory is the "nebular hypothesis" of Kant and Laplace. Whereas Buffon had assumed the prior existence of the sun, the Kant-Laplace theory supposes that the planets began to form at the same time as the sun, both arising as part of a single process of contraction and condensation of a whirling cloud. Laplace's version (1796) was somewhat more specific: he postulated that a ring of material first separates at the outer edge of the cloud, where shrinking has increased the speed of rotation (because of conservation of angular momentum) to the point where centrifugal force exceeds the gravitational attraction toward the center. The ring then accumulates to a single gaseous ball (spinning around its own axis), which later condenses to form a planet. A similar process of ring spin-off may produce satellites, which of course should revolve around the primary in the same direction that the primary itself rotates. Though Laplace did not carry through any quantitative calculations, his hypothesis seemed to offer the possibility of explaining the distances of planetary orbits.

When astronomers began to examine the distant objects known as "nebulae," they found that many of them could not be resolved into separate stars even with the best telescopes. William Herschel (see **12.1**) concluded that these must be clouds of gas still in the stage preceding the formation of stars and planets. Thus the term "nebular hypothesis" came to be applied to Laplace's theory; it implied that the formation of our own solar system was not a unique or rare event depending on the chance encounter of two celestial objects but was being continually duplicated elsewhere in the universe in a process that might even be observable with a sufficiently powerful telescope.

The universality of the nebular hypothesis has always been one of its great attractions: it promises the existence of intelligent life elsewhere in the universe, since the same process that forms a star may also form planets.

The Kant-Laplace hypothesis, like Buffon's theory, assumed that the planets had been formed in a gaseous or molten state and later solidified as they cooled. Thus both theories meshed with the idea of a hot, cooling Earth that dominated geological theory during most of the eighteenth and nineteenth centuries (**2.2, 7.3**). This led to a rather paradoxical situation: the nebular hypothesis was the first major example of an "evolutionary" theory in modern physical science and as such may have helped to pave a pathway for the later acceptance of Darwin's evolutionary biology. (This is suggested by Ronald Numbers in his 1977 book, *Creation by Natural Law.*) Early evolutionists like Robert Chambers and Herbert Spencer constructed schemes beginning with the formation of a nebula and running through the evolution of living forms including man. Yet Kelvin's calculation of the age of the Earth based on this theory became one of the major arguments *against* Darwin's theory.

Astronomical studies of nebulae influenced views about the nebular hypothesis throughout the nineteenth century. When William Parsons, the Earl of Rosse, succeeded in resolving some nebulae into distinct stars in the 1840s, this was taken as evidence against the hypothesis that our own system originated as a cloud of gas. When William Huggins discovered in the 1860s that the spectra of some nebulae consisted of a few isolated bright lines rather than a mixture of stellar spectra, he concluded that they were gaseous and this helped to revive the nebular hypothesis. When photography showed that some nebulae have a spiral structure, this was taken by T. C. Chamberlin as evidence that the solar system was formed by an encounter of two stars rather than by condensation of rings associated with the formation of the sun. Finally, in the twentieth century when the distances and sizes of nebulae were found to be on the order of hundreds of millions of light years, so that they are separate galaxies, the assumption of a direct connection between such nebulae and the formation of individual stars and planetary systems had to be abandoned, though the terms "nebular hypothesis" and "solar nebula" survive in modern discussions.

The advance of solar system astronomy also provided new evidence, though much of this evidence simply undermined the assumption of regularity on which the nebular hypothesis had been based without pointing to a specific alternative. Thus, it was assumed throughout the nineteenth century that the rotations of the first six planets are all "direct" (in the same direction as the motion of the planets in their orbits around the sun), though it was realized that reliable determinations of the rotation periods of Mercury and Venus were still lacking. (Venus was later found to have retrograde rotation.) By the middle of the nineteenth century it had been established that the Uranus system is a clear-cut exception; the rotation of the planet and the motions of all known satellites are around an axis nearly perpendicular to that of the orbital motion, so these motions are retrograde. However, the fact that the rotation of the primary agrees with the motion of the satellites could still be regarded as a confirmation of the Laplacian hypothesis for the formation of satellites, and the tilt of the entire system could be attributed to a single disturbing event. Hence when the orbital motion of Neptune's satellite Triton was found to be retrograde, it was generally assumed that the rotation of Neptune itself is also retrograde (reliable observations had not been made, and it was subsequently found that the rotation is actually direct). If this assumption were granted, one could say that the rotations and satellite motions for the first six planets are all direct, those of the Neptune system are retrograde, and those of the Uranus system are about halfway between direct and retrograde. This would suggest a trend toward retrograde motion as one recedes from the sun.

To account for this trend—based on "facts" which, in our superior wisdom, we may consider somewhat theory laden—it was generally assumed that there were two causes in operation during the formation of the system: one tending to produce direct rotation, the other retrograde. The first cause would have dominated inside the main part of the nebula, while the second dominated in its outer regions. For example, Laplace's original hypothesis implied that the inner parts of the nebula would be dense and viscous and thus might tend to move almost like a rigid body in which the linear speed of the outer parts would be greater than that of the inner parts. Accretion of particles from neighboring orbits would lead to direct rotation. On the other hand if the particles were moving independently in Kepler orbits, the opposite would be expected since linear speed would decrease with distance from the central body.

The nebular hypothesis also implied that the rotation of a central body would be greater than the speed of revolution of smaller bodies around it. This consequence was contradicted by the case of Phobos, the inner satellite of Mars discovered by Asaph Hall in 1877; it revolves around the primary three times for every rotation. More generally, if one assumes that the nebula was originally of uniform density and rotated with the same angular velocity throughout, there is a gross discrepancy with the present distribution of angular momentum, as was noted by J. Babinet in 1861; the sun is rotating much too slowly.

Geology

The state of geology at the end of the eighteenth century has usually been described in terms of a competition between two theories: Neptunism and Plutonism.

The Neptunists, led by Abraham Gottlob Werner (1750–1817) at Freiburg in Germany, assumed that all the rocks found on the Earth's surface had formed by precipitation from an ocean that once covered the entire globe. While Werner did not explicitly invoke the Biblical Flood to justify this assumption, the popularity of his theory is sometimes attributed to its implied connection with sacred history.

The Plutonists, following James Hutton (1726–1797) at Edinburgh, postulated that rocks were formed by heat and pressure inside the Earth. Although Hutton's theory assumed a "central fire" as did the cooling-Earth theories of Buffon and his followers, he rejected the view that the Earth had been much hotter in the past. Instead he postulated a cyclic process: upheavals powered by a *constant* internal heat alternating with erosion. As his disciple John Playfair expressed it in 1802, the Earth shows no signs of infancy or old age; just as in the solar system whose stability had recently been demostrated by Laplace, we cannot infer a beginning or end as long as the laws of nature are valid.

Recent historical research, presented most forcefully in a recent book by Rachel Laudan, *From Mineralogy to Geology* (1987), indicates that the supposed competition was quite one-sided. Werner and his followers actually dominated geology up to the 1830s. Their work addressed the outstanding problems of the science—the location and formation of rocks and minerals, crucial to the mining industry, which provided institutional support for geological research. Hutton's theories were not seen as relevant to these problems and he had few supporters until Lyell reformulated his views.

Charles Lyell (1797–1875) has likewise been seen as a major protagonist in another competition: Uniformitarianism versus Catastrophism. His importance has been inflated, especially in Anglo-American writings on the history of geology, for several reasons. First, Charles

Darwin followed Lyell's doctrines, and thus Lyellian geology became part of the prehistory of Darwinism. Second, in his *Principles of Geology* Lyell presented a version of the earlier history of the subject that denigrated or omitted the opposing theories, making it appear that he himself was the founder of modern geology, and his followers adopted this view in their own writings. Third, the term "Uniformitarian" has at least two different meanings, and the easy acceptance of one gives undeserved credibility to the other. It may simply mean that the laws of nature always remain the same, and perhaps also that we should explain the past in terms of causes or forces that can be seen to operate in the present. So far most geologists would agree. But Lyell also assumed that the *magnitude* of geological forces has never been significantly different than at present; this meaning of Uniformitarian would exclude cooling-Earth theories, since geological forces would have been stronger in a hotter Earth.

Lyell's opponents, such as Leonce Elie de Beaumont, were actually much more influential than Lyell on the Continent, as Mott Greene shows in his book *Geology in the Nineteenth Century* (1982). Their "catastrophes" were not arbitrarily invoked or supernatural events, but followed naturally from the model of a cooling, contracting Earth. For example, after the outer layer solidified it would not contract as rapidly as the still-liquid interior, so a gap would start to form between crust and core. Deprived of support, the crust would collapse, forming irregular bulges and wrinkles (continents and mountain chains).

Both sides (Uniformitarians and Catastrophists) relied on heat more than water to explain geological formations, and both rejected the assumption that geological history must be limited to the 6000 or fewer years of Biblical chronology. But where the Catastrophists linked geological history to the original formation of the Earth from a hot nebula, and thus could (at least in principle) estimate its age from its rate of cooling, the Uniformitarians pushed the present physical conditions back indefinitely far into the past. Lyell wanted to sever geology from cosmogony; the latter had acquired an excessively speculative character in the eighteenth century "theories of the Earth" and still seemed to threaten his goal of a rigorously scientific discipline.

Lyell was able to justify his neglect of any change in the Earth's internal temperature by citing the conclusions of the French mathematician Joseph Fourier (1768–1830). Fourier developed a theory of heat conduction, combined with a method for representing functions in terms of trigonometric series that allowed him to reconstruct the thermal history of an idealized Earth model. Assuming the Earth to be a homogeneous solid sphere, initially at a uniform high temperature in a space at zero temperature, Fourier derived an equation for the time required to cool down to the present temperature. The equation contained as factors the initial temperature, the heat capacity, thermal conductivity, and present value of the surface temperature gradient (rate at which temperature increases as you go down into the Earth). Using numerical values of these parameters suggested by Fourier one gets from 130 to 200 million years for the cooling time. Apparently he considered this such an absurdly long time that he didn't bother to state it explicitly. But he did state that the rate of decease in surface temperature would be only a fraction of a degree per century, and Lyell concluded that this was too small to have any geological consequences.

As long as geologists considered 100 million years an unimaginably long time, they could safely ignore the effect of the cooling of the Earth on its surface temperature. But by the middle of the nineteenth century, British geologists following Lyell had accepted the idea that geological history extends over "very long" periods, and some of them were willing to say that "very long" could mean "hundreds of millions of years." Charles Darwin, who found it convenient to suppose that physical conditions on the Earth's surface had been hospitable to life for such long periods, thus thought he could reasonably invoke geological periods of the order of 300 million years in the first edition of his *Origin of Species* (1859). At this point physical and biological evolution suddenly collided (**2.4**-*ii*).

BIBLIOGRAPHY

D. C. Lindberg and R. L. Numbers, eds. (1986), *God and Nature* (Berkeley: Univ. of California Press), includes articles by R. Hahn on Laplace and by M. J. S. Rudwick on Earth history.

Henry Faul and Carol Faul (1983), *It Began with a Stone: A History of Geology from the Stone Age to the Age of Plate Tectonics* (New York: Wiley), 85–143, could be used as a text for a course giving major emphasis to geology.

A. Hallam (1983), *Great Geological Controversies* (New York: Oxford Univ. Press).

Roy Porter (1983), *The Earth Sciences: an Annotated Bibliography* (New York: Garland); (1977), "Research in British Geology 1660–1800: A Survey and Thematic Bibliography," *Annals of Science* 34:33–42.

Wolf von Engelhardt (1982), "Kraefte der Tiefe—Zum Wandel Erdgeschichtlicher Theorien," *Janus* 69:119–40, is a survey of theories of the Earth from seventeenth to early twentieth centuries.

Topic *i:* Early Theories of the Origin of the Solar System

Stanley L. Jaki (1978), *Planets and Planetarians* (Edinburgh: Scottish Academic Press), Chaps. 1–4.

Topic *ii:* Nebular Hypothesis

Stephen G. Brush (1987), "The Nebular Hypothesis and the Evolutionary Worldview," *History of Science* 25:245–78.

Jacques Merleau-Ponty (1983), *La Science de l'Universe à l'Âge du Positivisme: Etude sur les Origines de la Cosmologie Contemporaine* (Paris: Vrin); (1977), "Laplace as a Cosmologist," in *Cosmology, History, and Theology,* ed. W. Yourgrau and A. D. Beck (New York: Plenum), 283–91; (1976), "Situation et Rôle de l'Hypothése Cosmogonique dans la Pensée Cosmologique de Laplace," *Revue d'Histoire des Sciences* 29:21–49.

Simon Schaffer (1980), "Herschel in Bedlam: Natural History and Stellar Astronomy," *British Journal for the History of Science* 13:211–39; (1978), "The Phoenix of Nature: Fire and Evolutionary Cosmology in Wright and Kant," *Journal for the History of Astronomy* 9:180–200.

J. H. Brooke (1979), "Nebular Contraction and the Expansion of Naturalism," *British Journal for the History of Science* 12:200–211 is an essay review of R. L. Numbers, *Creation by Natural Law* (cited below) and remarks on other recent historical studies.

Ronald L. Numbers (1977), *Creation by Natural Law: The Nebular Hypothesis in American Thought* (Seattle: Univ. of Washington Press).

S. L. Jaki (1976), "The Five Forms of Laplace's Cosmogony," *American Journal of Physics* 44:4–11.

Topic *iii:* Neptunism-Vulcanism

Rachel Laudan (1987), *From Mineralogy to Geology: The Foundations of a Science, 1650–1830* (Chicago: Univ. of Chicago Press).

Dennis R. Dean (1981), "The Age of the Earth Controversy: Beginnings to Hutton," *Annals of Science* 38:435–56.

Roy Porter (1980), "The Terraqueous Globe," in *The Ferment of Knowledge,* ed. G. S. Rousseau and R. Porter (New York: Cambridge Univ. Press), 285–324; (1979), "Creation and Credence: The Career of Theories of the Earth in Britain, 1660–1820," in *Natural Order,* ed. B. Barnes and S. Shapin (Beverly Hills, Calif.: Sage), 97–123; (1977), *The Making of Geology* (New York: Cambridge Univ. Press).

R. Grant (1979), "Hutton's Theory of the Earth," in *Images of the Earth,* ed. L. J. Jordanova and R. S. Porter (Chalfont St. Giles, Eng.: British Society for the History of Science), 23–38.

Topic *iv:* Cooling of Earth

Mott Greene (1982), *Geology in the Nineteenth Century* (Ithaca: Cornell Univ. Press), corrects the impression given by most Darwin-oriented historians of science that Lyell's views dominated the subject; Greene shows that continental European and American geologists were going in rather different directions.

Stephen G. Brush (1979), "Nineteenth-Century Debates about the Inside of the Earth: Solid, Liquid or Gas?" *Annals of Science* 36:225–54.

P. J. Lawrence (1977), "Heaven and Earth: The Relation of the Nebular Hypothesis to Geology" in *Cosmology, History, and Theology,* ed. W. Yourgrau et al. (New York: Plenum), 235–81.

Topic *v:* Lyell, etc.

Roy Porter (1982), "Charles Lyell: The Public and Private Faces of Science," *Janus* 69:29–50.

Martin J. S. Rudwick (1982), "Cognitive Styles in Geology," in *Essays in the Sociology of Perception,* ed. Mary Douglas (Boston: Routledge and Kegan Paul), 219–41; (1979), "Transposed Concepts from the Human Sciences in the Early Work of Charles Lyell," in *Images of the Earth,* ed. L. J. Jordanova and R. S. Porter (Chalfont St. Giles, Eng.: British Society for the History of Science), 67–83; (1977), "Historical Analogies in the Geological Work of Charles Lyell," *Janus* 64:89–107.

Leonard Wilson (1980), "Geology on the Eve of Charles Lyell's First Visit to America, 1841" *Proceedings of the American Philosophical Society* 124:168–202; (1972), *Charles Lyell* (New Haven: Yale Univ. Press).

Dov Ospovat (1977), "Lyell's Theory of Climate," *Journal of the History of Biology* 10:317–39.

Roy Porter et al. (1976), Papers in Lyell Centenary Issue, *British Journal for the History of Science* 9:91–242.

2.2 LIFE SCIENCE BEFORE DARWIN

i. Theories of reproduction and heredity
ii. Taxonomy, Linnaeus to Georges Cuvier; fossils; Buffon
iii. Theology: the argument from design, religious teachings
iv. Nature philosophy
v. Lamarck's theory of evolution
vi. Cell theory (Schleiden, Schwann)
vii. The organism as a machine: physico-chemical analysis in physiology

This is an example of a subject I pass over fairly quickly in my one-semester course, hence the absence of a synopsis. Instead, I present here some recommendations of works on the history of biology for those who wish a more extensive treatment.

Topics *i–v* are intended to lead directly into section **2.3**. Topics *vi* and *vii* may be included in a full-year course or in a semester course that gives greater emphasis to biology. Some knowledge of cell theory will be helpful in understanding the later material on the role of chromosomes in heredity (section **4.4**; see also **4.3**-*iii*).

READINGS

Of the general textbooks I consider suitable for this course (see section **1.2**), Mason's *History of the Sciences,* 331–92, has the most extensive discussion of topics *i–ii* and *vi–vii.* Toulmin and Goodfield, *Discovery of Time,* 96–101, 171–88, present a brief survey of *i–iii* and *v;* their *Architecture of Matter,* 307–60, provides good coverage of *vi* and *vii.* McKenzie's *Major Achievements of Science* has only a few pages (200–202, 469–72) and is really inadequate for any course giving serious attention to biology, so it would have to be supplemented by other readings. Another general text, Gillispie's *Edge of Objectivity,* 260–90, is good on items *ii, iii,* and *v.* Dampier, *History of Science,* 184–87, 252–75, is a little better than McKenzie.

H. J. Fyrth and M. Goldsmith, *Science History and Technology,* Book 2, Part I (1969), Chap. 5, is a brief overview of the relation between geology, evolution, religion, socialism, and the fate of nineteenth-century optimism (52–66). The same book also surveys industry and technology.

Among books on the history of biology, several have good sections dealing with these topics. If you want to include topics *vi* and *vii* but do not need to cover the life sciences before or after the nineteenth century, William Coleman's *Biology in the 19th Century* is one possibility, though many students do not find it easy to read (1–71 and 18–59 for this section, and the rest of the book for evolution). Lois Magner, *A History of the Life Sciences,* 179–234, 288–373, concentrates on *i, ii,* and *v–vii* with brief mentions of *iii* and *iv.* An older but still useful work is Charles Singer's *History of Biology to About the Year 1900* (3d ed. 1959), 174–203, 215–27, 296–300, 329–411, 505– 47 on topics *i, ii, iv–vii* (not in the same order).

Dale L. Ross, *Studies in Biology* (1972), provides a treatment for high school students, including ideas about embryology from Aristotle to modern theories of generation; it does not seem to be available at present in the United States.

Of the books specifically on evolution (which might be chosen because the course will give major emphasis to that topic), I find Peter Bowler's *Evolution* the most comprehensive and up-to-date; for this section, 1–22, 46–112, 119–26, 131–41 are appropriate. Loren Eiseley's *Darwin's Century* is still useful, perhaps supplemented by John Greene's *Science, Ideology, and World View* (1981). Eiseley concentrates on *ii* and *v* (1–55) with a brief section on *iii* in a later chapter (175–78). Greene attempts to test T. S. Kuhn's theory of scientific revolutions by treating the theories of Linnaeus and Buffon as paradigms; he is one of the handful of historians who have taken Kuhn's model seriously, so the article he reprints here (30–55) should be useful for that reason.

Greene's earlier book, *The Death of Adam,* deals extensively with pre-Darwinian ideas relevant to evolution and has been successfully used in courses given by my colleagues; 89–247 provide useful background for *ii, iii* and *v*. Michael Ruse, *Darwinian Revolution,* 1–15, 63–155, contains additional discussion of the views of Lyell, Richard Owen, and Chambers on organic progression and emphasizes some of the philosophical-methodological aspects omitted by historians. Appleman, *Darwin* (1979), 3–31, reprints extracts from G. de Beer, Lyell, J. D. Hooker, M. Millhauser, and Darwin on his predecessors. Since I recommend H. Wendt, *In Search of Adam* (1973), if one is going to cover the evolution of man (section **2.5**), his first seven chapters (3–190) would be quite appropriate here.

David Oldroyd's *Darwinian Impacts* (1983), covers **2.1–4**, parts of **3.1, 3.3, 4.3**, and the influence of Darwinism on theology, literature, music, etc. in about 375 pages.

The Open University course, *Science and Belief: From Copernicus to Darwin* (1974) has an excellent presentation of topic *iii* in its units 9–10, "Natural Theology in Britain from Boyle to Paley" by J. H. Brooke (47 large pages, with several study questions throughout the text). Unit 12 in the same course, "Precursors of Darwin?" (also by Brooke) covers *v* and parts of *i* and *ii*). These texts are well worth looking at, even if it turns out not to be practical to use them as required readings for students (because of cost or availability factors).

WAS THERE A PARADIGM BEFORE DARWIN?

Despite the fact that Kuhn's theory of the history of science is more widely known than any other, there have been very few serious attempts by historians of science to submit it to any kind of rigorous test. One such test is reported in John C. Greene's article, "The Kuhnian Paradigm and the Darwinian Revolution in Natural History" (cited below under *ii*).

Greene argues that systematic natural history did have a Kuhnian paradigm, established in the eighteenth century by the work of John Ray, Joseph Pitton de Tournefort, and Carl Linnaeus. It was presented in its definitive form in Linnaeus's book *System of Nature* (1735), where the three "kingdoms" (animal, vegetable, and mineral) are divided into classes, orders, genera, and species, and systematic rules are given for classifying and naming each species (based for example on the sexual organs and method of reproduction). It is assumed that each species was created in its present form by God and is immutable, although Linnaeus retreated from this doctrine and suggested the possibility of hybridization in some of his later writings.

According to Kuhn's theory one might have expected that the Linnaean paradigm would have dominated the field of natural history until "anomalies" led to a "crisis" and a revolution calling forth a new paradigm. Instead, an alternative paradigm appeared almost immediately in the works of Buffon. Buffon approached natural history from the standpoint of Newtonian physics. He rejected the Linnaean assumption that the task of natural history is to find the most "natural" system for classifying a fixed set of species; he assumed that God creates not species but matter and motion subject to definite forces and laws, resulting in the generation of various kinds of plants and animals under different physical conditions. Moreover, Buffon embedded his speculations about the evolution of biological species in the framework of a grand scheme of the origin and physical development of the Earth itself (**2.1**).

Buffon and Linnaeus differed on a basic philosophical issue: Linnaeus accepted the Aristotelian doctrine of final causes (teleology) and assumed that God has created all living things (and their parts) for a definite purpose. Buffon rejected final causes and asserted all things that *can* exist *do* exist regardless of their usefulness to an organism. Thus Buffon pointed to men's nipples as an example of the failure of teleological explanation.

Historians have found it difficult to make sense of Buffon's ideas about the origin and transformation of organic species. He seems to have believed that organisms arose spontaneously by the combination of "organic molecules" that attract each other in a manner analogous to Newtonian gravity; but he sees no definite sequence of development from simple to complex. He recognized the stability of species, once formed, resulting from the inability of members of different species to breed with each other; yet they could degenerate, as they had arisen, by natural processes.

Despite their sharp differences on many points, both Linnaeus and Buffon considered humans to be animals; Linnaeus included Man with the apes and sloths in the same order in the general class of Quadrupeds; Buffon advocated comparing humans with other animals to find similarities and differences.

Buffon's theory did not arise as a response to anomalies within the existing Linnaean paradigm. Instead, as Greene notes, it was a conscious attempt to introduce into natural history concepts derived from physical science. It did not replace the Linnaean paradigm but coexisted with it; thus it might be interpreted as a rival "research programme" or "research tradition" in the sense of Imre Lakatos or Larry Laudan, respectively. Or it could be seen as the echo in biology of a more general theme (in Gerald Holton's sense), the evolutionary worldview that was coming to dominate cosmological speculation in the late eighteenth century. (See **11.5** for the theories of Lakatos, Laudan, and Holton). While any generalization of this kind overlooks nationalistic and other differences, I would suggest that Linnaean and Buffonian biology be seen not as self-contained paradigms but as representatives of two opposed views of nature. The modern descendants of these views are Creationism and Evolution, but it is almost as long a path from the vague naturalism of Buffon to the precise neo-Darwinian synthesis of Theodosius Dobzhansky and Ernst Mayr (**4.6**) as from the respectable scientific taxonomy of Linnaeus to the antiscientific biblical creationism of Henry Morris and Duane Gish (**4.7**).

The theory most often considered as an alternative to Darwinism in the late nineteenth and early twentieth centuries is the one proposed by Jean Baptiste Pierre de Monet, Chevalier de Lamarck. In his book on zoological philosophy published in 1809, Lamarck proposed a theory of evolution based on four laws:

1. Nature tends to increase the size of living individuals to a predetermined limit.
2. The production of a new organ results from a new need.
3. The development reached by the organs is directly proportional to the extent to which they are used.
4. Everything acquired by the individual is transmitted to its offspring.

The result of the operation of these laws would be progressive development of species from simple to complex forms, branching off in different directions depending on local environmental conditions. One result would be humans, evolving from apes as a result of a "need" for better locomotion ability (thus upright posture) and for better means of communication (thus speech). The term "Lamarckism" usually refers primarily to the fourth law, known as "inheritance of acquired characteristics."

Lamarck's theory was not popular before 1859 and certainly does not qualify as a Kuhnian paradigm. But after Darwin's work convinced biologists that evolution had occurred, Lamarckism was frequently invoked to explain *how* it could have occurred by those who thought Darwin's explanation was not adequate.

We will not understand the origin of Charles Darwin's theory if we merely search out evolutionary precursors in the history of biology or anomalies in the biological theories of his day. By his own testimony, the most important stimulus to the invention of his hypothesis of natural selection came not from science but from the writings of an economist, Thomas Malthus. In his *Essay on Population* (1798), Malthus pointed out that population tends to grow geometrically (2, 4, 8, 16 . . .) whereas the food supply cannot grow more than arithmetically at best (1, 2, 3, 4 . . .). Hence there is an inevitable competition for food among the members of the human race, and population is limited by famine, disease, or war. Malthus did not suggest that biological species could change or that the struggle for existence could improve a species, yet Darwin (and Herbert Spencer) got just those ideas by reading Malthus. Philosophical theories of scientific change, until recently, have ignored such interactions between disciplines as a source of progress because they have refused to consider the origins of ideas—the notorious distinction between the "context of discovery" and the "context of justification."

BIBLIOGRAPHY

William Coleman and Camille Limoges, eds., *Studies in History of Biology,* is an annual series that includes important scholarly articles on pre-Darwinian biology.

Judith Overmier (1986), *The History of Biology: An Annotated Bibliography* (New York: Garland).

Philip F. Rehbock (1983), *The Philosophical Naturalists: Themes in Early Nineteenth-Century British Biology* (Madison: Univ. of Wisconsin Press).

Margaret Pelling (1983), "Experimentalism and the Life Sciences since 1800," in *Information Sources in the History of Science and Medicine,* ed. P. Corsi and P. Weindling (Boston: Butterworth Scientific), 361–78.

Carl Jay Bajema, ed. (1983), *Natural Selection Theory* (Stroudsburg, Pa.: Hutchinson Ross), includes extracts from Paley's *Natural Theology,* and from writings of W. Whewell, Erasmus Darwin, W. C. Wells, P. Matthew and W. Blyth.

Ernst Mayr (1982), *Growth of Biological Thought* (Cambridge: Harvard Univ. Press), Chaps. 1–4, 7–8, 14–15; see also the essay review by P. R. Sloan (1985) in *Journal of the History of Biology* 18:145–53.

Timothy Lenoir (1982), *The Strategy of Life: Teleology and Mechanics in Nineteenth Century German Biology* (Boston: Reidel).

C. Leon Harris, ed. (1981), *Evolution: Genesis and Revelations* (Albany: State Univ. of New York Press), Chaps. 5 and 6.

Mirko D. Grmek (1980), "A Plea for Freeing the History of Scientific Discoveries from Myth," in *On Scientific Discovery,* ed. M. D. Grmek and R. S. Cohen (Boston: Reidel), 9–42. A good historiographic essay based mostly on biological examples.

J. T. Merz (1912), *A History of European Thought in the Nineteenth Century* (Edinburgh: Blackwood), 2:200–464, covers general background.

Topic *i:* Reproduction and Heredity

John Farley (1982), *Gametes and Spores: Ideas about Sexual Reproduction, 1750–1914* (Baltimore: Johns Hopkins Univ. Press). See the essay review by R. C. Olby, *History and Philosophy of the Life Sciences* 8(1986):99–105.

Topic *ii:* Taxonomy and Fossils; Buffon

James A. Secord (1986), *Controversy in Victorian Geology: The Cambrian-Silurian Dispute* (Princeton, N.J.: Princeton Univ. Press).

Tore Frängsmyr, ed. (1983), *Linnaeus: The Man and His Work* (Berkeley: Univ. of California Press).

J. H. Eddy, Jr. (1983), "Buffon, Organic Alterations, and Man," *Studies in History of Biology* 7:1–45.

J. Lyon and P. R. Sloan, eds. (1981), *From Natural History to the History of Nature: Readings from Buffon and his Critics* (Notre Dame, Ind.: Univ. of Notre Dame Press).

Peter J. Bowler (1976), *Fossils and Progress: Paleontology and the Idea of Progressive Evolution in the Nineteenth Century* (New York: Science History).

M. J. S. Rudwick (1976), *The Meaning of Fossils: Episodes in the History of Palaeontology* 2d. ed. (Chicago: Univ. of Chicago Press).

Sten Lindroth (1973), "Linnaeus, Carl," *Dictionary of Scientific Biography* 8:374–81.

John C. Greene (1971), "The Kuhnian Paradigm and the Darwinian Revolution in Natural History," in *Perspectives in the History of Science and Technology,* ed. D. H. D. Roller (Norman: Univ. of Oklahoma Press), 3–25.

Topic *iii:* Theology

Neal C. Gillespie (1984), "Preparing for Darwin: Conchology and Natural Theology in Anglo-American Natural History," *Studies in History of Biology* 7:93–145; (1979), *Charles Darwin and the Problem of Creation* (Chicago: Univ. of Chicago Press).

Dov Ospovat (1978), "Perfect Adaptation and Teleological Explanation: Approaches to the Problems of the History of Life in the Mid-19th Century," *Studies in History of Biology* 2:33–56.

Peter J. Bowler (1977), "Darwinism and the Argument from Design: Suggestions for a Reevaluation," *Journal of the History of Biology.* 10:29–43.

Topic *iv:* Nature Philosophy

L. S. Jacyna (1984), "The Romantic Programme and the Reception of Cell Theory in Britain," *Journal of the History of Biology* 17:13–48.

Timothy Lenoir (1981), "The Goettingen School and the Development of Transcendental Naturphilosophie in the Romantic Era," *Studies in History of Biology* 5:111–205; also his *Strategy of Life* (Boston: Reidel), Chap. 5.

Topic *v:* Lamarck

Jean Baptiste Lamarck (1984 Engl. translation), *Zoological Philosophy: An Exposition with Regard to the Natural History of Animals,* translated by Hugh Elliot, with introductory essays by David L. Hull and Richard W. Burkhardt, Jr. (Chicago: Univ. of Chicago Press).

L. J. Jordanova (1984), *Lamarck* (New York: Oxford Univ. Press).

Yves Delange (1984), *Lamarck: Sa Vie, Son Oeuvre* (Le Paradou: Actes Sud, H. Nyssen).

Madeleine Barthélemy-Madaule (1982 Engl. translation), *Lamarck the Mythical Precursor* (Cambridge: MIT Press). See the critical review by Dorinda Outram (1984) in *British Journal for the History of Science* 17:319–20.

Lamarck et Son Temps; Lamarck et Notre Temps (Paris: Vrin, 1981).

Richard Burkhardt (1977), *The Spirit of System, Lamarck and Evolutionary Biology* (Cambridge: Harvard Univ. Press).

Leslie J. Burlingame (1973), "Lamarck, Jean Baptiste Pierre Antoine de Monet de," *Dictionary of Scientific Biography* 7:584–94.

Topic *vi:* Cell Theory

William Bechtel (1984), "The Evolution of Our Understanding of the Cell: A Study in the Dynamics of Scientific Progress," *Studies in History and Philosophy of Science* 15:309–36.

L. S. Jacyna (1984), "The Romantic Programme and the Reception of Cell Theory in Britain," *Journal of the History of Biology* 17:13–48.

Lois Magner (1979), *A History of the Life Sciences* (New York: Dekker), Chap. 9, with references to earlier historical works.

Topic *vii:* Organism as Machine

Jacques Roger (1986), "The Mechanistic Conception of Life," in *God and Nature,* ed. D. C. Lindberg and R. L. Numbers (Berkeley: Univ. of California Press), 277–95.

F. L. Holmes (1985), *Lavoisier and the Chemistry of Life* (Madison: Univ. of Wisconsin Press).

William Coleman (1979), "Bergmann's Rule: Animal Heat as a Biological Phenomenon," *Studies in History of Biology* 3:67–88. The rule proposed by Carl Bergmann in 1847 relates the size of animals to the climate of regions they inhabit.

William Randall Albury (1977), "Experiment and Explanation in the Physiology of Bichat and Magendie," *Studies in History of Biology* 1:47–131.

P. F. Cranefield (1957), "The Organic Physics of 1847 and the Biophysics of Today," *Journal of the History of Medicine* 12:407–23. A seminal article on the program to reduce biology to physics and chemistry.

2.3 ORIGIN AND EARLY DEVELOPMENT OF DARWIN'S THEORY

i. Family background (Erasmus Darwin), education
ii. Voyage of the *Beagle:* Darwin's observations, reading of Lyell's *Principles*
iii. Influence of Malthus and relation to nineteenth-century British thought
iv. Development of the theory of natural selection
v. Wallace; publication of *Origin of Species*

READINGS

Mason, *History of the Sciences,* 412–21 (*i–v*); McKenzie, *Major Achievements,* 203–15 (*i–v*), 473–81 (extracts from *Origin*); Toulmin and Goodfield, *Discovery of Time,* 197–211 (*i–iv*); Marks, *Science,* 174–82.

The importance of this topic is such that one needs more extensive readings than any of the above general texts. The choice depends partly on which other topics in the history of biology one wants to cover in the course. Some attractive possibilities are: Bowler, *Evolution,* 142–75; William Irvine, *Apes, Angels, and Victorians* (1955), 1–122; Eiseley, *Darwin's Century,* 46–51, 118–204 (including Wells and Chambers); Greene, *Death of Adam,* 249–94, and *Science, Ideology, and World View,* 60–87 (on Comte and Spencer); H. Wendt, *In Search of Adam,* 235–59; Magner, *History of the Life Sciences,* 377–91; Ruse, *Darwinian Revolution,* 16–35, 48–63, 155–201; Coleman, *Biology in the Nineteenth Century,* 57–91. Appleman, *Darwin,* 35–131 is an extended selection from the *Origin*. Darwin's own writings, especially *The Voyage of the Beagle,* are quite readable and may help some students appreciate how he developed his ideas. Jonathan Miller's *Darwin for Beginners* (1982) is a readable popularization, with cartoons.

The Open University course *Science and Belief: Copernicus to Darwin* (AMST 283) has a unit on Darwin by J. H. Brooke that provides an excellent review and discussion (unit 13, 42 large pages, including some material relevant to **2.4**).

Of the many articles based on recent historical research, the following are readable enough to be assigned to students: E. Mayr, "Darwin and Natural Selection," *American Scientist* 65 (1977):321–27; S. J. Gould, "Darwin's Deceptive Memories," *New Scientist* 85 (Feb. 1980):577–79; and "In Praise of Charles Darwin," *Discover* 3, no. 2 (Feb. 1982):20–25; S. S. Schweber, "The Genesis of Natural Selection—1838 . . . ," *BioScience* 28 (1978):321–26.

Several biographies can be used for collateral reading and reports by students: *The Autobiography of Charles Darwin,* edited by his son Francis Darwin with "reminiscences," selected letters, and other historical material (1892; Dover reprint); Gavin de Beer, *Charles Darwin* (1964); Irving Stone, *The Origin: A Biographical Novel of Charles Darwin* (1980).

Synopsis

Charles Robert Darwin came from a distinguished family: one grandfather was Erasmus Darwin, who had written on evolution in a Lamarckian fashion; the other was Josiah Wedgwood, known today for his pottery and one of the leaders of the British Industrial Revolution in the late eighteenth century. Darwin's family had enough money so that he did not have to work for a living, yet his father was annoyed when he seemed to be wasting his time on chemical experiments and on hunting with his friends. Perhaps fascination with horses and with animal breeding in general in Darwin's social circle contributed something to his later work on the theory of selection.

Darwin did originally intend to follow the family profession, medicine, but changed his mind after watching two operations performed (without anesthetics). He decided to study for the clergy in the Church of England instead and transferred to Cambridge University where he got through without distinction. But by this time he had acquired an interest in natural history and found one professor—J. S. Henslow, a botanist—who provided some inspiration to pursue that subject.

In 1831 Henslow recommended Darwin for an unpaid position on the HMS *Beagle,* which was to sail around the world in two years (it turned out to be five) primarily for the purpose of charting the coast of South America for the British Navy. This was Darwin's principal experience in collecting scientific data in geology, botany, and zoology. For light reading in the cabin he brought along Lyell's *Principles of Geology.*

Among the mass of observations that Darwin made on this trip, three seem to have been especially important in leading him to the theory of evolution:

1. *Variation of species in space at the same time.* He noted that as one proceeds southwards in the South American continent, there is a gradual change in the forms of animals, suggesting local modification of the same original species rather than separate creations of different types.
2. *Variation of species in time at the same place.* He observed the fossil remains of animals in the Pampas having skin armor similar to that of modern armadillos. Again, it seemed more plausible to suppose a gradual modification of a single original type, rather than to postulate a series of successive creations of slightly different species.
3. *Variation at the same time and almost the same place with special local conditions.* In the Galapagos Archipelago (600 miles west of Ecuador, on the equator) were a group of several islands isolated from each other where he observed many kinds of birds. In particular the birds now called "Darwin's finches" have long been thought to be strong evidence for evolution. There were at least fourteen distinct species, none of which existed on the mainland; they differed in the shape of their beaks or other features apparently associated with the kinds of food available (seeds, insects, etc.). Thus not only gross climatic and geographical features but very specific features of the environment could produce observable adaptations.

The usual story is that it was his observations of these finches that led Darwin to consider the possibility that species could change in response to environmental conditions. But, according to Frank J. Sulloway (1982), Darwin failed to record the specific islands from which each of his finch specimens had come; only afterwards did he reconstruct this information in the light of this theory. Rather than finches it was mockingbirds that first led him to notice the variations in species from one island to another.

The expedition returned to England in 1836 and Darwin retired to his country home to write up and think about his observations. He became convinced by July 1837 that species had changed but needed to explain *how*. According to his own account, the crucial stimulus came from reading (or rather rereading) on 28 September 1838 the *Essay on Population* by Thomas Malthus.

In his *Essay* (1798) Malthus pointed out that population tends to grow geometrically (e.g., doubling in each generation) whereas the food supply cannot grow more than arithmetically at best (constant increase with time). This is illustrated by the United States, where with ample food up to now the population has managed to double itself every twenty-five years. But any population tends to expand beyond the available food, producing an inevitable competition and struggle for existence among people. Population growth is restrained by famine, disease, or war; the only alternative is to limit reproduction but, as Malthus remarks, "toward the extinction of the passion between the sexes no progress whatever has hitherto been made."

This was a sour critique of the Enlightenment belief in unlimited progress and the perfectibility of man and society, as expressed most recently by Condorcet in his *Outline of the Intellectual Progress of Mankind* (1795). As Malthus noted, Condorcet wrote his book while sitting in jail, a victim of the French Revolution. It should be recalled that the excesses of violence in that revolution had frightened the English and led many of them to a conservative reaction against democracy.

It is not surprising that Darwin read the *Essay* since the Malthusian theory was widely discussed in England. Indeed, some historians have argued that Darwin's theory is nothing more than a transposition into biological terms of the ideology of nineteenth-century capitalism. But that interpretation fails to explain why no one else proposed such a theory before 1858.

According to S. S. Schweber, an important stimulus to Darwin's thinking came from a review of August Comte's *Cours de Philosophie Positive,* published anonymously in the *Edinburgh Review* in July 1838. The author, apparently unknown to Darwin, was the Scottish physicist

David Brewster. Of particular significance was Comte's discussion of Laplace's nebular hypothesis. In 1833 William Whewell had pointed to the regularities of the solar system as evidence of design. But Whewell, in another work, had slighted Scottish science, and Brewster became a fierce enemy for this and other reasons. Brewster pointed out in his review of Comte that these regularities could be explained as the result of an evolutionary history governed by physical laws. Comte and Brewster also emphasized the need for quantitative tests of the predictions of any such theory; Comte had in fact invented his own quantitative test of the nebular hypothesis though it was subsequently shown to be spurious.

Darwin was thus presented with a paradigm—in the original sense of that word, an example to be followed. Others like Robert Chambers and Spencer also tried to exploit the analogy (or continuity) between the nebular hypothesis for the origin of the Earth and the evolution of biological species but did not succeed in finding quantitative confirmations for the latter. Darwin was (according to Schweber) impelled to look up Malthus's essay with special attention to the mathematical law that population grows geometrically.

When Darwin did look at Malthus he was impressed by the conclusion that a population reproducing without limitation of resources would double in a fairly short time, such as twenty-five years. But the limitation on available food would force a competition in which only those organisms that are more "fit" (better adapted to the circumstances) can survive to reproduce themselves. This is the process that Darwin, going beyond Malthus, called natural selection.

There are two clarifications needed here. First, the word "selection" implies conscious choice by someone, an unfortunate connotation which Darwin did not intend; the selection occurs automatically by the mere fact that there is competition between organisms for a limited food supply, so that only a fraction of the total population can be successful, i.e., be "selected" to reproduce. Evolution depends on a brutal struggle for survival (long enough to produce offspring); it does not occur in a peaceful, harmonious world, as pre-Darwinian evolutionists may have hoped or believed.

The second point: in order for selection to operate, there must exist a *diversity* of organisms at the same time and place, some better and some worse adapted to the environmental conditions. This means that Darwin has to make the postulate of *random variation* or *random mutation* of species. It would appear that Darwin is thereby introducing a major component of the new worldview: *indeterminism*. But Darwin himself seems to have used the term "random" in the epistemological rather than the ontological sense—we treat events *as if* they happened by chance because we do not *know* their causes, but we may still maintain that those events do have causes. This was in fact the justification used for the statistical approach by Laplace, in the same essay on probability in which he gave the classic formulation of "Laplacean determinism." Darwin claimed that variations do have causes but was unable to find a plausible explanation of those causes and therefore simply postulated that variations occur. Similarly in physics Maxwell and Boltzmann postulated that molecules move *as if* by chance even though initially it was assumed that molecular motions are "really" deterministic; the probabilistic approach became so familiar that in the twentieth century physicists could slide into a theory that denied true determinism (**11.2**).

Ernst Mayr insists on the importance of two other elements in Darwin's thinking. First, he abandoned "essentialism"—the belief that all individuals of a species are essentially alike, as approximations to an ideal "type"—for "population thinking"—the belief that a species is a collection of different unique individuals. Second, he shifted from "soft" inheritance—the assumption that the material basis of inheritance can be modified by environmental influences, use or disuse of organs, or some inherent tendency to progress—toward "hard" inheritance, the denial of such changes. Darwin did not consistently maintain the modern (i.e., Mayr) views on these points, but his statements were generally more advanced than those of almost all other biologists before the 1930s. (Cf., **4.6**)

In 1839 Darwin married his cousin, Emma Wedgwood, and moved to London. He was elected a Fellow of the Royal Society and began to move in society (both scientific and otherwise). But he suffered considerably from ill health—historians are still arguing about the cause of Darwin's mysterious disease—and in 1842 retired to his country home, where he devoted himself to scientific work. In that year he wrote out for himself a short (35-page) summary of his theory of natural selection but refrained from publishing it; he wanted to be sure he had enough evidence to prove his case, since he realized that there would be strong opposition from both experts and laymen.

By 1856 Darwin had developed and documented his theory well enough to begin writing a treatise on natural selection. His friends and colleagues knew about his ideas and urged him to publish them, but he was reluctant to do so until 1858, when a communication from Alfred Russell Wallace shocked him into action. Wallace had reached the exact same conclusion independently of Darwin but by a very similar route (including the reading of the same passage in Malthus's essay). He had not, however, compiled such a large quantity of evidence as had Darwin to support his theory. Darwin and Wallace published extracts of their work together in 1858, and Darwin rushed to complete a longer version. This was said to be only a summary of the projected treatise but was itself a book of over 400 pages: *The Origin of Species by Means of Natural Selection of the Preservation of Favoured Races in the Struggle for Life* (1859).

BIBLIOGRAPHY

Frederick Burkhardt and Sydney Smith, eds. (1985–), *The Correspondence of Charles Darwin* (New York: Cambridge Univ. Press).

Ronald W. Clark (1984), *The Survival of Charles Darwin: A Biography of a Man and an Idea* (New York: Random House).

Carl Jay Bajema, ed. (1983), *Natural Selection Theory* (Stroudsburg, Pa.: Hutchinson Ross), 130–295.

George Gaylord Simpson, ed. (1982), *The Book of Darwin* (New York: Washington Square Press), contains extended extracts from Darwin's writings, with commentary.

R. J. Berry, ed. (1982), *Charles Darwin: A Commemoration, 1882–1982: Happy Is the Man That Findeth Wisdom* (London: Academic Press).

Roger G. Duval and Cleveland T. Duval, eds. (1982), *Charles Darwin, 1809–1882: A Centennial Commemorative* (Wellington, N.Z.: Nova Pacifica, 1982).

C. Leon Harris (1981), *Evolution* (Albany: State Univ. of New York Press), Chap. 7.

J. C. Greene (1975), "Reflections on the Progress of Darwin Studies," *Journal of the History of Biology* 8:243–73, is a review of earlier research.

Gavin de Beer (1971), "Darwin, Charles Robert," *Dictionary of Scientific Biography* 3:565–77.

Topic *i:* Family Background

Ralph Colp, Jr. (1985), "Notes on Charles Darwin's *Autobiography, Journal of the History of Biology* 18:357–401; (1977), *To Be an Invalid: The Illness of Charles Darwin* (Chicago: Univ. of Chicago Press).

Desmond King-Hele (1977), *Doctor of Revolution: The Life and Genius of Erasmus Darwin* (London: Faber and Faber).

Topic *ii:* Voyage of the Beagle

Sandra S. Herbert (1986), "Darwin as a Geologist," *Scientific American* 254, no. 5:116–23.

Paul Tasch (1986), "Geology and Zoology—A Symbiosis. Darwin's Beagle Voyage and Galapagos Experience," *Earth Sciences History* 4, no. 2:98–112.

M. J. S. Hodge (1983), "Darwin and the Laws of the Animate Part of the Terrestrial System (1835–1837): On the Lyellian Origins of His Zoonomical Explanatory Program," *Studies in History of Biology* 6:1–106.

F. J. Sulloway (1982), "Darwin and His Finches: The Evolution of a Legend," *Journal of the History of Biology* 15:1–53.

Richard D. Keynes, ed. (1979), *The Beagle Record* (New York: Cambridge Univ. Press).

Topic *iii:* Malthus, British Thought

S. S. Schweber (1980), "Darwin and the Political Economists: Divergence of Character," *Journal of the History of Biology* 13:195–289.

Edward Manier (1978), *The Young Darwin and His Cultural Circle* (Boston: Reidel).

Susan Cannon (1978), *Science in Culture* (New York: Science History).

Topic *iv:* Development of Theory

Frank Burch Brown (1986), "The Evolution of Darwin's Theism," *Journal of the History of Biology* 19:1–45.

Ralph Colp, Jr. (1986), "'Confessing a Murder': Darwin's First Revelations about Transmutation," *Isis* 77:9–32.

John F. Cornell (1986), "Newton of the Grassblade? Darwin and the Problem of Organic Teleology," *Isis* 77:405–21.

J. H. Brooke (1985), "The Relations between Darwin's Science and His Religion," in *Darwinism and Divinity,* ed. John Durant (New York: Blackwell), 40–75.

David Kohn (1985), "Darwin's Principle of Divergence as Internal Dialogue," in *The Kaleidoscope of Science,* ed. E. Ullmann-Margalit (Hingham, Mass.: Kluwer Academic); (1980), "Theories to Work by: Rejected Theories, Reproduction, and Darwin's Path to Natural Selection," *Studies in History of Biology* 4:67–170.

David Kohn, ed. (1985), *The Darwinian Heritage* (Princeton, N.J.: Princeton Univ. Press), Part One, "The Evolution of a Theorist."

David R. Oldroyd (1984), "How Did Darwin Arrive at His Theory? The Secondary Literature to 1982," *History of Science* 22:325–74.

C. J. Cela-Conde (1984), "Nature and Reason in the Darwinian Theory of Moral Sense," *History and Philosophy of Life Sciences* 6:3–24.

L. T. Evans (1984), "Darwin's Use of the Analogy between Artificial and Natural Selection," *Journal of the History of Biology* 7:113–40.

John F. Cornell (1984), "Analogy and Technology in Darwin's Vision of Nature," *Journal of the History of Biology* 17:303–44.

Jixing Pan (1984), "Charles Darwin's Chinese Sources," *Isis* 75:530–34.

H.-J. Rheinberger and Peter McLaughlin (1984), "Darwin's Experimental Natural History," *Journal of the History of Biology* 17:345–68.

R. Alan Richardson (1984), "Biogeography and the Genesis of Darwin's Ideas on Transmutation," *Journal of the History of Biology* 14:1–41.

Frank J. Sulloway (1984), "Darwin and the Galapagos," *Biological Journal of the Linnean Society* 21:29–59; (1979), "Geographic Isolation in Darwin's Thinking: The Vicissitudes of a Crucial Idea," *Studies in History of Biology* 3:23–65.

A. J. Cain (1984), "Islands and Evolution: Theory and Opinion in Darwin's Earlier Years," *Biological Journal of the Linnean Society* 21:5–27.

Ernst Mayr (1983), "Darwin, Intellectual Revolutionary," in *Evolution from Molecules to Men,* ed. D. S. Bendall (New York: Cambridge Univ. Press), 23–41; (1982), *Growth of Biological Thought* (Cambridge: Harvard Univ. Press), Chaps. 9–11, 16.

M. J. S. Hodge (1983), "The Development of Darwin's General Biological Theorizing," in *Evolution from Molecules to Men,* ed. D. S. Bendall (New York: Cambridge Univ. Press), 43–62.

Lindley Darden (1983), "Artificial Intelligence and Philosophy of Science: Reasoning by Analogy in Theory Construction," in *PSA 1982,* ed. P. D. Asquith and T. Nickles (East Lansing, Mich.: Philosophy of Science Association), 2:147–65, includes Darwin's theory of natural selection as a case study.

Dov Ospovat (1981), *The Development of Darwin's Theory: Natural History, Natural Theology, and Natural Selection, 1838–1859* (New York: Cambridge Univ. Press), argues that "between 1844 and 1859 the theory of natural selection was substantially transformed. . . . As late as 1844, the structure of Darwin's theory was to a large extent determined by . . . natural theological ideas or assumptions. The transformation of the theory that occurred in the 1850s eliminated some of these assumptions. . . . At the same time, it produced some of the most characteristically Darwinian ideas that we associate with the theory of natural selection. The theory of relative adaptation, for instance, is a product of the 1850s. . . ."

Scott A. Kleiner (1981), "Problem Solving and Discovery in the Growth of Darwin's Theory of Evolution," *Synthese* 47:119–62, tests H. A. Simon's thesis about the process of law discovery.

Howard E. Gruber (1980), "Cognitive Psychology, Scientific Creativity, and the Case Study Method," in *On Scientific Discovery,* ed. M. D. Grmek et al. (Boston: Reidel), 295–322, takes Darwin as a prime example because of his "images of wide scope" such as "war" and "tree of nature."

Sandra Herbert, ed. (1980), *The Red Notebook of Charles Darwin (Bulletin of the British Museum, Natural History,* Historical ser., vol. 7), indicates his 1837 change of views on species.

S. S. Schweber (1977), "The Origin of the *Origin* Revisited," *Journal of the History of Biology* 10:229–316.

Michael Ruse (1975), "Darwin's Debt to Philosophy: an Examination of the Influence of the Philosophical Ideas of John F. W. Herschel and William Whewell on the Development of Charles Darwin's Theory of Evolution," *Studies in History and Philosophy of Science* 6:159–81.

Topic *v:* Wallace; Publication of *Origin*

John Langdon Brooks (1986), "Development of Wallace's Perceptions of Biogeography, 1848–1859," *Earth Science History* 4:113–17; (1984), *Just Before the Origin: Alfred Wallace's Theory of Evolution* (New York: Columbia Univ. Press). See the review by M. Ruse in *Annals of Science* 42(1985):351–52.

Scott A. Kleiner (1985), "Darwin and Wallace's Revolutionary Research Programme," *British Journal for the Philosophy of Science* 36:367–92.

H. Lewis McKinney (1976), "Wallace, Alfred Russel," *Dictionary of Scientific Biography* 14:133–40.

2.4 RECEPTION OF *THE ORIGIN OF SPECIES*

i. Initial reactions, reviews, Huxley-Wilberforce debate, religious objections, public response
ii. Scientific criticisms: Kelvin and geological time scale, Fleeming Jenkin and "dilution" of mutants
iii. Biological/paleontological research inspired by or relevant to evolution (other than human)—fossils, discoveries of Marsh and Cope
iv. Neo-Lamarckism in the late nineteenth century, other modifications of Darwinian theory to avoid objections
v. The evolutionary worldview, Darwin's impact on Western culture

READINGS

Mason, *History of the Sciences,* 421–33, 497–98, 584–85 (*i–iv*). McKenzie, *Major Achievements,* 216–17, 244 (*i, iii*); Toulmin and Goodfield, *Discovery of Time,* 212–31 (*i, ii*), 232–72 (*v*); Gillispie, *Edge of Objectivity,* 320–28, 337–51 (*iii–v*). Marks, *Science,* 182–84. W. C. Dampier, *History of Science* (1965), 305–20. Richard Olson, *Science as Metaphor* (1971), 109–58. F. M. Turner, *Between Science and Religion* (1974), 8–37 (good concise exposition of "Victorian scientific naturalism").

Articles: David B. Wilson, "Shaping Modern Perspectives: Science and Religion in the Age of Darwin," in *Did the Devil Make Darwin Do It?* ed. D. B. Wilson (1983), 3–18. Stephen Jay Gould, "False Premise, Good Science," *Natural History* 92, no. 10(1983):20–26 (on Kelvin vs. geologists on age of Earth); "Fleeming Jenkin Revisited," ibid. 94, no. 6(June 1985):14–20 (on dilution of mutants). John Durant, "Darwinism and Divinity: A Century of Debate," in *Darwinism and Divinity,* ed. John Durant (1985), 9–39.

Books on history of biology/evolution: Bowler, *Evolution,* 176–216, 224–56; Irvine, *Apes, Angels and Victorians,* 123–216; Eiseley, *Darwin's Century,* 134–35, 216–23, 233–53, 325–52 (*ii, iv, v*); Michael Ruse, *Darwinian Revolution* (1979), 202–41, 248–73 (*i–v*); Greene, *Science . . . ,* 128–97 (*v*); Appleman, *Darwin,* 211–43, 297–367, 513–26 (*i, v*); Wendt, *In Search of Adam,* 259–90 (*i, iii*).

"The Crisis of Evolution" in the Open University course *Science and Belief: From Copernicus to Darwin* (AMST 283), 85–125, covers the impact of Darwin's theory on religious thought, with brief remarks on the parallel "revolution in historical thinking." "The Metaphysics of Evolution" in another Open University course, *Science and Belief from Darwin to Einstein* (A381), Unit 2, by James R. Moore and Colin Chant, is a 33-page commentary on the writings of Spencer, Huxley, and Tyndall (identified as "positivism" and "materialism") with discussion questions and illustrations. The accompanying collection of primary sources, *Darwin to Einstein,* ed. N. G. Coley and V. M. D. Hall (1980), includes selections from these authors. "Thermodynamics and Time" by Colin Russell (Unit 4 in the same series) provides an excellent account of the debate on the age of the Earth (topic *ii*) and of "Religion and Science in Kelvin's Thought."

Note that several aspects of the Darwinian impact will be covered in later sections—especially those dealing with the evolution of man (**2.5**), racial theories (**3.1, 3.2**), social Darwinism and eugenics (**3.3**), anthropology (**3.5**), and modern evolutionary theory (**4.6**). The readings listed here deal mainly with the initial and generalized response to the *Origin* and also make the important point that in the first few decades after

1859 the weighty *scientific* objections to the efficacy of natural selection seemed at times to be leading biologists away from Darwinism. This is especially clear in Emanuel Radl's book, *The History of Biological Theories* (English translation, 1930), which begins with the advent of Darwinism and ends with its decline. In between Radl gives a fascinating and almost contemporary account of most of the topics listed above. A similar view, with less detail, is found in Erik Nordenskioeld's *History of Biology* (English translation, 1928), also still worth reading.

Arthur Koestler's *Case of the Midwife Toad* (1972) describes an attempt to revive Lamarckism in the twentieth century; students may find this interesting to read although Koestler's presentation of the historical context is somewhat misleading [see S. J. Gould's review in *Science* 176(1972):623–5]. Another even more notorious attempt, with strong political overtones, is that of the Russian biologist T. D. Lysenko; two excellent sources that could be suggested for student reports are Zhores Medvedev, *The Rise and Fall of T. D. Lysenko* (1969) and David Joravsky, *The Lysenko Affair* (1970). (For further details on Lysenko see **4.5.**)

SYNOPSIS

The first major battle over Darwin's theory came at the 1860 meeting of the British Association for the Advancement of Science, at Oxford. Darwin himself was not a fighter; he left the job of defending his theories to Thomas Henry Huxley (1825–1895), another English biologist/geologist. Though Huxley had known Darwin since the early 1850s he was not converted to Darwin's theory until he read the *Origin.* To Huxley, the mechanism of natural selection seemed so obviously right that he is said to have exclaimed, "How extremely stupid not to have thought of that!" He quickly integrated his own extensive knowledge of natural history into the evolutionary framework and became known as "Darwin's bulldog" because of his campaign to establish evolution by numerous lectures and popular essays.

At the 1860 Oxford meeting, a violent attack on Darwin's theory came from Bishop Samuel Wilberforce, a crowd-pleasing orator widely known as "Soapy Sam." At the end of a witty and sarcastic tirade, which seemed to be winning over even the sober scientists who had carefully scrutinized Darwin's evidence, Wilberforce turned to Huxley and "begged to know, was it through his grandfather or his grandmother that he claimed his descent from a monkey?" Huxley rose and gave a brief but lucid account of Darwin's theory and then remarked that he would not be ashamed to have a monkey for his ancestor, but he *would* be ashamed (looking at Wilberforce) to have any relation to a man who used his great gifts to obscure the truth. This created a sensation, the popular report being that Huxley had said he'd rather be an ape than a bishop. The immediate effect was to publicize Darwin's theory and discredit his opponents, who had made the grave tactical error of choosing as their spokesman a man who knew absolutely nothing about science.

Such is the standard account of this famous episode; before using it, one should read the revisionist account of Lucas cited below.

More substantial objections were soon raised. Two of them are worth mentioning:

1. There has not been enough *time* for evolution to occur in the very slow manner that seems to be implied by Darwin's theory. This attack was made by William Thomson, Lord Kelvin, on the basis of his calculations of the rate of cooling of the Earth. Kelvin assumed that the Earth had simply been cooling down from a hot liquid ball, with no additional heat being supplied to replace that lost to space. From Fourier's theory of heat conduction, using data on the thermal conductivities of rocks and the rate of temperature increase with depth, he estimated that the Earth must have been liquid as recently as 100 million years ago. Darwin had carelessly mentioned 300 million years as a time scale available for evolution, though he was criticized by geologists and withdrew this estimate from later editions of the *Origin* even before Kelvin's attack. Huxley pointed out, quite correctly, that the validity of the theory did not depend on *any* particular time scale since no one had yet established the rate at which biological evolution would occur through natural selection. Nevertheless, Kelvin's great prestige as a physicist, and the inability of Huxley and Darwin to understand mathematics, meant that the evolutionists were continually on the defensive and had to produce somewhat specious arguments to support more and more rapid evolution as the physicists contracted the time scale.

The retreat of the evolutionists and geologists ended around 1900 when radioactivity was discovered. This changed the argument in two ways: first, by showing that there *are* heat sources in the Earth that can replace the heat lost to space by conduction—thus there is no need to assume that the crust has been much hotter in the past than it is now. Second, the method of radioactive dating led to the result that rocks on the Earth's surface had been solid for something like 3 *billion* years. Thus the age of the Earth was at least ten times longer than Darwin's original figure.

Part of the interest of this dispute is that it was a conflict between two sciences, physics and geology (and, indirectly, biology). For Kelvin it was obvious that in any such conflict physics must win because it is more "fundamental." There seems to be a "pecking order" within the sciences—physics at the top, next chemistry, then biology, psychology, and finally sociology at the bottom. Astronomy and geology are at the side as descriptive rather than experimental sciences—astronomy just slightly below physics, and geology perhaps just below chemistry. The basic idea is that each science is in principle "reducible" to the more fundamental ones. This idea is called "reductionism" and features prominently in modern discussions of the philosophy of science.

The "pecking order" itself is not something inherent in the nature of science but has evolved historically. In the nineteenth century, for example, what we now call "planetary science" was a perfectly respectable part of physics (or "natural philosophy"). Many major discoveries and theories in what we now call "pure" science originated in or were motivated by problems in planetary science, and most of the major scientists of the nineteenth century (including Kelvin) moved easily between pure and planetary science without recognizing that there was any significant difference between them. Yet in the twentieth century physics narrowed itself down to a few specialties like atomic physics, and the presumption (until the space program started around 1958) was that anyone who goes into planetary science must not be good enough to succeed in one of the "fundamental" fields.

To return to the objections to Darwin's theory:

2. In 1867 the British Engineer Fleeming Jenkin pointed out that a single chance variation, even if favorable to the survival of an individual member of a species when it occurred and thus transmitted to its individual offspring, had mathematically only an infinitesimal chance of being adopted by a *population*. It would simply be swamped by the process of interbreeding of descendants of the mutant with nonmutants. This of course was based on the assumption, shared by Darwin and Jenkin, that inheritance proceeds by "blending"—i.e., the offspring simply has the average value of any character of its parents.

Darwin had already considered this objection even before Jenkin's paper was published but was not able to find a satisfactory answer to it within the framework of his original theory, so he was forced to assume that habits or environmental factors common to an entire population would *produce* variations that would then be inherited; in other words, a relapse into Lamarckian theory.

The problem was solved—or at least put into a more satisfactory state—by Mendel's theory of heredity (**4.3**). This introduced the concept of "particulate" inheritance, according to which a variation, once introduced into the "gene pool," is never lost by "averaging" with other genes but remains, even if in a "recessive" form, available to cooperate with other similar variations. Mendel's work, though published in the 1860s, remained unknown to Darwin and other evolutionists until 1900, so it played no part in the debate in the nineteenth century.

Two American scientists played a major role in establishing the theory of evolution by finding sequences of fossil mammals showing clear lines of development: Othniel C. Marsh and Edward Drinker Cope. They popularized the dinosaurs of prehistoric times and thus encouraged the public as well as scientific colleagues to accept Darwin's theory.

Marsh found a series of fossil horses that very neatly displayed a gradual change in bone structure (e.g., going from the four-toed foot to the foot with the single toe of the living horse). Then in 1872–1873, he found fossil birds that possessed teeth and other reptilian characters, suggesting a genetic relationship between these two classes of animals.

When Huxley visited the United States in 1876 he examined Marsh's collection of horse fossils and recognized that for the first time the complete direct line of descent of an existing animal had been demonstrated. In his New York address a few days later, he predicted that a fossil horse with four complete toes in front and a rudiment of another, and probably with a rudimentary fifth toe on the hind foot, would eventually be found; within two weeks Marsh had discovered exactly this fossil. This is a good example to remember whenever it is claimed that evolutionary theory cannot make predictions!

The ultimate accolade for Marsh's work came from Darwin himself, who wrote to Marsh in 1880: "Your work on these old birds and on the many fossil animals of N. America has afforded the best support to the theory of evolution, which has appeared within the last 20 years."

In the public mind, Marsh was known not only as the man who found amazing prehistoric beasts, but also as one of the protagonists in a bitter contest to see who could collect the most and best fossil bones in the American West. The other leading vertebrate paleontologist, Marsh's hated rival, was E. D. Cope. Like Marsh, Cope had a considerable fortune at this disposal so he could pay collectors substantial prices for fossils. He was also one of the most prolific scientists of all time, publishing 1395 scientific articles and books.

Cope's discoveries of dinosaurs were of the same kind as Marsh's and reinforced support for Darwinism. But Cope also wrote at some length on theoretical problems and became one of the leaders of the "American school" of biology, otherwise known as the "neo-Lamarckians." Cope argued in 1871 that natural selection is not sufficient to explain evolution because it must wait for variations to arise before it can act. There must be some actively progressive principle to create the variations. So he postulated "originative" forces arising from the environment. These forces could operate on the organism's store of growth force or "bathmic energy" and cause the evolution of a particular organ to accelerate; but this would be compensated by taking the energy away from another organ, whose evolution would therefore be retarded.

Accelerated evolution didn't take as much time to do the job, and thus it seemed to offer a way out of the dilemma posed by Kelvin's estimates of the age of the Earth.

BIBLIOGRAPHY

David Kohn, ed. (1985), *The Darwinian Heritage* (Princeton, N.J.: Princeton Univ. Press).

Sydney Eisen and Bernard V. Lightman (1984), *Victorian Science and Religion: A Bibliography with Emphasis on Evolution, Belief, and Unbelief, comprised of works published from ca. 1900 to 1975* (Hamden, Conn.: Archon Books).

Richard J. Wassersug and Michael R. Rose (1984), "A Reader's Guide and Retrospective of the 1982 Darwin Centennial," *Quarterly Review of Biology* 59:417–37.

D. S. Bendall, ed. (1983), *Evolution from Molecules to Men* (New York: Cambridge Univ. Press).

Yvette Conry, ed. (1983), *De Darwin au Darwinisme: Science et Idéologie* (Paris: Vrin).

Topic *i:* Initial Reactions

D. C. Lindberg and R. L. Numbers, eds. (1986), *God and Nature* (Berkeley: Univ. of California Press), includes articles by A. H. Dupree and F. Gregory on the response of theologians.

Stephen Jay Gould (1986), "Soapy Sam's Logic: A True Scoundrel but with Redeeming Value," *Natural History* 95, no. 4:16–26; (1986), "Knight Takes Bishop? The Facts about the Great Wilberforce-Huxley Debate Don't Always Fit the Legend," *Natural History* 95, no. 5:18–22, 28–33.

Adel A. Ziadat (1986), *Western Science in the Arab World: The Impact of Darwinism, 1860–1930* (New York: St. Martin's Press).

Robert M. Young (1985), *Darwin's Metaphor: Nature's Place in Victorian Culture* (New York: Cambridge Univ. Press), reprints six essays on the social-intellectual context of Darwin's theory, first published

1969–1980, with updated references and a preface lamenting the failure of other scholars to pursue his themes further.

John Durant, ed. (1985), *Darwinism and Divinity* (New York: Blackwell), includes articles by J. Moore and A. Peacocke on the response of theologians to evolution.

Tess Cosslett, ed. (1984), *Science and Religion in the Nineteenth Century* (New York: Cambridge Univ. Press), consists of extracts from contemporary sources.

Alfred Kelly (1981), *The Descent of Darwin: the Popularization of Darwinism in Germany, 1860–1914* (Chapel Hill: Univ. of North Carolina Press).

Sheridan Gilley (1981), "The Huxley-Wilberforce Debate: A Reconsideration," *Studies in Church History* 17:325–40.

Sheridan Gilley and Ann Loades (1981), "Thomas Henry Huxley: The War between Science and Religion," *Journal of Religion* 61:285–308.

Hans Schwarz (1981), "The Significance of Evolutionary Thought for American Protestant Theology: Late 19th-Century Resolutions and 20th-Century Problems," *Zygon* 16:261–84.

R. J. Halliday (1981), "God and Natural Selection: Some Recent Interpretations of the Relation of Darwinism to Protestant Belief," *History of European Ideas* 2:237–46.

Neal C. Gillespie (1979), *Charles Darwin and the Problem of Creation* (Chicago: Univ. of Chicago Press).

James R. Moore (1979), *The Post-Darwinian Controversies: A Study of the Protestant Struggle to Come to Terms with Darwin in Great Britain and America, 1870–1900* (New York: Cambridge Univ. Press).

Harry Paul (1979), *The Edge of Contingency: French Catholic Reaction to Scientific Change from Darwin to Duhem* (Gainesville: Univ. Presses of Florida).

J. R. Lucas (1979), "Wilberforce and Huxley: A Legendary Encounter," *Historical Journal* 22:313–30, challenges the usual statement that Huxley demolished Wilberforce at Oxford.

Thomas F. Glick, ed. (1974), *The Comparative Reception of Darwinism* (Austin: Univ. of Texas Press).

Topic *ii:* Scientific Criticism

Claude C. Albritton, Jr. (1980), *The Abyss of Time: Changing Conceptions of the Earth's Antiquity after the Sixteenth Century* (San Francisco: Freeman, Cooper), Chaps. 14 and 15.

D. J. Hull, P. D. Tessner, and A. M. Diamond (1978), "Planck's

Principle," *Science* 202:717–23, is on the hypothesis that younger scientists are more likely to accept a new theory, applied to Darwin's theory.

Stephen G. Brush (1978), *The Temperature of History* (New York: Franklin), Chap. 3, discusses the controversy about the age of the Earth as a limit on the time available for evolution.

Joe D. Burchfield (1975), *Lord Kelvin and the Age of the Earth* (New York: Science History).

David L. Hull (1973), *Darwin and His Critics: The Reception of Darwin's Theory of Evolution by the Scientific Community* (Cambridge: Harvard Univ. Press). The essay review by R. Smith in *British Journal for the History of Science* 7(1974):278–85, discusses the problem of fitting the Darwinian "revolution" into Kuhnian or other schemes, and the relation of its historiography to current philosophy of science.

Topic *iii:* Research Inspired by Evolution

L. M. Cook, G. S. Mani, and M. E. Varley (1986), "Postindustrial Melanism in the Peppered Moth," *Science* 231:611–13, reports the latest developments in "one of the most fully documented cases of microevolutionary change."

Wolf-Ernst Reif (1986), "The Search for a Macroevolutionary Theory in German Paleontology," *Journal of the History of Biology* 19:79–130; (1983), "Evolutionary Theory in German Paleontology," in *Dimensions of Darwinism,* ed. M. Grene (New York: Cambridge Univ. Press), 173–203.

Ronald Rainger (1985), "Paleontology and Philosophy: A Critique," *Journal of the History of Biology* 18:267–87, concerns the alleged role of philosophical commitments in the response of late-nineteenth century paleontologists to Darwinism.

Ernst Mayr (1985), "Weismann and Evolution," *Journal of the History of Biology* 18:295–329; (1982), *The Growth of Biological Thought* (Cambridge: Harvard Univ. Press), 501–627.

Roger Lewin (1983), "Finches Show Competition in Ecology," *Science* 219:1411–12.

Michael Gross and Mary Beth Averill (1983), "Evolution and Patriarchal Myths of Scarcity and Competition," in *Discovering Reality,* ed. S. Harding and M. B. Hintikka (Boston: Reidel), 71–95, comments on the ideas of E. P. Odum, R. Ricklefs, J. Diamond, and E. Fisher; discusses the possibility of a feminist theory of evolution (including non-human evolution).

R. J. Berry et al. (1983), "Darwinism—A Hundred Years On," *Biological Journal of the Linnean Society* 20:1–135.

Mario A. Di Gregorio (1982), "The Dinosaur Connection: A Reinterpretation of T. H. Huxley's Evolutionary View," *Journal of the History of Biology* 15:397–418.

Everett C. Olson (1981), "The Problem of Missing Links: Today and Yesterday," *Quarterly Review of Biology* 56:405–42.

Topic *iv:* Neo-Lamarckism

Peter J. Bowler (1983), *The Eclipse of Darwinism: Anti-Darwinian Evolution Theories in the Decades around 1900* (Baltimore: Johns Hopkins Univ. Press). See also his article (1977), "Edward Drinker Cope and the Changing Structure of Evolutionary Theory," *Isis* 68:249–65.

F. B. Churchill (1977), "The Weismann-Spencer Controversy over the Inheritance of Acquired Characters," *Proceedings of the XV International Congress of History of Science* (Edinburgh, 1977), 450–68.

Topic *v:* Evolutionary Worldview

Robert M. Young (1985), *Darwin's Metaphor,* cited above under *i.*

Peter Morton (1984), *The Vital Science: Biology and the Literary Imagination, 1860–1900* (London: Allen and Unwin).

Ed Block, Jr. (1984), "Evolutionist Psychology and Aesthetics: *The Cornhill Magazine,* 1875–1880," *Journal of the History of Ideas* 45:465–75.

David Oldroyd and Ian Langham, eds. (1983), *The Wider Domain of Evolutionary Thought* (Boston: Reidel).

John Beatty (1982), "What's in a Word? Coming to Terms in the Darwinian Revolution," *Journal of the History of Biology* 15:215–39; also in (1983), *Nature Animated,* ed. M. Ruse (Boston: Reidel), 79–100, with comments by D. Hull and reply by Beatty.

Gerry Webster and Brian Goodwin (1981), "History and Structure in Biology," *Perspectives in Biology and Medicine* 25:39–62.

John C. Greene (1981), *Science, Ideology and World View: Essays in the History of Evolutionary Ideas* (Berkeley: Univ. of California Press).

D. P. Crooke (1981), "Darwinism: The Political Implications," *History of European Ideas* 2:19–34.

Hans Schwarz (1981), "The Significance of Evolutionary Thought for American Protestant Theology: Late Nineteenth-Century Resolutions and Twentieth-Century Problems," *Zygon* 16:261–84; (1980), "Darwinism Between Kant and Haeckel," *Journal of the American Academy of Religion* 48:581–602.

D. R. Oldroyd (1980), *Darwinian Impacts* (Atlantic Highlands, N.J.: Humanities Press), Chaps. 17–24, concerns Darwinism and politics, theology, philosophy, psychology, etc.

Robert Nisbet (1969), *Social Change and History, Aspects of the Western Theory of Development* (New York: Oxford Univ. Press).

2.5 HUMAN EVOLUTION

i. Ideas about the origin of man (Linnaeus, etc.) to 1800
ii. Evidence for prehistoric cultures and discovery of fossil men—Boucher de Perthes, Charles Lyell, J. K. Fuhlrott's "Neandertal Man" (1856), Huxley's *Man's Place in Nature* (1863)
iii. Darwin's *Descent of Man* and nineteenth-century theories of human evolution
iv. Survey of twentieth-century research, Piltdown fraud, etc., to 1970
v. Current views on the origin of man

READINGS

None of the general textbooks recommended for this course has an adequate treatment of the evolution of man; a rather surprising omission, in view of the historical and contemporary interest in the subject. Bowler, *Evolution,* 216–24, 303–8, gives a brief treatment. Eiseley, *Darwin's Century,* 255–324, concentrates on *iii* (especially the work of Wallace) and has brief remarks on *ii* and *iv*. Similarly for Greene, *Death of Adam,* 309–39, and remarks earlier in his book as one might expect from its title. Irvine, *Apes . . .,* deals only briefly with Darwin's ideas on man. Appleman, *Darwin,* 132–208, reprints extracts from Darwin's *Descent of Man.*

The most comprehensive discussion of all these topics (except *v*) is Herbert Wendt's *In Search of Adam* (1956). Although the book is more than 500 pages long, it covers to some extent the other aspects of evolution (as I have indicated in previous sections), and in my opinion is sufficiently readable that it should be seriously considered as a text for the entire subject of evolution, at least if one is willing to give that subject a prominent place in the course. However, being more than thirty years old, Wendt's book fails to cover developments after 1950, which need to be included even in a course whose primary focus is the period 1800–1850. I would recommend Donald Johanson and M. A. Edey, *Lucy* (1981), 27–96, for a lively personalized account of the period 1859–1959 from the perspective of a modern scientist.

Articles: Kenneth F. Weaver, Richard Leakey, and Alan Walker, "The Search for Early Man," *National Geographic* 168, no. 5 (Nov.

1985), 560–629, a profusely illustrated popular survey of recent research with a striking holographic image of the Taung child skull on the cover; Donald Johanson and Maitland Edey, "Lucy," *Science 81* (March) 44–55, an extract from their book of the same title; S. J. Gould, "We First Stood on Our Two Feet in Africa, Not Asia," *Discover 7,* no. 5 (May 1986):52–56; Philip V. Tobias, "From Linne to Leakey: Six Signposts in Human Evolution," in *Current Argument on Early Man,* ed. L. K. Königsson (New York: Pergamon Press, 1980), 1–12; C. Jolly, "Changing Views of Hominid Origins," *Yearbook of Physical Anthropology* 16 (1972):1–17; Tobias, "The Child from Taung," *Science 84,* 5, no. 9:99–100 (Nov. 1984) (on Raymond Dart's research); John Pfeiffer, "Early Man Stages a Summit Meeting in New York City," *Smithsonian* 15, no. 5:51–57 (Aug. 1984).

If you use Abram Kardiner and Edward Preble's book *They Studied Man* (1961) for anthropology (**3.4**) and Freud (**5.2**), their chapter on Darwin (17–36) could be used here for topic *iii.*

This may be an appropriate place to discuss Darwin's personal life and his reaction to the controversy started by his theories. In addition to the biographies mentioned in **2.3,** you may wish to look at the claim by Ralph Colp (1977) that Darwin's chronic but mysterious illness (which some historians say was fortunate for science because it let him concentrate on his work) resulted from the stress of developing a theory considered brutal and blasphemous. This is perhaps a more sophisticated version of the canard circulated by antievolutionists that Darwin recanted his theory on his deathbed and embraced religion. That story is so obviously wrong that even the creationists have rejected it; see W. R. Rusch (1975).

SYNOPSIS

The most direct evidence for the early evolution of humans comes from the discovery of fossil remains. For brevity one may skip earlier controversial findings (described in Wendt's book) and begin with the work of Edouard Lartet in the 1830s. Lartet, a French lawyer, became interested in prehistoric research when a farmer brought him a gigantic molar he had plowed up. Lartet identified it as belonging to an aboriginal elephant described by Georges Cuvier. He started digging, looking for bones of humans that might have lived at the same time. In 1837 he found the upper arm and skull fragments of a creature slightly smaller than a chimpanzee but with some human features. It was later named Dryopithecus, recognized as a relative of the South Asiatic gibbon and one of a group of anthropoid apes from which humans evolved (though probably not a direct ancestor).

In 1856 bones were discovered in the Neandertal Valley near Düsseldorf, Germany, by Johann Fuhlrott, a high school science teacher. They included a skull with massive brow ridges like those of a giant gorilla; and upper thighbones that were thick and curved so the creature must have walked with the upper part of its body thrust forward. Fuhlrott proposed that it was an intermediate between the gorilla and Homo sapiens. Hermann Schaafhausen, professor at the University of Bonn, supported this interpretation and considered the discovery a proof that humans had evolved from animals. That conclusion was rejected by several scientists, especially the influential physician Rudolf Virchow who argued that this was a human afflicted by rickets and arthritis and whose skull had been flattened by several blows in a fight.

Because of such skepticism and the inherent resistance to the idea that humans are descended from "lower" animals, the science of human evolution hardly existed as a legitimate subject as late as the middle of the nineteenth century. The breakthrough came in 1859 with two major events. The first is of course the publication of Charles Darwin's *Origin of Species,* which while having almost nothing to say directly about the origin of humans broke through the barrier of religious prejudice that had previously discouraged serious discussion of the subject. The second was the rather sudden acceptance of fossils, discovered earlier by Jacques Boucher de Perthes in France, as valid evidence of human antiquity. Initial skepticism about Boucher de Perthes's claim was dissolved by the discovery of similar fossils in England in 1858; British geologists went to inspect the French fossils and decided that humans had indeed existed much longer than the biblical six thousand years ago. Charles Lyell was one of those who abandoned the view that humans had been created recently; he published in 1863 an influential book on *Geological Evidences of the Antiquity of Man.*

Despite the coincidence in time there was no causal relation between these two events. Boucher de Perthes claimed that the humans whose remains he had found lived thousands or even millions of centuries ago, but rejected the idea that they had evolved from animals. Darwin's arguments for the animal origin of humans were not based on fossils known at the time; he could not point to any human fossils that were both *old* and *different from modern humans.* Until the middle of the twentieth century there was only a very loose connection between specific evolutionary hypotheses and the fossil record.

Lyell and other British geologists accepted "Neandertal Man" as a new species of human, distinct from Homo sapiens, and other fossils (including a skull found at Gibraltar in 1848) were included in this species. In the 1880s further specimens of Neandertal Man were found in the midst of bones of mammoths and other evidence showing that they lived during the Ice Age. Early in the twentieth century Marcellin Boule tried to show that the Neandertals were not actually ancestral to humans but were a side branch that became extinct. (The social history of Boule's work and its relation to the Piltdown fraud have been studied by Michael Hammond [1982].) But Neandertal Man remains a part of the popular image of our primitive "caveman" ancestors.

In the 1860s, after the British had validated the discoveries of Boucher de Perthes, Edouard Lartet again started to explore caves in southern France. In 1868 his son Louis Lartet excavated a cave found in the course of railway construction in the Vezère Valley. In this so-called "Cro-Magnon" cave (the term means simply "big hole") they found skeletons of a family of Ice Age humans together with tools, carved reindeer antlers, and other artifacts. The skulls were nearly identical to those of modern humans. Cro-Magnon Man was recognized as the immediate precursor of Homo sapiens, and was given the name Homo sapiens diluvialis.

The stereotype of the primitive "caveman" was significantly changed (though not completely eliminated from popular mythology) by the discovery of beautiful and (by nineteenth-century standards) artistically sophisticated paintings of animals on the walls of Ice Age caves in France. These paintings forced scientists to give serious attention to the cultural as well as the biological evolution of Homo sapiens.

These discoveries were made and publicized during the height of public interest and scientific controversy about Darwin's theory of evolution. Of course the main reason for such interest and controversy was the implication that the human race has evolved from another species. Darwin himself was reluctant to discuss this subject in the *Origin,* although it had been on his mind ever since he had first seen the savage inhabitants of Tierra del Fuego during his voyage on the *Beagle.* (At the end of *Descent* he recalls "the reflection at once rushed into my mind—such were our ancestors. These men were absolutely naked . . . they possessed hardly any arts, and like wild animals lived on what they could catch.") So it was T. H. Huxley rather than Darwin who first published a detailed analysis of the relation of man to animals.

In *Man's Place in Nature* (1863) Huxley surveyed reports on the appearance, anatomy, and behavior of "man-like apes" showing their similarity to humans. The bone structure of a human differs less from that of a gorilla than the gorilla does from the gibbon. Huxley shows sketches of the skulls of humans and various apes that vividly suggest by their placement on the page that the former somehow evolved from the latter. Another proof of the close relation of humans to animals can be derived from embryology: The development of the fetus of the dog, snake, frog, fish, and of man goes through precisely similar early stages.

Huxley boldly confronted the fact that many readers would be repelled by the view that humans are descended from apes. He asserted that it is not degrading to a human to say that he evolved from a "naked and bestial savage." One does not have to abandon the attempt to lead a noble life just because "the simplest study of man's nature reveals, at its foundations, all the selfish passions, and fierce appetites of the merest quadruped." On the contrary, by looking at how far we have come from lowly origins we gain hope for future progress.

There were many, in Darwin's time as well as more recently, who wanted to avoid the whole issue of human evolution by putting it beyond the reach of science. Darwin addressed these positivist critics (cf. **11.1-*iv***) in a memorable passage at the beginning of his *Descent of Man* (1871):

> It has often and confidently been asserted, that man's origin can never be known; but ignorance more frequently begets confidence than does knowledge: it is those who know little, and not those who know much, who so positively assert that this or that problem will never be solved by science.

Darwin presented only a brief sketch of the events that might have been involved in the transition from ape to human. Our "early progenitors" were "covered with hair, both sexes having beards; their ears were probably pointed, and capable of movement and their bodies were provided with a tail, having the proper muscles. . . . our progenitors, no doubts, were arboreal in their habits, and frequented some warm, forest-clad land. The males had great canine teeth, which served them as formidable weapons." [Darwin 1874, 164]

Darwin defined the paradigm or research programme for theories of human evolution. Six major changes, characterizing the transition from ape to human, must be explained: (1) "coming down from the trees" to live on land; (2) adoption of erect bipedal posture; (3) loss of most body hair; (4) reduction in size of teeth; (5) expansion of brain size, presumably associated with development of language and higher consciousness; (6) manufacture and use of tools. The explanation may involve adaptation to new environmental conditions through the mechanism of natural selection and development of features desired by prospective sexual partners through the mechanism of sexual selection. In a strictly Darwinian theory it may not involve inheritance of acquired characters (one aspect of "Lamarckism") or any overall *tendency* toward complexity, moral values, and other "human" features (orthogenesis), since evolution does not have any intrinsic direction. (Of course neither Darwin himself nor any of his supporters found it easy to stick to those rules!)

Darwin's theory, like those proposed afterwards, was a mixture of postulated environmental changes and interactions between the changes mentioned above. Loss of hair was a result of a hot climate; bipedalism was associated with coming down from the trees and free use of hands; reduction in tooth size was associated with use of tools (especially weapons). As with later theories, it was hard to determine which changes came first—which was cause and which was effect.

But Darwin did make a specific prediction from his theory: while we should not assume that the early progenitors of the simian stock, including humans, were identical with or even closely resembled any modern ape or monkey, we can still guess that our progenitors probably lived in Africa since the gorilla and the chimpanzee, our closest relatives, live there now. The absence (in 1871) of fossil remains linking apes and humans is not surprising since geologists haven't yet searched the place where such links are most likely to be found.

Ernst Haeckel, one of the first biologists to adopt Darwin's theory, was also one of the first to apply it to human evolution. His book *Natürlich Schöpfungsgeschichte* (1868) was highly praised by Darwin, who said that if it had appeared before he had written his own *Descent of Man* (1871) he probably wouldn't have bothered to finish it. In particular Darwin credited Haeckel with recognizing the importance of sexual selection. But Haeckel was really an evolutionist in the more general sense rather than a Darwinian; he accepted Lamarck's hypothesis about the inheritance of acquired characters, and popularized an anticlerical naturalistic worldview that did little to reconcile science and religion.

Haeckel postulated what he called the "biogenetic law," usually expressed by the phrase "ontogeny recapitulates phylogeny." This means that an individual organism, in its development as a fetus and later to maturity, goes through the same sequence of forms that the species itself has followed in its previous evolution from lower forms of life. If the biogenetic law were valid it would provide a fruitful method for studying evolution since embryological and developmental data on present-day organisms are much easier to obtain than paleontological data on prehistoric organisms.

Haeckel postulated an apelike man, to which he gave the name

Pithecanthropus, who evolved from the manlike apes (anthropoids) by walking erect, but did not yet possess speech and higher consciousness. Contrary to other Darwinians, he thought that humans were derived from gibbons on the basis of resemblance of embryos (applying his biogenetic law). This view inspired his disciple, the Dutch anatomist Eugène Dubois, to search for humanoid fossils in Java, where gibbons are prevalent. In 1892 he found a skull vaulting, teeth, and upper thighbone. From its skull volume he concluded that it was just halfway between the anthropoid apes and humans; thus it fit the description of Haeckel's Pithecanthropus. "Java Man," as he was popularly known, was more human than "Neandertal Man." In particular since his thighbone was not curved he apparently walked upright and therefore was called Pithecanthropus erectus.

Research in human paleontology has always been plagued with controversy about the validity of discoveries and claims of fraud. The most notorious case is "Piltdown Man," constructed from skull fragments found by an English lawyer, Charles Dawson, near Brighton in 1912. Pierre Teilhard de Chardin, a French Jesuit paleontologist who later became famous for his attempt to synthesize evolution and Christianity, participated in some of this work. Respected scientists like Arthur Smith Woodward, Keeper of Geology at the British Museum, vouched for the authenticity of Piltdown Man and considered it a "missing link" between human and ape. But several other scientists could not understand how the humanlike skull vaulting could have been joined to the chimpanzeelike lower jaw. Far from supporting the hypothesis that humans evolved from apes, Piltdown Man seemed to be an anomalous creature who did not fit into any plausible evolutionary scheme.

In 1949 Kenneth P. Oakley at the British Museum applied to the Piltdown Man recently developed techniques for estimating the ages of fossils from the amounts of fluorine, nitrogen, iron, and radioactive carbon they contained. He found that the skull was only 50,000 to 100,000 years old (late Ice Age), while the jaw actually belonged to a modern ape (orangutan or chimpanzee). The analysis, confirmed and extended by Wilfrid Le Gros Clark and J. S. Weiner, showed that the jawbone had been chemically treated to make it look like a fossil and the teeth skillfully ground down to resemble those of prehistoric humans. It was not a question of a "mistake" or "misinterpretation" by Dawson and his colleagues, but a deliberate forgery that could only have been done by an expert.

Since 1953 when Oakley, Le Gros Clark, and Weiner announced their discovery of the hoax, the question "who done it?" has fascinated the world of professional and amateur scientists. The finger has been pointed not only at the obvious suspect, Dawson, but at several others who could have concocted the fake jawbone and left it for Dawson to find: Teilhard de Chardin, geologist W. J. Sollas, the Australian anatomist Grafton Elliot Smith, and the creator of Sherlock Holmes, Arthur Conan Doyle. It is hard to imagine what could have been the motive of the culprit, since he or she never came forward to claim the credit for having fooled the experts. Perhaps he wanted to discredit the theory of evolution by sneaking tainted evidence into its empirical basis, in the hope that the eventual discovery of the hoax would undermine belief in the theory. If so, he found willing accomplices in the journalists and creationists who loudly proclaimed that an important "proof" of evolution had been demolished.

But the exposure of the fraud actually strengthened prevailing theories of human evolution. Piltdown Man, to the extent that his authenticity was accepted, had reinforced the idea that man's ancestors were similar to *modern* apes, and that the first stage in the transition from ape to human was the expansion of brain size rather than the development of erect posture. (This view had been advocated by Grafton Elliot Smith.) But most other scientists took their cue from Pithecanthropus with its apelike skull cap and humanlike thighbone, suggesting that erect posture came before a large brain. Some scientists argued, on the basis of the latter hypothesis, that the Piltdown skull and jaw did not belong to the same individual, before the fraud was discovered in 1953. [See for example L. S. B. Leakey (1960), v, 189, 226.] As in other areas of science, the power of the theory was demonstrated by its success in challenging an apparently contradictory piece of evidence; this is also yet another counterexample to the claim that evolutionary theory can't make predictions (cf. **4.7**).

Another prediction made by Darwin himself was that our earliest ancestors would be found in Africa. Before 1924 it was believed that Asia was the "cradle of mankind," but in that year Raymond Dart, professor of anatomy at the University of Witwatersrand, started to confirm Darwin's prediction. He was shown a fossilized skull blasted by workmen out of the face of a limestone cliff at Taung, 130 kilometers north of Kimberly in the Cape Province of South Africa. Dart identified the skull as belonging to a six-year-old child with features intermediate between apes and humans. In his announcement of the discovery in 1925, he claimed that the creature was a higher primate with rudimentary hominid (humanlike) features and that it represented a type ancestral to humans. He called it Australopithecus africanus. Despite much criticism Dart's interpretation is now generally accepted; some experts would go even farther and say that since the Taung skull is more humanlike than apelike, the name Australopithecus (meaning "southern ape") is incorrect.

Since 1925 many of the most important discoveries of fossils relating to human evolution in Africa have been made by the Leakey family: Louis S. B. Leakey, his wife Mary, and their son Richard. A major landmark was "Zinjanthropus," found by Mary Leakey in 1959, and known to the public as Nutcracker Man because of his huge molar teeth. Zinjanthropus was the first fossil hominid (generally classified now as a species of Australopithecus) that could be assigned an absolute age by radiometric dating: 1,750,000 years. This estimate substantially expanded the time scale for human evolution, previously thought to have taken place in much less than a million years. It also showed that the transition from ape to human was largely completed before the use of tools (or at least stone tools).

Louis Leakey, rejecting the view of most other experts, did not accept Zinjanthropus as a human ancestor, placing it instead in a line that coexisted with that leading to humans but became extinct. Another fossil, parts of which were first noticed by young Jonathan Leakey in 1960, was determined to belong to a more humanlike creature, relatively large brained, with upright posture; since the bones showed that his hand possessed considerable dexterity and could have made the tools found in the area, this hominid was named Homo habilis ("Handy Man" to the public).

In Ethiopia in 1974 Donald Johanson found a skeleton, complete enough to allow a fairly accurate reconstruction, of a female hominid, while playing a tape of the Beatles' song "Lucy in the Sky with Diamonds." "Lucy" was only about 3½ feet tall, walked erect, and died peacefully at the age of 25 or 30 years; she is about 3.5 million years old. According to a set of criteria for distinguishing between apes and humans established by Le Gros Clark in 1950, Lucy was not human but was advanced beyond the ape. Footprints found by several members of Mary Leakey's team in 1976–1978 showed that Lucy-like creatures were fully bipedal 3.5 million years ago. Johanson argued that bipedalism had now been proved to have evolved before brain enlargement and the use of tools.

Scientists have been more successful in finding evidence of the physical form of human ancestors than in establishing the reasons why those forms evolved to Homo sapiens. Several theories of human evolution have been proposed since the time of Darwin, but there is still little agreement among the experts about the selective factors and the behavioral developments associated with structural modifications.

One of the most popular theories during the first half of the twentieth century was that developed by Arthur Keith, professor at the Royal College of Surgeons of England. In 1899, as a medical officer assigned to a mining company in the Siamese jungle, he had a chance to observe gibbons and catarrhine monkeys. He noticed that the gibbon (unlike other monkeys) moves in trees with upright posture, grasping a branch with its feet and using its arms like a gymnast on a trapeze, grasping higher branches for support. When leaping from tree to tree it uses its arms for propulsion, whereas monkeys run along branches on all fours and use their hind limbs for propulsion. When he dissected gibbons he found that their back muscles were similar to those of humans (but unlike monkeys) having been modified to maintain upright posture. So he proposed that upright posture emerged when a monkey used its arms as the chief means of support and locomotion, becoming (in the course of time) a gibbon. From gibbons there evolved several species of which one modified its lower limbs to provide the chief means of support; this modification fitted the prehuman anthropoid for life outside the jungle.

Keith's mechanism for the evolution of bipedalism—use of arms for support and motion in and among trees—is known as "brachiation." It was adopted by several other theorists but eventually abandoned by Keith himself.

In addition to postulating that erect posture evolved before "coming down from the trees," Keith advocated two general principles of evolutionary change. First, isolation and inbreeding of a small group of organisms accelerates evolution because a mutation does not get swamped by mixing with a large "normal" population. This idea is elaborated in the "founder principle" of Ernst Mayr (**4.6**). Second, contrary to Haeckel's biogenetic law, a fetal characteristic of an organism often shows up as a mature characteristic in a *later* stage of evolution—for example, in humans, a brain which is large relative to the rest of the body, a small face, prominent forehead, and hairlessness. Thus the chimpanzee fetus goes through a stage in which its distribution of body hair is similar to that of humans. This principle is now called "neoteny."

Misia Landau (1981, 1984) has suggested that the major theories of human evolution popular in the first half of the twentieth century can all be described as variations on a simple narrative structure, using techniques for the analysis of folk tales. Each is a narrative featuring four important episodes: (1) terrestriality (shift from trees to ground); (2) bipedalism; (3) enlargement of the brain, development of language and intelligence; (4) civilization (emergence of technology, morals, society). These are essentially the same as the changes listed by Darwin (see above), leaving out loss of hair and reduction in tooth size for brevity. Keith's theory changes the sequence to 2–1–4–3; other theories can be summarized as permutations of these same four episodes. The structure of an evolutionary narrative consists of nine "functions" in Landau's terminology:

(1) The initial situation—a group of animals live in trees.
(2) Introduce the hero—he is smaller and weaker than the other animals.
(3) There is a change in the hero's environment (e.g., terrestriality) or in himself (brain enlargement).
(4) The hero leaves home because of this change.
(5) The hero is tested—by the environment or by qualities of his own character. The purpose of the test is to bring out the human character of the hero.
(6) He receives a gift from a beneficent power (intelligence, tools, moral sense, or bipedalism).
(7) He is transformed as a result of the gift (civilization or brain enlargement or terrestriality).
(8) He is tested again.
(9) He triumphs, becoming fully human (but perhaps with a foreboding of ultimate doom by technology or civilization).

In all of these theories the hero like the theorist is male; crucial steps in the transition from ape to human are ascribed to male activities such as hunting and fighting that call for upright posture and use of tools or weapons. Aggressiveness is seen as an innate quality of man, essential to his evolution. As Keith remarked in *Evolution and Ethics* (1947), war is part of the machinery of evolution; "Man is by nature competitive, combative, ambitious, jealous, envious and vengeful. These are the qualities which make men the slaves of evolution." In a passage written during World War II, he asserts: "Germany has drunk the vat of evolution to its last dregs."

The thesis that human evolution has necessarily developed aggression and male dominance, popularized by Konrad Lorenz, Robert Ardrey, and many other writers in the 1960s, finally provoked a counterthesis from feminist scientists. Adrienne Zihlman and Nancy Tanner argued that gathering of vegetables and small animals by females rather than hunting of large animals by males was the primary source of food for hominids; the mother, carrying her baby while gathering, had to evolve bipedalism and the use of tools. Females could have developed the first tools—biodegradable baskets for carrying food, rather than stone weapons that survive to be found by the archeologist. As the period of infant dependency lengthened, females would need more help from males in caring for offspring and would choose to mate with the more helpful males. Helen Fisher described *The Sex Contract* (1982) in which hominid females no longer limited their sexual receptivity to a short period of "estrus" but enticed males into long-term monogamous relationships. She saw the loss of body hair, enlargement of breasts and penis, and face-to-face copulation as aspects of a process of sexual selection initiated by females in order to enhance the survival chances of their offspring. Other experts such as Glynn Isaac and Owen Lovejoy incorporated similar factors such as food sharing and cooperation, and the development of a nuclear family, into their own theories of human evolution.

At present, no reputable scientist doubts that humans have evolved from some kind of apelike animal, but there is still considerable controversy about the sequence of steps and their causes. The history of the subject shows the difficulty of establishing objective facts free of preconceptions about what human nature is or should be like.

BIBLIOGRAPHY

Peter J. Bowler (1986), *Theories of Human Evolution: A Century of Debate, 1844–1944* (Baltimore: Johns Hopkins Univ. Press).

Matt Cartmill, David Pilbeam, and Glynn Isaac (1986), "One Hundred Years of Paleoanthropology," *American Scientist* 74:410–20.

John Reader (1981), *Missing Links: The Hunt for Earliest Man* (Boston: Little, Brown).

Glyn Daniel (1976), *A Hundred and Fifty Years of Archaeology,* 2d ed. (Cambridge: Harvard Univ. Press), 29–121.

Topic *i:* Ideas to 1840

J. Lyon (1968), "The Search for Fossil Man," *Isis* 61:68–84.

Topic *ii:* Discovery of Fossil Men

D. K. Grayson (1983), *The Establishment of Human Antiquity* (New York: Academic Press).

Topic *iii:* Darwin and Human Evolution

Elizabeth Knoll (1986), "The Science of Language and the Evolution of Mind: Max Müller's Quarrel with Darwinism," *Journal of the History of the Behavioral Sciences* 22:3–22.

Joel S. Schwartz (1984), "Darwin, Wallace, and the *Descent of Man,*" *Journal of the History of Biology* 17:271–89.

Martin Fichman (1984), "Ideological Factors in the Dissemination of Darwinism in England 1860–1900," in *Transformation and Tradition in the Sciences,* ed. E. Mendelsohn (New York: Cambridge Univ. Press), 471–85.

William F. Bynum (1984), "Charles Lyell's *Antiquity of Man* and Its Critics," *Journal of the History of Biology* 17:153–87.

Howard E. Gruber (1981), *Darwin on Man: A Psychological Study of Scientific Creativity,* 2d ed. (Chicago: Univ. of Chicago Press).

Sandra Herbert (1974, 1977), "The Place of Man in Darwin's Theory of Transmutation," *Journal of the History of Biology* 7:217–58; 10:155–227.

Ralph Colp, Jr. (1977), *To Be an Invalid: The Illness of Charles Darwin* (Chicago: Univ. of Chicago Press).

W. R. Rusch (1975), "Darwin's Last Hours," *Creation Research Society Quarterly* 12:99–102.

Topic *iv:* Twentieth Century

Charles Blinderman (1986), *The Piltdown Inquest* (New York: Prometheus Books).

Misia Landau (1984), "Human Evolution as Narrative," *American Scientist* 72:262–68; (1981), "The Anthropogenic: Paleoanthropological Writing as a Genre of Literature," Ph.D. diss., Yale University. A structural analysis of accounts of the "story of human evolution" (Darwin and early twentieth century).

Mary Leakey (1984), *Disclosing the Past* (Garden City, N.Y.: Doubleday). An autobiography of the wife of Louis Leakey who herself made several important discoveries of hominid fossils.

D. J. Greenwood (1984), *The Taming of Evolution: The Persistence of Nonevolutionary Views in the Study of Humans* (Ithaca: Cornell Univ. Press).

Kathleen J. Reichs, ed. (1983), *Hominid Origins: Inquiries Past and Present* (Washington, D.C.: Univ. Press of America). Includes Charles A. Reed, "A Short History of the Discovery and Early Study of the Australopithecines: The First Find to the Death of Robert Broom (1924–1951)," 1–77, and Beck A. Sigmon, "Turning Points in Early Hominid Studies," 79–100 (a review of developments after 1950).

Stephen Jay Gould (1983), *Hen's Teeth and Horse's Toes* (New York: Norton). Includes "The Piltdown Conspiracy" (reprint of his August 1980 essay, with new footnotes); "A Reply to Critics" (of the preceding essay).

John Hathaway Winslow and Alfred Meyer (1983), "The Perpetrator at Piltdown," *Science 83* 4, no. 7:33–43. Their suggestion that the hoaxer was Arthur Conan Doyle is criticized by Don R. Cox and Stephen J. Gould, "Piltdown Debate: Not so Elementary," *Science 83* 4, no. 9:18–26.

Michael Hammond (1982), "The Expulsion of the Neanderthals from Human Ancestry: Marcellin Boule and the Social Context of Scientific Research," *Social Studies of Science* 12:1–36; (1979), "A Framework of Plausibility for an Anthropological Forgery: The Piltdown Case," *Anthropology* 3:47–58.

P. V. Tobias (1977), "A Century of Research in Human Biology and Palaeo-anthropology in South Africa," in *A History of Scientific Endeavour in South Africa,* ed. Alexander C. Brown (Cape Town: Royal Society of South Africa), 214–39.

L. S. B. Leakey (1960), *Adam's Ancestors: The Evolution of Man and His Culture,* 4th ed. (New York: Harper).

Topic *v*: Current Views

Roger Lewin (1987), *Bones of Contention: Controversies in the Search for Human Origins* (New York: Simon and Schuster).

Randall R. Skelton, Henry M. McHenry, and Gerrell M. Drawhorn (1986), "Phylogenetic Analysis of Early Hominids," *Current Anthropology* 27:21–43, with comments by A. Bilsborough et al., and reply by Skelton et al.

J. Ebling (1985), "The Mythological Evolution of Nudity," *Journal of Human Evolution* 14:33–41, is a review of explanations for the loss of hair.

John H. Langdon (1985), "Fossils and the Origin of Bipedalism," *Journal of Human Evolution* 14:615–35.

John Devine (1985), "The Versatility of Human Locomotion," *American Anthropologist* 87:550–70.

Jeffrey A. Kurland and Stephen J. Beckerman (1985), "Optimal Foraging and Hominid Evolution," *American Anthropologist* 87:73–93.

R. C. Lewontin, Steven Rose, and Leon J. Kamin (1984), *Not in Our Genes* (New York: Pantheon).

Jeffrey H. Schwartz (1984), "Hominid Evolution: A Review and a Reassessment," *Current Anthropology* 25:655–64. With comments by R. B. Eckhardt et al., 664–68, and reply by Schwartz, 668–72.

Boyce Rensberger (1984), "A New Ape in Our Family Tree," *Science 84* 5, no. 1:16; (1984), "Bones of Our Ancestors," *Science 84* 5, no. 3:28–39.

Ruth Bleier (1984), *Science and Gender: A Critique of Biology and Its Theories on Women* (Elmsford, N.Y.: Pergamon Press), Chaps. 2 and 5.

David Pilbeam (1984), "The Descent of Hominoids and Hominids," *Scientific American* 250, no. 3:84–96.

Adrienne L. Zihlman (1983), "A Behavioral Reconstruction of Australopithecus," in *Hominid Origins,* ed. K. J. Reiche (Washington, D.C.: University Press of America), 207–38.

Allen L. Hammond (1983), "Tales of an Elusive Ancestor," *Science 83* 4, no. 9:37–43, reports recent views on Ramapithecus, and C. Jolly's new theory of evolution.

Glynn Isaac (1983), "Aspects of Human Evolution," in *Evolution from Molecules to Men,* ed. D. S. Bendall (New York: Cambridge Univ. Press), 509–43.

Helen E. Fisher (1982), *The Sex Contract: The Evolution of Human Behavior* (New York: Morrow).

A. J. Jelinek (1982), "The Tabun Cave and Paleolithic Man in the Levant," *Science* 216:1369–75.

C. Owen Lovejoy (1981), "The Origin of Man," *Science* 211:341–50; (1982), letters to editor, commenting on this article, by G. Isaac et al., *Science* 217:295–306.

Constance Holden (1981), "The Politics of Paleoanthropology," *Science* 213:237–40, is on the dispute between Richard Leakey and Donald Johanson and Tim White.

Donald C. Johanson and Maitland A. Edey (1981), *Lucy: The Beginnings of Humankind* (New York: Simon and Schuster).

Steven M. Stanley (1981), *The New Evolutionary Timetable* (New York: Basic Books); (1979), *Macroevolution, Pattern and Process* (San Francisco: Freeman). Stanley argues that recent fossil discoveries support the functional model of evolution and undermine (though they don't disprove) gradualism, by showing the coexistence of different hominid species over a million years.

Nancy M. Tanner (1981), *On Becoming Human* (New York: Cambridge Univ. Press).

L. K. Königsson, ed. (1980), *Current Argument on Early Man* (New York: Pergamon Press), includes the Tobias article suggested above under *Readings* and D. R. Pilbeam, "Major Trends in Human Evolution," 261–85.

Ruth Hubbard (1979), "Have Only Men Evolved?" in *Women Look at Biology Looking at Women,* ed. Ruth Hubbard et al. (Boston: Hall), 7–35; reprinted in *Discovering Reality,* ed. S. Harding and M. B. Hintikka (Boston: Reidel, 1983), 45–69. A critique of Darwinian "sexual selection" theories of human evolution and recent sociobiological theories.

E. Delson (1978), "Models of Early Hominid Phylogeny," in *Early Hominids of Africa,* ed. Clifford Jolly (New York: St. Martin's Press), 517–41. A technical review of current research.

Donna Haraway (1978), "Animal Sociology and a Natural Economy of the Body Politic, Part II: The Past is the Contested Zone: Human Nature and Theories of Production and Reproduction in Primate Behavior Studies," *Signs 4,* no. 1:37–60, is a feminist analysis of theories of reproduction and evolution proposed by S. Zuckerman, T. Rowell, S. Washburn, A. Zihlman, N. Tanner.

Richard E. Leakey and Roger Lewin (1977), *Origins: What New Discoveries Reveal About the Emergence of Our Species and Its Possible Future* (New York: Dutton).

3

EVOLUTION OF RACES AND CULTURES

3.1 RACE THEORIES IN EUROPE

i. Pre-1800 ideas about race; impact of geographical exploration
ii. Theories and observations before 1870
iii. Darwin's *Descent of Man* and the impact of evolutionary views

READINGS

Aside from the two or three pages in Stephen F. Mason's *History of the Sciences* (1962), 422–23, 433–34, there is essentially nothing on the topics of Chapter 3 in the standard texts available for a general course on history of science. Peter Bowler's *Evolution* (1984), a more specialized text, does only a little better (282–88). Yet one might well argue that it is precisely these topics—theories about racial differences, social Darwinism, eugenics, and cultural relativism—that provide the most striking influence of nineteenth-century science on modern society, and we need a better understanding of the history of science to help us straighten out our thinking on current problems. Thus for example a reviewer of John S. Haller's *Outcasts from Evolution: Scientific Attitudes of Racial Inferiority, 1859–1900* (1971) wrote: "In order to put the allegedly ludicrous claims of black militants into a much-needed perspective, one that makes them eerily and irresistibly unludicrous, this book is required reading." [*American Scientist* 60(1972):387].

The gap in the textbook literature cannot be filled by a single specialized book, but my recommendation is to start with one of the more elementary works on the history of anthropology and supplement it with articles on various topics. *Images of Man* by Annemarie De Waal Malefijt

(1974), 94–292, or H. R. Hays, *From Ape to Angel* (1979), 1–62, 207–68, 309–12, 340–54, could be used to build the foundation. Less comprehensive on anthropology but useful elsewhere in this course is Abram Kardiner and Edward Preble, *They Studied Man* (1961), 17–77, 134–203. Stephen Jay Gould's *The Mismeasure of Man* (1981), which I recommend for later sections, has a chapter on Paul Broca's head measurements (72–112) that could be used here. The article by J. S. Haller, "The Species Problem: Nineteenth-Century Concepts of Racial Inferiority in the Origin of Man Controversy," *American Anthropologist* 72(1970):1319–29, provides a good introduction to the major issues in *ii* and *iii* of this section and to **3.2;** his monograph mentioned above can be used by the instructor to prepare lectures going into some points in greater depth. If you want to put more emphasis on the role of the race concept in European and cultural history, Leon Poliakov's *The Aryan Myth: A History of Racist and Nationalist Ideas in Europe* (1974) or Jacques Barzun's *Race: A Study in Superstition* (1937, rev. ed. 1965) might be suitable for students who have some background in modern history.

BIBLIOGRAPHY

Topic *i:* Pre-1800 Ideas

R. H. Popkin (1974), "The Philosophical Bases of Modern Racism," in *Philosophy and the Civilizing Arts,* ed. C. Walton and J. P. Anton (Athens: Ohio Univ. Press), 126–65.

J. H. Elliott (1972), "The Discovery of America and the Discovery of Man," *Proceedings of the British Academy* 58:101–25, concerns sixteenth-century Europe's discovery of non-European men.

R. H. Osborne (1971), "The History and Nature of Race Classification," in *The Biological and Social Meaning of Race,* ed. Richard H. Osborne (San Francisco: Freeman), includes a summary of modern classification schemes.

Topic *ii:* 1800–1870

Nancy Stepan (1982), *The Idea of Race in Science: Great Britain, 1800–1960* (Hamden, Conn.: Shoe String Press), Chaps. 1–3.

Francis Schiller (1979), *Paul Broca, Founder of French Anthropology, Explorer of the Brain* (Berkeley: Univ. of California Press).

Popkin (1974), "Philosophical Bases," cited above.

J. S. Haller, Jr. (1970), "Concepts of Race Inferiority in Nineteenth-Century Anthropology," *Journal of the History of Medicine* 25:40–51.

Topic *iii:* Darwin's *Descent of Man*

Charles Darwin (1981), *The Descent of Man, and Selection in Relation to Sex* (reprint of the 1871 ed.), introduction by J. T. Bonner and R. M. May (Princeton, N.J.: Princeton Univ. Press).

Nancy Stepan (1986), "Race and Gender: The Role of Analogy in Science," *Isis* 77:261–77; (1985), "Biological Degeneration: Races and Proper Places," in *Degeneration,* ed. J. E. Chamberlin and S. L. Gilman (New York: Columbia Univ. Press), 97–120; (1982), *Idea of Race* (cited above), Chap. 4.

Allan Chase (1977), *The Legacy of Malthus: The Social Costs of the New Scientific Racism* (Urbana: Univ. of Illinois Press).

Bernard Campbell, ed. (1972), *Sexual Selection and the Descent of Man 1871–1971* (Chicago: Aldine), contains articles by Loren Eiseley, G. G. Simpson, T. Dobzhansky, E. Mayr, etc.

3.2 EVOLUTION AND RACISM IN THE UNITED STATES

i. Pre-1800 views about blacks and Indians
ii. Polygenist vs. monogenist theories, 1800–1870
iii. Impact of evolution on racial attitudes

READINGS

In addition to the background readings on the history of anthropology mentioned in **3.1**, I suggest using Stephen Jay Gould's *The Mismeasure of Man* as a text for parts of Chapters 3 and 4 of this course. His chapter on "American Polygeny and Craniometry before Darwin" (30–77) includes his own detailed scrutiny of the data of Morton. The chapters "Measuring Heads" (73–112) and "Measuring Bodies" (113–45) cover the work of Broca and Lombroso, the latter being also relevant for our **3.4**. If Robert C. Bannister's book *Social Darwinism* (1979) is used (see **3.3**), his Chapter 9, 188–200, would be appropriate here for *iii.*

This section is intended to take the story of "scientific racism" up to 1900, though some of the readings go into the early twentieth century. The critique of racism from the viewpoint of cultural anthropology by Franz Boas and others is covered in **3.5**; the more recent debates on heredity vs. environment and the correlation between race and IQ are discussed in **6.8**.

SYNOPSIS

Long before Darwin's theory was published, Americans discussed whether the differences between races are due to heredity or environment. According to the philosophy of the eighteenth-century Enlightenment, which inspired the basic documents of the United States, all human differences are due to environment. If this were true, it would seem that the African background and subsequent enslavement of the blacks was the only cause of their apparent mental inferiority, and there could be no justification for denying them the same rights as were enjoyed by whites. On the other hand, if blacks had been created as a separate race or had become separate through a process of biological evolution, the differences could be attributed to heredity and would not be expected to disappear even if the environment could be made the same as that of whites. "Evolution," in the most general sense, need not imply a simple progression from simple to complex or "advanced" forms. Indeed, the French scientist Buffon had suggested that under some conditions evolution might go in reverse, what he called *degeneration*. That is what he thought had happened on the American continent: as a result of a colder and wetter climate compared with Europe, the animal and plant forms that arose or migrated there had become smaller and less diverse.

The writings of Thomas Jefferson on racial differences are of interest not only because of his influence on American thought but also because they provide a response to European scientific theories such as Buffon's degeneration hypothesis. Jefferson's *Notes on Virginia* (1774) were written in part to refute Buffon; they also present his views on blacks and Indians. Whereas he seemed to regard Indians as closely akin if not identical to white Europeans, he believed that blacks are a completely different and probably inferior race. Miscegenation was perceived as a threat to the status of the white race. The accusation that Jefferson himself had fathered several children by one of his slaves was especially damaging, not so much because of the question of immorality but rather because it meant mixing the races, which Jefferson had publicly condemned.

There were two kinds of race theories in the nineteenth century. The *monogenist* theory asserted that all races had evolved from a common ancestor, while the *polygenist* theory postulated separate creation. The latter might seem to be the simplest way to account for racial differences and also to justify discrimination (e.g., "God made the Negroes to be servants for the whites"), but it did not seem to be consistent with the biblical account of creation. Hence people with fundamentalist religious views would hesitate to claim publicly that blacks are a different species from whites, however convenient that belief might be in supporting their concept of what society should be like.

Polygenist race theory was given "scientific" support in the period 1830–1860 by Samuel George Morton, a Philadelphia physician; Josiah Clark Nott, a physician in Mobile, Alabama; and Louis Agassiz, the well-known Swiss naturalist who emigrated to the United States and became a Harvard professor.

Morton developed a large collection of skulls and tried to show that different races had different head shapes. Using ancient Egyptian skulls he argued Negroes had been slaves, then as now, and that there had been no change in racial characteristics over several millenia. Southern politicians used Morton's conclusions to defend the continuation of slavery. Nott, more openly than Morton, used race theory to support slavery.

Agassiz believed that all existing species had been created separately and had not changed since their creation. He became one of the major scientific critics of Darwinian theory in the United States. He saw blacks for the first time when he visited the United States in 1846, as a guest of Morton at a Philadelphia restaurant employing black waiters. Agassiz jumped to the conclusion that blacks are a different species from whites, if indeed they are men at all. In 1850 he supported the polygenist theory at a meeting in Charleston of the American Association for the Advancement of Science (AAAS). When criticized for giving ammunition to the proslavery forces he replied that he was only a scientist, not responsible for how politicians used his results.

Agassiz was also subject to criticism from theologians for abandoning the literal interpretation of Genesis; he was confident of his own theory of God's plan for creating species. He pointed out with justice that the monogenist race theory implied some kind of evolution in order to account for race differences, a theory at least equally dangerous to religious beliefs. Indeed, in the subsequent controversy over Darwin's theory, Agassiz's leadership on the antievolutionary side did make him much more popular with theologians.

Asa Gray, Agassiz's colleague at Harvard in the 1850s and 1860s, was the first major promoter of Darwin's ideas in America. He had long been a monogenist on the race question, and since his religious beliefs coincided with his scientific ones he opposed those like Nott who asserted the inferiority of blacks. Gray realized that Darwinism could be used to refute polygenism, and thus he could appeal to the religious community to accept Darwinism because it derived all men from the same origin. But the theologians now saw Darwinism as the most serious threat to Christianity and so Agassiz became their hero.

BIBLIOGRAPHY

Topic *i:* Pre-1800

Winthrop D. Jordan (1968), *White over Black: American Attitudes toward the Negro, 1550–1812* (Chapel Hill: Univ. of North Carolina Press).

Topic *ii:* Polygenist vs. Monogenist Theories

Reginald Horsman (1986), *Race and Manifest Destiny: The Origins of American Racial Anglo-Saxonism* (Cambridge: Harvard Univ. Press).

Stephen Jay Gould (1979), "Agassiz' Later, Private Thoughts on Evolution: His Marginalia in Haeckel's *Natürliche Schöpfungsgeschichte* (1868)," in *Two Hundred Years of Geology in America,* ed. Cecil J. Schneer (Hanover, N.H.: Univ. Press of New England), 277–82.

Mary Pickard Winsor (1979), "Louis Agassiz and the Species Question," *Studies in History of Biology* 3:89–117.

Ronald T. Takaki (1979), *Iron Cages: Race and Culture in 19th-Century America* (New York: Knopf), discusses a wide variety of ethnic groups.

Richard H. Popkin (1978), "Pre-Adamism in 19th-Century American Thought: 'Speculative Biology' and Racism," *Philosophia* 8:205–39.

Edward Lurie (1974), *Nature and the American Mind: Louis Agassiz and the Culture of Science* (New York: Science History).

Topic *iii:* Impact of Evolution

Nancy Stepan (1985), "Biological Degeneration: Races and Proper Places," in *Degeneration,* ed. J. E. Chamberlin and S. L. Gilman (New York: Columbia Univ. Press), 97–120.

Dwight W. Hoover (1981), "A Paradigm Shift: The Concept of Race in the 1920s and 1930s" *Conspectus of History* I, no. 7:82–100.

John S. Haller, Jr. (1971), *Outcasts from Evolution: Scientific Attitudes of Racial Inferiority 1859–1900* (Urbana: Univ. of Illinois Press).

Guy Percy Griggs (1971), "The White American Social and Physical Scientists' Views of the Negro 1877–1920" (Ph.D. diss., University of Kansas).

George W. Stocking, Jr. (1968), *Race, Culture, and Evolution: Essays in the History of Anthropology* (New York: Free Press), Chap. 3: "The Persistence of Polygenist Thought in Post-Darwinian Anthropology," and Chap. 6: "The Dark-skinned Savage: The Image of Primitive Man in Evolutionary Anthropology."

3.3 "SOCIAL DARWINISM" AND EUGENICS

i. Spencer and Darwin
ii. Social Darwinism in American thought
iii. Francis Galton and the eugenics movement to 1914
iv. Theory of degeneration
v. Twentieth-century consequences

"I believe in social decency, not social Darwinism," said Walter Mondale in opening his campaign for the 1984 Democratic presidential nomination (*New York Times,* 22 Feb. 1983). Clearly today's voters are expected to recognize "social Darwinism" as a term of opprobrium. Where did it come from?

"Social Darwinism" was identified (and, some scholars would say, invented) in Richard Hofstadter's classic work, *Social Darwinism in American Thought* (1944, rev. ed. 1955). As Hofstadter and others have pointed out, it is based on ideas articulated most clearly not by Darwin himself but by Herbert Spencer. While there was certainly a lot of discussion of Darwinian ideas among intellectuals, the thesis that it inspired or justified businessmen in a propensity for cut-throat competition should not be accepted without some skepticism.

The "theory of degeneration" refers to the ideas proposed by B. A. Morel (1857), Cesare Lombroso, and Max Nordau (1892), which assumed a specific effect of environmental factors such as alcoholism on heredity. The theory was presupposed in Emile Zola's Rougon-Macquart novels and reinforced fin-de-siècle pessimism; it also provided part of the"scientific" justification for the Prohibition movement and other "Progressive" reforms in early twentieth-century America. The background of Freud's psychoanalysis (**5.1**) includes degeneration as one possible explanation of mental disorders.

Eugenics, on the other hand, was and is a real phenomenon, not merely a construction of intellectual historians. It was consciously advocated by a well-defined group of people and played a significant role in twentieth-century history. Its association with some of the horrors of Nazi Germany reacted back on science and discouraged legitimate research into human genetics.

READINGS

As with most aspects of the social history of science it is difficult to find a text that is brief, accurate, and readable enough for students. One possibility is Robert C. Bannister's *Social Darwinism,* 34–180, starting with a comprehensive account of Spencer's views and then discussing his American followers, especially William Graham Sumner. A chapter on eugenics (164–79), though brief, does show its relation to social Darwinism. In addition there is a chapter on late nineteenth-century racial theories inspired by evolution (180–200), which could be used for **3.2**, topic *iii.* Another good general book is Paul F. Boller's *American Thought in Transition: The Impact of Evolutionary Naturalism 1865–1900* (1969, reprinted 1981).

Among books on evolution mentioned in earlier sections, Bowler's *Evolution* deals with this topic (225–27, 266–81).

The topics discussed in **6.5** and **6.8** (heredity vs. environment, race-IQ controversy) are closely related and might well be combined with **3.3**. Hamilton Cravens's book, *The Triumph of Evolution: American Scientists and the Heredity-Environment Controversy 1900–1941* (1978), gives a good overview of much of this general subject. My book *The Temperature of History* (1978) discusses degeneration in the context of nineteenth-century theories of energy dissipation and the cooling of the earth, as well as Henry Adams and Oswald Spengler's ideas about history. Gould's *Mismeasure of Man* has a good section on Lombroso, 123–45, stressing his influence on criminal anthropology but omitting the connection with Morel and Nordau. Degeneration as an anthropological concept (J. Lubbock) is treated briefly by Hays, *Ape to Angel,* 50–55. Social aspects of twentieth-century eugenics are treated in Loren Graham's *Science and Values* (1981), 217–56.

Charles Rosenberg's *No Other Gods: On Science and American Social Thought* (1976) is a collection of case studies, several of which pertain to this section or to sections **4.1** and **4.5**, with a general introduction on "Science, Society, and Social Thought," 1–21. "The Bitter Fruit: Heredity, Disease and Social Thought," 25–54, and "George M. Beard and American Nervousness," 98–108, cover *iv;* "Charles Davenport and the Irony of American Eugenics," 89–97, touches on one aspect of *iii.*

The article by C. H. Corning, "Francis Galton and Eugenics," *History Today* 23(1973):724–32, is suitable for students, as are sections of Daniel Kevles's book *In the Name of Eugenics* (1985), published separately in *The New Yorker* under the title "Annals of Eugenics" (Oct. 1984).

SYNOPSIS

The eugenics movement was initiated in England by Francis Galton, though some of its ideas go back to Plato. After publishing influential works on heredity such as *Hereditary Genius* (1869) and experimenting with various tests of mental ability, Galton founded the Eugenics Education Society in 1908; this society published the widely read *Eugenics Review*. In the eugenics laboratory Galton endowed at the University of London, his student Karl Pearson carried on important work oriented especially toward the use of statistical methods. The journal *Biometrika* was founded by Galton and Pearson in 1901.

Galton believed that intelligence is primarily determined by heredity; he thought that the superior people in a society had a responsibility to marry other superior people and reproduce their kind. Like the earlier Spencerian Social Darwinists, Galton criticized philanthropy and welfare programs that helped the weak to survive at the expense of the strong and enabled the unfit to reproduce themselves irresponsibly. He and his followers attempted to identify various factors in the environment or in society that could be considered "racial poisons" tending to favor the reproduction of stupidity rather than intelligence. For example, like other nineteenth-century writers, he decried the bad effects of the movement of population from rural areas into cities.

By eliminating racial poisons and encouraging the breeding of better humans, the eugenists proposed to control the evolution of the human race. This program turned out to be very attractive in the United States, where it was incorporated into the thinking of many of the Progressives in the early decades of the twentieth century. It also provided "scientific" support for the prejudices of people who were interested in passing restrictive legislation (e.g., Prohibition and immigration laws) even though they did not accept the theory of evolution.

The leader of the American eugenics movement was Charles Davenport, a zoologist who pioneered the use of Pearson's statistical (biometric) methods. He collected data on the inheritance of traits in families and set up a Eugenics Record Office. Another leader was Paul Popenoe, author of popular books on eugenics and marriage. Popenoe considered socialism a menace (it ignored the basic and natural inequality of mankind), and questioned the wisdom of child-labor laws, minimum wage laws, mothers' and old-age pensions, and trade unionism, all of which (he claimed) helped preserve the biologically and mentally inefficient members of the race.

The American eugenics movement eventually became best known for its advocacy of sterilization, and this led to its downfall. By 1931, thirty states had passed laws calling for sterilization of persons judged to be insane, idiotic, or incurably criminal. When the Nazis came to power in Germany in the 1930s, they used sterilization to carry out their program of "race hygiene." Those judged hereditarily defective included the Jews. After World War II, revulsion against the Holocaust also tainted eugenics, and even today any proposals for sterilization, voluntary or otherwise, are likely to be condemned as "genocide."

The collapse of the eugenics movement in the late 1930s had two unfortunate effects. First, the public came to be highly suspicious of even those aspects of the program that might have been of some value if based more firmly on modern theories of heredity. Second, many scientists chose not to do research in *human* genetics because that field had acquired a bad reputation by its association with eugenics.

BIBLIOGRAPHY

Charles Rosenberg, ed. (1985–), *The History of Hereditarian Thought* (New York: Garland) is a reprint series of thirty-two books by or about Galton, C. B. Davenport, A. de Gobineau, C. Lombroso, other works on eugenics.

Linda L. Clark (1984), *Social Darwinism in France* (University: Univ. of Alabama Press), covers the period from 1859 to World War I.

Peter Morton (1984), *The Vital Science: Biology and the Literary Imagination, 1860–1900* (London: Allen and Unwin), discusses responses to Darwinism, fears of human degeneration, eugenics, and pre-Mendelian speculations on heredity.

Greta Jones (1980), *Social Darwinism and English Thought: The Interaction between Biological and Social Theory* (Atlantic Highlands, N.J.: Humanities Press), includes chapters on Lamarckism, eugenics, race, and class.

Topic *i:* Spencer and Darwin

Herbert Spencer: Structure, Function and Evolution (1971), ed. and with an introductory essay by Stanislav Andreski (New York: Scribner), includes extracts from his writings on evolution (biological, social, and cultural), 53–104.

Michael Ruse (1980), "Social Darwinism: The Two Sources," *Albion* 12:23–36, concerns the relations between Malthus, Darwin, and Spencer.

Scott F. Gilbert (1979), "Altruism and Other Unnatural Acts: T. H. Huxley on Nature, Man, and Society," *Perspectives in Biology and Medicine* 22:346–58.

Topic *ii:* American Social Darwinism

Robert C. Bannister (1979), *Social Darwinism: Science and Myth in American Social Thought* (Philadelphia: Temple Univ. Press).

Cynthia Eagle Russett (1976), *Darwin in America: The Intellectual Response, 1865–1912* (San Francisco: Freeman).

Topic *iii:* Galton and Eugenics

Carl J. Bajema, ed. (1976), *Eugenics, Then and Now* (Stroudsburg, Pa.: Hutchinson Ross), is an anthology of original sources, late nineteenth and early twentieth centuries.

Daniel J. Kevles (1985), *In the Name of Eugenics: Genetics and the Uses of Human Heredity* (New York: Knopf).

Timothy L. Alborn (1985), "Eugenics and the Meritocracy: Francis Galton's Educational Model," *Synthesis* 6, no. 1:16–37.

Raymond E. Fancher (1983), "Biographical Origins of Francis Galton's Psychology," *Isis* 74:227–33.

Karl H. Metz (1984), "The Survival of the Unfittest": Die sozialdarwinistische Interpretation der britischen Socialpolitik vor 1914," *Historische Zeitschrift* 239:565–601.

Michael Freeden (1983), "Eugenics and Ideology," *Historical Journal* 25:717–28; (1979), "Eugenics and Progressive Thought: A Study in Ideological Affinity," *Historical Journal* 22:645–71.

Donald A. MacKenzie (1981), *Statistics in Britain, 1865–1930* (New York: Columbia Univ. Press), Chaps. 2 and 3; the author's approach is that of the "social construction of knowledge" school.

L. A. Farrall (1979), "The History of Eugenics: A Bibliographical Review," *Annals of Science* 36:111–23.

Rosaleen Love (1979), " 'Alice in Eugenics-Land': Feminism and Eugenics in the Scientific Careers of Alice Lee and Ethel Elderton," *Annals of Science* 26:145–58.

R. S. Cowan (1977), "Nature and Nurture: The Interplay of Biology and Politics in the Work of Francis Galton," *Studies in History of Biology* 1:133–208.

G. R. Searle (1976), *Eugenics and Politics in Britain, 1900–1914* (Leyden: Noordhoff), contains chapters on Galton, Pearson, racial degeneration, and the race issue.

Topic *iv:* Degeneration

Max Nordau (Trans. from 2d German ed. 1895; reprinted 1968, with intro. by G. Mosse), *Degeneration* (New York: Fertig).

J. Edward Chamberlin and Sander L. Gilman (1985), *Degeneration: The Dark Side of Progress* (New York: Columbia Univ. Press).

Robert A. Nye (1984), *Crime, Madness, and Politics in Modern France: The Medical Concept of National Decline* (Princeton, N.J.: Princeton Univ. Press); (1976), "Heredity or Milieu: The Foundations of Modern European Criminological Theory," *Isis* 67:335–55.

Stuart C. Gilman (1983), "Degeneracy and Race in the 19th Century: The Impact of Clinical Medicine," *Journal of Ethnic Studies* 10, no. 4:27–50.

Francis Schiller (1982), *A Moebius Strip: Fin-de-Siècle Neuropsychiatry and Paul Moebius* (Berkeley: Univ. of California Press).

D. P. Crook (1981), "Darwinism—The Political Implications," *History of European Ideas* 2:19–34.

Stephen G. Brush (1978), *The Temperature of History* (New York: Franklin), see bibliography for Chap. 7, 180–90.

Peter R. Morton (1976), "Biological Degeneration: A Motif in H. G. Wells and other late Victorian Utopianists," *Southern Review: An Australian Journal of Literary Studies* 9:93–112.

Topic *v:* Twentieth-Century Consequences

Sheila Faith Weiss (1987), "The Race Hygiene Movement in Germany," *Osiris,* ser. 2, 3: 193–236; (1986), "Wilhelm Schallmayer and the Logic of German Eugenics," *Isis* 77:33–46.

Garland E. Allen (1986), "The Eugenics Record Office at Cold Spring Harbor, 1910–1940: An Essay in Institutional History," *Osiris,* series 2, 2:225–64; (1983), "The Misuse of Biological Hierarchies: The American Eugenics Movement, 1910–1940," *History and Philosophy of the Life Sciences* 5:105–28; (1980), "Genetics as a Social Weapon," in *Science and Liberation,* ed. Rita Arditti et al. (Boston: South End Press), 48–62; (1976), "Genetics, Eugenics and Society: Internalists and Externalists in Contemporary History of Science," *Social Studies of Science* 6:105–22, is an essay review of books by K. M. Ludmerer, D. K. Pickens, and W. Provine.

Steven Selden (1985), "Education Policy and Biological Science: Genetics, Eugenics, and the College Textbook, c. 1908–1931," *Teachers College Record* 87:35–51; (1978), "Biological Determinism and the Normal School Curriculum," *Journal of Curriculum Theorizing* 1:105–22, analyzes the influence of eugenics on American education before 1925.

Diane Paul (1984), "Eugenics and the Left," *Journal of the History of Ideas* 45:567–90, discusses a group of English and American scientists in the 1920s, 1930s, and early 1940s.

Gisela Bock (1983), "Racism and Sexism in Nazi Germany: Motherhood, Compulsory Sterilization, and the State," *Signs* 8:400–21.

Stephen Jay Gould (1983), *Hen's Teeth and Horse's Toes* (New York: Norton), includes an article on "Science and Jewish Immigration," 291–302.

W. Schneider (1982), "Toward the Improvement of the Human Race: The History of Eugenics in France," *Journal of Modern History,* 54:268–91.

Greta Jones (1982), "Eugenics and Social Policy between the Wars," *Historical Journal* 25:717–28.

J. Woodhouse (1982), "Eugenics and the Feeble-Minded: The Parliamentary Debates of 1912–14," *History of Education* 11:127–37.

Bentley Glass (1981), "A Hidden Chapter of German Eugenics between the Two World Wars," *Proceedings of the American Philosophical Society* 125:357–67, concerns a book by Erwin Baur, Eugen Fischer, and Fritz Lenz, used by the Nazis.

Loren R. Graham (1981), *Between Science and Values* (New York: Columbia Univ. Press); (1977), "Science and Values: The Eugenics Movement in Germany and Russia in the 1920s," *American Historical Review* 82:1133–64.

George L. Mosse (1966, 1981), *Nazi Culture: A Documentary History,* includes "racism," extracts from works by H. F. K. Günther et al. on heredity and biology, 57–91.

Roy Lowe (1980), "Eugenics and Education: A Note on the Origins of the Intelligence Testing Movement in England,"*Educational Studies* 6:108.

G. R. Searle (1979), "Eugenics and Politics in Britain in the 1930s," *Annals of Science* 36:159–79.

Georg Lilienthal (1979), "Rassenhygiene im Dritten Reich: Krise und Wende," *Medizinhistorisches Journal* 14:114–34.

Z. Suzuki (1975), "Geneticists and the Eugenics Movement in Japan," *Japanese Studies in the History of Science* 14:157–64.

3.4 ANTHROPOLOGY: THE TRANSITION FROM EVOLUTIONISM TO RELATIVISM

i. Early history
ii. Evolutionary anthropology
iii. Franz Boas and cultural relativism

LITERATURE

Readings: De Waal Malefijt, *Images of Man,* 3–115 (on *i*), 115–59 (on *ii*), 215–33 (on *iii*); or Hays, *From Ape to Angel,* 3–62 (mostly on *ii*), 227–69 (on *iii*); or Kardiner and Preble, *They Studied Man,* 17–77 (on *ii*), 134–59 (on *iii*). Dampier, *History of Science*, has a very brief section, 283–87.

SYNOPSIS

In my course anthropology is treated mainly in connection with the impact of evolution, race, and sex theories, the heredity-environment issue, and psychoanalysis. In the nineteenth century, "evolutionary anthropology" implied a one-dimensional approach in which all cultures were put in a sequence, starting with the most primitive and ending with the most advanced, i.e., Western European society as defined by white males; the assumption is that every society naturally passes through this sequence but some are further along than others at the moment. The scheme provided a convenient justification for colonial domination, while reassuring liberals that the "backward" people could eventually qualify for the rights and benefits of civilized societies if allowed to progress gradually under the tutelage of the white man.

The German-born anthropologist Franz Boas (1858–1942) had a major influence on the reorganization of anthropology as an academic discipline in the United States at the beginning of the twentieth century; he and his students have had a considerable impact on contemporary ideas about racial differences and the variations of human behavior in different cultures and have liberated us from many of the older ideas about biologically determined roles of people of different races and sexes. All of this may be summed up in the phrase "cultural relativism." As applied to the elementary-school curriculum in the 1970s through the MACOS ("Man: A Course of Study") project, cultural relativism offended many Americans who wanted children to accept their own moral values as absolute; the resulting backlash in Congress led to the withdrawal of federal support for science education. [See Dorothy Nelkin (1977), *Science Textbook Controversies and the Politics of Equal Time* (Cambridge: MIT Press), Chap. 7.] If you want to go into Boas more deeply, I recommend George Stocking's *Race, Culture, and Evolution,* 133–233.

BIBLIOGRAPHY

Murray J. Leaf (1979), *Man, Mind, and Science: A History of Anthropology* (New York: Columbia Univ. Press).

Topic *i:* Early History

Edward Evans-Pritchard (1981), *A History of Anthropological Thought* (New York: Basic Books), contains chapters on Montesquieu,

Henry Home, Lord Kames, Ferguson, Miller, Condorcet, other 18th-century writers.

Topic *ii:* Evolutionary Anthropology

Adam Kuper (1985), "The Development of Lewis Henry Morgan's Evolutionism," *Journal of the History of the Behavioral Sciences* 21:3–22.

Frank Spencer, ed. (1982), *A History of American Physical Anthropology, 1830–1930* (New York: Academic Press).

Neal C. Gillespie (1977), "The Duke of Argyll, Evolutionary Anthropology, and the Art of Scientific Controversy," *Isis* 68:40–54.

George W. Stocking, Jr. (1974), "Some Problems in the Understanding of Nineteenth Century Evolutionism," in *Readings in the History of Anthropology,* ed. Regna Darnell (New York: Harper and Row), suggests that the transition to evolutionary anthropology around 1860 and the reaction against it around 1900 both resemble Kuhnian revolutions.

Marvin Harris (1968), *The Rise of Anthropological Theory* (New York: Crowell), is a provocative analysis of major nineteenth- and twentieth-century theories from the viewpoint of "cultural materialism."

Topic *iii:* Boas

Franz Boas (1966), *Race, Language, and Culture* (1940), reprinted (New York: Free Press), is his own selection of essays; includes skeptical comments on eugenics and degeneration theories, race differences in IQ. See also (1974), *The Shaping of American Anthropology, 1883–1911: A Franz Boas Reader,* ed. George W. Stocking (New York: Basic Books).

Sydel Silverman, ed. (1981), *Totems and Teachers: Perspectives on the History of Anthropology* (New York: Columbia Univ. Press), contains essays on Boas and others.

Joan Mark (1981), *Four Anthropologists: An American Science in Its Early Years* (New York: Science History), says Boas has been given too much credit for ideas suggested first by others.

Dwight W. Hoover (1981), "A Paradigm Shift: The Concept of Race in the 1920s and 1930s," *Conspectus in History* 1, no. 7:82–100, discusses the influence of Boas on racism.

William W. Speth (1978), "The Anthropogeographic Theory of Franz Boas," *Anthropos* 73:1–31.

Hamilton Cravens (1978), *The Triumph of Evolution* (Philadelphia: Univ. of Pennsylvania Press), Chap. 3.

F. W. Voget (1970), "Boas, Franz," *Dictionary of Scientific Biography* 2:207–13.

4

GENDER AND GENETICS

4.1 NINETEENTH-CENTURY VIEWS OF SEX DIFFERENCES

i. European views before Darwin
ii. Spencer, Darwin, and the evolution of sex differences
iii. Anthropologists on sex roles and the history of the family (Henry Maine, John McLennan, J. J. Bachofen, L. H. Morgan)
iv. Women's education and role in science
v. American feminism and arguments about sex differences; physicians and female sexuality

READINGS

Charles Rosenberg, *No Other Gods* (1976), 54–88 (includes the article with Carroll Smith-Rosenberg, "The Female Animal: Medical and Biological Views of Women," and "Sexuality, Class and Role"). Annemarie De Waal Malefijt, *Images of Man* (1974), 123–39 (on *iii*); H. R. Hays, *From Ape to Angel* (1979), 32–49, 161–82 (on *iii*). Paul Robinson, *The Modernization of Sex* (chapter on Havelock Ellis) (1976). Lorna Duffin, "Prisoners of Progress: Women and Evolution," 57–91, in *The Nineteenth-Century Woman: Her Cultural and Physical World,* ed. Sara Delmont and Lorna Duffin (1978). Barbara J. Harris, "The Power of the Past: History and the Psychology of Women," 1–25, in *In the Shadow of the Past,* ed. Miriam Lewin (1984). Jill Conway, "Stereotypes of Femininity in a Theory of Sexual Evolution," 140–54, in *Suffer and Be Still: Women in the Victorian Age,* ed. Martha Vicinus (1972).

Stephen Jay Gould, *The Mismeasure of Man* (1981), 103–7, discusses Broca's attempt to establish differences in brain sizes of men and women, and Maria Montessori's attempt to invert his conclusions.

SYNOPSIS

This section is intended to provide a link between **2.5, 3.4,** and **4.2, 4.9, 5.4, 6.7.** The chapter as a whole introduces the nature/nurture issue in the context of sex differences, starting with the nineteenth-century view that these differences are primarily inborn, then swinging to the opposite extreme with Mead's claim that they are primarily cultural. Mendel's genetic theory provides the modern basis for biological sex determination; the discovery of the role of X and Y chromosomes was a major advance in the development of Mendelian genetics. The rest of the chapter follows the history of genetics in the first half of the twentieth century, including the discovery of the structure of DNA, and brings in the connection with Darwinian evolution through the "synthetic" theory, ending with a look at the current "creation-evolution controversy."

This section also provides an introduction to the topic "women in science" by examining the social barriers to women in higher education, based on Victorian preconceptions about their intellectual qualities and the alleged deleterious effects of intense mental exertion. These preconceptions were reinforced by the authority of science, for example Darwin's statements in *The Descent of Man* (1871) that male superiority has resulted from the process of evolution through "sexual selection." Darwin also accepted the view (attributing it to Galton) that the variability of traits is greater in men than in women, a view that survives up to the present time. The physical principle of energy conservation was also invoked by writers such as Herbert Spencer (*Principles of Biology,* 1867) in support of the claim that "mental labour carried to excess" causes a "deficiency in reproductive power" in women. This included the ability to suckle their infants, which "most of the flat-chested girls who survive their high pressure education are incompetent to do" (*ibid.* 2:486). Thus, according to the received scientific wisdom of the day, any woman who undertook a serious study of science would be endangering the future of the race!

The works of Havelock Ellis close the Victorian period and begin the modern period in the systematic study of sex differences. His *Man and Woman* (1894) assembled a mountain of evidence on the subject and concluded that women, though different from men, are not inferior.

Just two years before Darwin published *The Descent of Man,* the suffragist Lydia Ernestine Becker argued that women were just as capable of scientific achievement as men, according to the results of science examinations at the Royal College of Science for Ireland, one of the few which women were even allowed to take; yet women were excluded from the right to earn the best university degrees and to join the more prestigious scientific societies. (In 1906, Hertha Ayrton won an unrestricted prize offered by the Royal Society of London, which nevertheless refused to elect her to membership.)

In anthropology it was generally supposed (e.g., by Sir Henry Maine) that the family was always patriarchal until the 1860s when J. J. Bachofen in Germany, John McLennan in Scotland, and L. H. Morgan in the United States postulated a matriarchal origin. Karl Marx and Friedrich Engels adopted Morgan's theory (as presented in his *Ancient Society,* 1877), and Marxists promoted Morgan's theory as their own long after everyone else had abandoned it. (One wonders what Morgan, a middle-class Republican, would have thought of this kind of immortality!)

Many of the same issues were debated in the United States, with perhaps a greater emphasis on the views of female sexuality as expressed by physicians (at least recent historical studies seem to have stressed this aspect). The writings of Antoinette Blackwell provide an interesting source for the views of an articulate woman on the implications of Darwinism for sex differences; see *The Sexes Through Nature* (1875). Her sister-in-law Elizabeth Blackwell opened up medicine as a profession for women; see her *Essays in Medical Sociology* (1902, reprinted 1975). Other important authors were Lester Frank Ward and Charlotte Perkins Gilman.

BIBLIOGRAPHY

Nancy Leys Stepan (1986), "Race and Gender: The Role of Analogy in Science," *Isis* 77:261–77.

Mary Ann Warren (1980), *The Nature of Woman: An Encyclopedia and Guide to the Literature* (Inverness, Calif.: Edgepress).

S. Barbara Kanner (1972), "The Women of England in a Century of Social Change, 1815–1914: A Select Bibliography," in *Suffer and Be Still: Women in the Victorian Age,* ed. Martha Vicinus (Bloomington: Indiana Univ. Press), 173–206.

Topic *i:* Pre-Darwin

Lynda Birke (1982), "Cleaving the Mind: Speculations on Conceptual Dichotomies," in *Against Biological Determinism,* ed. Steven Rose (New York: Allison and Busby), 60–78, is on gender-associated dualities in Western thought.

Edward Shorter (1982), *A History of Women's Bodies* (New York: Basic Books).

Carol P. MacCormack and Marilyn Strathern, eds. (1980), *Nature,*

Culture, and Gender (New York: Cambridge Univ. Press), includes articles by L. J. Jordanova, "Natural Facts: A Historical Perspective on Science and Sexuality" (eighteenth- and early nineteenth-century concepts of women in French and British biomedical sciences) and M. Bloch and J. H. Bloch, "Women and the Dialectics of Nature in Eighteenth-Century French Thought."

Lorenne M. G. Clark and Lynda Lange, eds. (1979), *The Sexism of Social and Political Theory: Women and Reproduction from Plato to Nietzsche* (Toronto: Univ. of Toronto Press), includes articles by Lynda Lange on Plato and Rousseau, by Lorenne Clark on John Locke, by Steven Macleod and Louise Marcil-Lacoste on David Hume, by Patricia Mills on G. W. F. Hegel, by Mary O'Brien on Marx, and by Christine Allen on Friedrich Nietzsche.

Carol C. Gould and M. W. Wartofsky, eds. (1976), *Women and Philosophy* (New York: Putnam), includes Anne Dickason, "Anatomy and Destiny: The Role of Biology in Plato's Views of Women," 45–53; Caroline Whitbeck, "Theories of Sex Difference," 54–80; Carolyn Korsmeyer, "Reason and Morals in the Early Feminist Movement: Mary Wollstonecraft," 97–111.

Topic *ii:* Evolution and Sex Differences

Ruth Bleier (1984), *Science and Gender: A Critique of Biology and Its Theories on Women* (Elmsford, N.Y.: Pergamon Press).

Carl Jay Bajema, ed. (1984), *Evolution by Sexual Selection Theory: Prior to 1900* (New York: Van Nostrand Reinhold).

Evelleen Richards (1983), "Darwin and the Descent of Woman," in *The Wider Domain of Evolutionary Thought,* ed. D. Oldroyd and I. Langham (Boston: Reidel), 57–111.

Sandra Harding and Merrill B. Hintikka, eds. (1983), *Discovering Reality* (Hingham, Mass.: Reidel), contains articles by Ruth Hubbard, "Have Only Men Evolved?" and by Michael Gross and Mary Beth Averill, "Evolution and Patriarchal Myths of Scarcity and Competition."

Stephanie A. Shields (1982), "The Variability Hypothesis: The History of a Biological Model of Sex Differences in Intelligence," *Signs* 7:769–97.

M. J. Kottler (1980), "Darwin, Wallace, and the Origin of Sexual Dimorphism," *Proceedings of the American Philosophical Society* 124:203–26.

Phyllis Grosskurth (1980), *Havelock Ellis: A Biography* (New York: Knopf).

Vincent Brome (1979), *Havelock Ellis: Philosopher of Sex. A Biography* (London: Routledge and Kegan Paul).

Ruth Hubbard et al., eds. (1979), *Women Look at Biology Looking at Women* (Boston:Hall).

Susan S. Mosedale (1978), "Science Corrupted: Victorian Biologists Consider 'The Woman Question,'" *Journal of the History of Biology* 11:1–55.

Flavia Alaya (1977), "Victorian Science and the 'Genius' of Woman," *Journal of the History of Ideas* 38:261–80.

Paul Robinson (1976), *Modernization of Sex* (New York: Harper and Row), Chap. 1 on Ellis.

Elisabeth Fee (1976), "Science and the Woman Problem: Historical Perspectives," in *Sex Differences: Social and Biological Perspectives*, ed. Michael S. Teitelbaum (New York: Anchor Press), 175–223.

Topic *iii:* Anthropology

Lewis H. Morgan (1877, reprinted 1964 with Introduction to reprint ed. by L. A. White), *Ancient Society* (Cambridge: Harvard Univ. Press).

Bernice A. Carroll, ed. (1976), *Liberating Women's History* (Urbana: Univ. of Illinois Press), includes Sarah B. Pomeroy, "A Classical Scholar's Perspective on Matriarchy," and Ann J. Lane, "Woman in Society: A Critique of Frederick Engels," on Engels's use of L. H. Morgan's anthropological theories.

Topic *iv:* Women's Education and Role in Science

Margaret Rossiter (1982), *Women Scientists in America* (Baltimore: Johns Hopkins Univ. Press).

Joan N. Burstyn (1981), *Victorian Education and the Ideal of Womanhood* (Totowa, N.J.: Barnes and Noble).

Deborah Warner (1978), "Science Education for Women in Ante-Bellum America," *Isis* 69:58–67.

Carol Dyhouse (1976), "Social Darwinistic Ideas and the Development of Women's Education in England," *History of Education* 5:41–58.

Audrey B. Davis (1974), *Bibliography on Women: With Special Emphasis on Their Roles in Science and Society* (New York: Science History).

Topic *v:* American Views of Sex Differences and Sexuality

Louise Michele Newman, ed. (1984), *Men's Ideas/Women's Realities: Popular Science, 1870–1915* (Elmsford, N.Y.: Pergamon Press).

Ann Palmeri (1983), "Charlotte Perkins Gilman: Forerunner of a Feminist Social Science," in *Discovering Reality*, ed. S. Harding and M. B. Hintikka (Boston: Reidel), 97–119.

Rosalind Rosenberg (1982), *Beyond Separate Spheres: Intellectual Roots of Modern Feminism* (New Haven, Conn.: Yale Univ. Press), Chap. 1.

John Farley (1982), *Gametes and Spores: Ideas about Sexual Reproduction, 1750–1914* (Baltimore: Johns Hopkins Univ. Press).

Carl N. Degler (1980), *At Odds: Women and the Family in America from the Revolution to the Present* (New York: Oxford Univ. Press), Chaps. 11, 12.

Stephen Nissenbaum (1980), *Sex, Diet, and Debility in Jacksonian America: Sylvester Graham and Health Reform* (Westport, Conn.: Greenwood Press).

John D. Kasarda (1979), "How Female Education Reduces Fertility: Models and Needed Research," *Mid-American Review of Sociology* 4:1–22.

Sally Gregory Kohlstedt (1978), "In from the Periphery: American Women in Science," *Signs* 4:81–96.

Lois N. Magner (1978), "Women and the Scientific Idiom: Textual Episodes from Wollstonecraft, Fuller, Gilman, and Firestone," *Signs* 4:61–80.

Marie Tedesco (1978), "Science and Feminism: Conceptions of Female Intelligence and Their Effect on American Feminism, 1859–1920" (Ph.D. diss., Georgia State University), Chaps. 1–8.

G. J. Barker-Benfield (1976), *The Horrors of the Half-Known Life: Male Attitudes toward Women and Sexuality in 19th Century America* (New York: Harper and Row).

4.2 MARGARET MEAD AND THE ANTHROPOLOGY OF WOMEN

i. Mead's early research on adolescent girls and sex roles in primitive societies
ii. Other twentieth-century anthropologists on sex roles
iii. Women in anthropology
iv. Mead's later views and the nature/nurture controversy; evaluations of her work

READINGS

De Waal Malefijt, *Images of Man,* 293–314 (on psychological anthropology with very little on Mead). Abram Kardiner and Edward

Preble, *They Studied Man* (1961), 204–14 (on Ruth Benedict). Hays, *From Ape to Angel,* 340–54 (on Mead), 358–61 (on Benedict), 415–18 (WWII).

My focus on Margaret Mead to the neglect of other twentieth-century anthropologists reflects the fact that anthropology enters my course primarily because of its relation to other themes: race and sex differences, Darwinism, the nature/nurture controversy, and the role of women in science. Mead's early books popularized the view that female behavior and personality are not biologically determined but are conditioned by society. Her writings thus supported not only the "nurture" side but also provided scientific support for feminism, even though Mead herself seemed to retreat from the egalitarian position in her later writings.

Derek Freeman's recent attack, *Margaret Mead and Samoa,* produced a media blitz that demonstrated once again both the immense influence of Mead's ideas and the vitality of the nature/nurture debate. According to Freeman, Mead's reports on the easygoing character and sexual freedom of Samoan society were not objective observations but the product of her Boasian bias toward cultural determinism (cf., **3.4-***iii.*). Her defenders claim that Freeman's own bias in the opposite direction led him to overlook changes in Samoa between the time of her visits and his (e.g., interview with Leonora Foerstel and others in the Baltimore *Sun,* 15 Feb. 1983, C1–4).

SYNOPSIS

Margaret Mead, one of the first students of Franz Boas (**3.4**) at Columbia in the 1920s, wrote *Coming of Age in Samoa* (1928) as a popular account of her first field project, a study of the behavior and development of adolescent girls in a Polynesian community. It was an immediate best-seller and is still in print, though the sexy cover that graced the paperback edition some years ago has now been replaced by a more conservative illustration. The basic message of the book was that adolescence (cf., Hall's introduction of this concept, **6.1**) is not necessarily a time of neurotic conflict and confusion; it is only so in a society that prudishly represses the natural sexual behavior of young people before marriage. The Samoan girls, given the freedom to meet a new lover under the palm trees every night, grew up into emotionally stable adults without going through the various traumas that afflict young women in American society. Perhaps, as Freeman suggests, the girls were just telling the foreign lady what she wanted to hear, or it was all a big joke. Nevertheless, Mead's book reinforced the movement by Freud's followers to convince Americans that sexual repression is bad for your health (**5.6-*i***).

In 1931 Mead went to New Guinea to study the "conditioning of the social personalities of the two sexes." She claimed to find three tribes whose sex roles, taken in relation to the familiar stereotypes of Western society, exhibited a neat symmetry:

The Arapesh—both women and men display "feminine" personalities—cooperative, unaggressive, responsive to the needs and demands of others, maternal;

The Mundugumor—both men and women are "masculine"—ruthless and aggressive;

The Tchambuli—the women are dominant and managing ("masculine"), the men less responsible, emotionally dependent, etc. ("feminine").

In her report on these tribes, *Sex and Temperament in Three Primitive Societies* (1935), Mead suggested that there are no inherent biological reasons for personality differences between men and women; what we in Western society think of as masculine or feminine personality characteristics are entirely a product of social conditioning.

On the basis of studies of several primitive societies, Mead published in 1949 a major synthesis, *Male and Female,* in which she attempted to draw conclusions about American society. Contrary to the impression created by *Coming of Age,* promiscuity is not the norm in primitive society; and contrary to the impression created by *Sex and Temperament,* Mead now claimed that there *are* certain common (presumably biologically determined) features of the male and female roles in all societies. Inherent differences between the sexes should be recognized, but in a properly run society they should balance out so that neither sex feels inferior or wants to belong to the other. Although there is a waste of human potential when one sex is excluded from certain occupations or roles, it may not be advisable to change the situation if that means discouraging or hampering the effectiveness of the other sex. For example, Mead argued that the entrance of women into education has been a net loss because while women's qualities were needed in primary grades, men teachers have been driven out of higher grades also, so that boys have suffered by being taught only by women. Success in certain fields probably requires special abilities that are found more in one sex than the other: in the physical sciences, mathematics, and instrumental music, men "by virtue of their sex, as well as by virtue of their qualities as specially gifted human beings, [may] always have that razor-edge of extra gift that makes all the difference . . . while women may easily follow where men lead, men will always make the new discoveries." Conversely, women may have "a special superiority in those human sciences which involve that type of understanding which until it is analyzed is called intuition."

It is instructive at this point to read Betty Friedan's *The Feminine Mystique* (1963), where the critique of Mead's 1949 position is part of the inspiration for the Women's Liberation movement of the late 1960s and early 1970s.

In connection with Mead's last mentioned point, it is noteworthy that she and other women anthropologists have become well known for their contributions even though they are still in the minority. Some of the reasons for this may be found in a book edited by Peggy Golde, *Women in the Field* (1970), in which Mead and twelve other women anthropologists recall their own research experiences and point out their advantages (and disadvantages) as women in observing primitive cultures.

As a "role model" for women in science, Mead was perhaps less respectable than Marie Curie, but her work did suggest that in studying a primitive society a male observer can get only half the story. If anthropology is to be concerned with family structure and the roles of both men and women, then women must be included as observers and theorists in this science.

BIBLIOGRAPHY

Topic *i:* Early Research

Margaret Mead (1977), *Letters from the Field 1925–1972* (New York: Harper and Row); (1972), *Blackberry Winter: My Earlier Years* (New York: Morrow).

Marvin Harris (1968), *The Rise of Anthropological Theory* (New York: Crowell), Chap. 15, "Culture and Personality" (on Benedict and Mead).

Topic *ii:* Sex Roles

Peggy Reeves Sanday (1980), "Margaret Mead's View of Sex Roles in Her Own and Other Societies," *American Anthropologist* 82:340–48.

Sharon W. Tiffany (1980), "Anthropology and the Study of Women," *American Anthropologist* 82:374–80, is a review of recent books.

Rayna R. Reiter, ed. (1975), *Toward an Anthropology of Women*

(New York: Monthly Review Press), contains feminist critiques of male bias; attempts to interpret the roles of women in cultures.

Topic *iii:* Women in Anthropology

Shirley Ardener (1985), "The Social Anthropology of Women and Feminist Anthropology," *Anthropology Today* 1, no. 5:24–26.

Judith Modell (1983), *Ruth Benedict: Patterns of a Life* (Philadelphia: Univ. of Pennsylvania Press), includes her relations with Mead.

Rosalind Rosenberg (1982), *Beyond Separate Spheres* (New Haven, Conn.: Yale Univ. Press), Chap. 6 on E. C. Parsons.

Joan Mark (1980), *Four Anthropologists: An American Science in Its Early Years* (New York: Science History), has a chapter on Alice Cunningham Fletcher.

Peggy Golde, ed. (1970), *Women in the Field: Anthropological Experiences* (Chicago: Aldine).

Topic *iv:* Later Views of and about Mead

Derek Freeman (1987), letter to editor, *Scientific American* 265, no. 2 & 6; (1984), "The Truth Will Out," *Science 84* 5, no. 3:18; (1983), *Margaret Mead and Samoa: The Making and Unmaking of an Anthropological Myth* (Cambridge: Harvard Univ. Press). See the review by R. I. Levy (1983), *Science* 220:829–32; comments by Ward Goodenough (1983), *Science* 220: 906, 908; statement by the prime minister of Western Samoa that both Mead and Freeman are wrong about the sexual practices of Samoans, reported in *New York Times* 24 May 1983, C2; Ivan Brady, ed. (1983), "Speaking in the Name of the Real: Freeman and Mead on Samoa," *American Anthropologist* 85: 908–47.

Carleton Mabee (1986), "Margaret Mead's Approach to Controversial Public Issues: Racial Boycotts in the AAAS," *Historian* 48:191–208.

Mary Catherine Bateson (1985), *With a Daughter's Eye: A Memoir of Margaret Mead* (New York: Washington Square).

Jane Howard (1984), *Margaret Mead: A Life* (New York: Simon and Schuster).

Boyce Rensberger (1983), "The Nature-Nurture Debate I: Margaret Mead," *Science 83* 4, no. 3:28–37. See also the report by Eliot Marshall (1983), *Science* 219:1042–45.

Robert Cassidy (1983), *Margaret Mead: A Voice for the Century* (New York: Universe Books).

Rosalind Rosenberg (1982), *Beyond Separate Spheres* (New Haven, Conn.: Yale Univ. Press), Chap. 8.

"In Memoriam: Margaret Mead" (1980), *American Anthropologist* 82:261–373, is a collection of articles.

Edward Rice (1979), *Margaret Mead: A Portrait* (New York: Harper and Row).

4.3 MENDEL

i. Mendel's life and work; why it was ignored
ii. Other theories of inheritance in the late nineteenth century (Galton, etc.)
iii. Biological research on cellular reproduction (August Weismann, Oskar Hertwig, Hermann Fol)
iv. Revival of Mendel's theory in 1900

READINGS

A. E. E. McKenzie, *Major Achievements* (1973), 254–64, 499–501 (sources). John Marks, *Science* (1983), 276–79. Stephen F. Mason, *History of the Sciences* (1962), 531–32. Stephen Toulmin and June Goodfield, *Architecture of Matter* (1982), 352–67. W. C. Dampier, *History of Science* (1965), 321–29. Lois Magner, *History of Life Sciences* (1979), 405–32. Peter Bowler, *Evolution* (1984), 256–65. F. H. Portugal and J. S. Cohen, *A Century of DNA* (1977), 90–120. B. Norton, "Evolution after Darwin," Unit 13 in the Open University course *Science and Belief, from Darwin to Einstein* (A381), 33–57, and extracts from primary sources in the Noel G. Coley–Vance M. D. Hall anthology, *Darwin to Einstein* (1980). H. J. Fyrth and M. Goldsmith, *Science History and Technology,* Book 2, Part II, 33–41.

At this point, if one wants to go more deeply into modern biology, Garland Allen's *Life Science in the Twentieth Century* (1975, 1978) should be considered as a textbook. Some of my colleagues dislike Allen's attempt to categorize biological theories in terms of "materialism," etc., but he does cover much of the content of this chapter in a readable way. I would recommend 1–52 for general background on theories of heredity and the mechanistic conception of life as formulated at the turn of the twentieth century.

It is generally stated in historical accounts of Darwin's work that some of the difficulties encountered by his theory (**2.4**-*ii*) could have been overcome if only Mendel's 1865 paper had been known at the time. Thus historians have tried to explain why and to what extent Mendel's ideas were ignored by nineteenth-century biologists. Since Karl von Naegeli and a few others did in fact know about Mendel's theory, and copies of his paper were available in major scientific libraries, it is reasonable to suppose that prevailing biological thought was somehow hostile to this approach.

Another problem for historians is that Mendel's quantitative results are apparently "too good to be true." In 1936 British statistician R. A. Fisher concluded that these results were so close to the theoretical 3:1 frequency that they were unlikely to have been obtained without some theoretical bias. Comparison might be made with other cases in the history of science—John Dalton (**7.1**) and Cyril Burt (**6.8**) are well-known examples—where the reported "experimental" results that supposedly confirm a theory were apparently concocted *after* the theory was developed or were cleaned up in order to conform to theoretical expectations.

Galton's "biometric" school seemed to be more in accord with Darwinism and was regarded as an alternative to Mendel's theory in the early twentieth century.

Topic *iii* may be considered a sequel to **2.1**-*vi* and provides the technical background for the study of chromosomes in **4.4**.

BIBLIOGRAPHY

Robert C. Olby (1984), "Historiographical Problems in the History of Genetics," *Rivista di Storia della Scienza* 1:25–38.

John Farley (1982), *Gametes and Spores: Ideas about Sexual Reproduction, 1750–1914* (Baltimore: Johns Hopkins Univ. Press).

Wilma George (1982), "The Mendel Enigma, the Farmer's Son: The Key to Mendel's Motivation," *Archives Internationales d'Histoire des Sciences* 32:177–83 (pub. 1983), treats agriculture as a unifying theme in Mendel's work; his interest in meteorology.

Ernst Mayr (1982), *Growth of Biological Thought* (Cambridge: Harvard Univ. Press), Chaps. 12, 16–17.

Topic *i:* Life and Work

Gregor Mendel (1965), *Experiments in Plant Hybridisation,* Mendel's Original Paper in English Translation with Commentary and Assessment by Ronald A. Fisher, with a reprint of W. Bateson's Biographical Notice of Mendel (Edinburgh: Oliver and Boyd), includes Fisher's statistical critique of the credibility of Mendel's results.

R. C. Olby (1985), *Origins of Mendelism* (Chicago: Univ. of Chicago Press), is a 2d ed. of an authoritative work first published in 1966, about half of the book deals with work before Mendel; (1979), "Mendel No Mendelian?" *History of Science* 17:53–72.

Michael H. MacRoberts (1985), "Was Mendel's Paper on *Pisum* Neglected or Unknown?" *Annals of Science* 42:339–45.

Margaret Campbell (1985), "Pairing in Mendel's Theory," *British Journal for the History of Science* 18:337–40; (1982), "Mendel's Theory: Its Context and Plausibility," *Centaurus* 26:38–69; (1976), "Explanations of Mendel's Results," *Centaurus* 20:159–74.

Iris Sandler and Laurence Sandler (1985), "A Conceptual Ambiguity that Contributed to the Neglect of Mendel's Paper," *History and Philosophy of the Life Sciences* 7:3–70, suggest that in the late nineteenth century "there was no conceptual distinction between the transmission of an hereditary trait from parent to offspring (genetics) and the subsequent development of that trait in the offspring (embryology)." Since Mendel's theory treated only the first, it was not seen as a new theory of heredity.

W. W. Piegorsch (1983), "Has J. G. Mendel Been 'Too Accurate' in His Experiments? The χ^2 Test and Its Significance to the Evaluation of Genetic Segregation," *Historia Mathematica* 10:99–100, is on the analysis of Franz Weiling that challenges R. A. Fisher's conclusion.

J.-L. Serre (1981), "Mendel's Rejection of the Concept of Blending Inheritance," *Fundamenta Scientiae* 2:55–66.

C. Leon Harris, ed. (1981), *Evolution: Genesis and Revelations. With Readings from Empedocles to Wilson* (Albany: State Univ. of New York Press), Chap. 8.

Augustine Brannigan (1979), "The Reification of Mendel," *Social Studies of Science* 9:423–54, argues that Mendel was neither ignored in the 1860s nor simply rediscovered in 1900; he was concerned with different problems than his 1900 "rediscoverers."

Alexander Weinstein (1977), "How Unknown Was Mendel's Paper?" *Journal of the History of Biology* 10:341–64.

Topic *ii:* Other Theories of Inheritance

James W. Tankard, Jr. (1984), *The Statistical Pioneers* (Cambridge, Mass.: Schenken), includes a chapter on Galton.

Carl Jay Bajema, ed. (1983), *Natural Selection Theory: From the Speculations of the Greeks to the Quantitative Measurements of the Biometricians* (Stroudsburg, Pa.: Hutchinson Ross), 298–365.

Donald A. McKenzie (1981), *Statistics in Britain, 1865–1930* (New York: Columbia Univ. Press), Chaps. 3–6 on Galton, Pearson, biometrics vs. Mendelism.

Nils Roll-Hansen (1983), "The Death of Spontaneous Generation and the Birth of the Gene: Two Case Studies of Relativism," *Social Studies of Science* 13:481–519; response by Barry Barnes (1985), "A Case of Amnesia," *Social Studies of Science* 15:175–76, comment by H. M. Collins, *Social Studies of Science* 15:176–77 and reply by Roll-Hansen, *Social Studies of Science* 15:178–80. See also Nils Roll-Hansen (1980), "The Controversy between Biometricians and Mendelians: A Test Case for the Sociology of Scientific Knowledge," *Social Science Information* 19:501–17; response by Barry Barnes, "On the Causal Explanation of Scientific Judgment," *Social Science Information* 19:685–95.

Gloria Robinson (1979), *A Prelude to Genetics: Theories of a Material Substance of Heredity: Darwin to Weismann* (Lawrence, Kans.: Coronado Press), has chapters on Darwin, Galton, E. Haeckel, G. Jaeger, W. K. Brooks, K. von Naegeli, Weismann, Hugo De Vries; highly recommended by Jane Oppenheimer (*American Scientist,* Jan.-Feb. 1981).

Ruth Schwartz Cowan (1977), "Nature and Nurture: The Interplay of Biology and Politics in the Work of Francis Galton," *Studies in History of Biology* 1:133–208.

Topic *iii:* Cellular Reproduction

Ernst Mayr (1985), "Weismann and Evolution," *Journal of the History of Biology* 18:295–329.

Frederick B. Churchill (1979), "Sex and the Single Organism: Biological Theories of Sexuality in Mid-Nineteenth Century," *Studies in History of Biology* 3:129–77.

Topic *iv:* 1900 Revival

Onno G. Meijer (1985), "Hugo de Vries No Mendelian?" *Annals of Science* 42:189–232.

Lindley Darden (1985), "Hugo de Vries's Lecture Plates and the Discovery of Segregation," *Annals of Science* 42:233–42; (1980), "Theory Construction in Genetics," in *Scientific Discovery: Case Studies,* ed. Thomas Nickles (Boston: Reidel), 151–70, which is a philosophical analysis of these historical cases; (1977), "William Bateson and the Promise of Mendelism," *Journal of the History of Biology* 10:87–106; (1976), "Reasoning in Scientific Change: Charles Darwin, Hugo de Vries, and the Discovery of Segregation," *Studies in History and Philosophy of Science* 7:127–69.

Augustine Brannigan, Richard A. Wanner, and James K. White (1981), "The Phenomenon of Multiple Discoveries and the Re-publication of Mendel's Work in 1900," *Philosophy of the Social Sciences* 11:263–76.

Margaret Campbell (1980), "Did de Vries Discover the Law of Segregation Independently?" *Annals of Science* 37:639–55.

Malcolm Jay Kottler (1979), "Hugo de Vries and the Rediscovery of Mendel's Laws," *Annals of Science* 36:517–38.

Alice Baxter and John Farley (1970), "Mendel and Meiosis," *Journal of the History of Biology* 12:137–73, states "it was only *after* the rediscovery of Mendel's laws that cytologists came to share a set of common assumptions which led them to agree on what they saw under the microscope and eventually to accept the link between chromosomes and Mendelian factors."

4.4 CHROMOSOME BASIS OF SEX

i. Pre-1900 theories of sex determination
ii. Research of Nettie Stevens and E. B. Wilson
iii. Sex determination in humans; acceptance of the chromosome theory
iv. Chromosome anomalies (XYY, etc.) and their influence on behavior

READINGS

Allen, *Life Science in the Twentieth Century* (1978), 53–58. Ashley Montagu, "Chromosomes and Crime," *Psychology Today* 2, no. 5 (1968): 43–49, is a good explanation of the theory, with diagrams, but some of the conclusions are challenged by the articles cited below under topic *iv*.

SYNOPSIS

Prior to 1900, biologists thought that the sex of an organism was determined during the development of the embryo by environmental factors. Microscopic research on cells and the process of reproduction led to the discovery of chromosomes and conjectures about their possible role in sex determination. Finally, in 1905 the American biologist Nettie M. Stevens proposed that the presence or absence of the "Y" chromosome was correlated with production of a male organism, and her work was supported by experiments of E. B. Wilson. It had already been suggested (e.g., by Mendel in 1870) that the approximately 1:1 sex ratio could be explained by the Mendelian mechanism of dominance (Y is dominant, X is recessive).

The establishment of the XY chromosome theory of sex determination, though resisted at first by T. H. Morgan, soon became a cornerstone of twentieth-century genetics and played a major role in the achievements of Morgan's group (**4.5**). I give more attention to it than is usually done, because (1) sex differences is a major theme of my course; (2) it is much easier to understand than other parts of genetics; (3) it involves a major discovery by a woman scientist who is rarely given much credit and seems to be generally unknown even to modern biologists.

A possible link between presence of an extra Y chromosome in males and tendency toward criminal behavior was first reported by Patricia Jacobs in 1965. It was rumored (incorrectly) that Richard Speck, who murdered eight nurses in Chicago in 1966, had the XYY pattern. In 1968 Stanley Walzer and Park Gerald started to examine chromosomes of babies born at the Boston Hospital for Women. This project aroused considerable publicity and discussion of the ethical issues involved (what does one do when one discovers that a baby has XYY chromosomes?); see for example the report by Barbara J. Culliton in *Science* 186 (1974): 715–17. While this particular controversy seems to have died down, the problem of genetic screening remains and is likely to pop up again.

A more orthodox treatment of the subject would incorporate this topic into the next one (**4.5**); I would urge that if this is done, one resist the usual tendency to give all the credit to Morgan and forget about Stevens and Wilson.

BIBLIOGRAPHY

Jane Maienschein (1984), "What Determines Sex? A Study of Converging Approaches, 1880–1916," *Isis* 75:457–80.

John Farley (1982), *Gametes and Spores* (Baltimore: Johns Hopkins Press).

Topic *i:* Pre-1900

T. H. Morgan (1903), "Recent Theories in Regard to the Determination of Sex," *Popular Science Monthly* 64:97–116, is an original source for Morgan's views before he accepted the chromosome theory, as well as a good survey of the situation in 1900; see also Churchill's article cited above in **4.3-*iii***.

Alice Baxter and John Farley (1979), "Mendel and Meiosis," *Journal of the History of Biology* 12:137–73.

Topic *ii:* Stevens and Wilson

Marilyn Bailey Ogilvie and Clifford J. Choquette (1981), "Nettie Maria Stevens (1861–1912): Her Life and Contributions to Cytogenetics," *Proceedings of the American Philosophical Society* 125:292–311.

Stephen G. Brush (1978), "Nettie M. Stevens and the Discovery of Sex Determination by Chromosomes," *Isis* 69:163–72.

Scott Gilbert (1978), "The Embryological Origins of the Gene Theory," *Journal of the History of Biology* 11:307–51, explains why E. B. Wilson did and T. H. Morgan did not quickly accept the Sutton-Boveri synthesis of Mendelism and cytology; role of experiments on sex determination.

Alice Levine Baxter (1976), "Edmund B. Wilson as a Preformationist: Some Reasons for His Acceptance of the Chromosome Theory," *Journal of the History of Biology* 9:29–57.

Topic *iii:* Sex Determination in Humans; Acceptance of Theory

A. G. Cock (1983), "William Bateson's Rejection and Eventual Acceptance of the Chromosome Theory," *Annals of Science* 40:19–59.

Malcolm J. Kottler (1974), "From 48 to 46: Cytological Technique, Preconception, and the Counting of Human Chromosomes," *Bulletin of the History of Medicine* 48:465–502.

Topic *iv:* Anomalies

Jeremy Green (1985), "Media Sensationalism and Science: The Case of the Criminal Chromosome," in *Expository Science,* ed. T. Shinn and R. Whitley (Boston: Reidel), 139–61, argues scientists themselves were partly responsible for the popularization of the "XYY syndrome."

Herman A. Witkin (1976), "Criminality in XYY and XXY Men," *Science* 193:547–55.

Geoffrey Beale (1972), "Social Effects of Research in Human Genetics," in *The Biological Revolution: Social Good or Social Evil,* ed. Watson Fuller (Garden City, N.Y.: Doubleday/Anchor), 101–10.

4.5 THE GENE FROM MORGAN TO WATSON AND CRICK

i. Thomas Hunt Morgan and his group
ii. Ideological issues: Loeb and Lysenko
iii. Nucleic acids and proteins as genetic agents; the work of Avery, MacLeod, and McCarty
iv. Influences from physics: Bohr, Delbrück, Schrödinger
v. Watson, Crick, and the double helix structure of DNA; the role of Rosalind Franklin

READINGS

Mason, *History of the Sciences,* 533–40. McKenzie, *Major Achievements,* 265–73, 502–5. Magner, *History of Life Sciences,* 437–68. Bowler, *Evolution,* 292–93. Toulmin and Goodfield, *Architecture,* 369–72. Marks, *Science,* 280–81. Allen, *Life Science,* 50–81, 133–34, 187–228.

Edward D. Garber, ed., *Genetic Perspectives in Biology and Medicine* (1985), the articles by Arnold Ravin, H. V. Wyatt, Scott F. Gilbert, and Bruce Wallace present an excellent account of the history of modern genetics and molecular biology. Franklin H. Portugal and Jack S. Cohen, *A Century of DNA* (1977), 120–58, 230–71.

Articles: A. W. Ravin, "Genetics in America: A Historical Overview," *Perspectives in Biology and Medicine* 21(1978):214–31. Garland Allen, "The Rise and Spread of the Classical School of Heredity, 1910–1930: Development and Influence of the Mendelian Chromosome Theory," in *The Sciences in the American Context,* ed. N. Reingold (1979), 209–28. J. T. Edsall, "Progress in Our Understanding of Biology," in *Progress and Its Discontents,* ed. G. A. Almond et al. (1982). Lewis Thomas, "Oswald Avery and the Cascade of Surprises," *Esquire* 100, no. 6 (1983):74–78. H. F. Judson, "Annals of Science: The Legend of Rosalind Franklin," *Science Digest* 94, no. 1(1986):56–59, 78–83.

Many instructors choose the Watson-Crick work on the structure of DNA as an extended case history because of the readable and provocative account provided by J. D. Watson in his short book *The Double Helix*. This is available in a "Norton Critical Edition" (1980), which includes commentary and related material. Watson's version of the treatment of Rosalind Franklin, whose important X-ray work was used without her knowledge by Watson and Crick, is sometimes cited as an example of sexism in science.

Students can read extracts from sources pertaining to *ii* in N. G. Coley and V. M. D. Hall, *Darwin to Einstein* (1980).

SYNOPSIS

Much of the recent historical writing on early twentieth-century genetics has concentrated on the dispute between the "biometric" school (Francis Galton, Karl Pearson, W. F. R. Weldon) and the "Mendelian" school (William Bateson and others). This dispute was largely limited to Britain and, while clearly important in the development of modern ideas about heredity, seems somewhat peripheral to the main line of research that had already shifted to the United States by 1905. For simplicity I therefore focus on the work of T. H. Morgan and his group at Columbia, which established the function of the gene as a component of the chromosome; I then follow the attempts to determine the nature of the gene up to the discovery of the double helix structure. Subsequent developments such as the "genetic code" are beyond the scope of this *Guide*.

Biologists were at first reluctant to accept Mendelism because it seemed to be a reversion to discredited preformation theories and emphasize "structure" over "function." The trend around 1900 (the "revolt against morphology" as Garland Allen calls it) was toward Wilhelm Roux's "developmental mechanics."

But Wilhelm Johannsen's distinction between genotype and phenotype (1911) helped separate Mendelism from preformationism: what was inherited was not the character itself but only the "potential" for developing the character, not always realized. In this respect it might have turned out to be like the wave function in quantum mechanics (**11.2**), but instead it was eventually shown to be based on a concrete physicochemical object, the gene.

Morgan became a Mendelian as a result of his experiments with the fruit fly, Drosophila melanogaster, starting in 1908. In 1910 he found a curious white-eyed male fly and bred it to a normal red-eyed female. All the offspring (first generation) showed the normal red-eyed condition but white eyes appeared again in the next generation. They occurred in the 1:3 ratio predicted by Mendel's theory—but only in males. If a white-eyed male was mated with females from the first generation one found that 50 percent of the males and 50 percent of the females had white eyes. Morgan could explain this result by Mendel's theory if he assumed that the Mendelian factor for eye color always segregated with the factor for sex (i.e., is on an X chromosome). This is now called "sex-linked inheritance."

Since Drosophila could produce a new generation every ten to fourteen days it was easy to do experiments with them. Morgan assembled a team of students at Columbia University who quickly made important advances. In 1911 A. H. Sturtevant, following a suggestion from Morgan, assumed that the Mendelian factors are each at a definite position along a line in a chromosome, and that their (relative) positions can be found experimentally from the frequencies of "crossover"—when a chromosome breaks and exchanges parts with another chromosome. The farther apart two factors are on the chromosome, the more frequently exchanges occur between them. Sturtevant, starting out as an undergraduate assistant in Morgan's laboratory, eventually established his own reputation through the construction of maps showing the locations of genes.

Herman Muller, another student of Morgan, specialized in the effects or radiation on gene mutation and eventually (1947) received the Nobel Prize for this work.

Calvin Bridges, the third of Morgan's young collaborators in the period 1910–1915, was highly skilled in cytological research—the study of chromosomes in cells.

One of the first problems faced by Morgan's group was the accumulation of evidence that inheritance is more complicated than Mendel's theory assumed. For example, there were many cases of "incomplete dominance," in which the offspring has a character that is a mixture of those of its parents (red flower + white flower produces pink flower). The character is not determined by a single factor that is either dominant or recessive.

A related phenomenon was "quantitative inheritance," first observed by H. Nilsson-Ehle in Sweden. Certain traits like color can have a continuous range of variation; moreover, by selective breeding, it was possible to convert for example a dark red strain of wheat to successively lighter shades and eventually to a colorless strain. (This phenomenon was eventually explained by postulating that the color is produced by the joint action of several factors or genes.) Similarly W. E. Castle at Harvard, in a series of selection experiments on "hooded rats" (having a pigmented area on the head and back), was able to produce a strain of completely white rats and a strain of completely pigmented rats. Castle interpreted this result to mean that selection could produce individuals having traits outside the limits of variation of the original stock because the genes for a trait like hoodedness were altered ("contaminated") by proximity to other genes.

Castle's concept of "genetic contamination" was a radical departure from the Mendelian principle that the genes retain their individual properties unchanged during the process of reproduction. Morgan's group, especially Sturtevant and Muller, proposed an alternative hypothesis to explain similar phenomena that they had observed with Drosophila: there is a major gene for the character, which can be acted on quantitatively by other "modifier" genes in producing the variable phenotype. (This was an extension of the multiple factor hypothesis mentioned above.) Thus they preserved the Mendelian variability of genes at the cost of proliferating the number of genes needed to explain the phenomenon. Only a *mutation*—caused by random fluctuations or outside influences such as radiation, but not by other genes—was allowed to change the nature of the gene. Castle abandoned his theory when later experiments refuted it and accepted the modifier-gene hypothesis.

In 1915 Morgan, Sturtevant, Miller, and Bridges published a book on *The Mechanism of Mendelian Heredity* in which they surveyed the principles of their theory and its applications to many plants and animals as well as Drosophila. The book played a major role in persuading other biologists to accept the theory and in calling attention to further research opportunities. It strongly advocated the view that genes and chromosomes are specific physical entities, not just abstract concepts in a formal theoretical system as some other scientists still believed. Moreover, Morgan and his colleagues stressed the distinction between genotype and phenotype—the former being the collection of genes passed on unchanged (except for mutations) from one generation to the next, the latter being the character of the organism developed as a result of the interaction between genotype and environment.

Just as physicists and chemists needed an atom that would preserve its identity through chemical and physical transformations and could, as a first approximation, be ascribed an existence and properties independent of other atoms, so biologists wanted a gene that was impervious to environmental and somatic influences and could be associated with a specific trait.

In a second approximation genes, like atoms, could be assumed to interact with each other in order to generate more complicated phenomena. One example was the "modifier gene" hypothesis mentioned above—genes that produced small adjustments in a character largely determined by a "primary gene." Another example was the "position effect" postulated by Sturtevant in 1925 to explain the unusual hereditary behavior of the bar-shaped eye found as a mutation in Drosophila: the phenotype seemed to depend on the location of the bar gene on the chromosome in relation to other genes. More generally, biologists began to theorize that many traits were determined not by single genes but by the interaction of two or more genes, in order to explain continuous variations in phenotypes in terms of discrete unchangeable but mobile genes.

A third approximation, somewhat analogous to the introduction of nuclear reactions in atomic physics, was the hypothesis that *parts* of genes could be interchanged or "transposed" to produce new genes; this hypothesis was developed in the 1940s by Barbara McClintock in connection with her experiments on maize, but other geneticists were not ready to accept it until the 1960s. (She received the Nobel Prize in 1983.)

Opposition to Morgan's gene-chromosome theory came from those who (like Richard Goldschmidt) accepted the importance of chromosomes in inheritance but rejected their subdivision into discrete genes associated with specific characters; and from others (such as William Bateson) who rejected the assumption that heredity depends on any material substance at all. Bateson was apparently influenced by physicists who preferred to explain matter in terms of motion (e.g., wave vibrations) rather than substance—a view that was in and out of favor in physics at various times between 1850 and 1930.

Mechanism and Reduction

Reducing biology to physics was often criticized as "mechanistic" at the same time that physics itself was abandoning mechanistic conceptions. But the alternatives to mechanism—speculations based on vitalism or various forms of philosophical idealism—seemed even more out of place in the pragmatic world of experimental biology. Most twentieth-century biologists avoided the mechanism/vitalism issue by banishing it to the realm of philosophy, just as most twentieth-century physicists avoided the philosophical debates about the interpretation of quantum mechanics (**11.2**). The issue of reductionism has arisen again in connection with sociobiology, a science that claims to explain animal and human behavior in terms of inherited biological factors.

One biologist who explicitly advocated a mechanistic approach was Jacques Loeb, who came from Germany to America in 1891. He worked briefly with Morgan at Bryn Mawr and later in New York City. Loeb argued biological problems could and should be reduced to physics and chemistry and biologists should adopt the methods of controlled experimentation and quantitative measurement that had proved so effective in the physical sciences.

Loeb's legacy was the department he founded at Rockefeller Institute (now Rockefeller University) that provided a home, in both the physical and intellectual senses, for scientists like Oswald Avery who made contributions to unravelling the physicochemical nature of the gene. The discovery of the double helix structure of DNA, which is the culmination of the Second Scientific Revolution as far as biology is concerned, seems to represent among other things the triumph of the mechanistic, reductionist approach. At the same time modern biochemical genetics is in some ways less reductionist than early twentieth-century concepts, since it invokes concepts like the "operon," a complex interacting system.

Lysenko

The case of T. D. Lysenko provides a good example of the interactions between science and politics, not only in the Soviet Union but in the West where some left-wing biologists have been reluctant to criticize a person generally regarded as a charlatan.

In 1929 Lysenko announced that he had solved the problem of the killing of winter wheat in the Ukraine, in the USSR, by soaking and chilling the grain, then sowing it in the spring. This technique was called "vernalization." Because of the severe scarcity of food that accompanied forced industrialization and attempts to collectivize the peasants, Lysenko's "discovery" attracted considerable attention among Communist

Party officials. It was soon extended to the vernalization of *spring* wheat (a process that seems semantically if not biologically redundant), with loud but generally unverified claims of increased yields.

Lysenko rose to power in the Soviet agricultural establishment by ignoring or ridiculing scientists who had so far failed to solve the critical problems of Russian agriculture—low yields, disease, and vulnerability to harsh winters. By 1935 he had gained the favor of Joseph Stalin who saw Lysenko's work as a vindication of his preference for "practice over theory."

Lysenko's doctrines were sharply criticized in 1936 by Russian geneticists and by the visiting American scientist, Herman Muller. Lysenko denied the existence of discrete particles (genes) governing heredity but claimed to be a Darwinist. At that time the official philosophy of Marxism was held to be consistent with Darwinism but not with Lamarckism. It was only just becoming clear to scientists that Darwinism is consistent with Mendelian genetics (see **4.6**), but the theoretical basis for their connection may have been too subtle for Lysenko to grasp even if he had wanted to. Instead he used his power to suppress modern genetics by pinning on it the label of bourgeois foreign science. "Morganism" became a pejorative term.

Lysenko's opponents accused him of supporting Lamarckism, a charge which he initially denied for political reasons. Indeed one historian of the Lysenko affair, David Joravsky, has argued that Lysenko's approach was so completely unscientific that it does not deserve such a respectable label as Lamarckism, which is at least a scientific concept though a discredited one. But in 1946, after the publication of some pro-Lamarckist remarks by Stalin, Lysenko admitted what was already generally known: that his doctrines assumed that heredity could be manipulated directly by changing the environment, without waiting for the slow process of natural selection to work.

Following Stalin's death in 1953, Soviet agricultural officials stopped ignoring the fact that Lysenko's practice of "vernalization" and other techniques designed to increase yields had completely failed to do so, while costing billions of rubles. Some of his claims to transform one species into another were proved to be self-deception or deliberative fraud. By the early 1960s Russian biologists were allowed to publicly revive genetics, and the major journals were purged of Lysenkoite editors. But Nikita Khrushchev, during his regime, continued to support Lysenko, so it was not until after Khrushchev's ouster in 1964 (partly because his Lysenkoist agricultural schemes had failed) that biological science could displace pseudoscience.

DNA

Before the Avery-MacLeod-McCarty experiment in 1943, which clearly pointed to the genetic role of deoxyribonucleic acid, research on DNA was part of the general development of organic chemistry but was not for the most part motivated by the desire to understand the nature of inheritance. Thus it may be purely coincidental that both the abstract gene theory of Mendel and the discovery of DNA by Friedrich Miescher occurred at about the same time. Miescher isolated a substance from cell nuclei, later called "nucleic acid," in 1869. Major advances in genetics and in knowledge about DNA occurred during the period 1910–1916, again with little apparent connection between them. Finally, the period 1950–1965 saw the flowering of molecular biology, based on the recognition that genes are composed of DNA, the Watson-Crick model for the structure and replicating mechanism of DNA, and the cracking of the "genetic code." The fact that important discoveries seem to have been concentrated in three spurts of less than two decades each, separated by longer periods of relative inactivity, is perhaps best explained by the hypothesis that research on genetics and DNA tends to be favored by an atomistic reductionist approach to biology. The same three periods were also times when atomism was especially popular in physics.

Early in the twentieth century chemical analysis of chromosomes showed that they had two major constituents; the nucleic acid isolated by Miescher in 1869, and protein. The structure of protein was unknown but thought to be extremely complex. Nucleic acid was thought to be a somewhat simpler substance built up from several smaller molecules: several bases, two sugars, and the phosphate (PO_4) group. The discovery of the base components was due primarily to the German chemist Albrecht in the 1890s; there are five of them that occur in large quantities, two of them called purines (adenine and guanine), the other three called pyrimidines (cytosine, thymine, and uracil).[1]

Phoebus Levene, a Russian-American chemist, showed in 1909 that

[1]The purine molecules are composed of a six-membered ring (consisting of four carbon atoms and two nitrogen atoms) joined along a carbon-carbon double bond to a five-membered ring (consisting of three carbons and two nitrogens). The six-membered ring has a NH_2 group attached to one of the carbons. The difference is that in guanine that ring also has a hydroxyl (OH) group attached to it, while in adenine there is a hydrogen atom at the corresponding position.

The pyrimidines have a single six-membered ring with four carbons and two nitrogens. One of the carbons has a hydroxyl group attached to it, and another has a hydrogen. They differ in whether the groups attached to the other two carbons are CH_3 and OH, H and OH, or H and NH_2.

a sugar-type molecule, ribose, is a component of nucleic acid (ribose had been synthesized in 1891 by Emil Fischer but not previously known to occur in nature).[2] Later he found that some nucleic acid molecules contain a closely-related molecule which differs from ribose by lacking one oxygen atom. Hence it was called deoxyribose, and that kind of nucleic acid was subsequently called DNA, while the kind that contains ribose was called RNA for ribonucleic acid. Thus the "D" in "DNA" means "lacking an oxygen."

Levene and others in the 1920s and 1930s advocated the "tetranucleotide hypothesis": each nucleic acid contains four bases (two purines and two pyrimidines) in equal proportions. For DNA these would be adenine, guanine, thymine, and cytosine; for RNA the same except with thymine replaced by uracil. While this hypothesis may have seemed the simplest plausible one on the basis of data available at the time, it created the misleading impression that nucleic acids have a relatively simple structure and thus cannot be the basis of heredity; that role was generally assigned to the protein component of the cell nucleus.

The proteins were found to consist of at least twenty different "amino acids," which have structures similar to glycine or phenylalanine.[3] A complex molecule is usually built up by combining two glycines through a carbon-nitrogen bond called a "peptide linkage." The resulting molecule, which can be indefinitely long, is called a "polypeptide."

In the 1930s it was suggested by Dorothy Wrinch and others that "the specificity of the genes resides in their [the protein's] amino acid sequences" (Olby 1975, 116). A large number of possible protein molecules having similar overall structure can be formed by attaching various amino acids such as phenylalanine in different combinations as side chains to polypeptide "backbone."

Evidence that nucleic acid rather than protein carried genetic information emerged as a result of observations by Frederick Griffith, a medical officer of the Ministry of Health in England, in 1928. Griffith

---

[2]The ribose molecule has a five-membered ring with four carbons and one oxygen; each carbon has other groups such as OH or CH_2OH joined to it, or a single hydrogen atom (H). In deoxyribose one of the OH groups attached to a carbon is replaced by H.

[3]Amino acids are formed from ammonia (NH_3) by replacing one of the H atoms by groups of other atoms. In glycine the other part is similar to acetic acid (vinegar); the molecule has the formula $(NH_2)(CH_2)(COOH)$. In phenylalanine a more complicated group containing a benzene ring (six carbons) and another CH_2 group has been attached to one of the carbons.

studied several different types of pneumococci in the sputum of pneumonia patients, and found that the incidence of one type increased markedly over a period of several years while that of another decreased. In order to understand this, he tried the following experiment: he inoculated mice with a nonvirulent "R" (for "rough") form of the pneumococcus, and a virulent "S" (for "smooth") form, the latter having first been killed by heat. He found that the R form was rapidly changed to the S form. Some component of the S bacteria, even though dead, was able to act on the R bacteria and convert them into the S form.

Griffith's experiment led Oswald Avery at the Rockefeller Institute to undertake a thorough investigation to determine which component was responsible for the transformation. The first evidence for the genetic role of DNA was presented in 1944 by Avery, Colin MacLeod, and Maclyn McCarty. They showed the substance that transforms bacteria in Griffith's experiment is DNA. Avery was well aware of the significance of his work and had written to his brother Roy a year earlier that the active substance "may be a gene." But the published paper was phrased more cautiously and attracted little attention at the time. Avery died in 1955 without gaining the Nobel Prize that he obviously deserved.

Why did biologists ignore the Avery-MacLeod-McCarty work? There have been some attempts to give a sociological answer to this question, for example by showing that Avery was not part of the "invisible college," the small network of specialists who communicated with each other on such problems. According to Horace Judson (1978, 112) this explanation is "false and blindingly trivial" because the key members of that invisible college, in particular Max Delbrück, did know about Avery's work. (In this case it was not really invisible since it acquired a recognized name: the "phage group," after "bacteriophage," a parasitic virus that eats bacteria. Delbrück and others considered phage especially suitable for experiments on heredity.) In fact Avery's brother Roy happened to be at Vanderbilt University, where Delbrück was working in 1943 and showed him the letter mentioned above. The problem was not a failure to communicate, but a failure to overcome the philosophical bias of the invisible college against the very idea that a gene could be identified with a particular chemical compound. As another member of the invisible college (Salvador Luria) explained it, "People like Delbrück and myself, not only were we not thinking biochemically but we were somehow—and probably unconsciously—reacting negatively to biochemistry. And biochemists. As such. As a result, I don't think we attached great importance to whether the gene was protein or nucleic acid . . . " (Judson 1978, 117–8).

Influence of Quantum Physics on Molecular Biology

In order to understand why Delbrück and his group were hostile to biochemistry, and also why another group of scientists pursued the biochemical approach with great success a few years later, we must look at the situation in theoretical physics during the 1920s and 1930s (**11.2**). For Delbrück was originally a quantum physicist who worked with Niels Bohr. In 1932 he heard Bohr's lecture on "Light and Life" in which the prospects of biology were expounded in an especially interesting way, from the viewpoint of Bohr's "complementarity principle." When applied to biology, Bohr argued, the complementarity principle meant that a complete physicochemical explanation was forever impossible. Bohr's philosophy of quantum mechanics thus translated into an antireductionist philosophy of biology: the thesis that biological phenomena can never be completely explained by the laws of physics and chemistry, but that new laws operate at a higher level of organization. Delbrück and other physicists were inspired by this philosophy to go into biology in the hope of finding these new laws of nature, since they believed that they had already found all the laws of physics.

Bohr's philosophy of quantum mechanics was opposed by a small group of physicists led by Albert Einstein and Erwin Schrödinger (**11.2**), and in one important case their more "realist" view had an important impact on biology. In 1944 Schrödinger published a short book, *What Is Life?* in which he argued that while biological phenomena may depend on laws different from those ordinarily used in physics and chemistry, those laws would not be qualitatively different from the ones already known. For example, he was impressed by the stability of inherited characteristics over many generations and speculated about how this stability could be explained in terms of a genetic substance. If the gene were a large complex molecule, it would be subject to thermal fluctuations associated with the continual impacts of smaller molecules. But whereas in classical physics those impacts would change the structure of the molecule, in quantum physics they would have no effect unless they had enough energy to kick the molecule up to the next quantum level. Quantum theory might thus account for the stability of the gene over many generations, combined with its vulnerability to occasional mutations caused by high-energy radiation.

What Is Life? seems to have had a strong influence on certain physicists, such as Francis Crick, persuading them that there were important and interesting questions to be solved in biology, and that the solutions could be found by physical and chemical methods. Although Schrödinger's specific suggestions about the structure and functioning of genes were rather naive from a biologist's viewpoint, they did emphasize the concept of a "genetic code"—Schrödinger argued that the gene must be some kind of "aperiodic crystal" that stored the genetic information in its molecular structure in such a way that it could be used to synthesize other molecules in a definite way.

Francis Crick was a 35-year-old graduate student at the Cavendish Laboratory in Cambridge when James D. Watson arrived as a 23-year-old postdoctoral fellow in 1951. Watson has told the story of the discovery of the structure of DNA in his book *The Double Helix* (1968). While numerous factual details of this account have been challenged by other participants and by historians of science who have examined the documents, it is the best source now available for conveying the flavor of scientific research as it is actually done.

The best evidence for the structure of DNA came from X-ray crystallography work being done in London by Maurice Wilkins and Rosalind Franklin. But Wilkins and Franklin did not approach the problem in the way that Watson and Crick considered most fruitful, i.e., by building molecular models from balls. The success of Watson and Crick's work depended on getting Wilkins and Franklin to collaborate with them, or on seeing their data, in particular Franklin's X-ray photographs that were better than anyone else could obtain at the time. Watson and Crick had reason to think that Linus Pauling, who had previously used similar methods to find the helical structure of the polypeptide chain in proteins, was also working on the DNA problem. They were able to get advance information about Pauling's progress through his son Peter who was studying in Cambridge. They also benefitted from quantum-mechanical calculations by John Griffith (nephew of the Frederick Griffith who had found the first evidence for the genetic role of DNA) showing that adenine attracts thymine and guanine attracts cytosine. This suggested that a DNA molecule could replicate itself in a complementary fashion by gathering together free bases in solution and assembling them along its length.

Rosalind Franklin's X-ray photographs provided evidence for a helical structure of DNA. Watson and Crick obtained these photographs without her knowledge although they stated in their published paper that their results depended on Franklin's data. Watson asserts in *The Double Helix* that Franklin rejected a helical structure and became annoyed whenever this interpretation of her photographs was suggested. Evidence from Franklin's notebooks and other sources suggests that she was favorably considering a helical form at this time although she thought it had not been proved. One gets the impression of a cautious empirical scientist (Franklin) being skeptical of naive rash theoreticians (Watson and Crick).

The "moment of discovery" came on 21 February 1953, when Watson noticed that "an adenine-thymine pair held together by two hydrogen bonds was identical in shape to a guanine-cytosine pair held together by at least two hydrogen bonds." This was the key to constructing the double-helix model, which was immediately accepted by almost all of the experts. Crick, Watson, and Wilkins shared the 1962 Nobel Prize for the discovery; Franklin was not eligible because she had died in 1958.

BIBLIOGRAPHY

Raphael Falk and Gerrit A. M. Van Balen (1986), "New Perspectives in the History of Genetics," *Studies in History and Philosophy of Science* 17:133–73.

Richard J. Blackwell, comp. (1983), *A Bibliography of the Philosophy of Science, 1945–1981* (Westport, Conn.: Greenwood Press), 514–16.

Ernst Mayr (1982), *The Growth of Biological Thought* (Cambridge: Harvard Univ. Press), Chaps. 17–19.

John A. Fuerst (1982), "The Role of Reductionism in the Development of Molecular Biology: Peripheral or Central?" *Social Studies of Science* 12:241–78.

Franklin H. Portugal and Jack S. Cohen (1977), *A Century of DNA: A History of the Discovery of the Structure and Function of the Genetic Substance* (Cambridge: MIT Press). Parts of this book can be read by students in a course devoted to the history of biology.

Topic *i:* Morgan and His Group

A. H. Sturtevant (1965), *A History of Genetics* (New York: Harper and Row), gives the flavor of life in the "fly room" as recalled by one of its inhabitants.

Jane Maienschein (1987), "Heredity/Development in the United States, circa 1900," *History and Philosophy of the Life Sciences* 9:79–93.

Garland Allen (1984), "Thomas Hunt Morgan: Materialism and Experimentalism in the Development of Modern Genetics," *Social Research* 51:709–38; (1980), "The Evolutionary Synthesis: Morgan and Natural Selection Revisited," in *The Evolutionary Synthesis,* ed. Ernst Mayr and William Provine (Cambridge: Harvard Univ. Press), 356–82; (1979), "Naturalists and Experimentalists: The Genotype and the Phenotype," *Studies in History of Biology* 3:179–209; (1978), *Thomas Hunt Morgan: The Man and His Science* (Princeton, N.J.: Princeton Univ. Press).

Jonathan Harwood (1984), "The Reception of Morgan's chromosome theory in Germany: Inter-War Debate over Cytoplasmic Inheritance" *Medizinhistorische Journal* 19:3–32.

Jan Sapp (1983), "The Struggle for Authority in the Field of Heredity, 1900–1932: New Perspectives on the Rise of Genetics," *Journal of the History of Biology* 16:311–42.

Isabel Morgan Mountain et al. (1983), "The Place of Thomas Hunt Morgan in American Biology," *American Zoologist* 23:825–76, includes papers by G. E. Allen, J. M. Oppenheimer, J. A. Moore, and K. Keenan.

Lindley Darden (1982), "Aspects of Theory Construction in Biology," in *Logic, Methodology and Philosophy of Science VI* (Proceedings of the Sixth International Congress of Logic, Methodology and Philosophy of Science, Hannover, 1979) (Amsterdam: North-Holland), 463–77; (1980), "Theory Construction in Genetics," in *Scientific Discovery: Case Studies,* ed. T. Nickels (Boston: Reidel), 151–70.

Elof Axel Carlson (1981), *Genes, Radiation, and Society: The Life and Work of H. J. Muller* (Ithaca: Cornell Univ. Press).

Nathan Reingold and Ida Reingold, eds. (1981), *Science in America* (Chicago: Univ. of Chicago Press). Chapters 6, 7, and 11 include correspondence relating to genetics.

D. J. Kevles (1980), "Genetics in the United States and Great Britain, 1890–1930: A Review with Speculations,"*Isis* 71:441–55, reprinted in *Biology, Medicine and Society 1840–1940,* ed. Charles Webster (New York: Cambridge Univ. Press, 1981).

Scott F. Gilbert (1978), "The Embryological Origins of the Gene Theory," *Journal of the History of Biology* 1:307–51.

Nils Roll-Hansen (1978), "Drosophila Genetics: A Reductionist Research Program," *Journal of the History of Biology* 11:159–210, focuses on Morgan, Muller, Wilson, Bateson, and Johannsen.

Topic *ii:* Loeb and Lysenko

Philip J. Pauly (1987), *Controlling Life: Jacques Loeb and the Engineering Ideal* (New York: Oxford Univ. Press).

Loren R. Graham (1987), *Science, Philosophy, and Human Behavior in the Soviet Union* (New York: Columbia Univ. Press), Chapter 4.

Douglas R. Weiner (1985), "The Roots of 'Michurinism': Transformist Biology and Acclimatization as Currents in the Russian Life Sciences," *Annals of Science* 42:243–60.

Nils Roll-Hansen (1985), "A New Perspective on Lysenko?" *Annals of Science* 42:261–78, points out that "Lysenko was not merely a pseudo-scientist forced upon the scientific community by political bosses. Some scientific recognition by his peers was a necessary condition for his rise in the 1930s."

Garland E. Allen (1983), "The Several Faces of Darwin: Materialism in Nineteenth and Twentieth Century Evolutionary Theory," in

Evolution from Molecules to Man, ed. D. S. Bendall (New York: Cambridge Univ. Press), 81–102; (1981), "Morphology and Twentieth-Century Biology: A Response," *Journal of the History of Biology* 14:159–76.

David L. Hull (1982), "Philosophy and Biology," in *Contemporary Philosophy: A New Survey,* vol. 2, *Philosophy of Science,* ed. G. Fløistad (Boston: Nijhoff), 281–316.

Johann Peter Regelmann (1980), *Die Geschichte des Lyssenkoismus* (Frankfurt am Main: Rita Ge. Fischer).

Donald Fleming (1973), "Loeb, Jacques," *Dictionary of Scientific Biography* 8:445–47.

David Joravsky (1970), *The Lysenko Affair* (Cambridge: Harvard Univ. Press).

Topic *iii:* Nucleic Acids and Proteins

Maclyn McCarty (1985), *The Transforming Principle: Discovering that Genes Are Made of DNA* (New York: Norton).

Rene J. Dubos (1976), *The Professor, the Institute, and DNA: Oswald T. Avery, His Life and Scientific Achievements* (New York: Rockefeller Univ. Press).

Topic *iv:* Influences from Physics

E. J. Yoxen (1979), "Where Does Schroedinger's 'What Is Life?' Belong in the History of Molecular Biology?" *History of Science* 17:17–52.

Max Delbrück (1970), "A Physicist's Renewed Look at Biology: Twenty Years Later" *Science* 168:1312–15.

Gunther S. Stent (1968), "That Was the Molecular Biology that Was," *Science* 160:390–95.

John Cairns et al., eds. (1966), *Phage and the Origins of Molecular Biology* (Cold Spring Harbor, N.Y.: Cold Spring Harbor Laboratory of Quantitative Biology), includes Delbrück's article "A Physicist Looks at Biology."

Topic *v:* Double Helix Structure

Edward Yoxen (1985), "Speaking Out about Competition: An Essay on 'The Double Helix' as Popularization," in *Expository Science,* ed. T. Shinn and R. Whitley (Boston: Reidel), 163–81, seems surprised by the favorable reception of Watson's book, and reluctantly accepts its "message" about the competitive nature of science.

Philip Kitcher (1982), "Genes," *British Journal for the Philosophy of Science* 33:337–59, is on the relation between molecular and classical genetics, in the context of a theory of conceptual change in science.

Gunther S. Stent (1980), "To the Stockholm Station—Makers of the Molecular-Biological Revolution," *Encounter* 54, no. 3:79–85, is a review of books by Watson, Olby, and Judson.

Horace F. Judson (1979), *The Eighth Day of Creation: The Makers of the Revolution in Biology* (New York: Simon and Schuster). See also J. T. Edsall (1980), "Essay Review: Horace Judson and the Molecular Biologists," *Journal of the History of Biology* 13:141–58; Judson (1980), "Reflections on the Historiography of Molecular Biology," *Minerva* 18:369–421.

Robert Olby (1975), *The Path to the Double Helix* (Seattle: Univ. of Washington Press), is a very detailed account, favoring Crick's rather than Watson's view of how it happened.

Ann Sayre (1975), *Rosalind Franklin and DNA* (New York: Norton).

4.6 THE "SYNTHETIC THEORY OF EVOLUTION"

i. Statistical theory and population genetics: Hardy-Weinberg law, work of S. S. Chetverikov, R. A. Fisher, J. B. S. Haldane, S. Wright

ii. Neo-Darwinism, field studies, species concept: T. Dobzhansky, J. Huxley, E. Mayr, G. C. Simpson

iii. Recent debates on the synthetic theory: "punctuated equilibrium" theory of S. J. Gould, N. Eldredge, S. Stanley

READINGS

Allen, *Life Science,* 126–45. McKenzie, *Major Achievements,* 273–75. Marks, *Science,* 282–83, 394–97. Bowler, *Evolution,* 289–99, 317–26.

Article: Fred Hapgood, "The Importance of Being Ernst," *Science 84* 5, no. 5(1984):40–46, on Ernst Mayr.

Ernst Mayr's "Prologue" to the conference proceedings edited by himself and W. B. Provine, *The Evolutionary Synthesis* (1980), gives the history from the viewpoint of a scientist who helped to make it. It may be hard reading for students without a background in biology but does provide a springboard for discussing the way scientists use history. See also Mayr's *Growth of Biological Thought* (1982), Chaps. 1, 12–13 (comments on the relation of the Gould-Eldredge theory to his own, 617).

SYNOPSIS

The establishment of the neo-Darwinian evolutionary synthesis in the 1930s and 1940s completed the Second Scientific Revolution in biology, during the period when the discovery and application of nuclear fission was performing a similar role in physics (**10.2**). But, while the events of August 1945 instantly made atomic science an object of worldwide respect and fear, the much less spectacular developments in evolutionary theory went largely unnoticed outside the biological community. Since, to a first approximation, the synthesis merely restored Darwin's original concept of natural selection acting on random variations as the principal mechanism of evolutionary change, it may seem surprising that this concept was *not* generally accepted by biologists as recently as 1930.

The "synthesis" was primarily a union of Darwinian evolution with Mendelian genetics, but when the latter was revived in 1900 biologists at first saw the two theories as incompatible. Before 1900 evolutionists were divided into two camps: those who believed in gradual change and those who who postulated discontinuous jumps, called "saltations" or "mutations." The split roughly corresponded to the dichotomy between naturalists (including "systematists") and experimentalists (including geneticists), which was developing in the biological community. A third, smaller group—the specialists in "biometrics"—might be considered as distinct from the other two. Mendelism was seen as a version of saltationism, especially since de Vries, a leading Mendelian, advanced an influential mutation theory based on his work with Oenothera (evening primrose) plants.

An important distinction between the two camps, according to Ernst Mayr, was that naturalists tended to think of species as "populations"—collections of unique individuals—whereas experimentalists were still "essentialists," who thought of a species as a "type" and ignored variations. A mutation of a single individual could have little effect on a population (cf., Jenkin's objection, **2.4**) but if it changed the type it would change every individual at the same time. Later the geneticists, following T. H. Morgan, used the term "mutation" to mean very small changes in individuals rather than transformations of an entire species to another one.

Before the 1930s many biologists accepted Neo-Lamarckism, or more generally what Mayr calls "soft inheritance"—changes in the hereditary material due to use or disuse, or inherent tendencies toward progressive change (orthogenesis), as well as inheritance of acquired characters. A major reason for the popularity of soft inheritance was the belief that natural selection alone was not effective enough to transform very small advantages in fitness into major evolutionary change. Darwin himself had stated (and tried to refute) this belief in his famous discussion of the evolution of the eye.

Mendelism itself was not clearly understood even by its advocates.

For example the concept of "dominance" was sometimes believed to imply that a rare gene could gradually take over a substantial part of a population if it were dominant, but would die out if it were recessive. This misconception was corrected in 1908 by W. Weinberg and G. H. Hardy, who pointed out the following consequences of Mendel's rules:

> If in any generation the numbers of pure dominants (AA), heterozygotes (Aa) and pure recessives (aa) are p:2q:r, with random mating the numbers in the next generation will be $(p + q)^2:2(p + q)(q + r):(q + r)^2$. These may be written as $p_1:2q_1:r_1$. It can be shown that (1) the ratios will be the same in the second generation as in the first (i.e., $p_1/q_1 = p/q$, etc.) if $q^2 = pr$; (2) the ratios will be the same in all future generations regardless of the initial values of p, q, and r since $q^2_1 = p_1r_1$.
>
> Thus if we start with a dominant gene that is very rare in the first generation, its relative frequency will approximately double in the second but will have no tendency to increase after that. If we start with a recessive gene that is very rare in the first generation, its relative frequency will drop rapidly in the second generation (e.g., from 1 in 10,000 to 1 in 10,000,000) but will have no tendency to decrease after that.

These results, known as the "Hardy-Weinberg law," are valid as statistical averages in large populations, on the assumption that there is no selection. From a mathematical viewpoint they are simple algebraic identities, hardly worthy of the name "law"; Mendel himself could have derived them if he had considered the question.

In the decade after 1905, Mendelians began to recognize that natural selection rather than discontinuous mutation was the major factor in evolutionary change. This change of opinion resulted from three advances in genetics. First, experiments by William Castle and others (e.g., on hooded rats) showed that selection could act on continuous variations to change a character to a new stable level beyond the original limits of variation of that character. Second, H. Nilsson-Ehle and E. East found that Mendelian theory could explain some continuous variations, and what appeared to be blending inheritance, by invoking multiple factors. Third, T. H. Morgan and his coworkers on Drosophila discovered that Mendelian characters might be identified with very small variations rather than large changes.

One outcome of this research, pointed out by Nilsson-Ehle and East, was the proposal that the evolutionary "purpose" of sexual reproduction is to increase the possibility of genetic recombinations; natural selection acting on a broad distribution of genotypes enables a population to adapt better to changing environments. This is still considered one possible answer (though not the only one) to the burning question, "why sex?"

Complementary to the above-mentioned advances was the conclusion of B. M. Davis that the mutations in Oenothera, on which de Vries had based his theory, constituted a special kind of hybrid species that cannot be considered representative of species in general. The saltationist theory was gradually abandoned by geneticists after 1915 in favor of hypotheses based on small mutations and continuous variation, but other biologists remained unaware of this development for another decade or two and continued to assume that Mendelism was incompatible with Darwinian evolution.

The major advance in theoretical genetics during this period was the demonstration that selection can be very effective even though a character has only a marginal advantage in fitness. This result was first suggested by tables computed by H. T. J. Norton at the request of R. C. Punnett, published in Punnett's book *Mimicry in Butterflies* (1915). The tables also showed that selection is *not* very effective in removing a rare recessive gene from a population, contrary to the claim of eugenicists that bad traits could be eliminated in a few generations.

The first major contribution to the new science of mathematical population genetics was a paper by R. A. Fisher in 1918. Fisher proposed a method for separating hereditary influences from environmental ones and showed that many continuously varying characters such as human stature are primarily determined by Mendelian factors rather than by the environment. The motivation for this research may have been Fisher's interest in eugenics as well as his purely scientific interest in the theory of inheritance (Norton 1978). Trained in mathematics and astronomy, Fisher had found time during his undergraduate years at Cambridge University to organize a Eugenics Society there. When he married a woman of similar views and respectable intellect, he put his principles into practice by raising a family of eight children despite conditions of financial stringency. He later held the Galton Chair of Eugenics at University College, London.

The next advances came in Russia, though they were not known to Western biologists until much later. There were several genetics research groups; the one founded by Nikolai Kol'tsov in Moscow made the most important contributions to evolutionary theory. Kol'tsov succeeded in overcoming the separation between experimentalists and naturalists that had retarded progress in other countries by demanding that biologists have advanced training in several disciplines. While World War I cut off contact with Western research, the Communist government that took over following the revolution was strongly committed to science and gave support even to the "bourgeois" scientists in the 1920s. Research in genetics was favored because of its practical applications in agriculture and did not require as much expensive equipment as other fields did. Darwinism had been popular with Russian intellectuals before the revolution, and afterwards it was seen as a good "materialist" theory. Marxist ideology also favored the enterprise (or at least the jargon) of producing a "synthesis" from opposing positions.

In 1922 H. Muller visited the Russian geneticists, bringing with him Morgan's ideas and Drosophila cultures that became valuable research materials for the staff of Kol'tsov's institute. Sergei Chetverikov, originally a naturalist rather than a geneticist, had become interested in the statistical fluctuations of the populations of butterflies he collected and soon launched a research program in population genetics. His 1926 paper proposed a synthesis of evolutionary biology based on Darwinian selection and Mendelian genetics; it presented views and results comparable to those published a little later by Fisher, Haldane, and Wright. Chetverikov's influence on later developments was primarily through Theodosius Dobzhansky, another Russian who had been a member of a different group and became interested in population genetics only after coming to the United States. Haldane, a Marxist as well as a mathematical evolutionist, learned of Chetverikov's work and arranged for English translations of his papers. Chetverikov's scientific career was abruptly halted in 1929 when he was arrested by the Soviet secret police and exiled—possibly because of his criticism of Lamarckism (which was becoming popular among Communist Party theoreticians) or perhaps because he was denounced by students who had been excluded from his research discussion group.

During the 1920s the mathematical theory of population genetics was developed by Fisher, J. B. S. Haldane, and Sewall Wright. Haldane, son of the physiologist J. S. Haldane, served in World War I and acquired an interest in the physiology of respiration when he and his father were asked to improvise gas masks after the first German poison gas attacks. He seems to have been a counterexample to the saying "there are no atheists in foxholes"; he became an avid Communist and, even after his own research had demonstrated the effectiveness of natural selection, defended Lysenko's theories until 1949. He taught at Cambridge University and later at University College, London.

Sewall Wright studied biology at the University of Illinois, then worked with Castle at Harvard, assisting in the selection experiment on hooded rats. He was later employed by the Animal Husbandry Division of the U.S. Department of Agriculture. Starting from the problems of interpreting data, he devised a mathematical technique of "path coefficients" to estimate the degree to which a given effect was determined by each of a number of causes (1917–1921). He became convinced that the interaction of genes was an important factor in evolution.

Fisher, Haldane, and Wright disagreed with each other on several points, but this in itself seems to have stimulated further research and experimental tests. The major outcome of their work was that Mendelism and natural selection, together with processes that plausibly occur in natural populations, are sufficient to account for microevolution (shifts of gene frequencies in a population). They showed that a very slight selective advantage could cause a gene to spread throughout a population—this had previously been asserted but not proved. Moreover, quantitative work with paleontological evidence showed that natural selection acting on small variations could account for the rate of evolutionary change found in the fossil record. Thus it was no longer necessary to invoke neo-Lamarckian hypotheses.

The role of Fisher, Wright, and Haldane in demolishing earlier objections to Darwin's theory may be compared with the role of Galileo in refuting the major objection to the Copernican heliocentric system—that we do not observe the physical effects of the Earth's motion—through his law of inertia (**1.4**). The mathematical geneticists did not prove that evolution does proceed by natural selection with hard inheritance, but they destroyed the criticism that it could not proceed that way.

In addition to showing that the inheritance of acquired characters—which had never been directly observed, though often suggested—was an unnecessary assumption, the mathematical geneticists helped to eliminate other competing theories. Saltationism was undermined because selection was found to be a much more effective agent of evolutionary change than mutation; "directed" evolution (orthogenesis) also seemed to require arbitrary, unverifiable, and now superfluous assumptions.

One particular example of selection effects to which Haldane had applied his theory as early as 1924 although definitive evidence was not available until the work of H. B. D. Kettlewell and others decades later, has become well known as an illustration of Darwinism. Over the past two centuries the light-colored moth Biston betularia has been replaced, in the industrial areas of England and other countries, by a dark-colored mutant form. The Darwinian hypothesis is that the dark moth can more easily escape predators because it can't be detected against the smoke-covered trees; thus "industrial melanism" has a definite survival value.

Fisher published in 1930 a definitive treatise, *The Genetical Theory of Natural Selection*. It includes what he called the "fundamental theorem of natural selection": the rate of increase in fitness of any organism at any time is equal to its genetic variance in fitness at that time. He compared this theorem to the second law of thermodynamics: both involve statistical properties of populations or aggregates that are largely independent of the nature of the individual units that compose them; each requires the continual increase of a measurable quantity, fitness or entropy. But he was equally concerned about the social consequences of another trend: the greater fertility of the lower classes. Higher-paid positions in society tend to be filled by persons whose parents were in lower-paid positions; Fisher warned that such "social promotion" was not necessarily beneficial to society as a whole.

Population theory changed the practice of genetic research by providing a quantitative framework for the interpretation of data, identifying relevant parameters, suggesting experimental designs, and generating testable hypotheses. Dobzhansky, a leader in experimental work during the 1930s and 1940s, relied heavily on Wright's advice, as Provine has shown (1978) by examining their correspondence. Dobzhansky attributed a major role to the mathematicians in founding population genetics, and it has generally been assumed that population genetics provided a major part of the new evolutionary synthesis.

Mayr has challenged this view of the history of the evolutionary synthesis, claiming that systematics (his own specialty) played a more important role. He asserted that evolution is not simply a result of changes in gene frequencies; the crucial step is the formation of new species. "A new species develops if a population which has become geographically isolated from its parental species acquires during this period of isolation characters which promote or guarantee reproductive isolation when the external barriers break down" (Mayr 1980, 35).

The unification of systematics and genetics meant a reconciliation of experimentalists and naturalists as they agreed to accept the neo-Darwinian evolutionary synthesis. This unification took place between 1936 and 1947. It began with the publication in 1937 of Dobzhansky's *Genetics and the Origin of Species,* a book which contains hardly any mathematics but was nevertheless strongly influenced by the work of Chetverikov and Wright. Other landmarks were C. Darlington's *Evolution of Genetic Systems* (1939), E. Mayr's *Systematics and the Origin of Species,* and J. Huxley's *Evolution: The Modern Synthesis.* The evidence of paleontology bearing on the rate of evolution was integrated into the synthesis in G. G. Simpson's *Tempo and Mode in Evolution;* botany's contribution came with *Variation and Evolution in Plants* (1950) by J. Ledyard Stebbins. By that time the general acceptance of the synthesis had been demonstrated at a conference at Princeton in 1947 (Mayr in Mayr and Provine 1980, 42).

Of course research and debate about the nature of evolution did not end with the adoption of the synthetic theory, but continues to this day. The issue is not whether evolution occurred, but how and at what rate. In view of the frequent statements by creationists that scientists such as N. Eldredge, S. J. Gould, and S. Stanley have rejected Darwin and hence that evolutionary theory has been discredited, it is important to explain accurately the thesis of "punctuated equilibrium." Its advocates do not reject evolution, rather they argue that its rate varies from one time to another, and they are effective opponents of the creationist claim that evolution never occurred at all.

BIBLIOGRAPHY

Lindley Darden (1986), "Relations among Fields in the Evolutionary Synthesis," in *Integrating Scientific Disciplines,* ed. W. Bechtel (Dordrecht: Nijhoff), 113–23.

Marjorie Grene, ed. (1983), *Dimensions of Darwinism: Themes and Counterthemes in Twentieth-Century Evolutionary Theory* (New York: Cambridge Univ. Press).

Richard J. Blackwell, comp. (1983), *A Bibliography of the Philosophy of Science, 1945–1981* (Westport, Conn.: Greenwood), 499–513.

Garland E. Allen (1983), "The Several Faces of Darwin: Materialism in Nineteenth and Twentieth Century Evolutionary Theory," in *Evolution from Molecules to Men,* ed. D. S. Bendall (New York: Cambridge Univ. Press), 81–102.

Topic *i:* Population Genetics

Sewall Wright (1986), *Evolution: Selected Papers,* ed. W. B. Provine (Chicago: Univ. of Chicago Press).

William B. Provine (1986), *Sewall Wright and Evolutionary Biology* (Chicago: Univ. of Chicago Press); (1983), *The Development of Wright's Theory of Evolution: Systematics, Adaptation, and Drift,"* in *Dimensions of Darwinism,* ed. M. Grene (New York: Cambridge Univ. Press), 43–70; (1978), "The Role of Mathematical Population Genetics in the Evolutionary Synthesis of the 1930s and 1940s," *Studies in History of Biology* 2:167–92.

James W. Tankard, Jr. (1984), *The Statistical Pioneers* (Cambridge: Schenken), includes chapters on K. Pearson, W. S. Gosset, R. Fisher.

J. H. Bennett, ed. (1983), *Natural Selection, Heredity, and Eugenics: Including Selected Correspondence of R. A. Fisher with Leonard Darwin and Others* (New York: Oxford Univ. Press).

Paul Thomson (1983), "The Structure of Evolutionary Theory: A

Semantic Approach," *Studies in History and Philosophy of Science* 14:215–29, uses the "semantic view of theories" in B. van Fraassen's version to describe the synthetic theory of evolution based on population genetics.

Donald A. MacKenzie (1981), *Statistics in Britain, 1865–1930* (New York: Columbia Univ. Press), Chap. 8, on R. A. Fisher.

B. Norton (1983), "Fisher's Entrance into Evolutionary Science: The Role of Eugenics," in *Dimensions of Darwinism,* ed. M. Grene, 19–29; (1978), "Fisher and the Neo-Darwinian Synthesis," *Proceedings of the 15th International Congress of History of Science* (Edinburgh: Edinburgh Univ. Press), 481–94.

Joan Fisher Box (1978), *R. A. Fisher: The Life of a Scientist* (New York: Wiley-Interscience).

David L. Jameson, ed. (1977), *Evolutionary Genetics* (Stroudsburg, Pa.: Dowden, Hutchinson and Ross), contains reprints of papers by Darwin and Wallace (1859), Galton (1897), Mendel, W. Castle (1903), G. U. Yule (1906), G. H. Hardy (1908), W. Weinberg (1908), K. Pearson (1909), R. A. Fisher, S. Wright, J. B. S. Haldane, S. S. Chetverikov.

Topic *ii:* Neo-Darwinism

L. M. Cook, G. S. Mani, and M. E. Varley (1986), "Postindustrial Melanism in the Peppered Moth," *Science* 231:611–13, notes that "the response of the peppered moth, *Biston betularia,* to environmental changes brought about by industrialization remains one of the most fully documented cases of microevolutionary change." But as a result of smoke control legislation the trend has reversed.

Jonathan Harwood (1985), "Geneticists and the Evolutionary Synthesis in Interwar Germany," *Annals of Science* 42:279–301.

Harris Bernstein, Henry C. Byerly, Frederic A. Hopf, and Richard E. Michod (1985), "Genetic Damage, Mutation, and the Evolution of Sex," *Science* 229:1277–81.

G. Ledyard Stebbins and Francisco J. Ayala (1985), "The Evolution of Darwinism," *Scientific American* 253, no. 1:72–82, asserts, "recent developments in molecular biology and new interpretations of the fossil record are gradually altering and adding to the synthetic theory, for 40 years the standard view of the process of evolution." They claim that "punctuated equilibrium" (see *iii*) does not require much modification of the standard view.

Francisco J. Ayala (1985), "Theodosius Dobzhansky, January 25, 1900—December 18, 1975," *Biographical Memoirs, National Academy of Sciences* 55:163–213.

Stephen Jay Gould (1983), "The Hardening of the Modern Synthesis," in *Dimensions of Darwinism,* ed. M. Grene, 71–93.

Ernst Mayr and William B. Provine, eds. (1980), *The Evolutionary Synthesis: Perspectives on the Unification of Biology* (Cambridge: Harvard Univ. Press); see also the review by Niles Eldredge, "Gentlemen's Agreement," *The Sciences,* 23, no. 4:20–23, 31 (April 1981), noting two contradictory accounts of the origin of the "synthesis" in this book.

William B. Provine (1981), "Origins of the Genetics of Natural Populations Series," in *Dobzhansky's Genetics of Natural Populations I–XLIII,* ed. R. C. Lewontin, J. A. Moore, W. B. Provine, and B. Wallace (Chicago: Univ. of Chicago Press) 1–83; (1979), "Francis B. Sumner and the Evolutionary Synthesis," *Studies in History of Biology* 3:211–40.

Mark B. Adams (1980), "Science, Ideology, and Structure: The Kol'tsov Institute 1900–1970," in *The Social Context of Soviet Science,* ed. L. Lubrano, S. Solomon (Boulder, Colo.: Westview Press), 173–204; (1979), "From 'Gene Fund' to 'Gene Pool': On the Evolution of Evolutionary Language," *Studies in History of Biology* 3:241–85; (1970), "Towards a Synthesis: Population Concepts in Russian Evolutionary Thought, 1925–1935," *Journal of the History of Biology* 3:107–29.

Authur Caplan (1978), "Testability, Disreputability, and the Structure of the Modern Synthetic Theory of Evolution," *Erkenntnis* 13:261–78.

B. Norton (1978), "Fisher and the Neo-Darwinian Synthesis," *Proceedings of the XV International Congress of History of Science, 1977* (Edinburgh: Edinburgh Univ. Press), 481–94.

David L. Jameson, ed. (1977), *Genetics of Speciation* (Stroudsburg, Pa.: Dowden, Hutchinson and Ross), is a reprint of papers by S. Wright (1940), J. B. S. Haldane, T. Dobzhansky, A. H. Sturtevant, E. Mayr, etc.

Topic *iii:* Punctuated Equilibrium

Robert G. B. Reid (1985), *Evolutionary Theory: The Unfinished Synthesis* (Ithaca, N.Y.: Cornell Univ. Press).

Benton M. Stidd (1985), "Are Punctuationists Wrong about the Modern Synthesis?" *Philosophy of Science* 52:98–109.

Richard E. Grant (1983), "Evolution: A Cyclical Argument?" *Science* 219:1170.

F. J. Ayala et al. (1983), "Recent Developments in Biology," *PSA 1982,* vol. 2 (East Lansing, Mich.: Philosophy of Science Association), 275–328.

Stephen Jay Gould (1982), "Darwinism and the Expansion of Evolutionary Theory," *Science* 216:380–87.

Roger Lewin (1980), "Evolutionary Theory under Fire: An Historic Conference in Chicago Challenges the Four-Decade-Long Dominance of the Modern Synthesis," *Science,* 210:883–87, and letters to editor (1981) by D. J. Futuyama et al., *Science* 222:770ff.

4.7 THE CREATION-EVOLUTION CONTROVERSY

i. Religious fundamentalism: The historical background of "creationism"
ii. The Scopes Trial and its effect on the teaching of biology in public schools
iii. Revival of creationism in the 1970s, connection with Moral Majority and New Right; use of K. Popper's criterion

READINGS

Bowler, *Evolution,* 327–34, 338–48. J. R. Moore, "The Future of Science and Belief: Theological Views in the Twentieth Century," Unit 15 in the Open University course A381, *Science and Belief, Darwin to Einstein* (1981), 4–55 (on C. E. Raven, Protestant liberalism, William Bell Riley, Reinhold Niebuhr, etc.; while the background of "fundamentalism" is discussed, there is nothing on the recent creationist movement). J. Peter Zetterberg, ed., *Evolution versus Creationism: The Public Education Controversy* (1983), includes articles by both creationists and evolutionists. Dorothy Nelkin, "Science, Rationality, and the Creation/Evolution Dispute," in *Science and Creation,* ed. R. W. Hanson (1986), 33–45. Eileen Barker, "Let There Be Light: Scientific Creationism in the Twentieth Century," in *Darwinism and Divinity,* ed. J. Durant (1985), 181–204. Willard B. Gatewood, Jr., "From Scopes to Creation Science: The Decline and Revival of the Evolution Controversy," *South Atlantic Quarterly* 83(1984):363–83.

SYNOPSIS

The conflict between evolutionary theories and certain religious doctrines was not resolved in the nineteenth century, or even in the 1920s with the Scopes Trial in the United States, but has become since about 1970 a problem of serious concern to scientists and the public. In

1981 the passage of laws in Arkansas and Louisiana mandating "equal time" or "balanced treatment" of "scientific" creationism in public schools forced scientists to pay attention to the issue. Since then, numerous books and articles have explained the theory of evolution and supporting evidence.

Historians of science have a chance to play a role in this current debate by showing how scientists confronted and resolved the problem in the past. For example, many of the alleged fallacies of Darwin's theory (e.g., the "circular" argument about survival of the fittest) are based on misunderstandings that were clarified long ago. The thesis that the universe was created only 6000 years ago was generally accepted before the nineteenth century but was discarded by geologists (many of whom wanted to believe it for religious reasons) because of overwhelming evidence of the antiquity of the Earth (**2.1**). It is important to point out that going back to the "young Earth" doctrine of the creationists would demand throwing out much of the research done in geology, geophysics, and astronomy during the past 150 years, to say nothing of the implications for biology and paleontology. (See also **12.2**)

The creationists do raise a crucial issue in the philosophy of science, thanks to the confusion of some philosophers themselves. Karl Popper, in particular, has proposed a criterion for drawing a line of demarcation between science and pseudoscience, arguing that a truly scientific theory makes predictions that can be tested and possibly falsified, whereas a pseudoscience merely explains whatever facts are known. He argued that Darwin's theory of evolution is a pseudoscience, or rather a "metaphysical research programme," because it is not falsifiable but only offers an explanation of singular events that happened in the past and cannot be repeated in the laboratory. The creationists made effective use of Popper's specious claim, arguing that since neither evolution nor creation can be scientifically tested, both should be presented in the classroom. Popper eventually retracted his statement and concluded that Darwin's theory *is* scientific, but the damage was done. More generally it should be pointed out that if one applies the falsifiability criterion in this way, not only evolution but all theories in historical geology, astronomy, etc., would fail since they deal with phenomena taking place over wide regions of space and time that by their very nature cannot be brought into the laboratory for controlled experimentation. The criterion itself must be faulty if it excludes disciplines that are clearly scientific. Conversely, other philosophers like Mary Williams have argued that evolutionary theory does make testable statements and thus does satisfy a more reasonable version of Popper's criterion.

The political aspects of the creationist movement can hardly be separated from its religious component. Jerry Falwell and his "Moral Majority" were strong supporters of Reagan Republicanism, and their attack on evolution is part of a social agenda that includes legislation against abortion and pornography, as well as defeat of the Equal Rights Amendment. They see evolution as a component of "secular humanism," a movement undermining traditional society.

BIBLIOGRAPHY

J. Shea (1984), "A List of Selected References on Creationism," *Journal of Geological Education* 32:43–49.

Topic *i:* Fundamentalism

George M. Marsden (1980), *Fundamentalism and American Culture: The Shaping of Twentieth Century Evangelicalism 1879–1925* (New York: Oxford Univ. Press).

Topic *ii:* Scopes Trial

Donald E. Chittick (1984), *The Controversy: Roots of the Creation-Evolution Conflict* (Portland, Oreg.: Multnomah Press).

Ferenc Morton Szasz (1982), *The Divided Mind of Protestant America, 1880–1930* (University: Univ. of Alabama Press), 107–35.

Judith V. Grabiner and Peter D. Miller (1974), "Effects of the Scopes Trial," *Science* 185:832–37.

William B. Gatewood, Jr., ed. (1969), *Controversy in the Twenties: Fundamentalism, Modernism, and Evolution* (Nashville, Tenn.: Vanderbilt Univ. Press), contains extracts form the writings of fundamentalists and scientists.

L. Sprague De Camp (1968), *The Great Monkey Trial* (Garden City, N.Y.: Doubleday).

Topic *iii:* Revival of Creationism

Ronald L. Numbers (1986), "The Creationists," in *God and Nature,* ed. D. C. Lindberg and R. L. Numbers (Berkeley: Univ. of California Press), 391–423; (1982), "Creationism in 20th-century America," *Science* 218:538–44.

Edward J. Larson (1985), *Trial and Error: The American Controversy over Creation and Evolution* (New York: Oxford Univ. Press).

Eugenie C. Scott and Henry P. Cole (1985), "The Elusive Scientific Basis of Creation 'Science,' " *Quarterly Review of Biology* 60:21–30, is a survey of 68 journals showing that creationists did not even try to publish evidence for their theory in standard scientific journals, with very few exceptions.

Laurie R. Godfrey, ed. (1985), *What Darwin Began* (Boston: Allyn and Bacon), articles by Alice B. Kehoe and Steven D. Schafersman; (1983), *Scientists Confront Creationism* (New York: W. W. Norton).

Michael A. Cavanaugh (1985), "Scientific Creationism and Rationality" *Nature* 315:185–89.

Philip L. Quinn (1984), "The Philosopher of Science as Expert Witness," in *Science and Reality,* ed. J. T. Cushing et al. (Notre Dame, Ind.: Univ. of Notre Dame Press), 32–53, discusses testimony at the Arkansas trial by M. Ruse.

Frank Awbrey and William Thwaites, eds. (1984), *Evolutionists Confront Creationists* (San Francisco: California Academy of Sciences), includes papers by creationists and evolutionists, and a long article by G. B. Dalrymple on the age of the Earth.

Stanley Weinberg, ed. (1984), *Reviews of Thirty-One Creationist Books* (Syosset, N.Y.: National Center for Science Education).

Marcel Chotkowski La Follette, ed. (1983), *Creationism, Science, and the Law: The Arkansas Case* (Cambridge: MIT Press).

Dorothy Nelkin (1982), *The Creation Controversy: Science or Scripture in the Schools* (New York: Norton).

Philip Kitcher (1982), *Abusing Science: The Case Against Creationism* (Cambridge: MIT Press).

Stephen G. Brush (1982), "Finding the Age of the Earth: By Physics or by Faith," *Journal of Geological Education* 30:34–58, is a survey of radiometric methods for estimating the age of Earth, and a critical review of creationist objections to those methods and of T. Barnes's method of justifying the "young earth" hypothesis.

J. Patrick Gray and Linda D. Wolfe (1982), "Sociobiology and Creationism: Two Ethnosociologies of American Culture," *American Anthropologist* 84:580–94; Wolfe and Gray (1981), "Creationism and Popular Sociobiology as Myths," *Humanist* 11, no. 4:43–50.

Michael Ruse (1982), *Darwinism Defended* (Reading, Mass.: Addison-Wesley); (1981), "Karl Popper and Evolutionary Biology," in his *Is Science Sexist?* (Boston: Reidel), 65–84.

Karl Popper (1980), "Evolution," *New Scientist* 87:611; (1978), "Natural Selection and the Emergence of Mind," *Dialectica* 32:339–55; (1976), *Unended Quest* (LaSalle, Ill.: Open Court), 167–80.

Catherine A. Callaghan (1980), "Evolution and Creationist Arguments," *American Biology Teacher* 42:422–27, with supplement available from the author, Department of Linguistics, Ohio State University, Columbus, Ohio 43210.

Henry Morris (1974), *Scientific Creationism* (San Diego, Calif.: Creation-Life Publishers), is a basic reference work presenting the most popular version of creationism.

For more recent news and literature see the quarterly journal *Creation/Evolution* and the bimonthly *Creation/Evolution Newsletter* (National Center for Science Education, Box 32, Concord College, Athens, W.Va. 24712).

5

FREUD AND PSYCHOANALYSIS

5.1 PSYCHIATRY AND THE UNCONSCIOUS BEFORE FREUD

i. Conscious vs. unconscious mind to 1850
ii. Hypnotism: F. A. Mesmer, J. Elliotson, James Easdaile; Anesthesia, James Braid
iii. Psychiatry before 1850
iv. The late nineteenth-century background for Freud: theories of hysteria, degeneration, unconscious

READINGS

R. E. Fancher, *Pioneers of Psychology* (1979), 170–204 (hypnotism, hysteria, crowd psychology); Thomas Leahey, *History of Psychology* (1979), 156–59 (hypnotism), 217–19. Robert Thomson, *Pelican History of Psychology* (1968), 194–211. Merle Curti, *Human Nature in American Thought* (1980), 313–40. For a more extensive treatment, one may use Franz G. Alexander and Sheldon T. Selesnick, *The History of Psychiatry* (1966), Chaps. 7–10, 89–178, for seventeenth through nineteenth centuries.

Articles: Frank J. Sulloway, "Freud and Biology: The Hidden Legacy," in *The Problematic Science,* ed. W. R. Woodward and M. G. Ash (1982), 198–227. Henri Ellenberger, "The Unconscious before Freud," *Bulletin of the Menninger Clinic* 21 (1957):3–15.

Freudian psychoanalysis would not ordinarily be the first or even a major topic in a course on the history of psychology. My reason for giving it such prominence is the view that Freud has had an impact on modern thinking, especially on our concept of human nature, that transcends academic psychology. The most extreme statement of that view would group together Copernicus, Darwin, and Freud as instigators of three revolutions, each of which radically changed our understanding of "Man's Place in the Universe"—and would identify Freud as the one who provoked the modern controversy about the Nature of Woman as distinct from Man.

Even if one does not accept Freud as a scientific revolutionary, and indeed if one maintains that he was not even a scientist properly speaking, there is much useful material on which to base a discussion of what it means to be scientific—where does one draw the line between science and pseudoscience? This discussion could be incorporated into a more traditional course in history of science by using only **5.2** and **5.5**, while omitting the rest of the chapter.

According to L. L. Whyte, Leibniz is regarded as the first European thinker to state the idea of an unconscious mind with precision. The concept was developed primarily by British and German thinkers; Cartesian influence discouraged the French from recognizing the unconscious.

Topic *iv* is directly connected with the theory of degeneration (**3.3-*iv***): many of the phenomena such as hysteria, which Freud later suggested could be treated by psychotherapy, were formerly regarded as incurable results of hereditary defects.

BIBLIOGRAPHY

W. F. Bynum et al., eds. (1985), *The Anatomy of Madness: Essays in the History of Psychiatry* (New York: Tavistock).

Gregory A. Kimble and Kurt Schlesinger, eds. (1985), *Topics in the History of Psychology,* volume 2 (Hillsdale, N.J.: Erlbaum), includes J. D. Matarazzo, "Psychotherapy," and B. A. Maher and W. B. Maher, "Psychopathology II. From the Eighteenth Century to Modern Times."

George Mora (1965), "The History of Psychiatry: A Cultural and Bibliographic Survey," *Psychoanalytic Review* 52, no. 2:154 [298]–184 [328], is a review with 341 references.

Leon Chertok and Raymond de Saussure (1979), *The Therapeutic Revolution: From Mesmer to Freud* (New York: Brunner/Mazel).

Topic *i*: Unconscious before 1850

L. L. Whyte (1978), *The Unconscious before Freud,* new ed. (New York: St. Martin's Press), Chaps. 1–7.

Henri F. Ellenberger (1970), *The Discovery of the Unconscious* (New York: Basic Books).

Topic *ii:* Hypnotism

Heinz Schott (1984), "Mesmer, Braid und Bernheim: Zur Entstehungsgeschichte des Hypnotismus," *Gesnerus* 41:33–48.

Ronald E. Shor (1979), "The Fundamental Problem in Hypnosis Research as Viewed from Historic Perspectives," in *Hypnosis,* 2d ed., ed. E. Fromm and R. E. Shor (New York: Aldine), 15–41.

Topic *iii:* Psychiatry before 1850

Jose M. Lopez Piñero (1983), *Historical Origins of the Concept of Neurosis* (New York: Cambridge Univ. Press).

George Mora and Jeanne L. Brand, eds. (1970), *Psychiatry and Its History: Methodological Problems in Research* (Springfield, Ill.: C. C. Thomas).

Norman Dain (1964), *Concepts of Insanity in the United States, 1789–1865* (New Brunswick, N.J.: Rutgers Univ. Press).

Topic *iv:* Late Nineteenth-century Background for Freud

Sander L. Gilman (1985), "Sexology, Psychoanalysis, and Degeneration: From a Theory of Race to a Race to Theory," in *Degeneration,* ed. J. E. Chamberlin and S. L. Gilman (New York: Columbia Univ. Press), 72–96.

Gerlof Verwey (1985), *Psychiatry in an Anthropological and Biomedical Context: Philosophical Presuppositions and Implications of German Psychiatry, 1820–1870* (Boston: Reidel).

W. F. Bynum (1983), "Themes in British Psychiatry: J. C. Prichard (1785–1848) to Henry Maudsley (1835–1918)," in *Nature Animated,* ed. Michael Ruse (Boston: Reidel), 225–42.

Dennis B. Klein (1981), *Jewish Origins of the Psychoanalytic Movement* (New York: Praeger).

Wayne Viney et al. (1979), *History of Psychology: A Guide to Information Sources* (Detroit: Gale Research). See pp. 197–214 for references on the origins of psychoanalysis.

Frederic V. Grunfeld (1979), *Prophets without Honour: A Background to Freud, Kafka, Einstein and Their World* (New York: Holt, Rinehart and Winston).

Hannah S. Decker (1977), *Freud in Germany: Revolution and Reaction in Science, 1893–1907* (New York: International Universities Press), includes the relation of Freud's theories to degeneration.

5.2 FREUD AND HIS THEORY

i. Freud's early career, sources of his ideas
ii. Development of psychoanalytic theory

READINGS

John Marks, *Science* (1983), 485–86. Richard Olson, *Science as Metaphor* (1971), 183–206. Fancher, *Pioneers of Psychology,* 205–49; Abram Kardiner and Edward Preble, *They Studied Man* (1961), 217–70 (including "the influence of psychodynamics on the study of culture"). Alexander and Selesnick, *History of Psychiatry,* Chaps. 11–13, 181–210. Freud's own account is in a 1914 essay, "The History of the Psychoanalytic Movement," Part 1 (about 15 pages). Robert Thomson, *The Pelican History of Psychology,* 211–15, 249–56. Merle Curti, *Human Nature,* 341–71. Robert Watson, *The Great Psychologists* (1977), 491–522.

SYNOPSIS

Since one of Sigmund Freud's main themes was the influence of early childhood experiences on later behavior, one might well pay more than ordinary attention to his own early life. Thus for example he was an uncle (to the son of his father's son by his previous wife) when he was born, and his nephew, actually a year or two older than Sigmund, was one of his childhood playmates. When he was two and one-half, a sister was born, a threatening event he blamed not on his father but his half-brother, according to his later remembered fantasy.

In 1873, at age seventeen, Freud entered the University of Vienna and came into contact with the mainstream of nineteenth-century European science, e.g., Ernst Bruecke, one of the famous group of four German biologists who had launched in 1848 a program to make biology into a rigorous science with the same standards as physics. (See Cranefield's article cited in **2.2-*vii.***) Sulloway's biography explains the influence of biology on Freud's theories.

Freud gained a medical degree in 1881 and entered a Vienna hospital as an intern; then in 1885 he won a traveling fellowship and went to Paris to work with Jean-Martin Charcot who was studying hysteria. Charcot found that, contrary to the origin of the word (meaning "womb"), it could afflict males, and its symptoms could be produced by hypnosis. This may have suggested to Freud a medical motive for studying the psychology of patients. Back in Vienna he collaborated with Joseph Breuer whose patient Bertha Pappenheim displayed hysterical symptoms that disappeared after she described them to Breuer. Thus she discovered "catharsis"—reduction of fears and anxieties by bringing them to consciousness and talking about them. Freud eventually abandoned Charcot's technique of hypnotism and settled on the method of "free association," invented by another patient, "Elisabeth von R.," who reproved him for interrupting her flow of thought by asking too many questions.

The first results of the free association method were that the patient's memories kept going back to early childhood and often involved sexual experiences. Freud concluded that the frustration of sexual impulses was at the root of many of his patients' problems, and this is perhaps his best known thesis. By claiming that hysteria and neuroses are caused by the memory of earlier events, Freud was challenging the prevailing (late nineteenth-century) views that such afflictions are symptoms of hereditary degeneracy (**3.3-*iv***). Moreover, he was opening up the possibility that there is no strict dividing line between "pathological" and "normal" behavior, since the repression of sexual wishes and memories is scarcely avoidable by anyone growing up in human society.

In 1896 Freud published an article stating that the basic cause of hysteria is a traumatic sexual seduction (especially of the little girl by her father) at the age of three or four. He soon retracted this view, arguing that no one, not even he himself, could believe that 30 percent of the fathers in Vienna were seducing their daughters (as his cases led him to infer). Instead, he concluded that his patients' frequent memories of early seduction must be fantasies, and he admitted his "error" in 1897. This episode might be considered an example of Freud's willingness to admit his mistakes and a refutation of the charge that psychoanalysis is such a flexible doctrine that it can be twisted to account for anything and is therefore nontestable (cf., **5.5**). On the other hand it may be considered (in the view of Jeffrey Masson and other recent critics) a cowardly retreat from a valid but socially unacceptable discovery. By proclaiming that children's reports of sexual abuse are only products of their own imagination, Freud may have helped to cover up a scandalous situation that we are only now beginning to understand.

But Freud's new explanation was even more unacceptable to the average person than his old one, for he now postulated that every infant has a basic sexuality that does not need any adult seduction for its arousal. This is not to say all aspects of this sexuality are biologically determined, but rather that the earliest relations of the child with his or her parents in the *normal* family situation lead to a characteristic pattern of development common to almost all children. The theory of this pattern was gradually worked out in Freud's writings from 1898 on.

In 1900 Freud published a long treatise on the *Interpretation of Dreams,* considered by some his most important work. He introduced a system of symbolism to translate the apparently innocent content of the dream into sexual terms. Thus a candle, knife, necktie, or any long thin object means a penis; an oven or ship means the female genitals; climbing stairs means sexual intercourse, etc.

Freud's 1904 book, *Psychopathology of Everyday Life,* became well known to the public and to generations of students who read it in psychology courses; it is the source of the theory of "Freudian slips" in which one says, apparently inadvertently, something that reveals one's real but improper wishes or thoughts. Freud's own typical example is someone who says: "Yes, a woman must be pretty if she is to please the men. A man is much better off. As long as he has five straight limbs, he needs no more." The main theme is that there are no such things as "chance" actions—every act has its psychic cause even though the conscious mind may not be aware of it. Moreover, as the title suggests, these phenomena are not restricted to abnormal people but affect everyone to some degree.

Freud's second major treatise, *Three Contributions to the Theory of Sex,* first appeared in 1905 but was revised several times later. It deals with "sexual aberrations" (homosexuality, fetishes, oral and anal sex, sadism, masochism, etc.), which Freud insists are not due to hereditary degeneration but are merely extreme manifestations of tendencies common to all "normal" people, guided in certain directions by early experiences. Another chapter on "Infantile Sexuality" introduced several famous concepts: the "period of latency" starting around age eight, following an earlier lively sexuality which is then forgotten; "erogenous zones" of the body; a progression from oral to anal and then genital phases; the "castration complex"; and "penis envy." The latter is claimed to be the origin of women's feelings of inferiority to males. *Libido* (sexual force or energy) is said to be possessed by both boys and girls, yet Freud calls it essentially male; this seems to be another form of the ancient theory that "masculine" equals "active" and "feminine" equals "passive."

According to Freud, the little girl first masturbates with her clitoris just as the little boy masturbates with his penis, but during puberty this erogenous zone has to be repressed in the girl, and sexuality is transferred to the vagina. The need for this transformation is, according to Freud, the key to the nature of neuroses and hysteria in women, and is the basic reason why women's psychic development is more complicated and subject to inhibitions and repressions than men's.

Up to this point Freud has been playing the role of the objective scientist, but in essays published in 1907 and 1908 he starts to publicize his opinions on how society should handle sexuality. In "The Sexual Enlightenment of Children" he argues that schools should provide sex education before age ten in order to satisfy the child's natural curiosity and avoid neurosis (don't leave this to parents because they don't realize that such young children have sexual feelings). In " 'Civilized' Sexual Morality and Modern Nervous Illness" Freud argues that many people cannot comply with the "unnatural" demand that sex be limited to reproduction within marriage and rejects the idea that abstinence "steels the character." Suppression of sexuality is bad, and since women are pressured not to even *think* about sex they are thereby "scared away from *any* form of thinking, and knowledge loses its value for them"—another reason for the intellectual inferiority of women, but one that could be removed.

Another short publication of 1908 caused almost as much controversy as the ones just mentioned: Freud argued that the combination of three character traits, orderly (including bodily cleanliness), parsimonious, and obstinate, is generally found in persons who experienced long delay in getting control of their bowels in childhood. They had been stuck at the anal stage in which the child derives pleasure from withholding his feces. Hence the term "anal retentive" for this character type. (Money is psychologically equivalent to feces, in this theory.)

Freud's later opinion of America was apparently determined by his own bathroom experiences a few months after that paper was published. G. Stanley Hall, a founder of American psychology (**6.1-*iii***), invited Freud to lecture at Clark University. The publicity given to Freud's views at this time helped to launch psychoanalysis in America but Freud developed an intense prejudice against America. He was having problems with his appendix and his prostate and had to urinate frequently, but somehow the nearest men's room was always miles away. So he was subjected to continual embarrassment and physical discomfort. Later he said "America is a mistake; a gigantic mistake it is true, but none the less a mistake." Some of these feelings came out in the posthumous book on Woodrow Wilson, generally regarded as tarnishing Freud's reputation because of its many unjust or untrue statements.

In the years 1913–1923 Freud developed his three-part structure of the mind: Id, Ego, Superego. *Totem and Taboo* (1912) is a venture into speculative anthropology. Later he wrote on religion (*Future of an Illusion,* 1927) and finally arrived at a theory of the origin of Judaism that postulated an immense cultural guilt complex deriving from a murder of Moses by the tribe that could not live with the restrictions he imposed on them (*Moses and Monotheism,* 1937–1939). The relations between society, religion, and sex are also discussed in *Civilization and Its Discon-*

tents (1927). His elaboration of the psychology of women is mentioned below (**5.4**). A rather peculiar work, which doesn't fit too well with his earlier theories, is *Beyond the Pleasure Principle* (1920) in which Freud postulates a universal "death instinct" complementary to the sex drive.

BIBLIOGRAPHY

Peter Gay (1985), *Freud for Historians* (New York: Oxford Univ. Press).

Martin Grotjahn (1979), "New Insight into the Life and Work of Sigmund Freud: An Overview of the Literature since 1975," *Journal of the American Academy of Psychoanalysis* 7:299–313, is a continuation of earlier reports by Grotjahn on the Freud literature, cited herein.

Topic *i*: Early Career, Sources of Ideas

W. F. Bynum et al. (1985), *The Anatomy of Madness: Essays in the History of Psychiatry* (New York: Tavistock), includes articles by I. Dowbiggin, J. Forrester, and A. Clare on the origins and historical context of Freud's theories.

Jeffrey M. Masson (1985), ed. and trans., *The Complete Letters of Sigmund Freud to Wilhelm Fliess, 1887–1904* (Cambridge: Harvard Univ. Press); (1984), *The Assault on Truth: Freud's Suppression of the Seduction Theory* (New York: Farrar, Straus and Giroux); (1984), "Freud and the Seduction Theory," *Atlantic Monthly* 253, no. 2:33–60. Preliminary report by Daniel Goleman and Ralph Blumenthal, "Freud: Secret Documents Reveal Years of Strife," *New York Times,* 24 Jan. 1984, C1ff. For the aftermath see Masson (1984), "The Persecution and Expulsion of Jeffrey Masson as Performed by Members of the Freudian Establishment and Reported by Janet Malcolm of *The New Yorker,*" *Mother Jones* 9, no. 10:34–37, 42–47.

Janet Malcolm (1984), *In the Freud Archives* (New York: Knopf), is on the controversy about Masson's thesis on Freud's seduction theory.

Martin Rubinstein (1983), "Freud's Early Theories of Hysteria," in *Physics, Philosophy and Psychoanalysis,* ed. R. S. Cohen and L. Laudan (Boston: Reidel), 169–90.

Yehuda Elkana (1983), "The Borrowing of the Concept of Energy in Freudian Psychoanalysis," in *Psicoanalisi e Storia delle Scienze* (Firenze: Olschki), 55–80.

Frank Sulloway (1982), "Freud and Biology: The Hidden Legacy," in *The Problematic Science,* ed. W. R. Woodward and M. G. Ash (New York: Praeger), 198–227; (1979), *Freud: Biologist of the Mind* (New York: Basic Books), Chaps. 1–8, 12–14.

Leon Chertok and Raymond de Saussure (1979), *The Therapeutic Revolution: From Mesmer to Freud* (New York: Brunner/Mazel).

Kenneth Levin (1978), *Freud's Early Psychology of the Neuroses* (Pittsburgh: Univ. of Pittsburgh Press).

Larry Stewart (1977), "Freud before Oedipus: Race and Heredity in the Origins of Psychoanalysis," *Journal of the History of Biology* 9:215–28.

Jean-Marc Dupeu (1977), "Freud and Degeneracy: A Turning Point," *Diogenes* 97:43–64.

Hannah S. Decker (1977), *Freud in Germany: Revolution and Reaction in Science, 1893–1907* (New York: International Universities Press), includes the relation of Freud's theories to degeneration theory.

Topic *ii:* Development of Psychoanalytic Theory

Patrick J. Mahoney (1986), *Freud and the Rat Man* (New Haven: Yale Univ. Press).

Paul E. Stepansky, ed. (1986), *Freud: Appraisals and Reappraisals* (Hillsdale, N.J.: Erlbaum Associates/Analytic Press).

Daniel Goleman (1985), "Freud's Mind: New Details Revealed in Documents," *New York Times,* Nov. 12, C1, C3, reports that the journals of Princess Marie Bonaparte have been made available for research.

Timothy McCarthy (1981), "Freud and the Problem of Sexuality," *Journal of the History of the Behavioral Sciences* 17:332–39.

Reuben Fine (1979), *A History of Psychoanalysis* (New York: Columbia Univ. Press).

Anthony F. Badalamenti (1979), "Entropy in Freudian Psychology," *Methodology and Science* 12:1–16 (1979).

Frank Sulloway (1979), *Freud: Biologist of the Mind* (New York: Basic Books), Chaps. 9–11.

[Note: additional references, dealing with Freud's theories of female psychology, will be found below in the Bibliography for section **5.4-***i.*]

5.3 GROWTH OF PSYCHOANALYSIS: FOLLOWERS AND COMPETITORS

i. Initial response to Freud in Europe: foundation of a movement and its breakup into rival schools. Systems of Adler, Jung, etc.

ii. Reception of Freud's theories in America. Establishment of psychoanalysis as a profession

READINGS

H. J. Fyrth and M. Goldsmith, *Science History and Technology* (1969), Book 2, Part II, 43–46. Olson, *Science as Metaphor,* 207–30. Alexander and Selesnick, *History of Psychiatry,* Chaps. 14–16 (211–52), on *i;* Jacques M. Quen and Eric T. Carlson, eds., *American Psychoanalysis: Origins and Development* (1978), chapters by Barbara Sicherman, "The New Psychiatry: Medical and Behavioral Science, 1895–1921," 20–34, and John C. Burnham, "The Influence of Psychoanalysis upon American Culture," 52–68. Freud, *The History of the Psychoanalytic Movement,* Parts 2 and 3 (about 35 pages). Thomson, *Pelican History,* 215–18, 256–65. Watson, *Great Psychologists,* 522–46. A. R. Gilgen, *American Psychology since World War II* (1982), 65–85 (on *ii*) and 167–202 (competition from clinical psychology).

SYNOPSIS

Freud soon became the leader of a "movement" that included several people who weren't content to be followers but had their own somewhat different ideas. The best known were Alfred Adler (1870–1937), who split with Freud in 1911, and C. G. Jung (1875–1961), who left the movement in 1914. Adler in particular gained a considerable following in America, and today there are several study groups of parents who are enthusiastic about applying his "individual psychology" methods to child raising. Adler popularized the term "inferiority complex" and replaced sex by aggressiveness and power struggle as the fundamental drive. He also stressed sociological aspects of consciousness rather than the repressed unconscious. By avoiding the element of determinism and inevitability in Freud's theory, Adler raised hopes that humans can be improved by rational education and changes in the social order. Jung, on the other hand, stressed the determination of psychic life, e.g., dream symbols, by the "collective unconscious" inherited from past generations, manifested in mythology and religion.

BIBLIOGRAPHY

John Burnham (1985), "The Encounter of Christian Theology with Deterministic Psychology and Psychoanalysis," *Bulletin of the Menninger Clinic* 49:321–52; (1982), "The Reception of Psychoanalysis in

Western Cultures: An Afterword on Its Comparative History," *Comparative Studies in Society and History* 24:603–10.

Peregrine Horden, ed. (1985), *Freud and the Humanities* (New York: St. Martin's Press).

Jeffrey Berman (1985), *The Talking Cure: Literary Representations of Psychoanalysis* (New York: New York Univ. Press).

Louise E. Hoffman (1981), "War, Revolution, and Psychoanalysis: Freudian Thought Begins to Grapple with Social Reality," *Journal of the History of the Behavioral Sciences* 17:251–69.

Topic *i*: Psychoanalytic Movement in Europe; Adler, Jung

Geoffrey Cocks (1985), *Psychotherapy in the Third Reich: The Göring Institute* (New York: Oxford Univ. Press), shows that although the Nazis condemned Freudian psychoanalysis as "Jewish science" they tolerated other versions of psychotherapy.

Steven Marcus (1984), *Freud and the Culture of Psychoanalysis: Studies in the Transition from Victorian Humanism to Modernity* (Boston: Allen and Unwin).

Paul E. Stepansky (1983), *In Freud's Shadow: Adler in Context* (New York: Analytic Press).

Hannah S. Decker (1982), "The Reception of Psychoanalysis in Germany," *Comparative Studies in Society and History* 24:589–602.

Irving E. Alexander (1982), "The Freud-Jung Relationship: The Other Side of Oedipus and Countertransference. Some Implications for Psychoanalytic Theory and Psychotherapy," *American Psychologist* 37:1009–18, is on Freud's role as an "Oedipal father."

John-Raphael Staude (1981), *The Adult Development of C. G. Jung* (Boston: Routledge and Kegan Paul).

Harry Slochower et al. (1981), "Jung's 'Secret' Confrontations with Freud: A Symposium," *American Imago* 38:3–95.

Reuben Fine (1979), *A History of Psychoanalysis* (New York: Columbia Univ. Press), Chaps. 4ff (organized by topics).

S. Grossman (1979), "C. G. Jung and National Socialism," *Journal of European Studies* 9:231–59.

Jan Ehrenwald (1976), *History of Psychotherapy* (New York: Aronson).

Vincenzo Cappelletti (1976), "The Latest Freud Research," *History of Science* 14:1–16, concerns the Freud-Jung letters, etc.

Topic *ii*: America

Michael Shortland (1987), "Screen Memories: Towards a History of

Psychiatry and Psychoanalysis in the Movies," *British Journal for the History of Science* 20:421–52.

Karen J. Winkler (1986), "Scholars Prescribe Freud's 'Talking Cure' for Problems," *Chronicle of Higher Education* 33, no. 8:4–6, reports that "while Freud's reputation is in serious decline among psychiatrists and psychologists, it is on the rise among literary and film critics, historians, anthropologists, and political scientists."

Rand B. Evans and William A. Koelsch (1985), "Psychoanalysis Arrives in America: The 1909 Psychology Conference at Clark University," *American Psychologist* 40:942–48.

Constance M. McGovern (1985), *Masters of Madness: Social Origins of the American Psychiatric Profession* (Hanover, N.H.: Univ. Press of New England).

Darrell Smith (1982), "Trends in Counseling and Psychotherapy," *American Psychologist* 37:802–9, reports a survey showing that the most influential psychotherapists are Carl Rogers, Albert Ellis, and Freud. Books considered "most representative of the current *Zeitgeist*" are *Handbook of Psychotherapy,* ed. S. L. Garfield et al. (2d ed. 1978), and *Multimodal Behavior Therapy* by A. A. Lazarus (1976).

J. C. Burnham (1979), "From Avant-garde to Specialism: Psychoanalysis in America," *Journal of the History of the Behavioral Sciences* 15:128–34.

Nathan G. Hale, Jr. (1978), "From Berggasse XIX to Central Park West: The Americanization of Psychoanalysis, 1919–1940," *Journal of the History of the Behavioral Sciences* 14:299–315.

John Demos (1978), "Oedipus and America: Historical Perspectives on the Reception of Psychoanalysis in the United States," *Annual of Psychoanalysis* 6:23–39.

Elizabeth M. R. Lomax, J. Kagan, and B. G. Rosenkrantz (1978), *Science and Patterns of Child Care* (San Francisco: Freeman), Chap. 3.

Jacques M. Quen and E. T. Carlson, eds. (1978), *American Psychoanalysis: Origins and Development* (New York: Brunner/Mazel).

5.4 FREUDIAN AND NEO-FREUDIAN THEORIES ABOUT WOMEN

i. Freud on the Oedipus complex, penis envy, and the inferiority of women
ii. Helene Deutsch
iii. Karen Horney
iv. Modern psychiatry

READINGS

Viola Klein, *The Feminine Character* (1972), 71–90, 163–82. Rona Cherry and Laurence Cherry, "The Horney Heresy," *New York Times Magazine,* 26 Aug. 1973, 12, 75–84. Thomson, *Pelican History,* 266–68 (on Horney). Miriam Levin, ed., *In the Shadow of the Past: Psychology Portrays the Sexes* (1984), articles by Barbara J. Harris, "The Power of the Past: History and the Psychology of Women," 1–25, and Gabriele Wickert, "Freud's Heritage: Fathers and Daughters in German Literature (1750–1850)," 26–38.

SYNOPSIS

As early as 1905 Freud had stated his basic assumptions about the early development of sexuality: first, that the child's discovery of the possession of a penis by boys but not by girls produces penis envy in girls and castration anxiety in boys and that these emotions condition the later development of the unconscious mind; second, that well before puberty, a child chooses the parent of the opposite sex as love object and the parent of the same sex as a rival (the *Oedipus complex* for boys, or *Electra complex* for girls). The theory was elaborated in later writings such as his *New Introductory Lectures on Psychoanalysis* (1937). A key postulate is that the girl must abandon the "masculine" sexuality associated with the clitoris to the more passive "feminine" sexuality centered in the vagina. The investigation of the alleged distinction between the clitoral and vaginal orgasm was a major focus of the research of William Masters and Virginia Johnson in the 1960s, and of considerable discussion in the recent Women's Liberation movement which regarded Freud as an historic enemy.

The famous phrase "anatomy is destiny," now taken to signify that a woman's personality and behavior patterns are completely determined by her biological nature (and in particular the fact that female sex organs are different from male) first appeared in Freud's 1912 paper, "The Most Prevalent Form of Degradation in Erotic Life," where it is described as a modification of "a well-known saying of Napoleon's." (The *Oxford Dictionary of Quotations,* however, attributes the phrase "character is destiny," which Freud apparently had in mind, to George Eliot, quoting Novalis.) In 1912 it seems to imply the animal nature of human sexuality, both male and female, a nature that is basically incompatible with the demands of civilization; but in 1924 it appears in something like its current meaning (paper on "The Passing of the Oedipus Complex").

Helene Deutsch (1884–1982) and Karen Horney (1885–1952) both started as Freudian analysts and then came to America. Deutsch remained a defender of Freudian theory in her treatise *Psychology of Women* (1944), though she rejected Freud's assumption that the girl has a phase of masculine sexuality before the development of penis envy; her concept of femininity was even more biologically determined than Freud's. She criticized the "masculinity complex" of women who reject their feminine role and try to develop independent careers in business or the professions, using their minds rather than their bodies. Deutsch provided the "scientific" basis for a number of popular books and articles in the 1940s and 1950s urging women not to compete with men but to accept their natural role as wife-mother. A good example is the book by Ferdinand Lundberg and Marynia Farnham, *Modern Woman: The Lost Sex* (1947).

A modern reprint of the writings of Clara M. Thompson (1893–1948), *On Women* (1971, selections from *Interpersonal Analysis*), touts her as a critic of Freud's negative attitudes toward women. But Thompson did not move much farther away from Freud than Lundberg and Farnham. While emphasizing the role of cultural conditioning she warns the woman with "average training and ability" not to compete with men, and characterizes the suffragettes and feminists as women who "usually [have] a hatred and contempt for man at the same time that they strove to be like him."

Horney, in her early writings, inverted Freudian doctrine and argued that men feel inferior because they cannot be mothers and fear women because they associate intercourse with death. But rather than continuing as an opponent of Freud, Horney found that experience with her American patients required her to develop a radically different theory in which anxiety rather than sexuality is the basis of neurosis. She credited Margaret Mead with recognizing the effect of culture on personality and noted the fallacy of generalizing from neurotic Austrians to all of humanity (see *The Neurotic Personality of Our Time,* 1937). The outcome of her work was not a new "psychology of women" but a system that did not depend on sex differences at all (*Neurosis and Human Growth,* 1950).

BIBLIOGRAPHY

Mary Anne Warren (1980), *The Nature of Woman: An Encyclopedia and Guide to the Literature* (Inverness, Calif.: Edgepress).

Topic *I:* Freud on the Oedipus Complex and on Women

Charles Bernheimer and Claire Kahane, eds. (1985), *In Dora's Case* (New York: Columbia Univ. Press).

Sarah Kofman (1985), *The Enigma of Woman* (Ithaca, N.Y.: Cornell Univ. Press).

Judith Van Herik (1982), *Freud on Femininity and Faith* (Berkeley: Univ. of California Press).

Lucy Freeman and Herbert S. Strean (1981), *Freud and Women* (New York: Ungar).

Janet Sayers (1979), "Anatomy is Destiny: Variations on a Theme," *Women's Studies International Quarterly* 2:19–32.

Janine Chasseguet-Smirgel and Burness E. Moore (1976), two articles on "Freud and Female Sexuality . . . " *International Journal of Psychoanalysis* 57:275–86, 287–30.

Topic *ii:* Deutsch

Paul Roazen (1985), *Helene Deutsch: A Psychoanalyst's Life* (New York: Doubleday).

Topic *iii:* Horney

(1980), *The Adolescent Diaries of Karen Horney* (New York: Basic Books).

Marcia Westkott (1986), *The Feminist Legacy of Karen Horney* (New Haven: Yale Univ. Press).

Dee Garrison (1981), "Karen Horney and Feminism," *Signs* 6:672–91.

Jack L. Rubins (1978), *Karen Horney: Gentle Rebel of Psychoanalysis* (New York: Dial Press).

Topic *iv:* Modern Psychiatry

Rachel T. Hare-Mustin (1983), "An Appraisal of the Relationship between Women and Psychotherapy: 80 Years after the Case of Dora," *American Psychologist* 38:593–601.

Jane Gallop (1983), *The Daughter's Seduction: Feminism and Psychoanalysis* (Ithaca, N.Y.: Cornell Univ. Press), is a survey of several recent theories.

Sandra Harding and Merrill B. Hintikka, eds. (1983), *Discovering Reality* (Boston: Reidel), contains articles by E. F. Keller, N. Scheman, J. Flaxx.

Margery Collins and Christine Pierce (1976), "Holes and Slime: Sexism in Sartre's Psychoanalysis," in *Women and Philosophy,* ed. Carol C. Gould and Marx W. Wartofsky (New York: Putnam), 112–27.

Jean Baker Miller, ed. (1973), *Psychoanalysis and Women* (New York: Brunner/Mazel), includes extracts from K. Horney, A. Adler, Clara Thompson, many others.

5.5 IS PSYCHOANALYSIS A SCIENTIFIC THEORY?

i. Popper's falsifiability criterion and philosophical discussions
ii. Empirical tests
iii. Effectiveness of therapy

READINGS

Karl Popper, *Conjectures and Refutations* (1962), 33–39. Elliott Marshall, "Psychotherapy Faces Test of Worth," and "Psychotherapy Works, but for Whom?" *Science* 207(1980):35–36, 506–8. Adolf Grünbaum, "Freud's Theory: The Perspective of a Philosopher of Science," *Proceedings and Addresses of the American Philosophical Association* 57(1983):5–32. Nikki Meredith et al., "Therapy under Analysis," *Science 86,* 7, no. 5(1986):29–52.

SYNOPSIS

In most parts of a history of science course one is dealing with theories that may be more or less valid, but there is no doubt that they are scientific even if they have been refuted. But when one gets into the behavioral and social sciences, the question sometimes arises: is this science at all? Rather than omit such dubious cases from the course, I would argue that a major function of a course on history of science is to convey some idea of what it means, and has meant in the past, to *be* scientific. For this purpose, controversial cases like psychoanalysis are well suited.

Karl Popper, one of the most influential modern philosophers of science, proposed a "criterion of demarcation" between science and pseudoscience that seems to have been explicitly designed to put psychoanalysis on the side of pseudoscience. In a 1953 lecture (part of which is recommended reading for this section), Popper recalls how, as a student in Vienna after World War I, he was very impressed by the confirmation of Einstein's general relativity theory by eclipse observations (**8.7**)—not that the theory was necessarily true (for its postulates were still very hard to believe) but that it could be submitted to a clear-cut test. By contrast, three other "revolutionary" theories of the day—Marx's theory of history, Freud's psychoanalysis, and Alfred Adler's "individual psychology"—were so flexible that they could explain everything. Popper worked with Adler for a short period in 1919 and reported to him a case "which to me did not seem particularly Adlerian, but which he found no difficulty in analysing in terms of his theory of inferiority feelings." Yet

Popper thought that any case could be explained equally well by Freudian or Adlerian theory; hence there was no way to choose between them or to test either of them. From this experience Popper developed his thesis that a scientific theory must be capable of disproof by some specified experiment whose result is not already known. A theory that is so flexible it can explain anything may or may not be *true,* but it is not scientific.

One might imagine three kinds of response to Popper's claim that psychoanalysis is not a scientific theory. (1) To accept the validity of Popper's criterion and design tests of particular hypotheses; this has been done in a number of cases. (2) To admit that psychoanalysis is not a scientific theory but claim that it is instead a therapy that has to be judged by its success in helping mentally ill people. This claim leads to a different kind of research, which must of course involve a "control group" of patients with similar symptoms who are treated by another kind of therapy or not treated at all; and some critics claim that the no-treatment alternative is just about as effective ("spontaneous remission" of symptoms). (3) To dispute the validity of the criterion on the grounds that even theories generally accepted as scientific are excluded by it. As we have seen in **4.8**, Popper used the same criterion to argue (until recently) that Darwinism is not a testable scientific theory, an argument gleefully adopted by the creationists; this fact may warn against accepting a simpleminded version of the falsifiability criterion. More generally, any theory that deals with phenomena that take place over large domains of space and time, which thus cannot be brought into the laboratory for controlled experimentation, would fail to meet Popper's test (e.g., geology, astronomy).

The issue has been raised again recently because of proposals that public and private health insurance plans pay the cost of psychotherapy, and the demand by the federal government that this form of treatment be shown to be scientifically valid, to justify expenditure of tax dollars on it. Marshall's articles (listed above) report an early stage of the controversy.

BIBLIOGRAPHY

Topic *i:* Popper's Criterion and Philosophical Discussions

Adolf Grünbaum (1986), "Precis of *The Foundations of Psychoanalysis: A Philosophical Critique,*" *Behavioral and Brain Sciences* 9:217–28, with commentary by A. Caplan, K. Popper, et al., 228–66, and author's response, 266–81; (1984), *The Foundations of Psychoanalysis: A Philosophical Critique* (Berkeley: Univ. of California Press); (1984), "The Hermeneutic Construal of Psychoanalytic Theory and Therapy: An Ill-Conceived Paradigm for the Human Sciences," in *Science and Reality*, ed. J. T. Cushing et al. (Notre Dame, Ind.: Univ. of Notre Dame Press), 54–82; (1979), "Is Freudian Psychoanalytic Theory Pseudoscientific by Karl Popper's Criterion of Demarcation?" *American Philosophical Quarterly* 16:131–41; (1978), "Psychological Explanations for the Rejection or Acceptance of Scientific Theories," *Humanities in Society* 1:293–304; (1977, 1978) "Is Psychoanalysis a Pseudoscience? Karl Popper versus Sigmund Freud," *Zeitschrift für Philosophische Forschung* 31(1977):333–53, 32(1978):49–69; (1976) "Is Falsifiability the Touchstone of Scientific Rationality? Karl Popper versus Inductivism," in *Essays in Memory of Imre Lakatos*, ed. R. S. Cohen et al. (Boston: Reidel), 213–52.

Larry Laudan, ed. (1983), *Mind and Medicine: Problems of Explanation and Evaluation in Psychiatry and the Biomedical Sciences* (Berkeley: Univ. of California Press), contains articles by A. Grünbaum, M. S. Moore, and M. N. Eagle.

Morris N. Eagle (1983), "The Epistemological Status of Recent Developments in Psychoanalytic Theory," in *Physics, Philosophy, and Psychoanalysis*, ed. R. S. Cohen and L. Laudan (Boston: Reidel), 31–55.

V. L. Jupp (1977), "Freud and Pseudo-Science," *Philosophy* 52:441–53.

Topic *ii:* Empirical Tests

Saul Rosenzweig (1985), "Freud and Experimental Psychology: The Emergence of Idiodynamics," in *A Century of Psychology as Science*, ed. S. Koch and D. Leary (New York: McGraw-Hill), 135–207.

Michael T. Motley (1985), "Slips of the Tongue," *Scientific American* 253, no. 3:116–27, notes that modern interest started with research by Victoria A. Fromkin in the 1960s; discusses validity of the "Freudian slip" concept.

Gerald W. Vogel (1978), "An Experimental Examination of Freud's Dream Theory," *Scientia* 113:519–40, concludes "that Freud's evidence was anecdotal rather than systematic" and his theory "remains a series of untested hypotheses"—but "modern empirical laboratory studies have ignored Freud's hypotheses and have failed—as he did not—to answer the central questions of dream psychology."

[See also the books by S. Fisher and R. P. Greenberg cited below.]

Topic *iii:* Effectiveness of Therapy

Patrick H. De Leon, Gary R. Van den Bos, and Nicholas A. Cummings (1984), "Psychotherapy—Is It Safe, Effective, and Appropriate? The Beginning of an Evolutionary Dialogue," *American Psychologist* 38:907–11, is followed by seven articles on public policy aspects of mental health care by Senator Daniel K. Inouye and others.

Kenneth F. Schaffner (1983), "Clinical Trials: The Validation of Theory and Therapy," in *Physics, Philosophy, and Psychoanalysis,* ed. R. S. Cohen and L. Laudan (Boston: Reidel), 191–208.

Edward Erwin (1980), "Psychoanalytic Therapy: The Eysenck Argument," *American Psychologist* 35:435–43.

Seymour Fisher and Roger P. Greenberg, eds. (1978), *The Scientific Evaluation of Freud's Theories and Therapy. A Book of Readings* (New York: Basic Books); (1977), *The Scientific Credibility of Freud's Theories and Therapy* (New York: Basic Books).

6 BEHAVIOR AND INTELLIGENCE

6.1 AMERICAN PSYCHOLOGY IN 1900

i. The European legacy: Wilhelm Wundt and the founding of scientific psychology
ii. Psychology comes to America: William James, Hugo Muensterberg, E. B. Titchener, J. M. Cattell
iii. G. Stanley Hall: the concept of adolescence

READINGS

John Marks, *Science* (1983), 480–88 (general background on social sciences). Raymond Fancher, *Pioneers of Psychology* (1979), 126–69. Thomas Leahey, *History of Psychology* (1979), 185–213, 261–68. R. Thomson, *Pelican History of Psychology* (1968), 67–74 (on *i*), 125–67 (on *ii, iii*). Robert I. Watson, Sr., *The Great Psychologists* (1977), 275–95, 367–434. David Hothersall, *History of Psychology* (1984), 83–129, 244–71. Merle Curti, *Human Nature* (1980), 186–215. William Kessen and Emily D. Cahan, "A Century of Psychology: From Subject to Object to Agent," *American Scientist* 74(1986):640–49.

Most courses on history of science do not cover psychology at all; those that do probably devote more time to European psychology before 1900. I favor the texts by Fancher, Leahey, and Thomson because, though they do deal with the earlier history, they include substantial sections on the topics in which I am interested. S. J. Gould's *Mismeasure of Man* (1981), though polemical in tone, is extremely useful for **6.4–6, 6.8** and for the earlier sections on racism.

SYNOPSIS

"Wundt is the founder [of psychology] because he wedded physiology to philosophy and made the outcome independent" (Leahey, 188). He is also the only psychologist (other than Freud) on whom a substantial secondary literature now exists. Thus I would spend a little time on Wundt and then go directly to the founders of twentieth-century American psychology—James, Muensterberg, Titchener, and Cattell.

G. Stanley Hall (1844–1924) is less important that James as a major thinker in psychological theory but had a more direct impact on American society, through (1) his concept of adolescence; (2) introducing psychoanalysis; (3) promoting the applications of psychology to education, and (4) helping psychology get started as a professional discipline in the United States. He studied with James at Harvard and obtained the first Ph.D. in psychology in the United States in 1878. He discovered the great potential demand for psychology in education and was a founder of the "child study" movement. In 1887 he started the *American Journal of Psychology,* which he restricted to experimental papers—he criticized the philosophical writings of other psychologists including James. His laboratory at the Johns Hopkins University was the first in the United States (except for that of James at Harvard, which was not used for teaching purposes). In 1892 he organized the American Psychological Association. Thus for a while it appeared that Hall would be the leader of the profession as it turned away from its philosophical background and became more "scientific" (i.e., experimental).

In 1888 Hall accepted the invitation to become founder-president of a new university being started in Worcester, Massachusetts, by the initiative of a wealthy merchant, Jonas Clark. He was able to assemble a first-rate faculty, including the physicist A. A. Michelson (**8.5**) and the anthropologist Franz Boas (**3.4**). The psychology department, dominated by Hall himself, was especially strong and turned out most of the American Ph.D.s in this field before 1900. But difficulties soon developed at Clark University because the financial support for the new institution was not forthcoming from the business community in the area. Hall became embroiled in controversies with other psychologists but continued to do important research in child development.

Hall's *magnum opus* was his two-volume treatise on *Adolescence* (1904). Although Hall did not invent the word as is sometimes claimed, he established its modern usage and launched the vogue of interest in youth, which has dominated much of twentieth-century American culture. Hall called it the "paradise of the race"—the time of sudden expansion of interest in feelings, of turning from the self to others—but also a time of great "storm and stress," of emotional instability caused by the conflict of emerging impulses, especially sexuality. He applied the concept in his theories of race and sex difference. Thus "woman at her best never outgrows adolescence as man does, but lingers in, magnifies and glorifies this culminating stage of life with its all-sided interests, its convertibility of emotions, its enthusiasm, and zest for all that is good, beautiful, true, and heroic." He discussed the status of "adolescent races," using the "recapitulation" model for the relation between the evolution of the race and the growth of the individual.

Hall helped prepare the ground for the introduction of Freud's theories in America by his willingness to talk and write about sex and his encouragement of the sex education movement around 1900. He anticipated many of Freud's ideas about repression and sublimation of the sexual instincts though he did not make a coherent system out of them. In 1909, for the twentieth anniversary celebration of Clark University, Hall invited Freud and Jung to lecture. Freud thus received his first recognition from the academic world and, thanks to Hall, was made to appear in America as a much more respectable and eminent figure than he did at the time in Europe. As a result of personal contacts with Freud, a few key American psychiatrists such as James Jackson Putnam became enthusiastic converts.

Two of Hall's students played major roles in modern behavioral science: Lewis M. Terman (**6.5**) and Arnold Gesell, who carried on Hall's child study program.

BIBLIOGRAPHY

Georg Eckardt, Wolfgang G. Bringmann, and Lothar Sprung, eds. (1985), *Contributions to a History of Developmental Psychology: International Wilhelm Preyer Symposium* (Amsterdam/Berlin/New York: Mouton).

Gregory A. Kimble and Kurt Schlesinger, eds. (1985), *Topics in the History of Psychology* (Hillsdale, N.J.: Erlbaum).

Kurt Danziger (1985), "The Origins of the Psychological Experiment as a Social Institution," *American Psychologist* 40:133–40.

Josef Brozek and Ludwig J. Pongratz, eds.(1980), *Historiography of Modern Psychology* (Toronto: Hogrefe), contains essays by W. Woodward and M. Wertheimer.

Wayne Viney, Michael Wertheimer, and Marilyn Lou Wertheimer (1979), *History of Psychology: A Guide to Information Sources* (Detroit: Gale Research).

Robert Watson has compiled extensive bibliographies, published as (1978), *The History of Psychology and the Behavioral Sciences: A Biblio-*

graphic Guide (New York: Springer) and (1974, 1976), *Eminent Contributors to Psychology* (New York: Springer). The former is shorter but has annotations; the second is a two-volume list of primary and secondary references, arranged by name of psychologist. He has written on American historiography of psychology in his (1977), *Selected Papers on the History of Psychology* (Hanover, N.H.: Univ. Press of New England), 25–94.

E. G. Boring (1950), *A History of Experimental Psychology,* 2d ed. (New York: Appleton-Century-Crofts), provides good perspective on this period, though superseded in many details by recent scholarship. See Chap. 16 (Wundt), 410–20 (Titchener), Chap. 21 (American pioneers—James, Hall, Cattell), Chap. 22 (functionalism at Chicago and Columbia, educational psychology).

Topic *i:* European Legacy; Wundt

Kurt Danziger (1983), "Origins of the Schema of Stimulated Motion: Towards a Pre-History of Modern Psychology,"*History of Science* 21:183–210; (1979), "The Positivist Tradition of Wundt," *Journal of the History of the Behavioral Sciences* 15:205–30.

William R. Woodward and Mitchell G. Ash, eds. (1982), *The Problematic Science: Psychology in Nineteenth-Century Thought* (New York: Praeger), has articles by R. J. Richards, M. Marshall, L. Daston, R. S. Turner, F. J. Sulloway.

David E. Leary (1979), "Wundt and After: Psychology's Shifting Relations with the Natural Sciences, Social Sciences, and Philosophy," *Journal of the History of the Behavioral Sciences* 15:231–41.

Wolfgang G. Bringmann and Ryan D. Tweney, eds. (1980), *Wundt Studies: A Centennial Collection* (Toronto: Hogrefe).

R. W. Rieber, ed. (1980), *Wilhelm Wundt and the Making of a Scientific Psychology* (New York: Plenum).

Mitchell G. Ash (1980), "Academic Politics in the History of Science: Experimental Psychology in Germany, 1879–1941," *Central European History* 13:255–86.

Conrad G. Mueller (1979), "Some Origins of Psychology as a Science," *Annual Review of Psychology* 30:9–29, argues that psychology was founded by Wundt at Leipzig in 1879, but notes the paradox that almost all of the experimental methods proposed in the late nineteenth century are now acceptable *except* that of Wundt.

Topic *ii:* Americans

Ernest Hilgard (1987), *Psychology in America* (San Diego: Harcourt Brace Jovanovich).

William James (1984), *Essays in Psychology,* reprint with new introduction by W. R. Woodward (Cambridge: Harvard Univ. Press); (1984), *Psychology: Briefer Course,* reprint with introduction by M. M. Sokal (Cambridge: Harvard Univ. Press); (1981), *Principles of Psychology,* reprint with introduction by R. Evans and G. E. Myers (Cambridge: Harvard Univ. Press); (1977), *The Writings of William James,* ed. J. J. McDermott (Chicago: Univ. of Chicago Press).

Josef Brozek, ed. (1984), *Explorations in the History of Psychology in the United States* (Lewisburg, Pa.: Bucknell Univ. Press), has articles by R. B. Evans, W. R. Woodward, J. A. Popplestone and M. W. McPherson, and M. M. Sokal.

David W. Bjork (1983), *The Compromised Scientist: William James in the Development of American Psychology* (New York: Columbia Univ. Press).

Mark R. Schwehn et al. (1982), "A William James Renaissance: Four Essays by Young Scholars," *Harvard Library Bulletin* 30:367–480.

Michael M. Sokal (1982), "James McKeen Cattell and the Failure of Anthropometric Mental Testing, 1890–1901," in *The Problematic Science,* ed. W. R. Woodward and M. G. Ash (New York: Praeger); Sokal, ed. (1981), *An Education in Psychology: James McKeen Cattell's Journal and Letters from Germany and England, 1880–1888* (Cambridge: MIT Press).

Donald S. Napoli (1981), *Architects of Adjustment: The History of the Psychological Profession in the United States* (Port Washington, N.Y.: Kennikat Press).

Matthew Hale, Jr. (1980), *Human Science and Social Order: Hugo Muensterberg and the Origins of Applied Psychology* (Philadelphia: Temple Univ. Press).

K. Danziger (1979), "The Social Origins of Modern Psychology," in *Psychology in Social Context,* ed. A. R. Buss (New York: Irvington), 27–45.

Hamilton Cravens (1978), *The Triumph of Evolution* (Philadelphia: Univ. of Pennsylvania Press), 56–78.

Dorothy Ross (1978), "American Psychology and Psychoanalysis: William James and G. Stanley Hall," in *American Psychoanalysis: Origins and Development,* ed. Jacques M. Quen and E. T. Carlson (New York: Brunner/Mazel), 38–51.

Topic *iii:* Hall

Stewart H. Hulse and Bert F. Green, Jr., eds. (1986), *One Hundred Years of Psychological Research in America: G. Stanley Hall and the Johns Hopkins Tradition* (Baltimore: Johns Hopkins Univ. Press).

J. G. Morawski (1982), "Assessing Psychology's Moral Heritage through Our Neglected Utopias," *American Psychologist* 37:1082–95, focuses on proposals by Hall and others for the role of psychology in the betterment of American society.

Philip J. Pauly (1979), "Psychology at Hopkins: Its Rise and Fall and Rise and Fall and . . . " *Johns Hopkins Magazine* 30, no. 6:36–41, discusses Hall, J. M. Baldwin, J. B. Watson, K. Dunlap.

David Muschinske (1977), "The Nonwhite as Child: G. Stanley Hall on the Education of Nonwhite Peoples," *Journal of the History of the Behavioral Sciences* 13:328–36.

Dorothy Ross (1972), *G. Stanley Hall: The Psychologist as Prophet* (Chicago: Univ. of Chicago Press).

6.2 WATSON'S BEHAVIORISM

i. Darwinian psychology: humans as animals. Ivan Pavlov and the "conditioned reflex"
ii. Life and work of J. B. Watson
iii. The Progressive movement: use of science for social control

READINGS

Fancher, *Pioneers of Psychology, 295–338. Leahey, History of Psychology,* 245–60, 280–97. Hothersall, *History of Psychology,* 333–65. Thomson, *Pelican History of Psychology,* 92–102, 160–67, 225–28. Watson, *Great Psychologists,* 435–66. Curti, *Human Nature,* 372–406.

Article: T. E. Weckowicz and Helen Liebel-Weckowicz, "Typologies of the Theory of Behaviorism since Descartes," *Sudhoffs Archiv* 66 (1982): 129–51.

SYNOPSIS

In one sense behaviorism is a reversion to the Newtonian clockwork-universe philosophy as applied by John Locke and some French Enlightenment philosophers to human nature. In particular, it seems to claim that "man is a machine"—not a machine assembled by heredity but rather one whose operation is almost entirely determined by environment. Moreover, while some behaviorists may assert that human behavior can be entirely explained on a biological-physical-chemical basis without introducing any special "mental" entities, they do not seem to be talking about the kind of machine that can be taken apart and analyzed in terms of its internal mechanisms, but rather a "black box" which is characterized only by "input-output" functions. Contrary to the classical mechanistic view of the natural world, behaviorists adopt the positivistic viewpoint advocated by Comte and Ernst Mach in the nineteenth century: one should not postulate theoretical entities that cannot be directly observed.

A more immediate influence on the origin of behaviorism was the adoption of the evolutionary worldview in the nineteenth century. Not only did Darwin's theory of evolution by natural selection encourage a more "naturalistic" (i.e., nonreligious, nonmetaphysical) approach to the study of man; it also promoted much greater interest in scientific studies of animals. If man is to be regarded as the highest form of animal life rather than as a completely different form of life, then it may be possible to derive plausible conclusions about humans from results established by systematic study of lower animals. Darwin himself was interested in the psychology of animals (e.g., *Expression of the Emotions in Man and Animals,* 1872).

At the beginning of the twentieth century American psychology was dominated by William James and E. B. Titchener with their method of "introspection," which relied on the human subject's reports of his/her own mental states (**6.1**-*ii*). This technique struck some psychologists in the younger generation as being too "subjective" and hence not properly scientific. Moreover, it did not seem to fit in very well with the study of animal psychology. Thus the Darwinian hypothesis of continuity between animals and humans led to a conflict between psychologists who claimed that psychology is the study of consciousness and psychologists who wanted to study animal and human behavior from a unified viewpoint.

Behaviorism began with J. B. Watson's 1913 paper "Psychology as the Behaviorist Views It." That title is continued in the text as follows: ". . . is a purely objective experimental branch of natural science. Its theoretical goal is the prediction and control of behavior. Introspection forms no essential part of its methods, nor is the scientific value of its data dependent upon the readiness with which they lend themselves to interpretation in terms of consciousness. The behaviorist, in his efforts to get a unitary scheme of animal response, recognizes no dividing line between man and brute."

While granting to the new-born baby a few "unlearned beginnings" (in place of the "instincts" postulated by James and other psychologists), Watson gave primary importance to conditioned stimulus-response relations like those studied by the Russian physiologist Ivan Pavlov.

In 1920 Watson, married since 1903, fell in love with his laboratory assistant Rosalie Rayner. The resulting divorce scandal led to his involuntary resignation from the faculty at the Johns Hopkins University, but Watson quickly found a new career in the advertising business. By 1924 he was vice-president of the J. Walter Thompson company, using his psychological expertise to sell rubber boots and Yuban coffee. Having insisted from the beginning that control of behavior is a major goal of his science, he found the advertising business as congenial as pure research but continued to publish books and articles on behaviorism. Meanwhile his new wife Rosalie regaled the readers of *Parents Magazine* with accounts of her attempts to apply behaviorist techniques to raising their sons.

Watson never seemed to have any qualms about putting science at the disposal of those who wanted to manipulate the behavior of others for public or private purposes. On the contrary, he was in tune with many contemporary "reform" movements such as Prohibition and Eugenics (**3.3**), which were advocating the control of individual behavior on behalf of the greater good of society, and with other psychologists who were promoting the application of intelligence tests to the classification of students and workers (**6.5**). This was all part of the "Progressive movement" in the first quarter of the twentieth century.

BIBLIOGRAPHY

John M. O'Donnell (1985), *The Origins of Behaviorism: American Psychology, 1870–1920* (New York: New York Univ. Press).

John Burnham (1985), "The Encounter of Christian Theology with Deterministic Psychology and Psychoanalysis," *Bulletin of the Menninger Clinic* 49:321–52.

Laurence D. Smith (1981), "Psychology and Philosophy: Toward a Realignment, 1905–1935," *Journal of the History of the Behavioral Sciences* 17:28–37.

Topic *i:* Darwinian Psychology

Robert Boakes (1984), *From Darwin to Behaviorism: Psychology and the Minds of Animals* (New York: Cambridge Univ. Press).

George Windholz (1983), "Pavlov's Position toward American Behaviorism," *Journal of the History of the Behavioral Sciences* 19:394–407.

R. J. Richards (1982), "Darwin and the Biologizing of Moral Behavior," in *The Problematic Science,* ed. W. R. Woodward and M. G. Ash (New York: Praeger), 43–64.

George E. Gifford, Jr., et al. (1981), "Pavlov's Legacy to Behavioral

Psychology, Physiology, and Psychiatry: A Program about Pavlov," *Journal of the History of the Behavioral Sciences* 17:236–50.

Philip J. Pauly (1981), "The Loeb-Jennings Debate and the Science of Animal Behavior," *Journal of the History of the Behavioral Sciences* 17:504–15.

Brian MacKenzie (1976), "Darwinism and Positivism as Methodological Influences on the Development of Psychology,"*Journal of the History of the Behavioral Sciences* 12:330–37.

[See also **4.5-*i*** for references on the mechanistic conception of life; Watson was probably influenced by Jacques Loeb at Chicago.]

Topic *ii:* Watson

J. B. Watson (1936), chapter in *A History of Psychology in Autobiography,* ed. Carl Murchison (New York: Russell and Russell/Atheneum), 3:271–82.

Rosalie Rayner Watson (1930), "I Am the Mother of a Behaviorist's Sons," *Parents Magazine* 5, no. 12:16ff.

Ben Harris (1984), " 'Give Me a Dozen Healthy Infants . . .': John B. Watson's Popular Advice on Childbearing, Women, and the Family," in *In the Shadow of the Past,* ed. M. Lewin (New York: Columbia Univ. Press), 126–54.

Kerry W. Buckley (1982), "The Selling of a Psychologist: John Broadus Watson and the Application of Behavioral Techniques to Advertising," *Journal of the History of the Behavioral Sciences* 18:207–21; (1982), "Behaviorism and the Professionalization of American Psychology: A Study of John Broadus Watson, 1878–1958" (Ph.D. diss., University of Massachusetts).

J. G. Morawski (1982), "Assessing Psychology's Moral Heritage through Our Neglected Utopias," *American Psychologist* 37:1082–95, discusses proposals of Watson and others for improving society.

Franz Samelson (1981), "Struggle for Scientific Authority: The Reception of Watson's Behaviorism, 1913–1920," *Journal of the History of the Behavioral Sciences* 17:399–425, analyzes why it was *not* rapidly accepted; there was no paradigm-exemplar; psychology had to accept social control as its goal.

David Cohen (1979), *J. B. Watson: The Founder of Behaviourism* (Boston: Routledge and Kegan Paul).

Elizabeth M. R. Lomax, J. Kagan, and B. G. Rosenkrantz (1978), *Science and Patterns of Child Care* (San Francisco: Freeman), Chap. 4.

David Cornwell and Sandy Hobbs (1976), "The Strange Saga of Little Albert," *New Society* 35:602–4, concerns a well-known experiment by Watson; the following articles continue this discussion: Ben

Harris (1979), "Whatever Happened to Little Albert?" *American Psychologist* 34:151–60, treats the experiment as "an example of myth making"; Franz Samelson (1980), "J. B. Watson's Little Albert, Cyril Burt's Twins, and the Need for a Critical Science," *American Psychologist* 35:619–25, discusses the issue raised by Harris in a broader context; see the comment by H. J. Eysenck (1981), "On the Fudging of Scientific Data," *American Psychologist* 36:692 and Samelson's reply, ibid.

William Harrell and Ross Harrison (1938), "The Rise and Fall of Behaviorism," *Journal of General Psychology* 18:367–421, claims that although some ideas of behaviorism have been absorbed into the mainstream of psychology, it is no longer a major force. Note that 1938 was the year of publication of B. F. Skinner's first comprehensive work, *The Behavior of Organisms* (New York: Appleton Century Crofts).

Topic *iii:* Progressive Movement

David Bakan (1966), "Behaviorism and American Urbanization," *Journal of the History of the Behavioral Sciences* 2:5–28.

J. C. Burnham (1960), "Psychiatry, Psychology and the Progressive Movement," *American Quarterly* 12:457–65.

6.3 SKINNER'S BEHAVIORISM

i. Life and work of B. F. Skinner; relation to logical positivism and P. W. Bridgman's operationism
ii. Applications to education: the teaching machine, programmed instruction, Keller method
iii. Behavior modification: therapy and social control; ethical problems in the use of science
iv. Critiques of behaviorism, emergence of new paradigm(s)

READINGS

Fancher, *Pioneers,* 355–71. Leahey, *History,* 303–51. Hothersall, *History of Psychology,* 394–407. Richard Olson, *Science as Metaphor* (1971), 244–53 (extract from Skinner's *Science and Human Behavior*). A. R. Gilgen, *American Psychology since World War II* (1982), 87–95 ("The Hullian Era, 1940–60") and 97–110 (Skinner). Thomson, *Pelican History,* 230–33, 298–300. Curti, *Human Nature,* 372–406. T. H. Leahey, "Behaviorism," in *Encyclopedia of Psychology,* ed. R. J. Corsini (1984), 1:131–33.

Skinner owes much of his influence to his effective writing style—

his undergraduate short stories were praised by Robert Frost—and I would therefore recommend that one of his articles be included on the reading list for students. (See for example the selection in *Cumulative Record.*) He has published an extract from his autobiography, "Origins of a Behaviorist," in *Psychology Today* 17, no. 9(1983):22–33. See also "My Experience with the Baby Tender," ibid. 12, no. 10(1979):29–40.

SYNOPSIS

According to his own account Skinner read the works of Jacques Loeb and Ivan Pavlov on physiological reactions, but it was especially a series of articles by Bertrand Russell, criticizing J. B. Watson's ideas, that got him interested in behaviorism. (Recall that Charles Darwin first started thinking seriously about evolution when he read Lyell's critique of Lamarck.) As a graduate student at Harvard (Ph.D. 1931) Skinner came in contact not only with contemporary psychological thinking but also with the positivistic philosophy associated with developments in physics (see **11.1**). He read Mach and Henri Poincaré, and presumably was influenced by Harvard physicist P. W. Bridgman's *The Logic of Modern Physics* (1927). Bridgman argued that one should define physical concepts strictly in terms of operations by which they could be measured, and this became known as *operationism.* In this environment Skinner and his fellow graduate students such as Fred Keller found ample justification for the behavioristic viewpoint, which rejected mental states that are not accessible to direct observation by the psychologist.

Going beyond Watson who stressed the relation between stimulus and response, Skinner pointed to behavior not elicited by any external stimulus but "spontaneously emitted" by the organism and then conditioned by the response of the environment. He called this *operant behavior* and devoted himself to detailed investigations of *operant conditioning*—one waits for the organism to "emit" the response at random and then reinforces it positively or negatively. The analogy with Darwinian natural (or artificial) selection is obvious.

Shortly after World War II Skinner's name started to appear in the popular press in connection with military technology, child care, and utopian literature. He had developed a method of training pigeons to guide missiles toward their targets but it was never actually used because military experts thought it was too ridiculous. He designed an air-conditioned temperature-controlled box when his second daughter, Deborah, was born, and thus became known as the inventor of the "Skinner box" for child rearing. He also composed a novel, *Walden Two,* about a communal society operated according to behavioristic principles; about a decade after its publication in 1948, the book suddenly started to pick up sales and a few people actually started communes based on it.

In 1953 Skinner visited his daughter's fifth-grade arithmetic class and was appalled at the teaching methods used—"I saw minds being destroyed," he told a reporter years later. Having succeeded in teaching pigeons to play Ping-Pong, he decided to use the same methods to teach academic subjects to children. The result was the "teaching machine" and the system of "programmed instruction," which was rapidly developed and adopted in many American schools during the 1960s. Another behavioristic method was developed by Fred Keller and became popular in college courses in the 1970s under the initials PSI (for "personalized system of instruction").

During the 1960s and early 1970s when behaviorism was being applied to education and tried out as a substitute for psychotherapy ("behavior mod"), its scientific basis was being attacked as a result of new research in psychology. Many psychologists now say that it has been replaced by "cognitive science," a synthesis of cognitive psychology and computer models of the brain.

BIBLIOGRAPHY

Topic *i*: Skinner

B. F. Skinner (1976), *Particulars of My Life* (New York: Knopf); (1979), *The Shaping of a Behaviorist* (New York: Knopf); (1983), *A Matter of Consequences* (New York: Knopf), is an autobiography in three volumes (so far). See also his (1980), *Notebooks: B. F. Skinner,* ed. Robert Epstein (Englewood, N.J.: Prentice-Hall), and (1972), *Cumulative Record: A Selection of Papers* (New York: Appleton-Century-Crofts).

Laurence D. Smith (1986), *Behaviorism and Logical Positivism: A Reassessment of the Alliance* (Stanford, Calif.: Stanford Univ. Press), reinterprets the neobehaviorists C. L. Hull, E. C. Tolman, and B. F. Skinner as drawing from pragmatism and neorealism and as relatively impervious to logical positivism.

Kristjan Gudmundsson (1984), "The Emergence of B. F. Skinner's Theory of Operant Behavior: A Case Study in the History and Philosophy of Science" (Ph.D. diss., University of Western Ontario).

Eckardt Scheerer (1983), *Die Verhaltenslehre* (New York: Springer-Verlag), gives a historical-philosophical analysis of Skinner's works.

William R. Woodward (1982), "The 'Discovery' of Social Behaviorism and Social Learning Theory, 1880–1980," *American Psychologist* 37:396–410, argues that the contention of two modern neobehaviorisms for dominance indicates a widespread agreement on three principles of learning—the principles of contiguity, effect, and imitation. All three are traced from Freud and James in the nineteenth century.

S. R. Coleman (1981), "Historical Context and Systematic Functions of the Concept of the Operant," *Behaviorism* 9:207–26.

Harold J. Allen (1980), "P. W. Bridgman and B. F. Skinner on Private Experience," *Behaviorism* 8:15–29, concerns the influence of P. W. B.'s *Logic of Modern Physics* (1927) on Skinner.

R. J. Herrnstein (1977), "The Evolution of Behaviorism," *American Psychologist* 32:593–603; reply by B. F. Skinner, "Herrnstein and the Evolution of Behaviorism," ibid., 106–12; rejoinder by Herrnstein, ibid., 1013–16.

Topic *ii:* Education

B. F. Skinner (1984), "The Shame of American Education," *American Psychologist* 39:947–54.

Paul Chance (1984), "The Revolutionary Gentleman: At 85 He Fights What He Fears Is a Losing Battle against Educational Dogma," *Psychology Today* 18, no. 9: 42–48 (Sept. 1984), reports an interview with Fred Keller.

Topic *iii:* Behavior Modification

Alan E. Kazdin (1978), *History of Behavior Modification: Experimental Foundations of Contemporary Research* (Baltimore: University Park Press).

Harvey Wheeler, ed. (1973), *Beyond the Punitive Society. Operant Conditioning: Social and Political Aspects* (San Francisco: Freeman), contains assessments of Skinner's work by J. R. Platt, A. Toynbee, M. Black, A. Jensen, etc., followed by Skinner's reply.

Topic *iv:* Critiques of the Theory

Barry Gholson and Peter Barker (1985), "Kuhn, Lakatos, and Laudan: Applications in the History of Physics and Psychology," *American Psychologist* 40:755–69, conclude that the battle between behavioristic and cognitive learning theories since 1930 is better described by the models of Lakatos and Laudan than by "popularized Kuhnian versions" (see **11.5**).

Amedeo Giorgi (1982), "Issues Relating to the Meaning of Psychology as a Science," in *Contemporary Philosophy: A New Survey,* vol. II, *Philosophy of Science,* ed. G. Fløistad (Boston: Nijhoff), 317–42.

John Caiazza (1981), "Analyzing the Social 'Scientist'," *Intercollegiate Review* 16, no. 2:91–98, is a conservative critique of social sciences and behaviorism.

Wayne Viney et al. (1979), *History of Psychology: A Guide to Information Sources* (Detroit: Gale Research), 169–73, contains an annotated list of thirty-seven critiques of behaviorism.

Leonard Gardner (1979), "Behaviorism and Dynamic Psychology: Skinner and Freud," *Psychoanalytic Review* 66:253–62, contrasts the "bankruptcy of behaviorism" with the vast influence of psychoanalysis.

Lewis W. Brandt (1979), "Behaviorism: The Psychological Buttress of Late Capitalism," in *Psychology in Social Context,* ed. A. R. Buss (New York: Irvington), 77–99, comments on the critiques by K. Holzkamp (1964–73), A. Mitscherlich (1954–75), K. Lorenz (1972–74).

Allan R. Buss (1978), "The Structure of Psychological Revolutions," *Journal of the History of the Behavioral Sciences* 14:57–64, discusses four revolutions: behavioristic, cognitive, psychoanalytic, humanistic.

Brian MacKenzie (1977), *Behaviorism and the Limits of Scientific Method* (Atlantic Highlands, N.J.: Humanities Press), argues that behaviorism is not a Kuhnian paradigm and, for philosophical reasons, was destined to fail. Leahey (1980) calls this book "the best single account of the conceptual origins of behaviorism."

J. R. Kantor (1976), "Behaviorism, Behavior Analysis, and the Career of Psychology," *Psychological Record* 26:305–12, is a brief defense of behaviorism against its critics.

David S. Palermo (1971), "Is a Scientific Revolution Taking Place in Psychology?" *Science Studies* 1:135–55, attempts to apply Kuhn's theory to the rise and fall of behaviorism. See comments on this article by Neil Warren and L. B. Briskman in ibid. 1(1971): 407–13 and 2(1972):87–97, respectively; also Erwin M. Segal and Roy Lachman (1972), "Complex Behavior or Higher Mental Process: Is There a Paradigm Shift?" *American Psychologist* 27:46–55.

6.4 EARLY ATTEMPTS TO MEASURE INTELLIGENCE

i. Francis Galton and J. M. Cattell
ii. Alfred Binet
iii. Further development of Binet's test by W. Stern and H. H. Goddard

READINGS

S. J. Gould, *Mismeasure of Man,* 75–77 (on Galton), 146–74 (Binet, Goddard). Fancher, *Pioneers,* 254–94 (Galton), 344–46 (Binet). Hothersall, *History of Psychology,* 301–16. Thomson, *Pelican History,* 104–14, 137–40, 190–93. Watson, *Great Psychologists,* 319–45 (Galton), 347–66 (Binet and other French psychologists). Curti, *Human Nature,* 272–312.

SYNOPSIS

In his book *Hereditary Genius* (1869) Galton introduced a method of cataloging eminent men, rating them as to intelligence, putting them on a quantitative scale, and asking how many of their blood relatives were also very bright. He also designed tests to measure reaction times to sounds and lights and other aspects of sensory perception. While determined to satisfy the physicists' criterion for being scientific—quantification—Galton did not succeed in finding a satisfactory numerical measure of intelligence. He did, however, advocate not only measuring intelligence but improving it through eugenics (**3.3**).

In America a climate of opinion favorable to the introduction of intelligence tests was encouraged by James McKeen Cattell (see **6.1** for other aspects of his career). In 1903 he published "A Statistical Study of Eminent Men" in which the one thousand most eminent people in history (including thirty-two women) were assigned a rank order on the basis of lengths of articles about them in standard reference works. He also launched the publication of *American Men of Science* in 1906 (a work which did include some women, as recognized by a change of title in recent editions), and indicated which were the most eminent (as determined by a vote of ten leaders in each field) by putting an asterisk next to their names.

The modern IQ test goes back to the attempt of the French psychologist Alfred Binet to help educators sort out the various kinds of retarded children—the standard categories at the time being "idiot," "imbecile," and "moron." Which children were "really abnormal" and should be sent to special schools, and which "just lazy"? Assuming the existence of an intrinsic capacity as distinct from actual performance in school work, Binet tried to design a test to measure this capacity. At the start his principal human subjects were his two daughters.

Binet's approach differed from that of Galton and others in that he insisted that intellectual differences could be uncovered only by tests involving complex processes, not simple sensory or motor tasks. He assumed that an individual child should be compared with "normal" children of various ages, and that such normal children pass through a series of stages characterized by the ability to pass more and more difficult tests. He claimed that it was impossible to distinguish a normal from an abnormal child unless the child's age is known, since according to his test results a retarded child of age twelve will get the same score as a normal child who is somewhat younger.

From this it followed that intelligence was a single numerical variable that could be measured in terms of *years of age,* i.e., "mental age." A moron is simply a child whose mental age is much less than his chronological age.

Binet was well aware of most of the defects of intelligence tests; for example he insisted that a child can be called "retarded" only in relation to his own environment, and "is normal when he can conduct himself without having need of the tutelage of others, earns sufficient remuneration for his needs," etc.—there is no absolute standard.

In considering the social impact of Binet's tests it should also be kept in mind that previously a child's achievements in school were assumed to depend on his moral qualities over which the child supposedly had voluntary control; many children were therefore given severe physical punishment for academic failures. As a result of Binet's work the onus of moral blame for poor academic work has been largely removed; instead, children are often supposed to be rewarded for the *effort* they put in, and their achievement is judged in relation to their own ability.

Whereas Binet himself suggested that degree of abnormality should be measured by the *difference* between chronological age and mental age, W. Stern in 1911 suggested that one should *divide* the mental by the chronological age, thus introducing what later came to be called the intelligence *quotient.*

Binet's test was introduced into the United States by H. H. Goddard, who was interested in the inheritance of intelligence; he is also known for his study of the "Kallikak" family as an example of the inheritance of feeblemindedness (see Gould's remarks on this study).

Independently of Binet, Charles Spearman in England published in 1904 a paper on "general intelligence" as a component of each mental faculty. "Spearman's G" and the later work of Cyril Burt is taken up in **6.8**.

BIBLIOGRAPHY

Michael M. Sokal, ed. (1987), *Psychological Testing and American Society, 1890–1930.* New Brunswick, N.J.: Rutgers Univ. Press.

W. G. Dahlstrom (1985), "The Development of Psychological Testing," in *Topics in the History of Psychology,* ed. G. A. Kimble and K. Schlesinger, vol. 2 (Hillsdale, N.J.: Erlbaum), 63–113.

Jonathan Harwood (1983), "The IQ in History," *Social Studies of Science* 13:465–77, is an essay review of H. Cravens, *Triumph of Evolution;* D. L. Eckberg, *Intelligence and Race;* and R. Marks, *The Idea of IQ.*

Daniel N. Robinson, ed. (1977), *Significant Contributions to the History of Psychology, 1705–1920,* series B: *Psychometrics and Educational Psychology,* vol. 4 (Washington, D.C.: University Publications of America), includes Galton on "Co-relations . . . ," (1889), three papers by Binet and the 1911 Binet-Simon paper, and Stern (1914) on "Psychological Methods of Testing."

Topic *i:* Galton and Cattell

Ronald C. Johnson, Gerald E. McClearn, Sylvia Yuen, Craig T. Nagoshi, Frank M. Ahern, and Robert E. Cole (1985), "Galton's Data a Century Later," *American Psychologist* 40:875–92.

Michael M. Sokal (1982), "James McKeen Cattell and the Failure of Anthropometric Mental Testing, 1890–1901," in *The Problematic Science,* ed. W. R. Woodward and M. Ash (New York: Praeger), 322–45.

Roy Lowe (1980), "Eugenics and Education: A Note on the Origins of the Intelligence Testing Movement in England," *Educational Studies* 6:1–8.

Topic *ii:* Binet

Theta H. Wolf (1973), *Alfred Binet* (Chicago: Univ. of Chicago Press).

Topic *iii:* Further Development

Gillian Sutherland (1977), "The Magic of Measurement: Mental Testing and English Education, 1900–1940," *Transactions of the Royal Historical Society* 27:135–53. Testing was started by G. Newman, ca. 1910.

6.5 TERMAN'S IQ TEST

i. Lewis Terman's career; development of the IQ test

ii. Later refinements of the IQ test; factor analysis and Charles Spearman's g *iii.* The College Board, the SAT, social impact of tests; the meritocracy

READINGS

Hothersall, *History of Psychology,* 316–30. Gould, *Mismeasure of Man,* 174–272 (includes an explanation of correlation and factor analysis). Thomson, *Pelican History,* 325–31.

Articles: James Fallows, "The Tests and the 'Brightest': How Fair Are the College Boards?" *Atlantic* 245, no. 2 (1980):37–48. George A. Miller, "The Test," *Science 84* 5, no. 9(1984):55–57. Kevin McKean, "Intelligence: New Ways to Measure the Wisdom of Man," *Discover* 6, no. 10(1985):25–41.

SYNOPSIS

In the United States it was Lewis M. Terman (1877–1956) who established the quantitative concept of intelligence and added the term "IQ" to the language. Since Terman was a professor at Stanford University, his revision of the Binet test was called the "Stanford-Binet" and became widely accepted.

In his first major publication, *The Measurement of Intelligence* (1916), Terman argued that there was a pressing social need to identify mentally retarded children by a reliable and simple test to bring them under the "surveillance and protection of society" so they will not commit crimes. At the same time society can benefit by detecting children of superior ability so they can be given an appropriately challenging education, enabling them to put their talents to the best possible use, rather than merely falling into habits of laziness because ordinary schoolwork is too easy for them. Industries now waste a lot of money employing persons whose mental ability is not equal to the tasks they are expected to perform; every large company should hire a psychologist to weed out unfit applicants. Schools should determine grade placement and promotions on the basis of intelligence tests rather than age; as soon as we know the minimum intelligence level needed for each occupation we can tell young people which jobs they might reasonably expect to hold. The test can also be used in eugenics.

Following Stern (**6.4**) Terman used the ratio of mental age to chronological age as the basic index of intelligence; this ratio, multiplied by 100, he called the "intelligence quotient." He standardized his test on 1000 white California boys and girls, excluding all foreign-born children. The average IQ of the original population is thus 100 by definition;

moreover, the level of difficulty of items at different ages from 3 to 16 was adjusted in such a way that the average would remain constant over this range. (One has to keep this in mind in view of occasional references to the "discovery" that a person's IQ remains approximately constant as he or she gets older.)

Contrary to earlier ideas about the distribution of intelligence in a population, Terman found that (1) the curve is symmetrical around the mean—extreme deviations above are about as frequent as extreme deviations below; (2) there are no sharp boundaries between the traditional categories (idiot, imbecile, etc.) but a continuous change from one level to the next.

An important outcome of the development of the IQ test (not original with Terman but definitely established by him) was to refute the common opinion that men are smarter than women. It does appear that boys do better on some kinds of items (especially arithmetic) but girls do better on others, and indeed the first results showed that the average IQ of girls was a few points higher than that of boys. Terman then revised the test to give more weight to items on which boys do better, so that the average would come out to 100 for each sex. Hence it is now meaningless to say that boys and girls have equal average IQs. Nevertheless the IQ test has had a major impact on popular ideas about male and female intelligence; no knowledgeable person can ever again say that women's intellectual achievements are less than those of men because of inferior innate intelligence. Terman himself tried to explain the achievement difference mostly in terms of social factors.

During World War I, a modified version of the Stanford-Binet test was given to 1,700,000 soldiers; it was known as the "Army Alpha" (to distinguish it from the "Beta" test given to illiterates). Public attention was directed to two results: (1) the average mental age of soldiers was found to be 13 years; (2) IQs of blacks were lower than those of whites. Less attention was given to the facts that (1) adult intelligence, as defined by the test, is essentially a mental age of 16; (2) northern blacks, especially those who had lived in the north for several years, had IQs much closer to those of whites than southern blacks, indicating a substantial effect of environment.

The impact of these results in the 1920s has to be viewed in the context of ideas and movements discussed in earlier sections: evolution, genetics, heredity, degeneration, and eugenics. Americans expressed concern about racial mixing after Emancipation and about immigration from Ireland and southern Europe. Low IQ scores of blacks and certain groups of immigrants could be used to justify excluding or isolating them.

As mentioned in **3.4**, anthropologists at this time were moving away from the evolutionary/hereditarian view toward a cultural/environmentalist view. Franz Boas and his colleagues attacked the use of IQ tests to make comparisons between people of different races, pointing out that the kinds of test items used would not be familiar to children who hadn't attended U.S. white middle-class schools; the items assumed familiarity with standard English and other aspects of American culture.

In view of the willingness of white male psychologists to adjust the fundamental test of intelligence so as to eliminate an apparent inferiority of males to females, one may ask why a similar adjustment could not be made to eliminate an apparent inferiority of blacks to whites (see **6.7**).

In 1926 the College Entrance Examination Board introduced the "Scholastic Aptitude Test"—supposedly a measure of intelligence rather than of knowledge. The "SAT" has been used ever since as one factor in determining admissions to the more prestigious colleges, and thus indirectly in deciding who gets a better chance at the better careers in American society. It has been defended as a way of offering bright students an opportunity to get into colleges that might not have admitted them on the basis of high school grades and teachers' recommendations. But the assumption that the SAT measures innate aptitude has been undermined in recent years by the recognition that short intensive "coaching" sessions can significantly raise scores. Students whose families can afford to pay for coaching thus have an advantage. Another criticism is that while scores on the SAT may be good predictors of grades in college they have little relation to success in later life and may count against those students who would benefit most from a college education. The test may also discriminate against women who tend to get lower scores than men on the SAT but get better grades in college. All of these issues need to be discussed by teachers and students in the light of the historical origins of the intelligence/aptitude concept.

BIBLIOGRAPHY

Michael M. Sokal, ed. (1987), *Psychological Testing and American Society, 1890–1930.* New Brunswick, N.J.: Rutgers Univ. Press.

Hamilton Cravens (1978), *The Triumph of Evolution* (Philadelphia: Univ. of Pennsylvania Press), 224–65.

Topic *i:* Terman

M. M. Sokal (1982), review of S. J. Gould's *Mismeasure of Man, Annals of Science* 39:629–30.

Paul D. Chapman (1980), "Schools as Sorters: Lewis M. Terman and the Intelligence Testing Movement," Ph.D. diss., Stanford University; see *Dissertation Abstracts International* 40(1980):5759-A.

Franz Samelson (1979), "Putting Psychology on the Map: Ideology and Intelligence Testing," in *Psychology in Social Context,* ed. A. R. Buss (New York: Irvington), 103–68.

Nicholas Pastore (1978), "The Army Intelligence Tests and Walter Lippmann," *Journal of the History of the Behavioral Sciences* 14:316–27.

Marie Tedesco (1978), "Science and Feminism: Conceptions of Female Intelligence and Their Effect on American Feminism, 1859–1920," Ph.D., diss., Georgia State University.

Topic *ii:* Refinements

Robert J. Sternberg (1985), "Human Intelligence: The Model Is the Message," *Science* 230:1111–18, is a survey of modern theories.

Paul M. Dennis (1984), "The Edison Questionnaire," *Journal of the History of the Behavioral Sciences* 20:23–37.

John Garcia (1981), "The Logic and Limits of Mental Aptitude Testing," *American Psychologist* 36:1172–80.

Brian Evans and Bernard Waites (1981), *IQ and Mental Testing: An Unnatural Science and Its Social History* (Atlantic Highlands, N.J.: Humanities Press), concerns Spearman, Binet, effects of testing in Britain.

Donald S. Napoli (1981), *Architects of Adjustment: The History of the Psychological Profession in the United States* (Port Washington, N.Y.: Kennikat Press).

Jeffrey M. Blum (1978), *Pseudoscience and Mental Ability: The Origins and Fallacies of the IQ Controversy* (New York: Monthly Review Press).

Topic *iii:* SAT

Lawrence Biemiller (1986), "Critics Plan Assault on Admissions Tests and Other Standard Exams," *Chronicle of Higher Education,* 8 Jan., 1, 4.

David Owen (1985), *None of the Above: Behind the Myth of Scholastic Aptitude* (Boston: Houghton Mifflin).

Gillian Sutherland (1984), *Ability, Merit, and Measurement: Mental Testing and English Education, 1880–1940* (New York: Oxford Univ. Press).

Mark Snyderman and R. J. Herrnstein (1983), "Intelligence Tests and the Immigration Act of 1924," *American Psychologist* 38:987–95, rejects the charge of L. Kamin, S. J. Gould (see below), and others that the act was passed with the help of the intelligence testing community.

Allan Nairn et al. (1981), *The Reign of ETS* (Washington, D.C.:

Learning Research Project), advertised as "The Ralph Nader Report on the Educational Testing Service."

Stephen Jay Gould (1980), "Science and Jewish Immigration," *Natural History* 89, no. 12:14–19. Reprinted in his (1983), *Hen's Teeth and Horses' Toes* (New York: Norton), 291–302.

6.6 PSYCHOLOGY OF GENIUS AND THE MIND OF THE SCIENTIST

i. The concept of genius in the nineteenth century; relation to insanity and degeneration
ii. Terman's study of bright children and their adult lives
iii. IQs of famous men
iv. Psychology of scientists

READINGS

Gould, *Mismeasure of Man,* 183–88. Fancher, *Intelligence Men,* 141–45.

SYNOPSIS

In 1921 Terman, in collaboration with Melita Oden, began a study of 1000 "gifted children" in California, selected by a combination of teacher recommendation and IQ tests. The average IQ of the group was about 151. This was one of the first major "longitudinal studies" in psychology. Terman and Oden followed this group as they went to college and into professions, married, and developed their careers. The purpose was to test popular ideas about child prodigies and see how this group differed from the rest of the population. Terman pointed out that before 1850 the youthful prodigy was generally admired and expected to become successful as an adult; later on the idea spread that there was something wrong with the "precocious" child, and that he was more likely to become insane or stupid when he grows up. Lombroso suggested that there is a close connection between genius and insanity and that anyone who is very different from the norm in *any* direction should be classed as "degenerate" (**3.3**-*iv*). Conversely, it was often said that many of the great geniuses had been dunces in childhood (e.g., stories about Einstein). Should the gifted child be encouraged to advance at his or her own rate in school or held back so as not to develop "social disabilities"? The results provided answers to some of these questions and raised others.

Another study conducted by Catherine Morris Cox (later Catherine Miles), a colleague of Terman, analyzed *The Early Mental Traits of Three Hundred Geniuses* (1926). Starting from Cattell's list of the 1000 most eminent people in history (**6.4**), she eliminated those who were eminent only by reason of birth or for whom adequate biographical data were not available. IQs were then estimated on the basis of information about the early development and achievements of the person; Terman had previously used this method to estimate that Francis Galton's IQ was about 200. Because of the way Cattell's list had been compiled, no one who attained eminence after about 1840 was included. The winner was Goethe (210), with Leibniz a close second (205).

While all eminent scientists are undoubtedly quite intelligent, the one-dimensional IQ scale does not seem to give much interesting information about them or even correlate with their achievements, as one might rank them within this select group. Similarly, Terman and Oden found that within the gifted group those with the highest IQs did not make the highest salaries. Mahoney (*Scientist as Subject,* 36–39) cites some recent evidence on this point and questions the use of IQ-type tests in graduate school admissions.

Research on the psychology of scientists within the last three decades, starting with Anne Roe's *The Making of a Scientist* (1953), has concentrated on establishing the characteristics of scientists as a group compared with nonscientists, or physical scientists compared with biologists.

BIBLIOGRAPHY

Topic *i:* Genius in the Nineteenth Century

George Becker (1978), *The Mad Genius Controversy: A Study in the Sociology of Deviance* (Beverly Hills, Calif.: Sage), is a survey of the controversy during the period 1840–1950.

G. Mora (1964), "One Hundred Years from Lombroso's First Essay, 'Genius and Insanity,'" *American Journal of Psychiatry* 121:562–71.

Topic *ii:* Bright Children

L. M. Terman (1954), "The Discovery and Encouragement of Exceptional Talent," *American Psychologist* 9:221–30 (includes recollections of his work with bright children); Terman and Melita H. Oden (1947), *The Gifted Child Grows Up* (Stanford: Stanford Univ. Press).

May V. Seagoe (1975), *Terman and the Gifted* (Los Altos, Calif.: Kaufmann).

Topic *iii:* IQs of Famous Men

Catherine Morris Cox et al. (1926), *The Early Mental Traits of Three Hundred Geniuses* (Stanford: Stanford Univ. Press).

Dean K. Simonton (1984), *Genius, Creativity and Leadership: Historiometric Inquiries* (Cambridge: Harvard Univ. Press).

Topic *iv:* Psychology of Scientists

Roger N. Shepard (1983), "The Kaleidoscopic Brain: Spontaneous Geometric Images May Be the Key to Creativity," *Psychology Today* 17, no. 6:62–68, includes the possible role of mental images in the discoveries of Albert Einstein, James Clerk Maxwell, and Nikola Tesla.

William Broad and Nicholas Wade (1983), *Betrayers of the Truth* (New York: Simon and Schuster), is a journalistic account of cases of fraud and deceit in modern science.

Norriss S. Hetherington (1983), "Just How Objective Is Science?" *Nature* 306:727–30, concerns historical cases in astronomy, where personal bias affected reports of observations.

Richard S. Mansfield and T. V. Busse (1981), *The Psychology of Creativity and Discovery* (Chicago: Nelson-Hall).

M. T. H. Chiu, P. Feltovich, and R. Glaser (1981), "Categorization and Representation of Physics Problems by Experts and Novices," *Cognitive Science* 5:121–52, is an example of research in the new tradition of cognitive science: it is found that experts assign a problem to a category based on abstract physical principles (e.g., energy conservation), whereas novices use a category based on literal terms in the problem statement.

Michael J. Mahoney (1979), "Psychology of the Scientist: An Evaluative Review," *Social Studies of Science* 9:349–75; (1977), *Scientist as Subject: The Psychological Imperative* (Cambridge, Mass.: Ballinger).

6.7 MASCULINE AND FEMININE THINKING: WOMEN IN SCIENCE

i. The Terman-Miles MF scale
ii. Research on sex differences in intellectual performance; spatial visualization
iii. Women in science: innate vs. social factors; what would science be like if it were dominated by women?

READINGS

Viola Klein, *The Feminine Character* (1971), 104–12, on Terman-Miles test; other chapters in this book provide useful background. E. E. Maccoby, "Feminine Intellect and the Demands of Science," *Impact of Science on Society* 20(1970):13–28; see also the article by Kathleen Lonsdale in this issue. Doreen Kimura, "Male Brain, Female Brain: The Hidden Difference," *Psychology Today* 19, no. 11(1985):50–58. Miriam Lewin, "'Rather Worse than Folly?' Psychology Measures Femininity and Masculinity, 1. From Terman and Miles to the Guilfords," in *In the Shadow of the Past,* ed. M. Lewin (1984), 155–78. Lise Meitner, "The Status of Women in Professions," *Physics Today* 13, no. 8(1960):16–21. Jane Butler Kahle, "Women Biologists: A View and a Vision," *BioScience* 35(1985):230–34. Vera Rubin, "Women's Work," *Science 86* 7, no. 6(1986):58–65, on women in astronomy with a statistical summary by Betty Vetter on the current status of women in science.

A set of filmstrips and cassettes on *Women in Science* has recently been issued (Madison, Wis.: Hawkhill Associates, 1986).

SYNOPSIS

Lewis Terman and Catherine Cox Miles developed a quantitative scale for masculinity-femininity, the "Terman-Miles MF Test," described in their book *Sex and Personality* (1936). The test included word associations (which goes most naturally with the word HOME: (a) expenses (b) happiness (c) house (d) sleep); information questions (baby gets its first tooth at: (a) 6 months (b) 12 months (c) 15 months (d) 18 months); emotional and ethical attitudes (is putting pins on the teacher's chair (a) extremely wicked . . . (d) not really bad); and interests (do you like or dislike cooking?). Agreeing with an answer most frequently given by men counts +1; by women, −1.

It is sometimes stated that intellectual brilliance or achievement is associated with "cross-sex typing"—feminine mental characteristics in boys, masculine in girls. The Terman-Miles results confirm this idea but in a more subtle way. They obtained separate measures of "scholarship" (achievement) and intelligence and correlated each with MF score. They found that for men there is a small negative correlation between scholarship and masculinity but no correlation between intelligence and masculinity; for women, there is no correlation between scholarship and MF score but a small negative correlation between intelligence and femininity.

In recent years the efforts to encourage participation of women in science have revived the question of innate sex differences in thinking. Several studies have found that boys are better at tasks involving spatial visualization, while girls are better at verbal tasks. The old idea that men have a greater range of variation in ability than women (so that there are more very bright and more very stupid men, while average IQ is the same for males and females) has been revived and rationalized in terms of sex chromosomes: since females have two X chromosomes, deviations in traits on one will be averaged out by the other, whereas males with one X and one Y are more likely to deviate from the average (J. O'Connor, *Structural Visualization* [1943]).

Whether sex differences are culturally or biologically determined is still a matter of controversy. Lack of progress may be due to parochialism in American psychological research: it is just too easy to get college students as subjects for psychological experiments and then to assume that the results of experiments on them are characteristic of the entire human race. Psychologists who argue about the relative importance of nature and nurture rarely go to another culture, as the anthropologists do, to find subjects for their tests; moreover they seem to have little interest even in reading the results of research published in other countries. (A recent study showed little interest in foreign languages among psychology graduate students.)

The debate about women's participation in science raises other interesting questions: In what sense is science, as it has been developed in the West since the sixteenth century, a masculine enterprise? Is the language of science sexist? Should women scientists simply accept the paradigms established by males or try to change them so as to make science an activity more congenial to women? What would "female science" be like?

BIBLIOGRAPHY

Topic *i:* Terman-Miles MF Scale

Ronald A. La Torre and W. E. Piper (1978), "The Terman-Miles MF Test: An Examination of Exercises 1, 2 and 3 Forty Years Later," *Sex Roles* 4:141–54.

Topic *ii:* Sex Differences in Intellect

Elsie G. Moore and A. Wade Smith (1986), "Sex and Race Differences in Mathematics Aptitude," *Sociological Perspectives* 29:77–100.

Anne Fausto-Sterling (1985), *Myths of Gender: Biological Theories About Men and Women* (New York: Basic Books).

Paula J. Caplan, Gael M. MacPherson, and Patricia Tobin (1985),

"Do Sex-Related Differences in Spatial Ability Exist? A Multi-Level Critique with New Data," *American Psychologist* 40:786–99.

Lynn H. Fox (1984), "Sex Differences among the Mathematically Precocious," *Science* 224:1292–94.

Marian Lowe and Ruth Hubbard, eds. (1983), *Woman's Nature: Rationalizations of Inequality* (Elmsford, N.Y.: Pergamon Press), includes E. Fee, "Women's Nature and Scientific Objectivity" and K. Messing, "The Scientific Mystique: Can a White Lab Coat Guarantee Purity in the Search for Knowledge about the Nature of Women?"

Daniel B. Hier and William F. Crowly, Jr. (1982), "Spatial Ability in Androgen-Deficient Men," *New England Journal of Medicine* 306:1202–5, finds a positive correlation between spatial ability and androgen level.

Stephanie Shields (1982), "The Variability Hypothesis: The History of a Biological Model of Sex Differences in Intelligence," *Signs* 7:769–97.

Rosalind Rosenberg (1982), *Beyond Separate Spheres: Intellectual Roots of Modern Feminism* (New Haven: Yale Univ. Press), Chap. 4.

Meredith M. Kimball (1981), "Women and Science: A Critique of Biological Theories," *International Journal of Women's Studies* 4:318–38.

John Eliot and Anna Hauptman (1981), "Different Dimensions of Spatial Ability," *Studies in Science Education* 8:45–66.

Camilla P. Benbow and Julian C. Stanley (1980), "Sex Differences in Mathematical Ability: Fact or Artifact?" *Science* 210:1262–64; (1983), "Sex Differences in Mathematical Reasoning Ability: More Facts," *Science* 222:1029–31, attracted wide publicity and criticism for its conclusion that sex differences are mostly innate. See for example C. Tomizuka et al. (1981), "Mathematical Ability: Is Sex a Factor?" *Science* 212:114–21; S. Tobias (1982), "Sexist Equations," *Psychology Today* 16, no. 1:14–17; J. Beckwith and M. Woodruff (1984), "Achievement in Mathematics," *Science* 223:1247–48; J. Meer (1984), "Mathematical Gender Gap: Narrowing or Inborn?" *Psychology Today* 18, no. 3:76–77; R. A. Beckman and L. Fraser (1985), "Inventing Gender Differences," *Science 85* 6, no. 5:14; C. P. Benbow and J. C. Stanley, eds. (1984), *Academic Precocity* (Baltimore: Johns Hopkins Univ. Press), gives the editors' evaluation, including conclusions on sex differences, 205–14; J. C. Stanley and C. P. Benbow (1985), "Why Are Girls Different?" *Science 85* 6, no. 7:12.

Barbara Schaap Starr (1979), "Sex Differences among Personality Correlates of Mathematical Ability in High School Seniors," *Psychology of Women Quarterly* 4:212–20.

Michele Andrisin Wittig and Anne C. Petersen, eds. (1979), *Sex-*

Related Differences in Cognitive Functioning (New York: Academic Press), reviews current research including the "sex-linked major gene hypothesis" of O'Connor.

Evelyn Fox Keller (1978), "Gender and Science," *Psychoanalysis and Contemporary Thought* 1:409–33, reprinted (1983), in *Discovering Reality,* ed. S. Harding and M. B. Hintikka (Hingham, Mass.: Reidel), 187–205.

Julia A. Sherman (1978), *Sex-Related Cognitive Differences: An Essay on Theory and Evidence* (Springfield, Ill.: Thomas).

Marie Tedesco (1978), "Science and Feminism: Conceptions of Female Intelligence and Their Effect on American Feminism, 1859–1920," Ph.D. diss., Georgia State University, includes a discussion of the mental tests developed by Helen T. Woolley at Chicago, 1898–1900, and others by Luella Pressey (1914).

Topic *iii:* Women in Science

Marilyn Bailey Ogilvie (1986), *Women in Science: Antiquity through the Nineteenth Century* (Cambridge: MIT Press), is an authoritative reference work.

Evelyn Fox Keller (1985), *Reflections on Gender and Science* (New York: Longman); (1983), "Feminism as an Analytic Tool for the Study of Science," *Academe* 69, no. 5:15–21; (1980), "Feminist Critique of Science: A Forward or Backward Move?" *Fundamenta Scientiae* 1:341–47, along with a "response" by M. R. Paty and reply by Keller; (1977), "The Anomaly of a Woman in Physics," in *Working It Out,* ed. S. Ruddick and P. Daniels (New York: Pantheon Books), 77–91.

Jane Butler Kahle, ed. (1985), *Women in Science: A Report from the Field* (Philadelphia: Taylor and Francis), includes M. Behringer, "Women's Role and Status in the Sciences: An Historical Perspective."

P. J. Siegel and K. T. Finley (1985), *Women in the Scientific Search: An American Bio-Bibliography, 1724–1979* (Metuchen, N.J.: Scarecrow Press).

Stephen G. Brush (1985), "Women in Physical Science: From Drudges to Discoverers," *Physics Teacher* 23:11–19.

Laurel Furumoto and Elizabeth Scarborough (1985), "Placing Women in the History of Psychology: The First American Women Psychologists," *American Psychologist* 41:35–42.

Barbara Lotze, ed. (1984), *Making Contributions: An Historical Overview of Women's Role in Physics* (College Park, Md.: American Association of Physics Teachers).

Vivian Gornick (1983), *Women in Science: Portraits from a World in Transition* (New York: Simon and Schuster).

Agnes N. O'Connell and Nancy Felipe Russo (1983), *Models of Achievement: Reflections of Eminent Women in Psychology* (New York: Columbia Univ. Press).

Margaret Rossiter (1982), *Women Scientists in America: Struggles and Strategies to 1940* (Baltimore: Johns Hopkins Univ. Press), is the first comprehensive work to be written by a professional historian of science on a topic where there is a large literature of dubious value.

Susan Raven and Alison Weir (1981), *Women of Achievement* (New York: Harmony Books), 217–45, presents half-page sketches of women scientists and physicians.

Carolyn Wood Sherif (1979), "Bias in Psychology," in *The Prism of Sex,* ed. J. A. Sherman and E. T. Beck (Madison: Univ. of Wisconsin Press), 93–133.

Else Hoyrup (1978), *Women and Mathematics, Science and Engineering, A Partially Annotated Bibliography with Emphasis on Mathematics and with References on Related Topics* (Roskilde, Denmark: Roskilde Univ. Press).

Lois N. Magner (1978), "Women and the Scientific Idiom: Textual Episodes from Wollstonecraft, Fuller, Gilman, and Firestone," *Signs* 4:61–80.

Paul J. Campbell and Louise Grinstein (1977), "Women in Mathematics: A Selected Bibliography," *Philosophia Mathematica* 13/14:171–203, and articles by J. Fang in this volume.

Susan Schacher et al. (1976), *Hypatia's Sisters: Biographies of Women Scientists, Past and Present* (Seattle: Feminists Northwest).

Alison Kelly (1976), "Women in Science: A Bibliographic Review," *Durham Research Review* no. 36:1092–1108.

N. Weisstein (1976), "Adventures of a Woman in Science," *Federation [American Societies of Experimental Biology] Proceedings* 35:2226–31.

Audrey B. Davis (1974), *Bibliography on Women: With Special Emphasis on Their Roles in Science and Society* (New York: Science History).

6.8 HEREDITY VERSUS ENVIRONMENT AND THE RACE-IQ CONTROVERSY

i. The Nature-Nurture problem in intelligence; twin studies
ii. Racial bias in IQ testing; the Arthur Jensen controversy
iii. Cyril Burt: his career and impact on British society; the meritocracy

READINGS

Gould, *Mismeasure of Man,* 273–334 (Burt, Thurstone, Jensen). Curti, *Human Nature,* 272–312.

Article: Jeff Howard and Ray Hammond, "Rumors of Inferiority," *New Republic* 193, no. 11(1985):17–21, is on obstacles to black intellectual success.

SYNOPSIS

During the 1930s and again after 1965 the mental testing movement in psychology came into intense conflict with the American liberal-egalitarian movement and its counterpart, the environmentalist school in the social sciences. The concept of intelligence itself, as well as the applications of the IQ test, have been strongly challenged—on the grounds that the concept either has no scientific existence as a real property of human beings, or else on the grounds that one should not try to put labels on people (however accurate) that will affect their access to the various benefits and opportunities offered by society.

As early as 1922, columnist-social critic Walter Lippmann issued a warning against the IQ testers: "They claim not only that they are really measuring intelligence, but that intelligence is innate, hereditary, and predetermined. . . . Intelligence testing in the hands of men who hold this dogma could not but lead to an intellectual caste system . . . it could turn into a method of stamping a permanent sense of inferiority upon the soul of a child. . . ."

A question of crucial importance today is: does the widespread use of IQ tests (especially in the form of the SAT as a criterion in college admissions) contribute to (or is it used to justify) discrimination against minority groups? Are there genetic differences between the average IQs of different races, or are all apparent differences due only to environmental factors or to a "cultural bias" of the tests themselves? Does the question even have any meaning if IQ is not an inherent attribute but only a construct of the psychologist?

The first quantitative study of the relative influence of nature and nurture was one by Barbara Stoddard Burks in 1928. She studied the correlations of parent-child IQ and the correlations of foster-parent, foster-child IQs. The conclusion was that about 80 percent of the variation in IQs between different people can be attributed to heredity, and about 20 percent to environment. Another way of stating this conclusion is: a radical change in environment (e.g., taking a child out of a home with stupid parents and putting him in a home with smart parents) can increase his IQ by up to 20 percent (though in most cases it is much less).

The average IQ difference between blacks and whites in America is between 11 and 15 points. Thus one could argue that even if the test is a fair indicator of capacity for succeeding in a white-dominated society, the difference could be attributed primarily to environmental factors. This reasoning has been disputed and should not be used or even discussed by an instructor who does not have a firm grasp of the statistical theory involved.

In 1969 the psychologist Arthur Jensen published an article, "How Much Can We Boost IQ and Scholastic Achievement?" in the *Harvard Educational Review*. The following issue of the *Review* contains a collection of short articles by Jensen's critics and his replies, and the whole thing is reprinted in a booklet titled *Environment, Heredity and Intelligence*. This is a good source for anyone who wants to see both sides presented in a fairly rational way—except that it turns out that there really aren't "two sides" in the way one might think from reading second-hand reports on the controversy.

Cyril Burt was a leading British authority on intelligence tests; he published a number of articles supporting the thesis that most of the variation in individual scores is attributable to heredity rather than environment. One result of the general acceptance of that view was the use of an IQ-type test, the so-called eleven-plus exam, to determine which children would go to high schools that would prepare them for college, and which would go to vocational schools. It was difficult though not impossible to transfer from one track to the other at a later age.

After his death, Burt's publications were severely criticized because he apparently had no research results to back up most of his conclusions; he continued to publish the same correlation coefficients while claiming ever-larger sample sizes. Like Mendel's data on peas, Burt's results were "too good to be true" and were probably fabricated to reinforce his own preconceived ideas. Unlike Mendel's data, they were not supported by later investigators.

The exposure of Burt's fraud has seriously undermined the credibility of the hereditarian position on intelligence, although its advocates can still claim that much of their evidence is independent of Burt's work. It does provide a good example of the general issue of cheating in science, a problem that has recently received much publicity and is worth some class discussion once the technical issues have been explained.

BIBLIOGRAPHY

Lee Ellis (1984), "Reputed Changes in Social Scientists' Sympathies Regarding the Nature-Nurture Controversy: An Exploratory Comparison," *Politics and the Life Sciences* 2:194–97.

Topic *i:* Nature-Nurture

Robert Plomin and John C. DeFries (1985), *The Origins of Individual Differences in Infancy: The Colorado Adoption Project* (Orlando, Fla.: Academic Press).

Leon J. Kamin et al. (1985), "Criminality and Adoption," *Science* 227:983–89, is a discussion of recent research by S. A. Mednick on genetically transmitted predispositions toward criminality.

Gerard Lemaine and Benjamin Matalon (1985), *Hommes Superieurs, Hommes Inférieurs? La Controverse sur l'Hérédité de l'Intelligence* (Paris: Colin).

Jonathan Harwood (1982), "American Academic Opinion and Social Change: Recent Developments in the Nature-Nurture Controversy," *Oxford Review of Education* 8, no. 1:41–67.

Hamilton Cravens (1978), *The Triumph of Evolution* (Philadelphia: Univ. of Pennsylvania Press).

Topic *ii:* Race and IQ—Jensen

Brian MacKenzie (1984), "Explaining Race Differences in IQ: The Logic, the Methodology, and the Evidence," *American Psychologist* 39:1214–33, is a review and critique of genetic theories; extensive bibliography.

Michel Schiff, Michel Duyme, Annick Dumaret, and Stanislaw Tomkiewicz (1982), "How Much *Could* We Boost Scholastic Achievement and IQ Scores? A Direct Answer from a French Adoption Study," *Cognition* 12:165–96, reports research done in response to Jensen's question; they conclude that IQs can be raised as much as 14 points by adoption into a family of different social class.

Alexander K. Wigdor and Wendell R. Garner, eds. (1982), *Ability Testing: Uses, Consequences, and Controversies* (Washington, D.C.: National Academy Press).

Sandra Scarr, ed. (1981), *Race, Social Class, and Individual Differences* (Hillsdale, N.J.: Lawrence Erlbaum Associates), is a reprint of papers by Scarr et al. on studies of black and white twins, interracial adoptions, and comments by L. Kamin and A. Jensen. She says: "I was prepared to emigrate if the blood-grouping study had shown a substantial relationship between African ancestry and low intellectual skills. I had decided that I could not endure what Jensen had experienced at the hands of colleagues" (525).

Jerry Hirsch (1981), "To 'Unfrock the Charlatans,' " *SAGE Race Relations Abstracts* 6, no. 2:1–67, is a critique of Shockley and Jensen, polemical but a useful historical presentation with a bibliography of 599 items.

James R. Flynn (1980), *Race, IQ, and Jensen* (Boston: Routledge and Kegan Paul).

Douglas Lee Eckberg (1979), *Intelligence and Race: The Origins and Dimensions of the IQ Controversy* (New York: Praeger).

Robert V. Guthrie (1976), *Even the Rat Was White: A Historical View of Psychology* (New York: Harper and Row), includes information on black psychologists and the study of racial differences.

M. A. B. Deakin (1976), "On Urbach's Analysis of the 'IQ Debate'," ibid., 27:60–65, comments on Urbach (1974).

Peter Urbach (1974), "Progress and Degeneration in the 'IQ Debate'," *British Journal for the Philosophy of Science* 25:99–135, 235–59, is an application of the "methodology of scientific research programmes" of Imre Lakatos (**11.5-*ii***).

Topic *iii:* Burt

Diane B. Paul (1985), "Textbook Treatments of the Genetics of Intelligence," *Quarterly Review of Biology* 60:317–26, complains that textbooks continue to rely on Burt's discredited results.

Gillian Sutherland (1984), *Ability, Merit, and Measurement: Mental Testing and English Education, 1880–1940* (New York: Oxford Univ. Press); (1977), "The Magic of Measurement: Mental Testing and English Education, 1900–1940," *Transactions of the Royal Historical Society* 27:135–53.

Richard L. Kellogg (1982), "The Case of the Suggestible Psychologist," *Psychology Today* 16, no. 8:69–70, concludes: "Cyril Burt's readings of Sherlock Holmes may have contributed to his intense belief in the genetic origins of human behavior. . . . Holmes, along with Darwin and Galton, was a strict hereditarian in his perspective on human behavior, a position that stems, no doubt, from the medical training that his physician creator, Arthur Conan Doyle, received at the University of Edinburgh."

Charles Webster, ed. (1981), *Biology, Medicine and Society 1840–1940* (New York: Cambridge Univ. Press), includes: Bernard Norton, "Psychologists and Class" (intelligence tests from Galton to Burt) and Gillian Sutherland, "Measuring Intelligence: English Local Education Authorities and Mental Testing 1919–1939."

Brian Evans and Bernard Waites (1981), *IQ and Mental Testing: An Unnatural Science and Its Social History* (Atlantic Highlands, N.J.: Humanities Press).

L. S. Hearnshaw (1979), *Cyril Burt, Psychologist* (Ithaca, N.Y.: Cornell Univ. Press), is a comprehensive account by an expert who started as a supporter of Burt and had access to all the remaining unpublished material; he eventually concluded that Burt obtained no new data after 1955 and that he wrote at least some of the papers published under the names of others to bolster his case.

Leon Kamin (1974), *The Science and Politics of IQ* (New York: Wiley/Halsted Press), was the first effective refutation of Burt's work.

7

ATOMS, ENERGY, AND STATISTICS

7.1 CHEMICAL ATOMIC THEORY

i. The revolution in chemistry: Joseph Priestley, Antoine Lavoisier, discovery of oxygen, stoichiometry, system of elements

ii. John Dalton and his theory; influence of Newton's ideas about gas structure; mixing of gases in the atmosphere; solubility of gases in water; table of atomic weights; law of multiple proportions

iii. Further development of the theory by Joseph Louis Gay-Lussac, Amedeo Avogadro; William Prout's hypothesis; atomic weights and the Karlsruhe Congress

iv. The periodic system of elements; D. I. Mendeleeff and others; prediction of new elements and their properties

READINGS

Stephen F. Mason, *History of the Sciences* (1962), 305–13, 449–67. A. E. E. McKenzie, *Major Achievements of Science* (1973), 91–106, 133–44, 276–77. Stephen Toulmin and June Goodfield, *Architecture of Matter* (1982), 207–46. John Marks, *Science* (1983), 144–52 (on *i*), 152–54 (on *ii*), 31–15 (on *iv*). L. P. Williams and H. J. Steffens, *The History of Science in Western Civilization* 3 (1978):108–24, 133–35, 164–72. Harold Sharlin, *Convergent Century* (1966), 48–79 (also a section on organic/physical chemistry in the late nineteenth century, 120–41). C. A. Russell and D. C. Goodman, "Atomism," Unit 5 in Open University course AST 281, *Science and the Rise of Technology since 1800* (1973), 8–36. C. J. Schneer, *The Evolution of Physical Science* (1960),

131–58; *Mind and Matter* (1969), 91–180. Robert E. Schofield, "Atomism from Newton to Dalton," *American Journal of Physics* 49(1981):211–16.

SYNOPSIS

My course deals with the history of chemistry rather superficially, touching only on those subjects relevant to the atomic structure of matter. Rather than start with alchemy, I begin this topic with Joseph Priestley (1733–1804), using his life and work to illustrate the eighteenth-century British background, the connection of chemistry with electricity and Newtonian ideas, and the prehistory of oxygen. Comparison of the "phlogiston" theory of combustion of Georg Ernst Stahl, accepted by Priestley, with the oxygen theory, leads to the preliminary conclusion that they are equivalent provided one is willing to entertain the assumption that a substance (phlogiston) may have negative weight, along with the (axiom?) that total weight or mass must be conserved in a reaction. One can then discuss whether Newton's laws permit mass to be negative or nonconserved.

Priestley is also known as the inventor of artificially carbonated drinks and as the first to mention the use of what he called "rubber" to erase pencil marks. He was hounded out of England as a suspected sympathizer with the French Revolution and went to America where (at that time) unorthodox political views were tolerated more than they were in England; he was one of the founders of the Unitarian Church in the United States.

Antoine Laurent Lavoisier (1743–1794) is sometimes credited with having discovered or established the law of conservation of mass in chemical reactions. Or did he simply *assume* it must be true in reporting his data? In any case he took over the discovery of oxygen from Priestley—an episode that can be used to discuss what one means by "discovery" in science. By establishing the modern terminology for chemical substances, Lavoisier became the "father of modern chemistry" almost by definition—very few people can understand the chemical works written before Lavoisier. Finally, Lavoisier's connections with the tax-farming system of the Old Regime in France, and his resulting execution by the Revolution, provide a dramatic case history in the relations of science and society.

The term "stoichiometry" was introduced by Jeremias Benjamin Richter (1762–1807) in 1792; he defined it as the mathematical analysis of undecomposable bodies (elements) in chemistry and foresaw the quantification of chemical reactions. Early examples were the law of constant composition proposed by Louis Joseph Proust (1754–1826, not to be confused with William Prout!) and the law of reciprocal proportions or law of equivalents proposed by Richter.

Proust's law was attacked by Claude-Louis Berthollet (1748–1822, not to be confused with Pierre Eugene Marcellin Berthelot!), in his *Essai de Statique Chimique* (1803). Berthollet's claim that some compounds have variable composition was generally abandoned by chemists by around 1810, although his ideas about incomplete reactions in solutions were reincarnated to some extent in the later "mass action law." Those chemical compounds that do not in fact have definite ratios—e.g., ionic crystals in which ions of similar size and valence can replace each other—are now called "berthollides." Fortunately they were not discovered until much later, since it appears that Berthollet's theory had to be at least temporarily defeated in order to allow Dalton's theory to be accepted.

It is difficult to explain Richter's law of reciprocal proportions without using an atomic theory (see e.g., Partington's *History of Chemistry* 3:680) and thus it seems as though there must be a direct line from Richter to Dalton. The line may have gone through E. G. Fischer's note on Richter's rules, added to a translation of Berthollet's book; the note appeared in an English version (1804) of the *Essai,* which may have been read by Dalton.

John Dalton (1766–1844) is one of the best examples in the history of science of how progress can be made by people who hold firmly, even obstinately, to obsolete doctrines. He insisted that the particles of gases repel each other with forces varying inversely as their distance (*not* the square of their distances), thinking erroneously that Newton had proved this in the *Principia.* (In fact Newton claimed only that *if* the pressure of a gas is due to repulsive forces between particles at rest, *then* that force must be as $1/r$.) Dalton adopted the view, held in the eighteenth century, that this repulsive force is associated with heat ("caloric"), together with the hypothesis (perhaps suggested by analogy with electrical force) that there is no repulsion if the particles are of different kinds. In his work on meteorology, Dalton encountered the question: how can gases of different densities remain mixed in the atmosphere? Answer: each oxygen repels other oxygens and thus tends to be surrounded by nitrogen atoms which it does not repel, and conversely. Moreover, the total pressure of the mixture is just the sum of the pressures that would be due to the individual components if each were by itself in that volume—because pressure is due to repulsive force, which only acts on atoms of the same kind. Result: Dalton's law of partial pressures! (I have never seen a textbook which explained this origin of the law.)

The assumption that atoms of different kinds do not repel was inconsistent with the view, held by Dalton and others at this time, that interatomic repulsion is caused by atmospheres of caloric condensed around each atom. This may be one reason why Dalton abandoned his first theory of mixed gases, mentioned above, and adopted a second theory based on attributing different sizes to atoms of different kinds.

Dalton's work on the proportions of oxygen and nitrogen in the atmosphere may have led him to his atomic theory, though that is disputed by some historians. (Arnold Thackray stresses his work on solubility of gases in water, generalizing William Henry's law to different kinds of gas.) Of particular interest is his report of data on the reactions of nitrogen and oxygen under different conditions, leading to a simple numerical ratio of 2:1 for the amounts of reactants. The report makes it appear that the atomic hypothesis is an induction from the experimental facts. But chemist-historians who tried to repeat Dalton's experiment in the twentieth century (J. R. Partington, L. K. Nash) found that one does not get such clean-cut results and concluded that Dalton must have used a previously conceived atomic theory to rectify his data. (Cf., the case of Mendel's 3:1 ratio, **4.3.**)

Dalton discovered what is usually called "Charles's law" for the expansion of gases by heat, but most of the credit for establishing this law should go to Joseph Louis Gay-Lussac (1778–1850). In the development of chemical atomic theory, Gay-Lussac is known for the law of combining volumes of gases (1808). Dalton realized that if all gases have the same number of atoms in the same volume, this law would be consistent with his own theory; but he explicitly denied that assumption. He assumed instead that atoms of different elements have different sizes, and since he also assumed that in a gas they are close together, they would occupy different amounts of space.

Gay-Lussac also performed an experiment on the *free* expansion of gases (into a vacuum), finding almost no change in temperature; this result was later seen as a refutation of the idea that temperature is simply proportional to the density of caloric (so that it ought to be inversely proportional to volume) and hence as an argument against a basic conclusion of the caloric theory. It could also be used as an objection to the premise that gas pressure is due to repulsive forces associated with heat, since those forces ought to be weaker when the atoms are farther apart.

In 1811 Avogadro proposed his hypothesis that all gases at the same pressure and temperature contain the same number of molecules in equal volumes. "Molecules" may contain more than one atom of the same kind. Dalton rejected this assumption on the grounds that atoms of the same kind repel each other. Thus Dalton's formula for water is HO, not H_2O, and many of his atomic weights differ from those determined by the analysis of Gay-Lussac and Avogadro. Avogadro justified his hypothesis by assuming, contrary to Dalton, that molecules are separated by great distances in gases, so the fact that they may have different sizes (perhaps due to varying attraction for caloric) does not prevent the same number from occupying the same space. But he was not able to estimate any numerical value for the number of molecules in a given volume, and thus the term "Avogadro's number" gives him a little too much credit.

In 1815 William Prout proposed that the atomic weights of all elements are integer multiples of that of hydrogen; and further that perhaps all the elements are constructed by putting together hydrogen atoms. Prout's hypothesis promised a return to the satisfying belief that the world is constructed out of only one (or a small number of) basic kinds of substance, whereas Lavoisier's system implied (by the end of the nineteenth century) that there are ninety-two qualitatively different kinds of matter. In addition to this metaphysical attractiveness, the hypothesis might be considered an example of a good scientific theory in the sense of Karl Popper, because it can be directly tested by experiment, and historically the more accurate data on atomic weights (found partly to test Prout's hypothesis) were ultimately useful in twentieth-century atomic theory, even though for several decades they were thought to refute the hypothesis. This subject is taken up again in **9.2** and **10.1**.

After a half-century of confusion and controversy among chemists about methods for determining atomic weights, an international conference was held at Karlsruhe in 1860. Avogadro's hypothesis and his method for determining atomic weights was revived by Stanislao Cannizzaro. The adoption of this method made possible systematic studies leading to the development of the periodic system of elements in the 1860s and 1870s.

According to J. W. van Spronsen, the periodic system may be regarded as an example of independent and nearly simultaneous discovery by six scientists: A.-E. Beguyer de Chancourtois, J. A. R. Newlands, L. Meyer, W. Odling, G. D. Hinrichs, and D. I. Mendeleev. But H. Cassebaum and G. Kauffman argue that, using a more stringent definition of what an acceptable periodic system must do, Odling was the first to arrive at a periodic system, while the general adoption of a system is due to Meyer and Mendeleev.

While recognizing that the usual practice of giving all the credit to Mendeleev is incorrect, I would suggest that in a general undergraduate course on history of science it is less important to worry about the distribution of credit than to explain the process by which the discovery was made and accepted in the context of nineteenth-century science. My impression is that the whole enterprise was somewhat outside the mainstream of nineteenth-century chemical *research,* even though it depended on recent developments within the mainstream such as Cannizzaro's rules for finding atomic weights. If one can put oneself in the frame of mind of the average chemist in the 1860s, not knowing how it was all going to turn out later, I think one would be quite skeptical of the scientific value of most of the writings on periodic systems. (Thus the Chemical Society of London refused to publish Newlands's paper simply because it *never* published purely theoretical work.) Often they seemed to be little more than pure numerology, comparable to attempts to explain the wavelengths of spectral lines (**9.3**) or the distances of planets from the sun (**12.1**). Indeed, Hinrichs applied the same approach to all three problems. The theoretical presupposition that atomic weights must fall into some simple numerical pattern led many of these theorists to ignore or manipulate inconvenient data and make arbitrary assumptions—practices that scientists consider reprehensible unless they lead to major breakthroughs.

From the writings of the developers of the periodic system it appears that one important motivation, other than the fascination of numerical patterns, was to organize the already formidable mass of information about the elements into a simple, easily remembered form for textbooks and lectures. Such a system would also be useful for those who were not doing chemical research themselves but needed an overview of the properties of the elements for other purposes—e.g., Beguyer de Chancourtois, a geologist interested in the distribution of elements in the Earth's crust.

The story of Mendeleev's work and the confirmation of his predictions of the existence and properties of new elements has been told many times, including the readings listed above. To exploit it as a case history in the interaction of theory and experiment, one should include two particular episodes. First, we know that a powerful theory can sometimes overcome apparent experimental refutation by forcing the experimenters to redo their measurements and find results in agreement with theory after all. Lecoq de Boisbaudran had initially found a density of 4.6 for gallium (Mendeleev's predicted "eka-aluminium") but after Mendeleev pointed out that it should be closer to 6, it turned out that the initial sample had been impure and a revised value, 5.96, was eventually obtained. But, thanks to such confirmations, Mendeleev's theory became a little *too* powerful, as shown by the second episode. He placed tellurium in group 6, row 7, just before iodine, which is in group 7, row 7, and hence announced that its atomic weight must be 125 (less than iodine, 127). B. Brauner, after considerable effort, obtained 127.64 (very close to the modern value) but the disagreement with Mendeleev's prediction convinced Brauner that his sample must have been contaminated and that the true atomic weight of tellurium is less than 127. (The tellurium-iodine case was eventually accepted as an "inversion"—an exception to the rule that the elements must be arranged in order of atomic weight.)

Anyone interested in the history of chemistry should subscribe to *CHOC News,* the Newsletter of the American Chemical Society's Center for History of Chemistry. For information write to Center for History of Chemistry, E. F. Smith Hall/D6, University of Pennsylvania, 215 South 34th Street, Philadelphia, PA 19104.

BIBLIOGRAPHY

Alan J. Rocke (1984), *Chemical Atomism in the Nineteenth Century, Dalton to Cannizzaro* (Columbus: Ohio State Univ. Press).

Topic *i:* Revolution in Chemistry

Alfred Nordmann (1986), "Comparing Incommensurable Theories," *Studies in History and Philosophy of Science* 17:231–46.

Evan Melhado (1985), "Chemistry, Physics, and the Chemical Revolution," *Isis* 76:195–211; (1983), "Oxygen, Phlogiston, and Caloric: The Case of Guyton," *Historical Studies in the Physical Sciences* 13:311–34.

David J. Rhees (1983), Catalogue of the Exhibition "Joseph Priestley, Enlightened Chemist," Publication No. 1 of the Center for History of Chemistry (see address above).

Maurice Crosland (1980), "Chemistry and the Chemical Revolution," in *The Ferment of Knowledge,* ed. G. S. Rousseau and R. Porter (New York: Cambridge Univ. Press), 389–416.

Derek de Solla Price (1980), "The Analytical (Quantitative) Theory of Science and Its Implications for the Nature of Scientific Discovery," in *On Scientific Discovery,* ed. M. D. Grmek et al. (Boston: Reidel), suggests that "the (Lavoisier) revolution in chemistry is not any sort of 'postponed' event, but a marker of a period when all of the modern movement would have had its 'natural' and expected beginning."

Alan Musgrave (1976), "Why Did Oxygen Supplant Phlogiston? Research Programmes in the Chemical Revolution," in *Method and*

Appraisal in the Physical Sciences, ed. C. Howson (New York: Cambridge Univ. Press), 181–209, is an attempt to apply the methodology of I. Lakatos (**11.5**).

Henry Guerlac (1975), *Antoine-Laurent Lavoisier, Chemist and Revolutionary* (New York: Scribner's); (1973), "Lavoisier, Antoine-Laurent," *Dictionary of Scientific Biography* 8:66–91.

Topic *ii:* Dalton

Kiyosha Fujii (1986), "The Berthollet-Proust Controversy and Dalton's Chemical Atomic Theory 1800–1820," *British Journal for the History of Science* 19:177–200, criticizes S. C. Kapoor's conclusion that Proust could not have been the forerunner of Dalton's law of constant and multiple proportions.

Donald Cardwell and Joan Mottram (1984), "Fresh Light on John Dalton," *Notes and Records of the Royal Society of London* 39:29–40.

Iyama Hiroyuki (1983), "A Case of Fabricated Discovery: The Law of Multiple Proportions," *Historia Scientiarum* 24:19–28.

Theron Cole, Jr. (1978), "Dalton, Mixed Gases, and the Origin of the Chemical Atomic Theory," *Ambix* 25:117–30.

W. A. Smeaton (1978), "Berthollet's *Essai de Statique Chimique:* A Supplementary Note," *Ambix* 25:211–12; (1977), "Betholleť's *Essai de Statique Chimique* and Its Translations: A Bibliographical Note and a Daltonian Doubt," *Ambix* 24:149–58.

Arnold Thackray (1973), "Dalton, John," *Dictionary of Scientific Biography* 3:537–47; (1972), *John Dalton* (Cambridge: Harvard Univ. Press).

Topic *iii:* Gay-Lussac, Avogadro, Prout

W. H. Brock (1985), *From Protyle to Proton: William Prout and the Nature of Matter 1785–1985* (Boston: Hilger), Chaps. 1–7.

Mario Morselli (1984), *Amedeo Avogadro: A Scientific Biography* (Boston: Reidel); (1980), "The Manuscript of Avogadro's *Essai d'une manière de déterminer les masses relatives des molécules élémentaries," Ambix* 27:147–72.

Nicholas Fisher (1982), "Avogadro, the Chemists, and Historians of Chemistry," *History of Science* 20:77–102, 212–31.

J. H. Brooke (1981), "Avogadro's Hypothesis and Its Fate: A Case-Study in the Failure of Case-Studies," *History of Science* 19:235–73.

(1980), *Gay-Lussac: La Carrière et l'Oeuvre d'un Chimiste Français durant la première Moitié du XIX Siècle* (Palaiseau, France: École Polytechnique), has its contents listed in *Isis Critical Bibliography* (1982), item 1883.

M. Fricke (1976), "The Rejection of Avogadro's Hypotheses," in *Method and Appraisal in the Physical Sciences,* ed. C. Howson (New York: Cambridge Univ. Press), 277–307.

Maurice P. Crosland (1972), "Gay-Lussac, Joseph Louis," *Dictionary of Scientific Biography* 5:317–27; (1970), "Avogadro, Amedeo," ibid. 1:343–50.

Topic *iv:* Periodic System

Bernadette Bensaude-Vincent (1986), "Mendeleev's Periodic System of Chemical Elements," *British Journal for the History of Science* 19:3–17.

Harold Goldwhite (1979), "Mendeleev's Other Prediction," *Journal of Chemical Education* 56:35, concerns polonium.

B. M. Kedrov (1974), "Mendeleev, Dmitry Ivanovich," *Dictionary of Scientific Biography* 9:286–95.

D. G. Rawson (1974), "The Process of Discovery: Mendeleev and the Periodic Law," *Annals of Science* 31:181–204.

H. Cassebaum and G. B. Kauffman (1971), "The Periodic System of the Chemical Elements: The Search for Its Discoverer," *Isis* 62:314–27.

J. W. van Spronsen (1969), *The Periodic System of Chemical Elements: A History of the First Hundred Years* (New York: American Elsevier).

[Note: numerous Russian-language publications on Mendeleev are listed in recent issues of the *Isis Critical Bibliography.*]

7.2 ENERGY AND THE KINETIC WORLDVIEW

i. Discovery of energy conservation law; background in early nineteenth-century science; role of steam engine, nature philosophy, wave theory of heat
ii. Electromagnetism: Hans Christian Oersted and André Marie Ampère
iii. Wave theory of light: Thomas Young and Augustin Fresnel
iv. Thermodynamics (first law)
v. Kinetic theory of gases revived by Rudolf Clausius and James Clerk Maxwell

READINGS

Mason, *History of the Sciences,* 468–78, 486–95. McKenzie, *Major Achievements,* 145–74, 442–56. Toulmin and Goodfield, *Architecture of*

Matter, 247–54. Marks, *Science,* 135–41 (steam engine), 165–70 (energy). Sharlin, *Convergent Century,* 1–37, 99–119, 142–60. Schneer, *Evolution of Physical Science,* 186–257; *Mind and Matter,* 188–219. Williams and Steffens, *History of Science* 3:150–59, 181–241. Einstein and Infeld, *The Evolution of Physics,* 35–121 (see comments below). E. Segre, *From Falling Bodies to Radio Waves,* 79–132, 168–74, 186–251.

More specialized books: C. A. Russell, "Time, Chance, and Thermodynamics," Open University course A381, *Science and Belief: Darwin to Einstein* (1981), Block III, 5–12, and corresponding extracts from Mayer, Clausius, and Maxwell in N. G. Coley and V. M. D. Hall, eds., *Darwin to Einstein: Primary Sources in Science and Belief* (1980). L. P. Williams, *The Origins of Field Theory* (1980), 3–72. S. P. Bordeau, *Volts to Hertz . . . the Rise of Electricity (1982), 60–85.* For topic *iii* one may use the section on the nature of light by R. Stannard and N. G. Coley in Block IV of this course, 70–76.

The short book by Joan Solomon, *The Structure of Space* (1974), is a good elementary introduction to several topics in the history of physics and astronomy. Chapter 6 (71–81) provides the seventeenth-century background for the wave and particle theories of light, and Chapter 9 (90–97) covers the revival of the wave theory by Young and Fresnel.

For more emphasis on the economic background and social-technological relations of science, I suggest H. J. Fyrth and M. Goldsmith, *Science, History and Technology,* Book 2, Part I (1969), 1–25.

My personal recommendation is the Einstein-Infeld book, though it has some defects that will be obvious to the historian of physical science. The reader is likely to be confused by the way the authors jump into an exposition of the "mechanical view" based on the kinetic theory of matter, mixing in the work of J. P. Joule, Hermann von Helmholtz, and Robert Brown interpreted more or less from the standpoint of 1900. This section of the book should not be read as an account of how the kinetic theory developed (as it might at first appear to be) but as an exposition of a general "mechanistic" attitude toward scientific explanation prevalent in the nineteenth century. The events in Chapter 2, most of which occurred *before* those mentioned in the last half of Chapter 1, can nevertheless be described as contributing to the "decline of the mechanical view" as the authors claim. The authors give a summary of what was known about electricity and magnetism before the remarkable discovery of the connection between them by H. C. Oersted in 1820. They present as much background information on this subject as one can expect nonscience students to absorb, and for a class of such students I would resist the temptation to inject any more technical material at this point. Instead I would follow Einstein and Infeld's policy of covering rather superficially various topics in early nineteenth-century physics, calling attention to the possible connections with philosophical ideas (see below).

SYNOPSIS

This section deals with several areas of physics in the period 1800–1860 associated with the establishment of the general law of conversion and conservation of energy, and in particular the change from the caloric theory (regarding heat as a substance) to the "mechanical" or "kinetic" theory (regarding heat as a form of molecular motion). Whereas scientists at the beginning of the century explained the various phenomena of heat, light, electricity, and magnetism by attributing them to various substances (including "imponderable fluids"), during this period they tried to reduce the number of different substances to only one ("ether") and to explain all phenomena as the result of motions of ether and ordinary matter.

According to T. S. Kuhn's influential article on the discovery of energy conservation (1959), a dozen men have been given some share of the credit in the discovery; but if one requires both a general conversion/conservation statement and a numerical computation of at least one conversion coefficient, the list can be reduced to four: J. P. Joule, J. R. Mayer, H. von Helmholtz, and L. A. Colding. Even with this short list, one still has to explain a phenomenon of "simultaneous discovery" since all four seem to have done their work independently within a short period of time in the 1840s. Thus one should look for common factors in the early nineteenth-century background that might have inspired energy conservation.

Kuhn argues that the major factors were: (1) the practical interest in steam engines and a bookkeeping approach to their efficiency in producing work from heat (even though this was not yet regarded as an actual transformation of heat to work); (2) the discovery of conversion processes between heat, electricity, and magnetism; (3) German "nature philosophy" (*Naturphilosophie*), which popularized the idea that all natural phenomena could be explained in terms of one or two basic forces.

Oersted's discovery of electromagnetism in 1820 is the best-documented example of the influence of metaphysical views, deriving from nature philosophy, on a scientific discovery (see Stauffer's paper cited below). In other respects the nature-philosophy approach seems to be quite unscientific, since it disdains the use of mathematics and quantitative experimentation. Nevertheless, it provides a possible connection between science and cultural movements such as Romanticism that seem to recur periodically in Western society; students who prefer the humanities to the sciences should be aware that the qualitative, intuitive, holistic approach can sometimes be fruitful in science as long as it is balanced against an analytical approach. [See my book *The Temperature of History* (1978) for further discussion.]

The revival of the wave theory of light by Young and Fresnel fits into the rise of the kinetic worldview if one recalls that it replaced the conception of light as a *substance* (particles) by the conception of light as *motion* (of ether). But the connection is even closer than that, because of the great interest in *radiant heat* in the early nineteenth century, and the general conviction among physicists that light and heat are "identical"—i.e., they are qualitatively the same phenomenon, differing only in some quantitative parameter. As long as the particle theory of light was dominant, the "identity thesis" supported the substance (caloric) theory of heat; but when scientists switched their allegiance to the wave theory of light, many of them also started to lean toward a "wave theory of heat." Instead of identifying heat with an imponderable fluid, they found it almost as easy to identify it with the *vibrations* of that fluid, without giving up any of the explanatory power of the caloric theory; and it was natural to make the further assumption that this is the same ethereal fluid that transmits the vibrations of light. But then, as A. M. Ampère pointed out in 1832, the vibrations of the ether should interact and perhaps come into equilibrium with the vibrations of the atoms of ordinary matter. Thus one could go indirectly from a caloric to a kinetic theory of heat and then set aside the ether vibrations for separate treatment. (The latter leads to the topic of black-body radiation, **9.3.**)

Ampère, who followed up Oersted's discovery of electromagnetism by discovering the force between currents—a force that is used to define the "ampere" as a unit of current—also proposed to reduce magnetism to electricity. Since a circulating current acts like a magnetic dipole, one may simply assume that a magnet is a collection of circulating molecular currents. This could be called a "kinetic theory of magnetism." It would imply the nonexistence of magnetic monopoles, a question that is still unsettled today.

The history of electromagnetism is continued with Faraday in **8.1**.

Following the establishment of energy conservation in the 1840s, thermodynamics in its modern form was developed by Rudolph Clausius, William Thomson, and W. J. M. Rankine in the early 1850s. The first law may be considered a special case of energy conservation, quantifying the convertibility of heat and mechanical work. The second law is discussed in **7.3**.

The kinetic theory of gases, originally proposed by Daniel Bernoulli in 1738 and independently by John Herapath in the 1820s and J. J. Waterson in the 1840s, was finally successfully revived in the 1850s by Clausius. In addition to showing how the billiard-ball model could be used to derive the ideal gas laws and explain, at least qualitatively, the nature of the solid, liquid, and gaseous states of matter, Clausius was impelled to introduce a finite diameter for his molecules in order to account for the slowness of diffusion. This was the origin of the "mean-free-path" concept, later used by Maxwell to give a quantitative account of transport processes (viscosity, heat conduction, and diffusion).

While the molecular diameter had originally been introduced somewhat arbitrarily as an ad hoc postulate, Maxwell's theory could be used to estimate it from experimental data; this was done first by Josef Loschmidt (1865). As William Thomson pointed out in 1870, the atom now seemed much more real when one could measure, weigh, and count it.

Maxwell's kinetic theory paper of 1859 also introduced (in a much more systematic way than Clausius had done) a statistical approach for dealing with molecular motions. This established an important link between atomic physics and the statistical tradition that had been developed primarily in the social sciences (see **7.5**). There was also a connection with the study of double stars in astronomy (**12.1**).

BIBLIOGRAPHY

Christa Jungnickel and Russell McCormmach (1986), *Intellectual Mastery of Nature,* vol. 1, *The Torch of Mathematics, 1800 to 1870* (Chicago: Univ. of Chicago Press), is the first volume of a projected comprehensive history of modern theoretical physics.

R. W. Home (1984), *The History of Classical Physics: A Selected Annotated Bibliography* (New York: Garland).

Rudolf Stichweh (1984), *Zur Entstehung des modernen Systems wissenschaftlicher Disziplin: Physik in Deutschland, 1740–1890* (Frankfurt: Suhrkamp Verlag). See the review by K. Olesko (1985), *Isis* 76:607–8.

P. M. Harman (1982), *Energy, Force, and Matter: The Conceptual Development of Nineteenth-Century Physics* (New York: Cambridge Univ. Press).

Lewis Pyenson (1981), "History of Physics," in *Encyclopedia of Physics,* ed. Rita G. Lerner and George L. Trigg (Reading, Mass.: Addison-Wesley), 404–14, is an excellent survey concentrating on the growth of physics and the physics discipline since the eighteenth century.

Susan Faye Cannon (1978), *Science in Culture: The Early Victorian Period* (New York: Science History), Chap. 4, "The Invention of Physics," argues that physics in the modern sense was invented by the French during the years 1810–1830, then transferred to Britain.

Maurice Crosland and Crosbie Smith (1978), "The Transmission of Physics from France to Britain: 1800–1840," *Historical Studies in the Physical Sciences* 9:1–61.

Crosbie Smith (1976), "'Mechanical Philosophy' and the Emergence of Physics in Britain: 1800–1850," *Annals of Science* 33:3–29.

Topic *i:* Energy Conservation

M. Norton Wise and C. Smith (1986), "Measurement, Work and Industry in Lord Kelvin's Britain," *Historical Studies in the Physical Sciences* 17: 147–73.

Frank A. J. L. James (1985), "Between Two Scientific Generations: John Herschel's Rejection of the Conservation of Energy in His 1864 Correspondence with William Thomson," *Notes and Records of the Royal Society of London* 40:53–62.

H.-G. Schöpf (1984), "Frühe Ansichten und Einsichten über die Wärme and die Gase," *NTM* 21:35–47.

Herbert Breger (1982), *Die Natur als arbeitende Maschine: Zur Entstehung des Energiebegriffs in der Physik 1840–1850* (Frankfurt and New York: Campus Verlag).

David Gooding (1980), "Metaphysics versus Measurement: The Conversion and Conservation of Force in Faraday's Physics," *Annals of Science* 37:1–29.

Henry John Steffens (1979), *James Prescott Joule and the Concept of Energy* (New York: Science History); see comments by E. Mendoza and D. S. L. Cardwell (1981), "On a Suggestion Concerning the Work of J. P. Joule," *British Journal for the History of Science* 14:177–80.

M. Norton Wise (1979), "William Thomson's Mathematical Route to Energy Conservation," *Historical Studies in the Physical Sciences* 10:49–84.

Jacques Merleau-Ponty (1979), "La Découverte des Principes de l'Energie: l'Itinéraire de Joule," *Revue d'Histoire des Sciences* 32:315–31.

Armin Hermann (1978), "Die Entdeckung des Energie-Prinzips: Wie die Arzt Julius Robert Mayer die Physiker belehrte," *Bild der Wissenschaft* 4:140–48.

Crosbie Smith (1978), "A New Chart for British Natural Philosophy: The Development of Energy Physics in the Nineteenth Century," *History of Science* 16:231–79; (1976), "Faraday as Referee of Joule's

Royal Society Paper 'On the Mechanical Equivalent of Heat,' " *Isis* 67:444–49.

P. M. Heimann [now Harman] (1976), "Mayer's Concept of 'Force': The 'Axis' of a New Science of Physics," *Historical Studies in the Physical Sciences* 7:277–96.

R. Bruce Lindsay, ed. (1975), *Energy: Historical Development of the Concept* (Stroudsburg, Pa.: Dowden, Hutchinson and Ross), is an anthology of original sources, Aristotle to J. R. Mayer, J. P. Joule, and L. A. Colding.

Stanley W. Jackson (1967), "Subjective Experiences and the Concept of Energy," *Perspectives in Biology and Medicine* 10:602–26.

T. M. Brown (1965), "Resource Letter EEC-1 on the Evolution of Energy Concepts from Galileo to Helmholtz," *American Journal of Physics* 33:759–65.

T. S. Kuhn (1959), "Energy Conservation as an Example of Simultaneous Discovery," in *Critical Problems in the History of Science,* ed. M. Clagett (Madison: Univ. of Wisconsin Press), 321–56; reprinted in Kuhn's (1977) *The Essential Tension* (Chicago: Univ. of Chicago Press), 66–104.

Topic *ii:* Electromagnetism

James R. Hofmann (1987), "Ampère, Electrodynamics, and Experimental Evidence," *Osiris,* 2d ser., 3: 45–76.

T. Hashimoto (1983), "Ampère vs. Biot: Two Mathematizing Routes to Electromagnetic Theory," *Historia Scientiarum* 24:29–51.

Christine Blondel (1982), *Ampère et la Creation de l'Electrodynamique 1820–1827* (Paris: Bibliothèque Nationale); (1978), "Sur les premières recherches de formula électrodynamique par Ampère (octobre 1820)," *Revue d'Histoire des Sciences* 31:53–65. L. P. Williams calls Blondel's *Ampère* " . . . the best published work to date on what Ampère was up to in those bursts of incredible creativity during the 1820s" [review in *Isis* 75(1984):591–92]. Yet author and reviewer differ on some points: see Blondel, "Ampère and the Programming of Research," *Isis* 76 (1985):559–61, and a reply from Williams, ibid., 561.

Walter Kaiser (1981), *Theorien der Elektrodynamik im 19. Jahrhunderts* (Gerstenberg: Hildesheim).

Ole Immanuel Franksen (1981), *H. C. Oersted: A Man of the Two Cultures* (Birkerød, Denmark: Strandbergs Forlag).

Geoffrey Sutton (1981), "The Politics of Science in Early Napoleonic France: The Case of the Voltaic Pile," *Historical Studies in the Physical Sciences* 11:329–66.

Kenneth L. Caneva (1981), "What Should We Do with the Monster? Electromagnetism and the Psychosociology of Knowledge," in *Sciences and Cultures,* ed. E. Mendelsohn and Y. Elkana (Boston: Reidel), 101–31; (1980), "Ampère, the Etherians, and the Oersted Connexion," *British Journal for the History of Science* 13:121–38; (1978), "From Galvanism to Electrodynamics: The Transformation of German Physics and Its Social Context," *Historical Studies in the Physical Sciences* 9:63–159.

Pierre-Gérard Hamamdjian (1978), "Repères pour une Biographie Intellectuelle d'Ampère" and "Contribution d'Ampère au 'Theorème d'Ampère'," *Revue d'Histoire des Sciences* 31:233–48, 249–68.

Alfred Kastler (1977), "Ampère at les Lois d'Electrodynamique," *Revue d'Histoire des Sciences* 30:143–57.

Eizo Yamazaki (1977), "Sur l'Électrodynamique d'Ampère," *Memoirs of the Institute of Sciences and Technology, Meiji University* 16, no. 12:1–31.

Topic *iii:* Light

Frank A. J. L. James (1984), "The Physical Interpretation of the Wave Theory of Light," *British Journal for the History of Science* 17:47–60.

Geoffrey N. Cantor (1984), "Was Thomas Young a Wave Theorist?" *American Journal of Physics* 52:305–8; (1983), *Optics after Newton: Theories of Light in Britain and Ireland, 1704–1840* (Dover, N.H.: Manchester Univ. Press); (1978), "The Historiography of 'Georgian' Optics," *History of Science* 16:1–21.

J. Z. Buchwald (1983), "Fresnel and Diffraction Theory," *Archives Internationales d'Histoire des Sciences* 33:36–111.

D. H. Arnold (1983), "The Mecanique Physique of Simeon Denis Poisson: The Evolution and Isolation in France of His Approach to Physical Theory (1800–1840). V. Fresnel and the Circular Screen," *Archive for History of Exact Sciences* 28:321–42.

G. N. Cantor and M. J. S. Hodge, eds. (1981), *Conceptions of Ether: Studies in the History of Ether Theories, 1740–1900* (New York: Cambridge Univ. Press), has an introduction by the authors, and papers by P. M. Heimann, L. Laudan, J. Z. Buchwald.

Jean Rosmorduc (1977), "Ampère et l'Optique: Une Intervention dans le Débat sur la Transversalité de la Vibration Lumineuse," *Revue d'Histoire des Sciences* 30:159–67.

Eugene Frankel (1976), "Corpuscular Optics and the Wave Theory of Light: The Science and Politics of a Revolution in Physics," *Social Studies of Science* 6:141–84.

John Worrall (1976), "Thomas Young and the 'Refutation' of Newtonian Optics: A Case-Study in the Interaction of Philosophy of Science and History," in *Method and Appraisal in the Physical Sciences,* ed. C. Howson (New York: Cambridge Univ. Press), 107–79.

Topic *iv:* Thermodynamics

Ernst Mach (1986), *Principles of the Theory of Heat, Historically and Critically Elucidated,* trans. from the German ed. (1900) (Norwell, Mass.: Kluwer) is an important source for the late-nineteenth-century conflict between the kinetic-mechanistic worldview and the empirical viewpoint.

Eri Yagi (1984), "Clausius's Mathematical Method and the Mechanical Theory of Heat," *Historical Studies in the Physical Sciences* 15:177–95.

Philip Lervig (1982–1983), "What Is Heat? C. Truesdell's View of Thermodynamics. A Critical Discussion," *Centaurus* 26:85–122.

C. Truesdell (1982), "The Disastrous Effects of Experiment upon the Early Development of Thermodynamics," in *Scientific Philosophy Today,* ed. J. Agassi and R. S. Cohen (Boston: Reidel), 415–23; (1980), *The Tragicomical History of Thermodynamics, 1822–1854* (New York: Springer-Verlag); Truesdell and S. Bharatha (1977), *The Concepts and Logic of Classical Thermodynamics as a Theory of Heat Engines: Rigorously Constructed upon the Foundation Laid by S. Carnot and F. Reech* (New York: Springer-Verlag).

Fabio Sebastiani (1981), "La Teorie Caloricistiche di Laplace, Poisson, Sadi Carnot, Clapeyron e la Teoria dei Fenomeni Termici nei Gas formulata da Clausius nel 1850," *Physis* 23:397–438.

Keith Hutchison (1981), "W. J. M. Rankine and the Rise of Thermodynamics," *British Journal for the History of Science* 14:1–26; (1976), "Mayer's Hypothesis: A Study of the Early Years of Thermodynamics," *Centaurus* 20:279–304.

Crosbie Smith (1977), "William Thomson and the Creation of Thermodynamics," *Archive for History of Exact Sciences* 16:231–88; (1976), "Natural Philosophy and Thermodynamics: William Thomson and 'The Dynamical Theory of Heat,' " *British Journal for the History of Science* 9:293–319.

D. S. L. Cardwell and Richard L. Hills (1976), "Thermodynamics and Practical Engineering in the Nineteenth Century," *History of Technology* 1:1–20.

Edward E. Daub (1971), "Clausius, Rudolf," *Dictionary of Scientific Biography* 3:303–11.

Topic *v:* Kinetic Theory

Peter Achinstein (1987), "Scientific Discovery and Maxwell's Kinetic Theory," *Philosophy of Science* 54:409–34.

Elizabeth Garber, Stephen G. Brush, and C. W. F. Everitt, eds. (1986), *Maxwell on Molecules and Gases* (Cambridge: MIT Press), contains published and unpublished papers by J. C. Maxwell on kinetic theory and atomic concepts, and surveys nineteenth-century research on gas properties.

Stephen G. Brush (1983), *Statistical Physics and the Atomic Theory of Matter* (Princeton, N.J.: Princeton Univ. Press), Chap. 1; (1976), *The Kind of Motion We Call Heat* (New York: North-Holland/American Elsevier).

Eric Mendoza (1983), "The Kinetic Theory of Matter, 1845–1855," *Archives Internationales d'Histoire des Sciences* 32:184–220.

Theodore M. Porter (1981), "A Statistical Survey of Gases: Maxwell's Social Physics," *Historical Studies in the Physical Sciences* 12:77–116.

M. A. El'yashevich and T. S. Prot'ko (1981), "Maxwell's Contribution to the Development of Molecular Physics and Statistical Methods," *Soviet Physics Uspekhi* 24:876–903.

David B. Wilson (1981), "Kinetic Atom," *American Journal of Physics* 49:217–22.

R. B. Lindsay (1979), *Early Concepts of Energy in Atomic Physics* (New York: Academic Press), 112–79, includes papers by Joule, A. Kroenig, Clausius, Maxwell.

Elizabeth Garber (1978), "Molecular Science in Late-Nineteenth-Century Britain," *Historical Studies in the Physical Sciences* 9:265–98.

7.3 ENTROPY, TIME, AND CHANCE

i. The second law of thermodynamics and the principle of dissipation of energy
ii. Entropy and disorder; the reversibility and recurrence paradoxes

READINGS

McKenzie, *Major Achievements,* 168–70, 173–74, 456–57. Mason, *History of the Sciences,* 488–90, 495–97. Marks, *Science,* 166–68 (Carnot). Toulmin and Goodfield, *Architecture of Matter,* 259–60. Schneer, *Evolution of Physical Science,* 258–65. C. A. Russell, "Time, Chance, and Thermodynamics," Open University course A381, *Science and Belief: From Darwin to Einstein,* Block III, 12–17, 45–72, and extracts from W. Thomson, Eddington, etc., in the Colin-Hall anthology of primary sources; E. Daub's article on "Maxwell's Demon" reprinted in the Chant-Fauvel anthology of historical studies.

This section should be preceded by the topic on cooling of the Earth and geological theory up to Lyell (**2.1**-*i*), and the later topic on the debate between Kelvin and the geologists about the age of the Earth (**2.4**-*ii*). If you prefer to put the physical sciences before the life sciences in your course, you may wish to insert those sections before **7.3**. The Fourier theory of heat conduction is taken up again in connection with mathematical physics in **7.5**.

SYNOPSIS

The major emphasis in this section is on two interacting themes in the transition from Newtonian to modern physics: (1) the replacement of the cyclic, reversible, clockwork universe of the eighteenth century by the evolutionary, irreversible, dissipative universe of the nineteenth century; (2) the replacement of the deterministic causal universe of the nineteenth century by the random acausal universe of the twentieth century.

The history of the first theme goes back to the seventeenth century or earlier, with the argument by Descartes and Robert Boyle that the world is like a perfect machine that has been created by God with a fixed amount of matter and motion in such a way that it never runs down and its parts never wear out. Newton's second law, $F = ma$, is time-reversible (provided that the force F does not change when one changes t to $-t$). When applied to celestial mechanics, Newton's laws led to the conclusion that all deviations from Kepler's laws were cyclic, not secular (continuing indefinitely in the same direction), contrary to Newton's opinion that the solar system would eventually run down and require rejuvenation (presumably by divine intervention). Thus Laplace could make his famous reply to Napoleon, when asked why his book about the universe did not mention its author, "Sir, I have no need of that hypothesis."

Laplace himself advanced the idea of progressive change in the universe through his nebular hypothesis, which seems to contradict the assumption that all changes in the solar system are cyclic. The nebular hypothesis was closely connected with the idea that the Earth has been cooling down since its formation, and thus with the doctrine of "progression" in geology (**2.1**). Fourier's theory of heat conduction took the obvious fact that heat flows from hot to cold and gave it mathematical expression, thus setting the stage for a confrontation between the principle of irreversibility in heat theory and the principle of reversibility in mechanics. The confrontation was delayed for several decades, however, because Fourier himself considered the science of heat independent of mechanics so the question of reducing one to the other did not arise.

Whereas in geology the natural flow of heat from hot to cold may be considered a "good thing"—it cools the Earth's surface to a temperature hospitable to life—in the heat engine, according to Sadi Carnot, it is a "bad thing"—it reduces the motive power of heat below its maximum possible value. Since any such flow might have been used to produce useful work by inserting a Carnot-cycle steam engine, failure to extract such work represents waste. This idea, briefly mentioned by Carnot, was made explicit by William Thomson in his 1852 paper on the principle of dissipation of energy. At the same time Thomson noted the implication that because of the irreversible cooling of the Earth, the time during which life could exist on Earth was limited: at some time in the past the Earth's surface was too hot, and at some time in the future it will be too cold for life.

The modern form of the second law of thermodynamics, taking account of the first law (heat is not conserved but is interconvertible with other forms of energy), was given by Rudolf Clausius with the help of his "entropy" concept. The concept was introduced in 1854 but not named until 1865; originally, it was defined so that the change in entropy is the amount of heat flow divided by the absolute temperature ($dS = dq/T$). Thus when a given amount of heat (dq) flows from a high temperature to a low temperature, the net change in S is positive. So in general the second law implies that the entropy of the universe, or of a closed system, increases to a maximum, while the energy remains constant. The final state of maximum entropy was known as the "Heat Death" of the universe.

It is important to note that the second law does *not* mean entropy always increases. It may decrease in one part of a system while increasing in another. Moreover, in a system maintained at constant temperature the equilibrium state is not the one with greatest entropy but rather the one with lowest "free energy" F, where $F = E - TS$. Thus at low temperatures a system of molecules that exert forces on each other will seek an arrangement that minimizes the potential energy of the forces; this will often be a state of low entropy. Ignorance of this aspect of thermodynamics (whether innocent or intentional) is responsible for the claim that complex molecules cannot spontaneously form by natural processes because their formation would decrease the entropy of the system—this is a common but fallacious argument against Darwinian evolution.

The macroscopic science of thermodynamics as formulated by Clausius left up in the air the real nature of entropy. The first clue was provided by Maxwell in his famous "demon" thought experiment. Maxwell's demon can violate the second law by sorting molecules of different speeds, thus causing heat to flow from cold to hot. Conversely the thought experiment shows that the irreversible process of heat flowing from hot to cold, which occurs if there are no demons able to discriminate and sort individual molecules, is equivalent to an irreversible process in which an ordered state becomes more disordered. This equivalence was quantified by Ludwig Boltzmann in his equation $S = k \log W$, where W is the probability of a state of the system, which is assumed to be proportional to the number of molecular arrangements corresponding to that state. Highly disordered states, such as those corresponding to thermal equilibrium, correspond to a very large number of possible arrangements and thus have the greatest entropy.

Thus physics seemed to imply that the universe as a whole, and any isolated system within it, must inevitably tend toward a final state of death and disorder. This result was strikingly consistent with the predictions of the "theory of degeneration" **(3.3)** and the pessimistic "fin de siècle" mood of late-nineteenth-century European and American culture, even though it is difficult to establish a causal connection between the scientific and cultural doctrines.

The connection between scientific and cultural decline was made most explicitly by the American historian Henry Adams, who in the 1890s began collecting evidence for all kinds of dissipation—physical, biological, geological, and social. In three essays, later published under the title *The Degradation of the Democratic Dogma,* he argued that a science of history could be based on the general properties of energy and entropy as discovered by the physical sciences. As his brother Brooks Adams pointed out in the introduction to this book, disillusion with American democracy lurked in the background of Henry Adams's speculations.

The inevitability of dissipation did not go unchallenged in the scientific community. William Thomson in 1874 and Josef Loschmidt in 1876 raised the question: how can a theory based on the assumption that Newton's reversible laws apply to individual molecules be consistent with a principle of irreversible increase of entropy? This became known as the "reversibility paradox."

Henri Poincaré in 1893 and Ernst Zermelo in 1896 raised the related question: how can such a theory be consistent with irreversibility, if any mechanical system in a bounded space is subject to periodic recurrences of any initial state (as Poincaré had shown in 1889)? This became known as the "recurrence paradox." (Friedrich Nietzsche's revival of the ancient "eternal return" doctrine suggests that the theme of cyclic recurrence can be elicited on a philosophical as well as a scientific level in response to pronouncements about tendencies toward decline.)

Ludwig Boltzmann's attempts to resolve the reversibility and recurrence paradoxes played a major role in developing the modern "statistical" interpretation of thermodynamics. One result was his relation between entropy and disorder ($S = k \log W$, where W is the probability of a physical state, expressed in terms of the number of atomic arrangements corresponding to it). He also postulated a condition of "molecular disorder" in his analysis of collisions. Thus physicists became accustomed to the idea of randomness on the molecular level—not merely as a convenient assumption for calculating properties of systems of many molecules, but as perhaps a necessary assumption for explaining why such systems behave irreversibly. The discussion of thermodynamic irreversibility by Maxwell, Thomson, Loschmidt, Poincaré, Zermelo, and Boltzmann was part of the overall change from determinism to indeterminism in physical science.

One of Boltzmann's ideas undermined the absolute character of time itself. Faced with the argument that entropy cannot always increase but may itself be subject to statistical fluctuations, he suggested half-seriously that one could retain the principle of increasing entropy if one simply defines the direction of time as the direction in which entropy increases. In a world with (from our viewpoint) decreasing entropy, irreversible processes including those in living organisms would be going backwards in time, and thus the subjective time perception of those organisms would be reversed; from their point of view entropy is increasing with time.

While Boltzmann's proposed link between entropy and time has not been generally accepted by physicists, it does indicate that the concept of time as an absolute independent variable was beginning to be questioned at the end of the nineteenth century. Philosophers (H. Lotze, F. H. Bradley, G. E. Moore, and others) doubted whether the past is real, much less the future; many agreed that time is not an ultimate reality but merely a relation between events.

In a more mundane context, the railway and the telegraph—especially the Atlantic Cable—made people aware that clocks tell different times in different places and created a need for "standard time zones." For Victorians on both sides of the Atlantic, any uncertainty about what was the "correct" time was a threat to society. As one American put it,

> The furnishing of correct time is educational in nature, for it inculcates in the masses a certain precision in doing the daily work of life which conduces, perhaps, to a sounder morality; and this idea will not seem farfetched if we consider how strikingly indicative of the character of a people in the scale of civilization is the promptness with which they transact their business. [L. Waldo, "The Distribution of Time," *North American Review* 131(1880):528–36]

Newton's smoothly flowing stream of time was about to be replaced by Einstein's observer-dependent fourth dimension (**8.6**). The continuity of temporal processes was to be broken up by other theories proposed around 1905: the Planck-Einstein quantum theory of radiation (**9.4**), the Chamberlin-Moulton theory of the origin of the solar system (**12.4**), the revival of Mendel's genetics (**4.3**) and the instantaneous determination of sex by chromosomes (**4.4**), and Rutherford's radioactive decay (**9.2**). Even Binet's concept of the development of intelligence by stages (**6.4**) and Freud's theory of the impact of childhood events on psychosexual development (**5.2**) could be considered examples of the twentieth-century tendency to see processes as quantized rather than gradual, and thus as part of the overall change in the conception of time.

BIBLIOGRAPHY

Topic *i:* Second Law of Thermodynamics

Sadi Carnot (1978), *Réflections sur la Puissance Motrice du Feu,* ed. R. Fox (Paris: Vrin); review of this book by Eric Mendoza (1981), "The Life and Work of Sadi Carnot," *British Journal for the History of Science* 14:75–78; E. Mendoza, ed. (1960), *Reflections on the Motive Power of Heat, and Other Papers on the Second Law of Thermodynamics by E. Clapeyron and R. Clausius* (New York: Dover), is a translation of the original memoirs with introduction and notes.

M. Bailyn (1985), "Carnot and the Universal Heat Death," *American Journal of Physics* 53:1092–99.

Philip Lervig (1985), "Sadi Carnot and the Steam Engine: Nicolas Clement's Lectures on Industrial Chemistry 1823–28," *British Journal for the History of Science* 18:147–96; (1982), "What Is Heat? C. Truesdell's View of Thermodynamics. A Critical Discussion," *Centaurus* 26:85–122; (1978), "Sadi Carnot and Nicolas Clement," *Proceedings of the XV International Congress of History of Science* (Edinburgh: Edinburgh Univ. Press), 293–304.

Guenter Bierhalter (1985), "Die mechanische Entropie- und Disgregationskonzepte aus dem 19. Jahrhundert: Ihre Grundlagen, ihr Versagen und ihr Entstehungshintergrund," *Archive for History of Exact Sciences* 32:17–41; (1981), Boltzmanns mechanische Grundlegung des zweiten Hauptsatzes der Wärmelehre aus dem Jahre 1866," *Archive for History of Exact Sciences* 24:195–205; (1981), "Clausius' mechanische Grundlegung des zweiten Hauptsatzes der Wärmelehre aus dem Jahre 1871," ibid. 24:207–19; (1981), "Zu Hermann von Helmholtzens mechanischer Grundlegung der Wärmelehre aus dem Jahre 1884," ibid. 25:71–84.

Janez Strand (1984), "The Second Law of Thermodynamics in a Historical Setting," *Physics Education* 19:94–100, advises on how to teach the law to sixteen-year-olds.

Yung Sik Kim (1983), "Clausius's Endeavor to Generalize the Second Law of Thermodynamics," *Archives Internationales d'Histoire des Sciences* 33:22–39.

Keith Hutchinson (1981), "W. J. M. Rankine and the Rise of Thermodynamics," *British Journal for the History of Science* 14:1–26; (1981), "Rankine, Atomic Vortices, and the Entropy Function," *Archives Internationales d'Histoire des Sciences* 31:72–134.

S. S. Wilson (1981), "Sadi Carnot," *Scientific American* 254, no. 2:134–45.

Georges Lochak (1981), "Irreversibility in Physics: Reflections on the Evolution of Ideas in Mechanics and on the Actual Crisis in Physics," *Foundations of Physics* 11:593–621.

Fabio Sebastiani (1981), "La teorie caloricistiche di Laplace, Poisson, Sadi Carnot, Clapeyron e la teoria dei fenomeni termici nei gas formulata da Clausius nel 1850," *Physis* 23:397–438.

C. Truesdell (1980), *The Tragicomical History of Thermodynamics, 1822–1854* (New York: Springer-Verlag); (1979), "Absolute Temperatures as a Consequence of Carnot's General Axiom," *Archive for History of Exact Sciences* 20:357–80.

André Fridberg (1978), *Sadi Carnot, Physicien et les "Carnot" dans l'Histoire* (Paris: La Pensée Universelle).

Edward E. Daub (1978), "Sources for Clausius' Entropy Concept: Reech and Rankine," *Proceedings of the XV International Congress of History of Science* (Edinburgh: Edinburgh Univ. Press), 342–58.

(1976), *Sadi Carnot et l'Essor de la Thermodynamique* (Paris: Centre National de la Recherche Scientifique), is a collection of papers by C. C. Gillispie, E. Mendoza, and others.

Topic *ii:* Entropy and Disorder; Concept of Time

Greg Myers (1986), "Nineteenth-Century Popularizations of Thermodynamics and the Rhetoric of Social Prophesy," *Victorian Studies* 29:35–66.

T. M. Porter (1986), *The Rise of Statistical Thinking 1820–1900* (Princeton, N.J.: Princeton Univ. Press).

O. B. Sheynin (1985), "On the History of the Statistical Method in Physics," *Archive for History of Exact Sciences* 33:351–82.

Ilya Prigogine (1984), "The Rediscovery of Time," *Zygon* 19:433–47, discusses the influence of modern theories of irreversibility.

Stephen Kern (1983), *The Culture of Time and Space 1880–1918* (Cambridge: Harvard Univ. Press).

Engelbert Broda (1983), *Ludwig Boltzmann: Man, Physicist, Philosopher* (Woodbridge, Conn.: Ox Bow Press), is a translation of his 1955 biography, with minor revisions; (1980), "Boltzmann, Einstein, Natural Law and Evolution," *Comparative Biochemistry and Physiology* 67B:373–78.

D. Flamm (1983), "Ludwig Boltzmann and His Influence on Science," *Studies in History and Philosophy of Sciences* 14:255–78.

Vincent S. Steckline (1983), "Zermelo, Boltzmann, and the Recurrence Paradox," *American Journal of Physics* 51:894–97.

Stephen G. Brush (1982), "Changes in the Concept of Time during the Second Scientific Revolution," in *Ludwig Boltzmann Internationale Tagung . . . 1981,* ed. R. Sexl and J. Blackmore (Graz, Austria: Akademische Druck- u. Verlagsanstalt), 305–28; (1981), "Nietzsche's Recurrence Revisited: The French Connection," *Journal of the History of Philosophy* 19:235–38; (1979), "Scientific Revolutionaries of 1905: Einstein, Rutherford, Chamberlin, Wilson, Stevens, Binet, Freud," in *Rutherford and Physics at the Turn of the Century,* ed. M. Bunge and W. R. Shea (New York: Science History), 140–71; (1978), *The Temperature of History* (New York: Franklin); (1976), *The Kind of Motion We Call Heat* (New York: North-Holland/American Elsevier), Chap. 14; (1976), "Irreversibility and Indeterminism," *Journal of the History of Ideas* 37:603–30, reprinted in (1983), *Statistical Physics and the Atomic Theory of Matter* (Princeton, N.J.: Princeton Univ. Press), Chap. 2.

Silvan S. Schweber (1982), "Demons, Angels, and Probability: Some Aspects of British Science in the Nineteenth Century," in *Physics as Natural Philosophy,* ed. A. Shimony and H. Feshbach (Cambridge: MIT Press), 319–63.

Michael Heidelberger and Lorenz Krüger, eds. (1982), *Probability and Conceptual Change in Scientific Thought* (Bielefeld, Germany: Universität Bielefeld/B. Kleine Verlag).

7.4 THE REACTION AGAINST MECHANISM AND ATOMISM

i. Philosophical movements in the late nineteenth century (positivism, neoromanticism, Idealism; critique of "mechanistic" science)
ii. The atomic debates in physics and chemistry: Mach, Wilhelm Ostwald, Pierre Duhem; Boltzmann's defense of atomism
iii. The resurrection of atomism after 1900: Einstein's theory of Brownian movement, Jean Perrin's experiments and "molecular reality"

READINGS

McKenzie, *Major Achievements,* 338–45, 528–30 (Mach and Poincaré). Mason, *History of the Sciences,* 499–502 (Mach). Schneer, *Evolution of Physical Science,* 288–97. Brush, *Temperature of History,* 84–101.

SYNOPSIS

In physics texts one often reads that scientists at the end of the nineteenth century thought they had discovered all the basic laws of nature, and nothing remained but to measure all the physical constants to another decimal place of accuracy. This alleged state of complacency, according to the textbook history, was shattered by the revolution in atomic physics and relativity theory at the start of the twentieth century. While one can indeed find statements suggesting such complacency (e.g., by A. A. Michelson), it seems to me that the peak of confidence in the success of classical physics occurred around 1870; the last quarter of the nineteenth century saw substantial loss of confidence in the adequacy of the mechanistic worldview, combined with strong criticism of its pretensions to reduce all other sciences to physics. (Some of this feeling may indeed have been fear that the reduction would be successful; antagonism to Darwin's theory of evolution may have escalated into a general hostility toward mechanistic theories.)

The change in attitude toward science is reflected in the ambiguous use of the term "positivism." On the one hand it stands for the attitude that science is the only route to knowledge; on the other hand it constitutes a criticism within science, directed against theoretical speculation and favoring empiricism. In particular, empiricist/positivist critics objected to the use of atomic models on the grounds (1) they illegitimately attributed reality to something that had never been and perhaps might never be observed; (2) they had become so complicated and inconsistent as to be no longer useful in organizing research.

While Mach, Ostwald, Duhem and a few others were able to make skepticism about atoms the most fashionable position around 1900, one may wonder whether the average scientist really accepted that position. Yet some historians have claimed that antiatomism was partly responsible for the decline of French science and, in particular, it may have prevented the Curies from recognizing the evidence for nuclear transmutation (Malley 1979).

It is curious that Boltzmann, the most vociferous defender of atomism and the most experienced practitioner of statistical methods in physics, failed to recognize how the phenomenon of Brownian movement could be used to support his case. That was left to Einstein, who showed in one of his famous 1905 papers that fluctuations in the molecular bombardment of microscopic particles would lead to observable consequences, and to Jean Perrin, who performed the experiments to test Einstein's theory. Perrin used Einstein's theory to "prove" the reality of the atom—i.e., he made such a good case that even Wilhelm Ostwald, a leading antiatomist before 1905, changed his mind and agreed that the kinetic-atomic hypothesis must be accepted in physical chemistry. (Mach, whose objections were more metaphysical, never did retract his criticism, at least not in print.)

Any course on the history of science dealing with atoms should at least mention the scientist who established their existence and explain how he did it. That this did not happen until after 1900, in spite of the rather large quantity of evidence about atomic properties available before that time (e.g., from kinetic theory), demands in turn an explanation of why scientists were skeptical or opposed to atomic theories, and that is one justification for the cultural-philosophical excursion suggested in this section.

Incidentally, the first person to actually "see" an atom was Erwin Mueller, in 1955, with the field-ion microscope he perfected.

BIBLIOGRAPHY

Mary Jo Nye, ed. (1984), *The Question of the Atom, from the Karlsruhe Congress to the First Solvay Conference, 1860–1911* (Los Angeles: Tomash), includes the major original sources for this topic.

Stephen G. Brush (1978), *The Temperature of History* (New York: Burt Franklin), Chaps. 6–8; (1976), *The Kind of Motion We Call Heat* (New York: North-Holland), Sect. 1.9 and Chaps. 8, 15.

Topic *i:* Philosophical Movements

John T. Blackmore (1985), "An Historical Note on Ernst Mach," *British Journal for the Philosophy of Science* 36:299–329, argues that his antiatomism was philosophical rather than scientific; (1982), "The Rise and Fall of Three Fashionable Expectations," *Indian Journal of History of Science* 17:279–88, concerns electromagnetic, energetic, and phenomenalist-reductionist doctrines.

Wolfram W. Swoboda (1982), "Physics, Physiology, and Psychophysics: The Origins of Ernst Mach's Empiriocriticism," *Rivista di Filosofia* 73:234–74.

Roy MacLeod (1982), "The 'Bankruptcy of Science' Debate: The Creed of Science and Its Critics, 1885–1900," *Science, Technology and Human Values* 7, no. 41:2–15.

R. E. Martin (1981), *American Literature and the Universe of Force* (Durham, N.C.: Duke Univ. Press).

Topic *ii:* Atomic Debates

Ernst Mach (1943, reprint of 5th ed. of 1895), *Popular Scientific Lectures* (La Salle, Ill.: Open Court), includes "On the Economical Nature of Physical Inquiry."

Wilhelm Ostwald (1896), "Emancipation from Scientific Materialism," (translation from 1895 lecture) *Science Progress* 4:419–36.

H. Krips (1986), "Atomism, Poincaré and Planck," *Studies in History and Philosophy of Science* 17:43–63.

Paul K. Feyerabend (1984), "Mach's Theory of Research and Its Relation to Einstein," *Studies in History and Philosophy of Science* 15:1–22.

D. Flamm (1983), "Ludwig Boltzmann and His Influence on Science," *Studies in History and Philosophy of Science* 14:255–78.

J. L. Heilbron (1982), "*Fin-de-siècle* Physics," in *Science, Technology, and Society in the Time of Alfred Nobel,* ed. C. G. Bernhard et al. (New York: Pergamon Press), 51–73.

Ulrich Hoyer (1982), "Boltzmanns Verhältnis zum Positivismus," *Rivista di Filosofia* 73:275–89.

Jaakko Hintikka et al., eds. (1981), *Probabilistic Thinking, Thermodynamics and the Interaction of the History and Philosophy of Science, Proceedings of the 1978 Pisa Conference on the History and Philosophy of Science,* vol. 2 (Boston: Reidel), Section V: "Thermodynamics and Physical Reality." Papers by L. Krüger, E. N. Hiebert, V. Kartsev, O. A. Lezhneva on Boltzmann and Mach.

Marjorie Malley (1979), "The Discovery of Atomic Transmutation:

Scientific Styles and Philosophies in France and Britain," *Isis* 70:213–23, shows how positivism prevented the Curie group from recognizing transmutation.

Jacques Petrel (1979), "La Negation de l'Atome dans la Chimie du XIXieme siècle—Cas de Jean-Baptiste Dumas,"*Cahiers d'Histoire et de Philosophie des Sciences* 3, 142 pp.

Mary Jo Nye (1976), "The Nineteenth-Century Atomic Debates and the Dilemma of an 'Indifferent Hypothesis,' " *Studies in History and Philosophy of Science* 7:245–68.

Peter Clark (1976), "Atomism versus Thermodynamics," in *Method and Appraisal in the Physical Sciences,* ed. C. Howson (New York: Cambridge Univ. Press), 41–105.

Topic *iii:* Resurrection of Atomism and Brownian Movement

J. D. Nightingale and R. E. Kelly (1984), "Note on Einstein's First Paper," *American Journal of Physics* 52:560–61, comments on Einstein's paper on capillarity.

David Fenby (1981), "Einstein and Molecular Reality," *Chemistry in Britain* 17:114–18.

R. B. Lindsay (1979), *Early Concepts of Energy in Atomic Physics* (New York: Academic Press), 342–49, has excerpts from papers by Einstein and Langevin.

Milton Kerker (1976), "The Svedberg and Molecular Reality," *Isis* 67:190–216.

Erwin W. Müller (1957), "Atoms Visualized," *Scientific American* 196, no. 6:113–22 and cover, is an account of the field emission technique that finally made it possible to see an individual atom on the tip of a tungsten needle, thereby answering Mach's objection.

7.5 MATHEMATICS: STATISTICS, SERIES, AND SETS

i. Probability theory and statistics: J. Bernoulli, Laplace, Carl Friedrich Gauss, Adolphe Quetelet; applications in the physical and social sciences

ii. The vibrating string problem; partial differential equations in physics; Fourier's trigonometric series; redefinition of "function"

iii. Georg Cantor's set theory, transfinite numbers; intuitionism and the axiom of choice; later impact of set theory on the "new math" of the 1960s

READINGS

Marks, *Science,* 324–70. Howard Eves, *An Introduction to the History of Mathematics* (4th ed. 1976), Chaps. 13–15. J. M. Dubbey, *Development of Modern Mathematics* (1972), Chap. 7. Carl B. Boyer, *A History of Mathematics* (1968), 536–37, 598–617. Edna E. Kramer, *The Nature and Growth of Modern Mathematics* (1970), 297–324 (on Cantor, Zermelo, set theory). J. W. Dauben, "Georg Cantor and the Origins of Transfinite Set Theory," *Scientific American* 248, no. 6(1983):122–31. I. Grattan-Guinness, "Joseph Fourier and the Revolution in Mathematical Physics," *Journal of the Institute of Mathematics and Applications* 5(1969):230–53. M. Kline, *Mathematics in Western Culture* (1953), Chaps. 19 (on *ii*), 22–23 (on *i*), 25 (on *iii*). Cecil J. Schneer, *Evolution of Physical Science* (1984), 250–56 (on *i*).

The Open University course AM 289, *History of Mathematics* (1976), provides good coverage of topics *i* and *iii* in Unit 8, "Non-Euclidean Geometry" by G. Flegg, 24 pp.; Unit 10, "Paradoxes and the Infinite" by G. Flegg, 19 pp; and Unit 11, "Mathematics and Man" by S. H. Hollingdale, R. L. Wilder, and G. Flegg, 46 pp. In the latter, 7–38 are on statistics, probability, and the computer and are keyed to Kline's *Mathematics in Western Culture,* Chaps. 22 and 23.

Introductory textbooks on the history of science (and, I suspect, introductory courses) rarely give much attention to the history of mathematics. It is a subject reserved for separate treatment and thought to be of interest only to mathematics students and professors. But I would urge the inclusion of a few topics selected either for their importance to the development of other sciences or because they can be understood by students without strong mathematical backgrounds. The first two topics recommended in this section fall in the first category, the third in the second. Statistical theory is obviously closely related to the atomic-molecular theory of matter presented in the immediately preceding sections, as well as to population genetics (**4.6**) and intelligence testing (**6.4–6.8**). Fourier's method for solving differential equations by trigonometric series arose in the context of heat theory (**7.3**) which in turn was partly inspired by geological problems (**2.2**). Periodic solutions of related differential equations are important in other branches of physics such as electromagnetic waves (**8.4**) and quantum theory (**9.6**).

COMMENT

When Fourier showed that trigonometric series could be used to represent functions that are discontinuous or have discontinuous slopes as well as the "smooth" functions familiar to earlier scientists and mathematicians, he forced the introduction of a more general definition of function and the development of a more powerful theory (set theory) to deal with such functions. Since the 1960s American schoolchildren have been forced to learned this highly nonintuitive definition: a function is a set of ordered pairs no two of which have the same first member. In my opinion the introduction of abstract mathematical concepts in this apparently arbitrary way has convinced many students, quite unjustifiably, that they cannot understand mathematics and should drop the subject as soon as possible. Explaining the historical basis for such concepts is one way (though perhaps not the best way) to alleviate the "math anxiety" that seems to affect most of us.

BIBLIOGRAPHY

David Abbott, ed. (1986), *The Biographical Dictionary of Scientists: Mathematicians* (New York: Bedrick).

Joseph W. Dauben (1985), *The History of Mathematics from Antiquity to the Present: A Selective Bibliography* (New York: Garland), has sections on "Mathematics in the 19th Century" (146–203), "Probability and Statistics" (363–71), "Mathematical Physics" (296–27) and "Set Theory" (378–84).

S. A. Jayawardene (1983), "Mathematical Sciences," in *Information Sources in the History of Science and Medicine,* ed. P. Corsi and P. Weindling (Boston: Butterworth Scientific), 259–84.

George W. Mackey (1978), "Harmonic Analysis as the Exploitation of Symmetry: A Historical Survey," *Rice University Studies* 64, nos. 2, 3:73–228, ties the topics of this section to the ideas of symmetry groups; it assumes that the reader has a background in modern mathematics.

Morris Kline (1972), *Mathematical Thought from Ancient to Modern Times* (New York: Oxford Univ. Press), is the most comprehensive single reference on the history of mathematics. Chapters 28, 40, and 41 deal with topics *ii* and *iii;* Chap. 43 presents an overview of mathematics in 1900.

Topic *i:* Statistics

Lorenz Kruger et al., eds. (1987), *The Probabilistic Revolution* (Cambridge, Mass.: MIT Press).

Stephen M. Stigler (1986), *The History of Statistics* (Cambridge: Harvard Univ. Press).

T. M. Porter (1986), *The Rise of Statistical Thinking 1820–1900* (Princeton, N.J.: Princeton Univ. Press).

Elliott W. Montroll and Michael F. Shlesinger (1984), "On the Wonderful World of Random Walks," in *Nonequilibrium Phenomena II, From Stochastics to Hydrodynamics,* ed. J. L. Lebowitz and E. W. Montroll (New York: North-Holland), 1–121, discusses several episodes from the history of probability theory.

Michael Heidelberger et al., eds. (1983), *Probability Since 1800* (Bielefeld, Germany: Universität Bielefeld).

S. S. Schweber (1982), "Demons, Angels, and Probability: Some Aspects of British Science in the Nineteenth Century," in *Physics as Natural Philosophy,* ed. A. Shimony and H. Feshbach (Cambridge: MIT Press), 319–63, explains connections with Darwinism and kinetic gas theory.

Michael Heidelberger and Lorenz Krüger, eds. (1982), *Probability and Conceptual Change in Scientific Thought* (Bielefeld, Ger.: Universität Bielefeld/B. Kleine Verlag).

Lorraine J. Daston (1981), "Mathematics and the Moral Sciences: The Rise and Fall of the Probability of Judgments, 1785–1840, in *Epistemological and Social Problems of the Sciences in the Early Nineteenth Century,* ed. H. N. Jahnke and M. Otte (Boston: Reidel), 287–309.

Donald A. MacKenzie (1981), *Statistics in Britain, 1865–1930: The Social Construction of Scientific Knowledge* (New York: Columbia Univ. Press), emphasizes social influences on the development of scientific theory.

O. B. Sheynin (1979), "C. F. Gauss and the Theory of Errors," *Archive for History of Exact Sciences* 20:21–72.

Edward R. Hogan (1977), "Robert Adrain: American Mathematician," *Historia Mathematica* 4:157–72.

Topic *ii:* Differential Equations, Series, Functions

W. Scharlau (1981), "The Origins of Pure Mathematics," in *Epistemological and Social Problems . . . ,* ed. H. N. Jahnke and M. Otte (Boston: Reidel), 331–47, is on the importance of Fourier's heat theory.

Pierre Dugac (1981), "Des Fonctions comme Expressions Analytiques aux Fonctions Representables Analytiquement," in *Mathematical Perspectives,* ed. Joseph Dauben (New York: Academic Press), 13–36.

I. Grattan-Guinness (1981), "Mathematical Physics in France, 1800–1840: Knowledge, Activity, and Historiography," in *Mathematical Perspectives,* ed. J. Dauben (New York: Academic Press), 95–138; (1980), "The Emergence of Mathematical Analysis and Its Foundational Progress, 1780–1880," in his *From the Calculus to Set Theory* (London: Duckworth).

Thomas Hawkins (1980), "The Origins of Modern Theories of Integration" in *From the Calculus to Set Theory,* ed. I. Grattan-Guinness (London: Duckworth), 145–80.

Topic *iii:* Set Theory

J. W. Dauben (1980), "The Development of Cantorian Set Theory" in *From the Calculus to Set Theory,* ed. I. Grattan-Guinness (London: Duckworth), 181–219; (1979), *Georg Cantor* (Cambridge: Harvard Univ. Press); (1977), "Georg Cantor and Pope Leo XIII: Mathematics, Theology, and the Infinite," *Journal of the History of Ideas* 38:85–108.

Howard Delong (1970), *A Profile of Mathematical Logic* (Reading, Mass.: Harvard Univ. Press), Chaps. 1, 2.

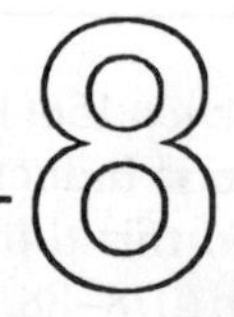

ELECTROMAGNETISM AND RELATIVITY

8.1 FARADAY

i. Life of Faraday; Humphrey Davy and the Royal Institution; alleged philosophical influences
ii. Discoveries: electric motor, electromagnetic induction, and the dynamo
iii. Field concept, impact on physics

READINGS

A. E. E. McKenzie, *The Major Achievements of Science* (1973), 176–85, 458–60. John Marks, *Science* (1983), 170–73. G. Holton and S. G. Brush, *Introduction to Concepts and Theories in Physical Science* (1973, 1985), 417–20. C. Schneer, *Evolution of Physical Science* (1984), 266–76. E. Segre, *From Falling Bodies to Radio Waves* (1984), 132–55.

Books that present electromagnetism as part of the prehistory of relativity: A. Einstein and L. Infeld, *Evolution of Physics* (1938, 1967), 125–42. Joan Solomon, *The Structure of Space* (1974), 98–107. B. Hoffmann, *Relativity and Its Roots* (1983), 64–69.

Books dealing specifically with electricity: Harold Sharlin, *Making of the Electrical Age* (1963), 133–41. D. C. Goodman, "Faraday and Electromagnetism," Unit 7 in the Open University course AST 281, *Science and the Rise of Technology since 1800* (1973), 87–98. Sanford P. Bordeau, *Volts to Hertz* (1982), 108–33. L. P. Williams, *The Origins of Field Theory* (1966, 1980), 72–117.

Articles: L. P. Williams, "Michael Faraday and the Physics of 100

Years Ago," *Science* 156(1967):1335–42. Samuel Devons, "The Search for Electromagnetic Induction," *Physics Teacher* 16(1978):625–31. A 16 mm color film on this topic is also available; see *American Journal of Physics* 49(1981):1078–88. Jose M. Sanchez-Ron, "The Problem of Interaction: On the History of the Action-at-a-Distance Concept in Physics," *Fundamenta Scientiae* 4 (1983):55–76.

SYNOPSIS

Michael Faraday (1791–1867) is a national hero to the British, not only because of the importance of his discoveries but because his career seems to show that a bright energetic lad from the lower classes can rise to the top of the heap. The romantic story of Faraday's "discovery" (in the show-biz sense) by Humphrey Davy and his remarkable feats in the laboratory has been told many times. I will note here only a few aspects of the story that call for discussion in a college-level course.

A valuable function of scientific journals, in addition to their primary function of publishing original research, is to provide for their readers comprehensive reviews of recent advances in science. In 1821 the editor of the British journal *Annals of Philosophy* asked Faraday to survey the flurry of work on electromagnetism that had appeared in the previous year, starting with Oersted's discovery (**7.2-*ii***). This seems to have provided the impetus for Faraday to start his own experiments on electricity—previously he had been working in chemistry. His first discovery was made on 3 September 1821, when he realized that the force exerted by a current on a magnet is circular in nature, so a magnetic pole that is free to move will be pushed in a circle around a fixed wire. But how to prevent the wire from getting tangled up with the apparatus that suspends the magnet? Solution: use mercury (a *liquid* conductor of electricity) to complete the circuit. Thus Faraday invented the first electric "motor."

Faraday was guided by the idea that for every effect of electricity on magnetism there must be a corresponding effect of magnetism on electricity (cf., Newton's third law); thus, by a slight modification, he designed a motor in which the magnet was fixed and the current-carrying wire rotated around it.

The French physicist A. M. Ampère criticized Faraday's report of his "electromagnetic rotator" on theoretical grounds, starting from the presupposition that all forces in nature must be similar to the Newtonian gravitational force. In particular they must act directly between particles of matter in a direction along the line between the centers of the particles ("central" forces). They may be either attractive or repulsive, and need not be inverse square, but certainly they cannot be "circular." The Oersted-Faraday description of the force exerted by a current-carrying wire made it neither attractive nor repulsive, but acting in a direction perpendicular to the line between the current element and the pole. Ampère argued that all electromagnetic interactions can be reduced theoretically to interactions between individual current elements, and that magnets are composed of molecular currents. Thus what seemed to be a circular force was really, according to Ampère, the total effect produced by a large number of direct forces between current elements.

One might interpret the disagreement between Faraday and Ampère as a clash of paradigms, and one might even argue that Ampère's insistence on seeing the world in terms of Newtonian central forces had prevented him (and other physicists) from discovering electromagnetism. Oersted and Faraday had to break out of that paradigm to make their discoveries. But, once the discoveries were made, Ampère was able to reinterpret them within his own scheme and make new discoveries such as the force between current elements (**7.2-*ii***). It should be noted, however, that Ampère's theoretical expression for the force between current elements, which assumes the forces are central although their magnitudes vary with relative orientation, is not the one now accepted; it does give the same result when integrated over the entire lengths of the wires.

Ampère preferred his explanation of electromagnetic interactions because it could be expressed mathematically with the equations familiar in Newtonian physics. Faraday, on the other hand, did not understand much mathematics, but he did have an amazing intuitive feeling for physical phenomena. He is sometimes cited as the last physicist who could work successfully (especially in developing theoretical concepts) without using mathematics, and this may be why he is such a popular figure. Others argue that he did use mathematics of a nonquantitative kind, i.e., topology, but this seems a little farfetched to me.

Faraday's most important discovery was electromagnetic induction in 1831; this provided a means for generating electric currents by moving a wire relative to a magnetic field. In its original form the discovery was that a current in one coil could induce a current in a nearby coil, but that induction occurs only when the primary current stops or starts. Faraday reasoned that this means that the magnetic lines of force around the wire are moving inwards or outwards at that time; it is essentially the relative motion of field and wire that induces the current, and thus the same effect can be produced by moving a magnet without having any current to start with.

Like many discoveries that have been preceded by a period of preliminary research and discussion among scientists, this one was made almost simultaneously by two scientists working independently in different countries. The American scientist Joseph Henry, who was teaching at an academy in Albany, New York, actually discovered induction a little before Faraday. Unfortunately for the reputation of American science, teachers at that academy were expected to spend all their time on teaching and administrative duties, with no time left for research. Henry had hardly any opportunity to follow up his discovery, which he made during a one-month summer vacation, and did not publish it until a year later.

In order to generate a continuous current, Faraday constructed a "disc dynamo" in which a copper disc could be rotated between the poles of a magnet; brass brushes maintained electrical contact with the axis on which the disc turned. This arrangement produced alternating current; the "commutator," invented by H. Pixii (a French technician working for Ampère) was later used to convert AC to DC.

The discovery of electromagnetic induction had enormous technological consequences, since it allowed electricity to be produced cheaply (by converting mechanical energy, which could be supplied from the steam engine, to turn the crank of the dynamo), rather than by using batteries that consumed expensive metals by chemical reactions. While Faraday realized the practical importance of his discoveries, his primary interest was in pure science; he left the development of the generator and motor to others. This development did not really "take off" until about fifty years later. The story is that Zenobe Gramme, a Belgian inventor, sent two of his dynamos for display at the Vienna Exhibition of 1873. An unknown workman happened to hook them up in such a way that the electricity generated by one dynamo made the other one run backwards, in effect as a motor.

In his biography of Faraday and in other publications, L. Pearce Williams has suggested that Faraday's ideas about forces and fields were influenced by I. Kant's theories and German *Naturphilosophie,* which reached Faraday through S. T. Coleridge and Humphrey Davy. This suggestion has been criticized by J. Brookes Spencer and others. The issue might well be discussed along with the influence of Nature Philosophy on the discovery of energy conservation (**7.2**) as an example of the possible role of metaphysics in guiding scientists.

See **8.3-*i*** for Faraday's discovery of the "magnetization of light" (rotation of plane of polarization).

BIBLIOGRAPHY

David Gooding and Frank A. J. L. James, eds. (1985), *Faraday Rediscovered: Essays on the Life and Work of Michael Faraday* (New York: Macmillan).

Topic *i:* Life of Faraday

L. Pearce Williams (1971), "Faraday, Michael," *Dictionary of Scientific Biography* 4:527–40; (1965), *Michael Faraday* (New York: Basic Books).

Topic *ii:* Discoveries

Michael Faraday (1965), *Experimental Researches in Electricity* (1839–1855, reprinted New York: Dover, 1965); (1932–1936), *Faraday's Diary,* ed. T. Martin (London: Bell).

David Gooding (1986), "How Do Scientists Reach Agreement about Novel Observations?" *Studies in History and Philosophy of Science* 17:205–30; (1982), "Empiricism in Practice: Teleology, Economy, and Observation in Faraday's Physics," *Isis* 73:46–67; (1982), "A Convergence of Opinion on the Divergence of Lines: Faraday and Thomson's Discussion of Diamagnetism," *Notes and Records of the Royal Society of London* 36:243–59; (1980), "Metaphysics versus Measurement: The Conversion and Conservation of Force in Faraday's Physics," *Annals of Science* 37:1–29; (1978), "Conceptual and Experimental Bases of Faraday's Denial of Electrostatic Action at a Distance," *Studies in History and Philosophy of Science* 9:17–49.

P. Tucci (1984), "Faraday contro Nobili: Un Episodio della Polemica Antiamperiana," *Giornale di Fisica* 25:347–64.

P. Marazzini and P. Tucci (1983), "The Physical Meaning of Faraday's Law from an Historical Point of View," *European Journal of Physics* 4:170–79.

Stanley M. Guralnick (1979), "The Contexts of Faraday's Electrochemical Laws," *Isis* 70:59–75.

Topic *iii:* Field Concept, Influence on Physics

Nancy J. Nersessian (1984), "Aether/Or: The Creation of Scientific Concepts," *Studies in History and Philosophy of Science* 15:175–212; *Faraday to Einstein* (Dordrecht: Nijhoff), 3–66.

Robert D. Friedel (1981), *Lines and Waves: Faraday, Maxwell, and 150 Years of Electromagnetism* (New York: Center for the History of Electrical Engineering).

Robert L. Oldershaw (1981), "Faraday, Maxwell, Einstein and Epistemology," *Nature and System* 3:99–107.

Enrico Bellone (1981), *La Relativita da Faraday a Einstein* (Turin, Italy: Editore Loescher).

David Gooding (1981), "Final Steps to the Field Theory: Faraday's Study of Magnetic Phenomena, 1845–1850," *Historical Studies in the Physical Sciences* 11:231–75; (1980), "Faraday, Thomson and the Concept of the Magnetic Field," *British Journal for the History of Science* 13:91–120.

Ole Knudsen (1976), "The Faraday Effect and Physical Theory, 1845–1873," *Archive for History of Exact Sciences* 15:235–81.

L. Pearce Williams (1975), "Should Philosophers Be Allowed to Write History?" *British Journal for the Philosophy of Science* 26:241–53, is a review of books by Agassi (on Faraday) and Berkson (on field theory); see replies by Berkson and Agassi (1978), ibid. 29:243–48, 248–52, and rejoinder by Williams, 252.

8.2 BEGINNINGS OF ELECTRICAL TECHNOLOGY

i. Invention of the electric telegraph
ii. The Atlantic Cable as a case history of a technological project; Kelvin's role
iii. The telephone: Alexander Graham Bell and Elisha Gray; growth of telephone systems and their social impact
iv. Electric power systems and their early applications, especially to lighting

READINGS

Marks, *Science,* 192–95. Harold I. Sharlin, *Making of the Electrical Age* (1963), 1–71, 141–230. D. C. Goodman, "Faraday and Electromagnetism," Unit 7 in the Open University course AST 281, *Science and the Rise of Technology since 1800,* 99–126 (on generators, incandescent lamp, power systems). G. R. M. Garratt, "The Communications Explosion in the Nineteenth Century: Some Contributions of Electrical Engineering," Unit 14 in ibid., 8–37 (telegraph, telephone). S. P. Bordeau, *Volts to Hertz,* 134–63, 208–19.

Articles: David Hounshell, "Two Paths to the Telephone," *Scientific American* 244, no. 1(1981):157–63 (on A. G. Bell and E. Gray). John Harris, "The Hairpin in the Bottle," *Physics Education* 14(1979):436–40 (development of light bulbs). Matthew Josephson, "The Invention of the Electric Light," *Scientific American* 201, no. 5(1959):98–114. Thomas P. Hughes, "How Did the Heroic Inventors Do It?" *American Heritage of*

Invention and Technology 1, no. 2(1985):18–25 (on Edison, Sperry, Tesla, De Forest). Noel Vietmeyer, "Inventing Charles Wheatstone," *Science 84,* 5, no. 5(1984):70–75. Robert Friedel, "New Light on Edison's Light," *American Heritage of Invention and Technology* 1, no. 1(1985):22–27.

This would be a good place for an excursion into the relations between science, technology, and the Industrial Revolution. The series of short texts by M. J. Fyrth and M. Goldsmith, *Science, History and Technology* (1965–1969), offers a good overview; see Book 2, Part I, Chap. 3, for the relations between electrical communications and other topics.

SYNOPSIS

One of the first practical applications of electromagnetic induction was the development of a practical electric telegraph. As early as 1809, Samuel von Soemmerring in Germany had constructed an electrochemical telegraph: using 35 wires attached to 35 gold electrodes in a tank of water, he could transmit any letter or number by sending current through one of the wires, producing gas bubbles at the electrode. While this device was not practical for long-distance communication, it stimulated other scientists and inventors to think about better methods. Soon after Oersted's discovery in 1820, Ampère suggested that messages could be sent by using electric current to deflect magnetic needles. J. S. C. Schweigger in Germany took one step toward realizing this proposal by inventing the "galvanometer," a device for magnifying the deflection of the needle by rolling the wire into a coil. Electromagnetic induction provided a cheaper and more powerful source of electric current than the galvanic battery and thus opened up the possibility of long-distance communication.

Several of the most eminent scientists of the early nineteenth century—Ampère, Joseph Henry, C. F. Gauss, and W. Weber—speculated and experimented on the electric telegraph but didn't quite manage to make a commercially successful invention out of it. Instead it was W. F. Cooke and Sir Charles Wheatstone in England, in 1837, building on the progress made by others, who first developed a practical system. Their telegraph was adopted for communication between railway stations in England in 1839 and later became an essential part of the system for controlling train movements to avoid collisions.

Another advantage of rapid communication was dramatically demonstrated in 1845 when John Tawell, disguised as a Quaker, jumped on a train to London after killing his mistress at Slough. The police sent a telegraph message ahead, giving his description so he could be arrested when he arrived. This was a nice illustration of the new balance of social forces brought about by technology: fast transportation permitted the criminal to make a getaway, but even faster communication permitted the long arm of the law to catch him anyway.

During this period Samuel F. B. Morse (1791–1872), an American portrait painter, was working on his own telegraph system. In contrast to the European and British systems requiring several different wires to send different letters, the essential feature of Morse's system was the "Morse code," which permitted the message to be sent on a single circuit by representing letters by a series of dots and dashes. Morse's system was therefore much cheaper than Cooke and Wheatstone's though it required more skill on the part of the operator who had to remember the code. Congress, after some delay, voted funds for a trial of the system in 1843, and the first experimental line was opened between Washington and Baltimore in 1844. Commercial success came with the Civil War, when military use of the telegraph brought prosperity to companies like Western Union whose lines did not happen to cross the boundary between North and South. In 1866 Western Union bought up its two chief rivals and soon had a fairly complete monopoly.

Increasing use of the telegraph led engineers and inventors to look for ways in which information could be sent more rapidly. To avoid the expense of constructing duplicate lines, they tried to devise ways of sending two or more messages at the same time on a single line. One person who tried to develop such a "multiple telegraph" was Alexander Graham Bell (1847–1922). Bell, whose father and grandfather had been professors of elocution in London and Edinburgh, came to the United States in 1872 to open a school in Boston for deaf mutes and others with defective speech. Continuing his father's interest in the science of speech, Bell studied the work of the German scientist Hermann von Helmholtz on sound waves and looked into methods of transmitting sound electrically by means of tuning forks. In the course of trying to develop a multiple telegraph on this basis, he discovered in 1874 that a miniature electric generator could be made from a vibrating magnet placed near a coil of wire. If the vibrations were caused by sound waves (according to the principle of "resonance" or "sympathetic vibration" as explained by Helmholtz) then this generator would send into the transmission line electrical currents varying in time in the same way as did the original sound waves. The variations in electric currents could be turned back into motions of a vibrating magnet at the other end of the line, and thus back into sound waves that reproduced the original sound waves.

Like many inventions, the telephone involved well-known scientific principles—in this case, Faraday's electromagnetic induction combined with the reverse action of generators and electric motors. But before Bell, it had not been realized that sound waves carried enough mechanical energy to make a magnet vibrate or that such vibrations would produce a strong enough current to go through a wire and be received at the other end.

Bell now realized that it was theoretically possible to make a device that would transmit the human voice, in recognizable form, across fairly long distances. Knowing little about electricity, he sought advice from Joseph Henry, who had left his professorship at Princeton in 1848 to become Secretary of the new Smithsonian Institution in Washington, D.C. With Henry's encouragement he developed a working model with a wire plunger that could be moved up and down in a cup of dilute sulfuric acid by a mechanical coupling with a diaphragm. He obtained his first patent for this "telephone" on 7 March 1876, and three days later actually sent the first sentence through it. Unlike Morse, who had inaugurated his telegraph in 1844 with the solemn declaration "What hath God wrought," Bell transmitted the first telephone message by accident, not realizing how sensitive his device actually was. He spilled some of the sulfuric acid on his clothes, and had to call quickly to his assistant Thomas Watson in the next room. Hence the historic message:

"Mr. Watson: come here, I want you!"

The story of the "simultaneous invention" of the telephone by Elisha Gray is told by Hounshell in the article mentioned above. Gray filed a caveat against Bell's patent in February 1876 but didn't think enough of the practical possibilities to carry through his own work and get a patent.

The rapid growth of the telephone system in the United States, and the monopoly the Bell Company was able to establish, raised important questions of public policy in dealing with technology. It was generally recognized by 1900 that some kind of state or federal government regulation would be needed to protect the public interest and keep rates down. The telephone was becoming an essential part of business and social life and seemed to be a "natural monopoly" in which competition meant wasteful duplication. Thus there arose the idea of a "public utility," which could be run as a monopoly by a single company under government regulation. This was a peculiarly American invention that has now become fairly well established, although other countries have not adopted it and even in the United States there is a continuing need to revise the scope of the permitted monopoly in the light of new technical developments. Comparison of the history of communication systems in the United States and other countries provides a fascinating case history in the interaction of technology and society, which would be an appropriate part of any course in the history of modern science.

THE ELECTRIC LIGHT

One of the most important electrical inventions of the nineteenth century was the development of a practical incandescent light by Thomas A. Edison (1847–1931). The large quantity of relevant documentary evidence, recently analyzed in detail by Friedel and Israel (1985), makes this a good case history for studying the nature of technological innovation in modern times.

Although Edison and others had for several years pondered the problem of using electricity to heat a filament and thereby produce light, Edison began his first serious efforts in 1878 at his laboratory in Menlo Park, New Jersey. He had several advantages over his competitors. Aside from his own talent, determination, and self-confidence, he had established a remarkable reputation for other inventions in telegraphy and telephony, and most recently for the invention of the phonograph. This success provided the resources to equip a laboratory and staff it with competent dedicated workers who could do the research needed to find the right materials, design the most efficient electrical devices, and test various ways of combining all the components into a workable system. It also attracted wealthy investors to put up the necessary cash, and sympathetic reporters to generate public interest and a demand for his product.

The early incandescent lamps failed because their filaments burned up too quickly at the high temperatures needed to produce a useful amount of light. At first Edison tried to solve this problem by designing an automatic regulator to shut off the electric current when the filament became dangerously hot. This line of work was not fruitful so he then turned to a systematic study of the properties of materials at high temperatures, with the hope of finding a substance that could survive repeated heating. Platinum seemed to be the best prospect but still burned up when heated in air.

Edison now returned to an earlier idea: enclose the filament in an evacuated bulb, so the absence of oxygen will prevent burning. Previously the vacuum pumps available had not been effective in removing enough air for this purpose, but Hermann Sprengel's invention of a new kind of pump in 1865 made this approach much more attractive. Note that the Sprengel pump also made possible the study of cathode rays and thus led to the discovery of the electron (**9.1**).

With the help of expert glassblowers and technicians, Edison's group built an improved version of the Sprengel pump suited to their own purposes, and were soon able to evacuate a bulb to a millionth of an atmosphere pressure in twenty minutes. A scientific instrument had been transformed into a useful engineering device.

In October 1879 Edison returned to another earlier idea: the use of carbon rather than platinum to make the filament. The reason may have been that he was still involved in the manufacture of his carbon transmitter for the telephone; there was lots of lampblack around the lab, and constant handling, rolling it between the fingers, etc., could have made him realize that this substance can easily be formed into filaments. (At this point the Freudian psychohistorian of technology would inquire into the childhood toilet training experiences of our hero.)

Charles Batchelor, following Edison's instructions at Menlo Park, tested lamps with various kinds of carbon filaments, and on 22 October 1879 found that one with carbonized cotton thread could burn more than fourteen hours. This is generally regarded as the crucial breakthrough, but other developments were essential to Edison's ultimate success. In addition to having a better vacuum pump than his competitors, he developed a generator that was 80 percent efficient when those currently in use achieved only about 60 percent. That made his light cheaper in comparison to gas, the primary means of illumination at the time. To make the lamp easy to use, his staff developed a simple switch, screw base, and socket for the bulb. Other features such as the high resistance of the filament and parallel circuits were designed to be part of a low-cost electrical distribution system.

Customers signed up for electricity so they could use the incandescent light, but once cheap electric power was available in homes and businesses it could be used to run other appliances. Thus Edison's light stimulated the growth of the electrical industry and played a major role in the rapid introduction of electricity throughout society and industry, especially in the United States, in the decades after 1880.

BIBLIOGRAPHY

Daniel R. Headrick (1981), *The Tools of Empire: Technology and European Imperialism in the Nineteenth Century* (New York: Oxford Univ. Press).

Bernard S. Finn (1976), "History of Electrical Technology: The State of the Art," *Isis* 67:31–35, is a review of historical studies.

Arthur P. Molella (1976), "The Electric Motor, the Telegraph, and Joseph Henry's Theory of Technological Progress,"*Proceedings of the Institution of Electrical and Electronic Engineers* 64:1273–78.

Daniel Boorstin, ed. (1972), *Technology and Society* (New York: Arno Press), is a collection of reprints of books including Arthur A. Bright, *The Electric-Lamp Industry . . . from 1800 to 1947* (1949); Henry M. Field, *The Story of the Atlantic Telegraph* (1893); J. E. Kingsbury, *The Telephone and Telegraph Exchanges: Their Invention and Development* (1915); Harold C. Passer, *The Electrical Manufacturers, 1875–1900* (1953); George B. Prescott, *Bell's Electric Speaking Telephone* (1884); Robert L. Thompson, *Wiring a Continent: The History of the Telegraph Industry in the United States, 1832–1866* (1947).

Topic *i:* Telegraph

Yasuo Nakagawa (1984), "The Development of Early Practical Electromagnetic Telegraphs and the Mechanization of Skilled Operation," *Historia Scientiarum* 27:77–89.

Jürgen Teichmann (1981), "Zwischen Physik und Technik: Elektrizität, Elektromagnetismus, und Gauss-Weberscher Telegraph," *Technikgeschichte* 48:298–307.

George Shiers (1977), *The Electric Telegraph: An Historical Anthology* (New York: Arno Press), has articles by W. J. King (1962), J. C. MacKechnie (1959), and several from the late nineteenth century.

Keith Dawson (1976), "Electromagnetic Telegraphy: Early Ideas, Proposals, and Apparatus," *History of Technology* 1:113–41.

Topic *ii:* Atlantic Cable

Vary T. Coates and Bernard Finn, eds. (1979), *A Retrospective Technology Assessment: Submarine Telegraphy. The Transatlantic Cable of 1866* (San Francisco: San Francisco Press).

Topic *iii:* Telephone

Neil H. Wasserman (1985), *From Invention to Innovation: Long-Distance Telephone Transmission at the Turn of the Century* (Baltimore: Johns Hopkins Univ. Press).

Robert W. Garnet (1985), *The Telephone Enterprise: The Evolution of the Bell System's Horizontal Structure, 1876–1909* (Baltimore: Johns Hopkins Univ. Press).

Ithiel de Sola Pool (1983), *Forecasting the Telephone: A Retrospective Technology Assessment* (Norwood, N.J.: Ablex); Pool, ed. (1977), *The Social Impact of the Telephone* (Cambridge: MIT Press).

George Shiers (1977), *The Telephone: An Historical Anthology* (New York: Arno Press).

John Brooks (1976), *The Telephone: The First Hundred Years* (New York: Harper and Row).

Topic *iv:* Power and Lighting

Robert Friedel and Paul Israel (1985), *Edison's Electric Light: Biography of an Invention* (New Brunswick, N.J.: Rutgers Univ. Press).

James R. Chiles (1985), "Learning from the Big Blackouts," *American Heritage of Invention and Technology* 1, no. 2:26–30. "Two nights of darkness, in 1965 and 1977, showed how fragile the nation's power system could be."

Thomas P. Hughes (1983), *Networks of Power: Electrification in Western Society, 1880–1930* (Baltimore: Johns Hopkins Univ. Press); (1979), "The Electrification of America: The System Builders," *Technology and Culture* 20:124–61.

Wyn Wachhorst (1981), *Thomas Alva Edison: An American Myth* (Cambridge: MIT Press).

William J. Broad (1979), "Rival Centennial Casts New Light on Edison," *Science* 204:32–33, 35–36, reports on claims for J. Swan as inventor of the light bulb.

8.3 THE NATURE OF LIGHT AND THE MAXWELLIAN SYNTHESIS

i. Pre-Maxwellian theories (W. Weber, etc.); hints about the connection between light and electromagnetism; Faraday's discovery of magnetic rotation of the plane of polarization; William Thomson's contributions to mathematizing the field concept

ii. Maxwell's development of electromagnetic theory; the displacement current; role of mechanical models and Maxwell's attitude toward them; light as electromagnetic waves

READINGS

Marks, *Science,* 251–55. Sharlin, *The Convergent Century* (1966), 81–98; *Making of the Electrical Age* (1963), 72–86. Schneer, *Evolution of Physical Science,* 276–87. M. Kline, *Mathematics in Western Culture* (1964), 304–10. A. E. E. McKenzie, *Major Achievements of Science* (1973), 185–89, 460. Holton and Brush, *Introduction to Concepts and Theories,* 420–23. Einstein and Infeld, *Evolution of Physics,* 142–53. L. S. Swenson, *Genesis of Relativity* (1979), 26–41. Solomon, *Structure of Space,* 107–10. Williams, *Origins of Field Theory,* 117–37.

Article: A. Einstein, "Maxwell's Influence on the Development of the Conception of Physical Reality," in *James Clerk Maxwell: A Commemoration Volume* (1931), 66–73.

SYNOPSIS

Wilhelm Weber (1804–1891), who worked with Gauss on the electric telegraph, developed an influential theory of electrodynamics in 1846, based on a velocity-dependent formula for the force between electric charges. It contained a constant c_W, the relative speed at which the force between two moving charges would be zero. In 1856 Weber and Rudolph Kohlrausch measured c_W by comparing electrostatic and electrodynamic forces. Unfortunately Weber had defined his constant so that it differs from the now standard definition of c by a factor of the square root of 2, and thus it was not immediately obvious from their data that it is numerically equal to the speed of light. It does follow from Weber's theory that a periodic disturbance will be propagated in a perfectly conducting circuit at the speed $c_W/2^{1/2}$, a quantity that *is* equal to the speed of light, as was first pointed out by G. R. Kirchhoff in 1857. Yet Kirchhoff did not follow up what now seems to be the obvious consequence of this coincidence and develop an electromagnetic theory of light.

The Danish physicist L. Lorenz did attempt to derive an electromagnetic theory of light from Weber's action-at-a-distance approach in 1867, assuming the electrical force is propagated through space at speed c, and was able to arrive independently at some of Maxwell's results. From Maxwell's viewpoint any such theory is unsatisfactory because it fails to say *where the energy is located* in the interval of time when the disturbance passes from one charge to another; the action-at-a-distance viewpoint makes no provision for energy to be carried by anything other than particles of matter.

Maxwell was influenced by Faraday's experiment on the rotation of the plane of polarization of light by a magnetic field. In 1856 William Thomson suggested that magnetism is associated with rotational motions, either of matter or ether. Thomson had in the 1840s begun the process of expressing Faraday's ideas in mathematical language, pointing out the analogy between the phenomena of electrostatics, hydrodynamics, and heat conduction: all can be described by the same differential equation (Laplace-Poisson equation) by assigning appropriate physical meanings to the mathematical symbols. Maxwell pursued this kind of analogy, pointing out how certain relations between physical quantities may be independent of the ultimate nature of those quantities. Thus the equation $V{\sim}e/r$ for the electrostatic potential of a point charge can be derived from two completely different assumptions: a "field" theory in which the potential is postulated to obey the Laplace-Poisson equation, as if one were dealing with a space-filling fluid; or an "action-at-a-distance" theory in which the force between two charges is postulated to be $F \sim e_1 e_2 / r_{12}^2$. Yet historically those two kinds of theories had been associated with completely different views of the physical world.

Maxwell followed up Thomson's suggestion about magnetism by assuming that a magnetic line of force can be represented by a vortex. In order to explain how neighboring vortices rotate in the same sense without interference at their boundaries, Maxwell postulated that the space between vortices is filled with smaller particles which serve as "idle wheels" or ball bearings to transmit the rotation. It then turned out that the *linear* motion of these particles could be identified with electric current. Here his familiarity with machines and his interest in "governors" (feedback devices) came in handy, for he knew that in some machines the idle wheels do not have to revolve around a fixed point.

The vortex-idle wheel model immediately allowed Maxwell to "explain" Ampère's law for the magnetic force around currents, which says that the x component of the current j is proportional to the difference between the y derivative of the z component of the magnetic field H and the z derivative of the y component of H. These derivatives simply represent the motion of the vortices at particular points in space compared to the motion at neighboring points; the idle wheels (electric currents) must move in the appropriate way to link the motions mechanically.

Maxwell then argued that in order to explain the action of an electric field on a *dielectric* substance (which does not ordinarily transmit current) we may imagine that each electric charge is displaced by a certain amount, but there is an elastic restoring force that keeps it within the same molecule and returns it to its original position when the external electric field is removed. If the ether itself can be regarded as supporting magnetic fields, then it follows from the model that it must have such elastic restoring forces that act on the ball bearings—otherwise one could make currents flow through it by applying an arbitrarily small constant electric field. The "elasticity" of the ether is, according to the model, a constant relating the electric field E to the charge displacement it produces.

The motion of any charge is a current. That seemingly innocent statement is a key step in Maxwell's theory. In particular the rate of change of displacement with time is a "displacement current" that must be added to the ordinary current, or indeed (in a "vacuum" or dielectric ether) is the entire current. If it behaves like a current with respect to magnetic fields, then it must appear in Ampère's law.

Following and extending this line of reasoning from his mechanical model, Maxwell arrived at his famous set of four equations relating the electric and magnetic fields with their space and time derivatives. He then showed from the equations that the fields must be propagated as waves moving through space at a speed c whose numerical value can be determined entirely from electric and magnetic experiments and is just the speed of light, $c = 3 \times 10^8$ m/sec.

In a letter to William Thomson in 1861, Maxwell claimed that he had arrived at this result completely unexpectedly, before he knew that the Weber-Kohlrausch experiment provided numerical values suggesting a relation between light and electromagnetism.

Maxwell stated his conclusion as follows: "light consists in the tranverse undulations of the same medium which is the cause of electric and magnetic phenomena." Note that the waves are *transverse* in the sense that the **H** and the **E** vectors must oscillate in directions perpendicular to each other and to the direction in which the wave itself moves. Maxwell's theory thus resolves the problem of the transversality of light without having to postulate a solid ether, and relates it to the peculiar "circular" nature of electromagnetic interactions.

Maxwell did not, in his 1862 paper announcing this result or in later publications, go beyond his statement that light consists of waves in the electromagnetic ether to say explicitly that electromagnetic waves of other frequencies can exist and ought to be detectable. Even though one can say that his *theory* predicts such waves, it is not strictly true to say that Maxwell himself predicted them. The closest he came to a prediction was a statement in 1865: "we have strong reason to conclude that light itself (including radiant heat, and other radiations if any) is an electromagnetic disturbance. . . ."

In his 1873 *Treatise* Maxwell showed that electromagnetic waves exert a mechanical pressure against a surface they strike. This result was used by Boltzmann in deriving the relation between energy and temperature of black-body radiation (**9.3**).

It is sometimes asserted that Maxwell abandoned the program of mechanical explanation of electromagnetic fields when he later reformulated his theory in terms of abstract Lagrangian dynamics. While this approach avoided the need to adopt a specific mechanical model for fields and currents, Maxwell still maintained that the energy of fields and currents is mechanical, and that an ether obeying the laws of mechanics does exist. The use of the Lagrangian formulation of mechanics does not mean that Newtonian mechanics has been abandoned, only that it has been refined in such a way that the maximum number of conclusions can be drawn from the minimum number of assumptions.

BIBLIOGRAPHY

P. M. Harman, ed. (1985), *Wranglers and Physicists: Studies on Cambridge Mathematical Physics in the Nineteenth Century* (Dover, N.H.: Manchester Univ. Press), contains articles by J. J. Cross, O. Knudsen, D. M. Siegel, P. M. Harman, and J. Z. Buchwald.

Salvo d'Agostino (1981), "Weber and Maxwell on the Discovery of the Velocity of Light in Nineteenth Century Electrodynamics," in *On Scientific Discovery,* ed. M. D. Grmek et al. (Boston: Reidel), 281–93.

Topic *i:* Pre-Maxwell Theories

David Gooding (1982), "Final Steps to the Field Theory: Faraday's Study of Magnetic Phenomena, 1845–1850," *Historical Studies in the Physical Sciences* 11:231–75; (1982), "A Convergence of Opinion on the Divergence of Lines: Faraday and Thomson's Discussion of Diamagnetism," *Notes and Records of the Royal Society of London* 36:243–59.

M. N. Wise (1981), "The Flow Analogy to Electricity and Magnetism, Part I: William Thomson's Reformulation of Action at a Distance," *Archive for History of Exact Sciences* 25:19–70.

Walter Kaiser (1981), *Theorien der Elektrodynamik im 19. Jahrhundert* (Hildesheim: Gerstenberg).

Kenneth L. Caneva (1981), "What Should We Do with the Monster? Electromagnetism and the Psychosociology of Knowledge," in *Sciences and Cultures,* ed. E. Mendelsohn and Y. Elkana (Boston: Reidel), 101–31; (1978), "From Galvanism to Electrodynamics: The Transformation of German Physics and Its Social Context," *Historical Studies in the Physical Sciences* 9:63–159.

Jed Z. Buchwald (1977), "William Thomson and the Mathematization of Faraday's Electrostatics," *Historical Studies in the Physical Sciences* 8:101–36.

Ole Knudsen (1976), "The Faraday Effect and Physical Theory, 1845–1873," *Archive for History of Exact Sciences* 15:235–81.

Salvo d'Agostino (1976), "La scopertà di una Velocità quasi Uguale alla Velocità della Luce nell'Electtrodinamica di Wilhelm Weber (1804–1891)," *Physis* 18:297–318.

Topic *ii.* Maxwell

James Clerk Maxwell (1873), *A Treatise on Electricity and Magnetism* (reprinted 1977, Oceanside, N. Y.: Dabor Science); (1890), *The Scientific Papers of James Clerk Maxwell* (reprinted 1952, New York: Dover); (1865), *A Dynamic Theory of the Electromagnetic Field* (reprinted 1982, ed. T. F. Torrance, New York: Columbia Univ. Press).

Alan Chalmers (1986), "The Heuristic Role of Maxwell's Mechanical Model of Electromagnetic Phenomena," *Studies in History and Philosophy of Science* 17: 415–27.

John Hendry (1986), *James Clerk Maxwell and the Theory of the Electromagnetic Field* (Accord, Mass.: Adam Hilger).

Daniel Siegel (1986), "The Origin of the Displacement Current," *Historical Studies in the Physical Sciences* 17:99–146.

Jed Z. Buchwald (1985), *From Maxwell to Microphysics: Aspects of Electromagnetic Theory in the Last Quarter of the Nineteenth Century* (Chicago: Univ. of Chicago Press).

Nancy J. Nersessian (1984), "Aether/Or: The Creation of Scientific Concepts," *Studies in History and Philosophy of Science* 15:175–212; (1984), *Faraday to Einstein* (Dordrecht: Nijhoff), 69–93.

C. W. F. Everitt (1983), "Maxwell's Scientific Creativity," in *Springs of Scientific Creativity,* ed. R. Aris et al. (Minneapolis: Univ. of Minnesota Press), 71–141, is a detailed analysis of family and psychological background.

Martin Goldman (1983), *The Demon in the Aether: The Story of James Clerk Maxwell* (Philadelphia: Heyden).

Peter Galison (1983), "Re-Reading the Past from the End of Physics: Maxwell's Equations in Retrospect," in *Functions and Uses of Disciplinary Histories,* ed. L. Graham et al. (Hingham, Mass.: Reidel), 35–51.

Hans-Georg Schöpf (1982), "Maxwell's Aethertheorien," *Astronomische Nachrichten* 303:29–37.

M. L. Levin and M. A. Miller (1981), "Maxwell's *Treatise on Electricity and Magnetism,*" *Soviet Physics Uspekhi* 24:904–13.

Mauro La Forgia and Carlo Tarsitani (1981), "Analogia, Modelli e Teoria Dinamica nel Contributo di Maxwell all'Elettromagnetismo," *Physis* 23:525–54.

Ivan Tolstoy (1981), *James Clerk Maxwell: A Biography* (Chicago: Univ. of Chicago Press). " . . . the book may offer a readable account of Maxwell's science, life, and humor, but it is rather marred by the author's disregard of the benefits to be derived from modern historical scholarship," review by C. Smith, *Isis* 73(1982):480. See also the review by C. W. F. Everitt (1984), in *Physics Teacher* 22:264–66.

R. L. Oldershaw (1981), "Faraday, Maxwell, Einstein and Epistemology," *Nature and System* 3:99–107.

Max Jammer and John Stachel (1980), "If Maxwell Had Worked between Ampère and Faraday: An Historical Fable with a Pedagogical Moral," *American Journal of Physics* 48:5–7.

M. Norton Wise (1979), "The Mutual Embrace of Electricity and Magnetism," *Science* 203:1310–18.

Salvo d'Agostino (1978), "Experiment and Theory in Maxwell's

Work. The Measurements for Absolute Electromagnetic Units and the Velocity of Light," *Scientia* 113:469–80.

Donald Franklin Moyer (1978), "Continuum Mechanics and Field Theory: Thomson and Maxwell," *Studies in History and Philosophy of Science* 9:35–50; (1977), "Energy, Dynamics, Hidden Machinery: Rankine, Thomson and Tait, Maxwell," ibid. 8:251–68.

Pamela Mack (1977), "Physical Reality in Maxwell's Electrodynamics," *Synthesis* [Cambridge, Mass.] 3, no. 4:44–58.

8.4 ELECTROMAGNETIC WAVES

i. Attempts to detect electromagnetic waves before Heinrich Hertz
ii. Hertz's experiments on electromagnetic waves and his writings on Maxwell's theory
iii. Guglielmo Marconi and the invention of radio; its effects on society
iv. The invention of television; its social effects

READINGS

McKenzie, *Major Achievements,* 190, 461–62. Holton and Brush, *Introduction to Concepts and Theories,* 424–25. Swenson, *Genesis of Relativity,* 63–64, 81–92. G. R. M. Garratt, "The Communications Explosion . . . ," Unit 14 in Open University course AST 281, 38–54 (on radio). Solomon, *Structure of Space,* 111–14. Sharlin, *Making of the Electrical Age,* 87–130. Kline, *Mathematics in Western Culture,* 311–21. Segre, *Falling Bodies,* 174–81.

Articles: Philip and Emily Morrison, "Heinrich Hertz," *Scientific American* 197(1957):98–106. C. Suesskind, "Guglielmo Marconi 1874–1937," *Endeavour* 33(1974):67–72. Leonard S. Reich, "From Edison's Wastebasket," *Science 84* (Nov. 1984):73, 75, is on Lee De Forest's vacuum tube. Thomas S. W. Lewis, "Radio Revolutionary," *American Heritage of Invention and Technology* 1, no. 2(1985):34–41 (on Edwin Armstrong).

SYNOPSIS

The first successful attempt to generate and detect electromagnetic waves (other than light) was made by the German physicist Heinrich Hertz (1857–1894). In 1887 Hertz noticed a peculiar effect produced by the sparking of an induction coil. It had already been observed by other scientists that sparks sometimes jumped through air between terminals of an induction coil. Hertz observed that when a piece of wire was bent around so that there was a short gap between its two ends and held near an induction coil, a spark would jump across the air gap in the wire when a spark jumped across the terminals of the coil.

Each spark produced by an induction coil is actually a series of many sparks jumping rapidly back and forth between the terminals. Hertz could control the jumping frequency by changing the size and shape of the 16-inch square plates he used for terminals. He reasoned that as sparks jump back and forth they must be setting up rapidly changing electric and magnetic fields in the air gap, and these fields, according to Maxwell's theory, will propagate through space as electromagnetic waves. When they pass over the air gap in the bent piece of wire, they will set up rapidly changing electric and magnetic fields there, thus producing a spark.

If Hertz's interpretation is correct there must be a short time delay between the two sparks, corresponding for the time required for the electromagnetic waves to go from one place to the other. In 1888, Hertz was able to measure the speed of these waves and found that it is the same as the speed of light. He then showed that electromagnetic radiation from his induction coil has all the usual properties of light waves: reflection, refraction, diffraction, etc.

Following work by E. Branly and O. Lodge, Guglielmo Marconi in 1894 started to develop a sensitive receiver for radio waves and improved Hertz's transmitter by grounding one of the plates and raising the other on a mast. Marconi went to England where he expected to find a market for his invention (the Italian government was not interested). A successful transmission of signals across the English Channel was followed by the astonishing feat of sending a message across the Atlantic Ocean in 1901. In a few years the safety value of having radio on board passenger ships was recognized; the wireless was credited with saving 700 lives when the *Titanic* sank in 1912.

In 1904 J. A. Fleming invented a valve or diode, based on the Edison effect (emission of electrons from a hot filament). Fleming's valve "rectified" the incoming radio waves, converting them from high-frequency oscillations to direct current. Lee De Forest developed a practical vacuum-tube detector by adding a grid to control the flow of electrons; the vacuum tube was found to be exceptionally useful in amplifying and generating currents and thus formed the basis of the new science of "electronics." (Further technical and commercial developments of radio up to 1927 are described in Sharlin's book.)

BIBLIOGRAPHY

Topic *i:* Pre-Hertz

C. Suesskind (1964), "Observations of Electromagnetic Radiation before Hertz," *Isis* 55:32–42.

Topic *ii:* Hertz

H. Hertz (1893), *Electric Waves* (reprinted 1962, New York: Dover); (1977), *Memories, Letters, Diaries,* 2d ed. (San Francisco: San Francisco Press).

Bruce J. Hunt (1983), " 'Practice vs. Theory': The British Electrical Debate, 1888–91," *Isis* 74:341–55, relates "how Maxwell's followers in Britain used Hertz's discovery . . . to promote their theory and push it into practical applications."

Jean Cazenobe (1982), "Les incertitudes d'une decouverte: L'onde de Hertz de 1888 à 1900," *Archives Internationales d'Histoire des Sciences* 32:236–65.

Salvo d'Agostino (1975), "Hertz's Researches on Electromagnetic Waves," *Historical Studies in the Physical Sciences* 6:261–323.

Topic *iii:* Marconi and Radio

H. G. J. Aitken (1985), *The Continuous Wave: Technology and American Radio, 1900–1932* (Princeton, N.J.: Princeton Univ. Press); (1976), *Syntony and Spark: The Origins of Radio* (New York: Wiley).

Reynold M. Wik (1981), "The Radio in Rural America during the 1920s," *Agricultural History* 55:39–50.

Susan J. Douglas (1979), "Exploring Pathways in the Ether: The Formative Years of Radio in America, 1896–1912" (Ph.D. diss., Brown University).

George Shiers, ed. (1976), *Development of Wireless to 1920* (New York: Arno Press), is a reprint of twenty articles published from 1890 to 1967.

Topic *iv:* Television

Joshua Meyrowitz (1986), *No Sense of Place: The Impact of Electronic Media on Social Behavior* (New York: Oxford Univ. Press).

James M. Wall et al. (1985), *Violence and Sexual Violence in Film, Television, Cable and Home Video. Report of a Study Commission* (New York: National Council of the Churches of Christ).

Joseph H. Udelson (1982), *The Great Television Race: A History of the American Television Industry, 1925–1941* (University, Ala.: Univ. of Alabama Press).

Jeff Greenfield (1978), *Television: The First Fifty Years* (New York: Harry N. Abrams).

H. J. Eysenck and D. K. B. Nias (1978), *Sex, Violence, and the Media* (New York: St. Martin's Press).

8.5 EINSTEIN AND THE ETHER

i. Concepts of space and time from Newton to Mach; non-Euclidean geometry
ii. The Michelson-Morley experiment and other attempts to detect the earth's motion; controversy about whether the result of this experiment motivated Einstein to develop relativity
iii. Electromagnetic theories of Lorentz, Poincare, and others; the "electromagnetic worldview"
iv. Einstein's early life (up to 1905)

READINGS

McKenzie, *Major Achievements,* 300–303. Schneer, *Evolution of Physical Science,* 296–307. Holton and Brush, *Introduction to Concepts and Theories,* 502–8. Marks, *Science,* 263–64. Segre, *Falling Bodies,* 181–85. Segre, *X-Rays to Quarks,* 78–80. Paul M. Clark, "Einstein: Philosophical Belief and Physical Theory," Unit 6 in the Open University course A381, *Science and Belief from Darwin to Einstein* (1981), 9–34. Einstein and Infeld, *Evolution of Physics,* 153–77. Swenson, *Genesis of Relativity,* 59–81, 97–127, 145–67. Kline, *Mathematics in Western Culture,* 410–31 (on non-Euclidean geometry). S. Goldberg, *Understanding Relativity* (1984), 1–102. Solomon, *Structure of Space* (1974), 115–21. Hoffman, *Relativity and Its Roots,* 1–88. H. Pagels, *The Cosmic Code* (1982), 17–31. F. P. Dickson, *The Bowl of Night* (1968), 71–86. C. Lanczos, *Space through the Ages* (1970), Chap. 3, "The Evolution of Metrical Geometry," 47–98. B. M. Casper and R. J. Noer, *Revolutions in Physics* (1972), 291–416.

Articles: John Stachel, "Einstein and Ether Drift Experiments," *Physics Today* 40, no. 5 (May 1987): 45–47. Loyd S. Swenson, Jr., "Measuring the Immeasurable," *American Heritage of Invention and Technology* 3, no. 2 (Fall 1987): 42–49 (on Michelson's experiments).

SYNOPSIS

For students who have just learned Copernicus and Galileo established that the Earth really does rotate around its own axis and revolve in orbit around the Sun, it may be confusing to be told that Einstein proved that there is no such thing as absolute motion—no absolute space relative to which the Earth could move. Thus it is important to explain the difference between relativity of linear motion and of circular motion.

Newton's *Principia* (1687) asserts the reality of absolute space, time and motion. His rotating bucket experiment demonstrates the existence of absolute *rotational* motion—the water in the bucket rises at the sides showing the effects of centrifugal force even when it has no motion relative to the bucket itself [but see Laymon (1978)]. His understanding of the dynamics and kinematics of circular motion enabled him to see the fallacy of Galileo's statement that the rotation of the Earth would have no effect on the observed motion of falling objects; he pointed out to Robert Hooke that instead of being "left behind" by the motion of the Earth (as Aristotelians argued and Galileo denied) the object would actually run ahead of it (since its linear momentum in the direction parallel to the Earth's surface would be greater if it started at a greater distance from the center of the rotating Earth). Despite Hooke's claim to have verified this prediction by experiment, it was not conclusively confirmed until around 1800.

Other phenomena that depend on the rotation of the Earth are the patterns of winds and ocean currents, the equatorial bulge of the solid part of the Earth, the reduction in the gravitational acceleration at the equator (partly due to rotation and partly to the bulge), motion of the Foucault pendulum, and (probably) the Earth's magnetic field. All of these came too late to have much effect on the Copernican Revolution, but they do raise difficulties for the simpleminded idea that *all* motion is relative. The Earth rotates—but relative to what?

An important consequence of the Earth's motion in orbit around the sun is *parallax,* the apparent change in position of stars due to the shift in our point of observation. But parallax could not be observed in the seventeenth century, so astronomers who wanted to accept the heliocentric system had to postulate that all the stars are so far away that the parallax angle is too small to measure. The first well-confirmed observation of stellar parallax was not until 1838 (**12.2**).

In 1728 James Bradley, looking for parallax, discovered an apparent motion of stars he correctly explained as being due to the Earth's motion around the sun. This effect, called "stellar aberration," seems to imply that the Earth moves with respect to the medium in which light is propagated. But an experiment by François Arago in 1810 showed that this interpretation is not valid. Augustin Fresnel interpreted Arago's results to mean that the moving telescope lens drags the ether along with it to such an extent that it prevents one from directly measuring the motion of the Earth through the ether, at least to first order in v/c.

In the meantime philosophers and mathematicians were debating whether space itself must have the properties indicated by Euclidean geometry. The controversy centered around Euclid's fifth postulate, the "parallel postulate," which asserts that through a given point one and only one line can be drawn parallel to another line. It is equivalent to the statement that the sum of the angles in a plane triangle is 180°. This seemed so obvious that it should be possible to derive it from the others rather than have to state it as an independent postulate.

Immanuel Kant, in his *Critique of Pure Reason* (1781), argued that our minds are constructed in such a way that they compel us to view the external world in only one way. Thus Euclidean geometry is what Kant called an "a priori synthetic truth" we possess before we acquire any experience of the world and governs the way we interpret out experience. If one accepts Kant's argument, it would be impossible for the physical world to be other than Euclidean.

By the late eighteenth century, the failure of all attempts to derive Euclid's fifth postulate from the others made some mathematicians suspect that it came from experience rather than logic, and that one could create a logically consistent geometry without it, though such a geometry might have no relevance to the real world. The great German mathematician Carl Friedrich Gauss (1777–1855) was probably the first to work out a non-Euclidean system of geometry, but he hesitated to publish his ideas. There is a story, often repeated in textbooks and popular histories, that Gauss tried to test the Euclidean character of space by measuring the angles of a triangle; if the sum was different from 180°, that would mean space is "curved." It was pointed out by Arthur Miller (1972) that this story is based on a misunderstanding: Gauss was not trying to measure the curvature of space in that sense, but simply studying a three-dimensional surface which we already know to be curved *in* Euclidean space, namely the surface of the Earth. This was part of Gauss's work on a geodetic survey.

Nikolai Ivanovich Lobatchevsky (1793–1856), a Russian mathematician, is generally regarded as the first discoverer of a non-Euclidean geometry. (Contrary to Tom Lehrer's song, he did not plagiarize it from others, though it is clear that he was putting together ideas some of which had been common property in mathematics for centuries). Lobatchevsky presented his theory in an unpublished lecture in 1826, in a series of papers in 1829–1837, and finally in two books in 1840 and 1855.

Janos Bolyai (1802–1860), an Hungarian army officer, is often considered as coinventor of the theory because he wrote a paper outlining similar ideas, which was published as an appendix to his father's mathematics textbook in 1832–1833.

In Lobatchevsky's geometry the sum of the angles in a triangle is less than 180°; the trigonometric relations are the same as for a sphere of imaginary radius, and the space is not closed in real coordinates. Since the ordinary sine, cosine, and tangent functions transform into "hyperbolic" functions for imaginary variables, and for other reasons, Lobatchevsky's geometry is sometimes called "hyperbolic" geometry.

Another version of non-Euclidean geometry was developed by the German mathematician G. F. B. Riemann (1826–1866), in a lecture given at Göttingen University in 1854 but not published until 1868. In Riemann's geometry the sum of the angles in a triangle is greater than 180°. The space has a positive curvature; it is finite but has no boundary. The distance between two points differing by an infinitesimal amount is determined by "metric coefficients" g_{ij} which, Riemann suggested, may depend on the "binding forces which act on it."

Riemann's ideas were popularized in England by W. K. Clifford, who suggested that "small portions of space are of a nature analogous to little hills on a surface which is on the average flat" and "that this property of being curved or distorted is continually passed on from one position to another after the manner of a wave. . . ."

Felix Klein saw the connection between the two different kinds of non-Euclidean geometry developed by Lobatchevsky and Riemann. He proposed a general theory to unite them (1871), using the terminology "hyperbolic" for Lobatchevsky's geometry, "parabolic" for Euclidean, and "elliptic" for Riemannian.

The French mathematician Henri Poincaré (1854–1912) suggested around 1890 that Euclidean geometry may not be *true* but it is the most *convenient.* In fact he claimed that one geometry cannot be more true than another, it can only be more convenient; we adopt a certain geometry and then *adapt* our physical laws to it. If the sum of angles in a triangle were found to be different from 180°, we should assume that Euclidean geometry still describes physical space but that light does not travel in "straight" lines. This attitude of Poincaré came to be known as *conventionalism* and is part of the general movement of skepticism toward the possibility of finding the real nature of the world that arose in the late nineteenth century. It is perhaps the reason why Poincaré failed to discover relativity theory even though he had developed all the equations for it; this is an ongoing controversy, which might be discussed as an example of the role of philosophical preconceptions in science.

Returning to the problem of measuring the Earth's motion through space, we may locate the origin of the late-nineteenth-century interest in this problem in a letter from J. C. Maxwell to D. P. Todd of the U. S. Nautical Almanac Office in 1879, discussing possible methods. Albert Abraham Michelson (1852–1931), generally called an American physicist, was actually born in a place which is now Strzelno, Poland, and came to the United States with his parents two years later. Thus while his work must be considered an integral part of American science, one might also note that he was following the tradition of the Polish astronomer Nicolaus Copernicus who produced a radical change in our views about the structure of the universe without really meaning to do so. Michelson's collaborator was E. W. Morley, a chemist. The experiment was first done in 1881 but definitive results were not reported until 1887.

The failure of Michelson and Morley to observe the expected time difference in the transit of light beams along different paths, due to a supposed motion of the Earth through the ether, was discussed by the Irish physicist G. F. FitzGerald in 1889. FitzGerald suggested that the solid arm of the interferometer contracts by a factor depending on the speed v relative to the ether, by an amount just sufficient to nullify the time difference. The contraction was made somewhat plausible by the remark that forces between atoms (which determine the size of a macroscopic object) are electrical in nature and thus would be affected by motion through an electromagnetic ether. The Dutch physicist H. A. Lorentz arrived at a similar hypothesis a few years later and used it to develop his electrodynamic theory.

Until recently, textbooks and histories of modern physics introduced Einstein's theory of relativity as if it were an attempt to account for the negative result of the Michelson-Morley experiment—i.e., the failure to observe any motion of the Earth relative to the ether. The notion that relativity theory was invented to account for an anomalous experimental result was convenient for physics teachers and empiricist philosophers of science but did not stand up under detailed analysis of the available evidence about the origins of Einstein's ideas. Gerald Holton (1969) showed that the problem of the Michelson-Morley experiment played little if any part in Einstein's early development of his theory, though of course he was aware of the general problem of ether-drift and of the FitzGerald-Lorentz contraction through his reading of the works of Lorentz and others.

Einstein *was* probably influenced by Ernst Mach's book *Science of Mechanics* (1883), which criticized Newton's postulate that absolute space and time exist. Mach argued that the rotating bucket experiment did not prove rotation with respect to absolute space but only with respect to the "fixed stars," i.e., the matter in the universe as a whole. This assumption, later used in some versions of general relativity theory, is now known as "Mach's principle." More generally, Einstein attributed to Mach the idea that one could legitimately be skeptical of ideas associated with the name of Newton.

As for specific influences on Einstein's early thinking, Holton identifies a textbook by August Foeppl, which Einstein may have used to learn Maxwell's electromagnetic theory. Foeppl raised some of the same issues about the symmetry between electric and magnetic fields (and how this relates to the question of which is considered at rest, which in motion) that Einstein discussed in his 1905 paper.

Another feature of late nineteenth-century physics that has lately attracted the attention of historians is the "electromagnetic worldview." According to this view, advocated by Lorentz, Planck, and others, one should abandon mass as a fundamental entity and replace it by electric and magnetic fields. This was one way to get rid of the mechanical worldview but it did not survive as a competitor to relativity.

BIBLIOGRAPHY

Gerald Holton (1988), *Thematic Origins of Scientific Thought, Kepler to Einstein,* rev. ed. (Cambridge: Harvard Univ. Press), includes "On the Origins of the Special Theory of Relativity," "Poincare and Relativity," "Influences on Einstein's Early Work," "Einstein, Michelson and the Crucial Experiment," and "On Trying to Understand Scientific Genius."

H. Nelkowski et al., eds. (1979), *Einstein Symposion Berlin . . .*, includes Imre Toth, "Spekulation ueber die Moeglichkeit eines nicht euklidischen Raumes vor Einstein," 46–83; David Cassidy, "Biographies of Einstein," 490–500.

Richard Schlegel (1976), "'Bohr and Einstein': A Course for Nonscience Students," *American Journal of Physics* 44:236–39, is an outline of a course given by Schlegel for nonscience students, presenting ideas of relativity and quantum physics in association with the lives of Einstein and Bohr. List of possible readings.

Topic *i:* Space and Time Concepts, Non-Euclidean Geometry

Rosine van Oss (1983), "D'Alembert and the Fourth Dimension," *Historia Mathematica* 10:455–57, concludes "the priority of connecting time to space as a fourth dimension seems to belong to d'Alembert in one of the articles he wrote for the *Encyclopedie* in 1754."

Michael Bradie and Comer Duncan (1982), "A Course on the Philosophy and Physics of Space and Time," *Teaching Philosophy* 5:109–16.

Miodrag Cekic (1981), "Mach's Phenomenalism and Its Consequences in Physics," *International Philosophical Quarterly* 21:249–59, claims that Mach is the originator of the theory of relativity.

Dietrich Fliedner (1981), "Zum Problem des vierdimensionalen Raums: Eine theoretische Betrachtung aus historischgeographischer Sicht," *Philosophia Naturalis* 18:388–412.

Jeremy Gray (1979), *Ideas of Space: Euclidean, Non-Euclidean, and Relativistic* (New York: Oxford Univ. Press).

Ronald Laymon (1978), "Newton's Bucket Experiment," *Journal of the History of Philosophy* 16: 399–413.

Arthur Miller (1972), "The Myth of Gauss' Experiment on the Euclidean Nature of Physical Space," *Isis* 63: 345–48.

Topic *ii:* Michelson-Morley Experiment

Bruce Hunt (1986), "Experimenting on the Ether: Oliver J. Lodge and the Great Whirling Machine," *Historical Studies in the Physical and Biological Sciences* 16:111–34.

G. H. Keswani and C. W. Kilmister (1983), "Intimations of Relativity. Relativity before Einstein," *British Journal for the Philosophy of Science* 34:343–54, discusses ideas of Maxwell and Poincaré on the Earth's motion through the ether and the relativity principle.

Robert S. Shankland et al. (1982), "Michelson in Potsdam," *Astronomische Nachrichten* 303, is a special issue containing articles on the Michelson-Morley experiment and related topics.

H. J. Haubold and R. W. John (1982), "Albert A. Michelsons Aetherdriftexperiment 1880/81 in Berlin und Potsdam," *NTM* 19, no. 1:31–45.

Horst Melcher (1982), "Aetherdrift und Relativität: Michelson, Einstein, Fizeau und Hoek," *NTM* 19, no. 1:46–67.

M. A. Handschy (1982), "Reexamination of the 1887 Michelson-Morley Experiment," *American Journal of Physics* 50:987–90.

Ronald Laymon (1980), "Independent Testability: The Michelson-Morley and Kennedy-Thorndike Experiments," *Philosophy of Science* 47:1–37.

Tsuyoshi Ogawa (1979), "Japanese Evidence for Einstein's Knowledge of the Michelson-Morley Experiment," *Japanese Studies in the History of Science* 18:73–81.

Topic *iii:* Electromagnetic Theories—Lorentz, etc.

Nancy J. Nersessian (1986), "Why Wasn't Lorentz Einstein: An Examination of the Scientific Method of H. A. Lorentz," *Centaurus* 29: 37–73; (1984), *Faraday to Einstein* (Dordrecht: Nijhoff), 95–119.

Jed Z. Buchwald (1985), *From Maxwell to Microphysics: Aspects of Electromagnetic Theory in the Last Quarter of the Nineteenth Century* (Chicago: Univ. of Chicago Press); (1985), "Modifying the Continuum: Methods of Maxwellian Electrodynamics," in *Wranglers and Physicists,* ed. P. M. Harman (Dover, N.H.: Manchester Univ. Press), 225–41; (1981), "The Abandonment of Maxwellian Electrodynamics: Joseph Larmor's Theory of the Electron," *Archives Internationales d'Histoire des Sciences* 31:135–80, 373–438.

R. M. Nugayev (1985), "The History of Quantum Mechanics as a Decisive Argument Favoring Einstein over Lorentz," *Philosophy of Science* 52:44–63. Starting from the Lakatos scheme as applied by Zahar (see below), Nugayev modifies it to take account of the unity of Einstein's work in quantum theory and relativity.

Dennis Dieks (1984), "The Reality of the Lorentz Contraction," *Zeitschrift für allgemeine Wissenschaftstheorie* 15:330–42.

Wilfried Schröder (1984), "Hendrik Antoon Lorentz and Emil Wiechert (Briefwechsel und Verhältnis der beiden Physiker)," *Archive for History of Exact Sciences* 30:167–87, reproduces twelve letters, 1899–1921, on the Michelson experiment, electrodynamics, and relativity.

Arthur I. Miller (1984), *Imagery in Scientific Thought* (Boston: Birkhäuser), Chap. 1, "Poincaré and Einstein"; (1981), *Albert Einstein's Special Theory of Relativity* (Reading, Mass.: Addison-Wesley), Chaps. 1 and 2. This material is discussed more briefly in Miller's article (1980), "On Some Other Approaches to Electrodynamics in 1905," in *Some Strangeness in the Proportion,* ed. H. Woolf (Reading, Mass.: Addison-Wesley), 66–93.

E. G. Zahar (1983), "Poincaré's Independent Discovery of the Relativity Principle," *Fundamenta Scientiae* 4:147–75; (1973), "Why Did Einstein's Programme Supersede Lorentz's?" *British Journal for the Philosophy of Science* 24:95–123, 223–62; reprinted in *Method and Appraisal in the Physical Sciences,* ed. C. Howson (New York: Cambridge Univ. Press, 1976), 211–75, is an attempt to apply the "methodology of scientific research programmes" of I. Lakatos; claims that Lorentz's programme, including the contraction hypothesis, was a strong competitor even after 1905. See the critical remarks by A. I. Miller (1974), "On Lorentz's Methodology," *British Journal for the Philosophy of Science* 25:29–45, and further discussion by W. Brouwer (1980), "Einstein and

Lorentz: The Structure of a Scientific Revolution," *American Journal of Physics* 48:425–31.

Jose M. Sanchez Ron (1982), "Einstein y Lorentz: Del Significado de la Relatividad Especial a la Incommensurabilidad entre Paradigmas," *Pensiamento* 38:441–53.

Ole Knudsen (1978), "Electric Displacement and the Development of Optics after Maxwell," *Centaurus* 22:53–60.

Topic *iv:* Einstein's Early Life

John Stachel, ed. (1987), *The Collected Papers of Albert Einstein,* Vol. I: *The Early Years,* 1879–1902 (Princeton, N.J.: Princeton Univ. Press).

Albert Einstein (1949, New York: Library of Living Philosophers), "Autobiographical Notes" in *Albert Einstein Philosopher-Scientist,* ed. P. A. Schilpp (reprinted 1959, New York: Harper), 1–95.

Lewis Pyenson (1985), *The Young Einstein: The Advent of Relativity* (Boston: Hilger); (1980), "Einstein's Education: Mathematics and the Laws of Nature," *Isis* 71:399–425.

Ronald W. Clark (1984), *Einstein: The Life and Times—An Illustrated Biography* (New York: Abrams).

8.6 SPECIAL THEORY OF RELATIVITY

i. Einstein's 1905 papers
ii. Response of scientific community of relativity
iii. Later applications and experimental tests

READINGS

Marks, *Science,* 256–62. McKenzie, *Major Achievements,* 304–7, 513–16. Swenson, *Genesis of Relativity,* 173–82. Schneer, *Evolution of Physical Science,* 307–20. Holton and Brush, *Introduction to Concepts and Theories,* 508–25. Segre, *X-Rays to Quarks,* 80–86. Paul M. Clark, "Einstein: Philosophical Belief and Physical Theory," (cited in **8.5**), 43–47, and accompanying readings from Einstein (1933) and Holton (1973) in the anthologies on *Darwin and Einstein* by Coley and Hall and Chant and Fauvel, respectively. Solomon, *Structure of Space,* 122–40. Hoffmann, *Relativity and Its Roots,* 88–127. Goldberg, *Understanding Relativity*. Pagels, *Cosmic Code,* 31–39. Kline, *Mathematics in*

Western Culture, 432–52. Einstein and Infeld, *Evolution of Physics,* 177–208. Dickson, *Bowl of Night,* 87–109. Casper and Noer, *Revolutions in Physics,* 291–416.

Articles: Gerald Holton, "On Trying to Understand Scientific Genius," *American Scholar* 41(1971–1972), no. 1:95–110. Stanley Goldberg, "Albert Einstein and the Creative Act: The Case of Special Relativity," in *Springs of Scientific Creativity,* ed. R. Aris et al. (1983), 232–53. Arthur Miller, "On Einstein's Invention . . . ," in *PSA 1982,* ed. P. D. Asquith and T. Nickles, 2(1983):377–402. William Eamon, "Inventing the World: Einstein and the Generation of 1905," *Antioch Review* 43(1985):340–51.

While I have not had much success in getting students to read original sources, an exception can be made in this case. Aside from the Einstein-Infeld book, which was written mostly by Infeld, Einstein's exposition in his book *Relativity: The Special and General Theory* (1916, 1931), 1–68, is as clear as that by any of the popularizers or textbook writers.

SYNOPSIS

Einstein's first paper on relativity (1905) appears to have emerged from a prolonged period of reflection on Maxwell's electromagnetic theory rather than from any special concern with the results of the Michelson-Morley experiment (see **8.5**). At the age of sixteen he had discovered the paradoxical consequences that would follow from pursuing a beam of light and actually catching up with it. In the opening paragraph of his 1905 paper he discussed another inconsistency in the conventional interpretation of electromagnetic theory as applied to the reciprocal action of a magnet and a conductor. He located the difficulty in the assumption that absolute space exists and suggested that this assumption might be discarded.

Einstein postulated (1) that the laws of nature are the same for observers in any inertial frame of reference; (2) the speed of light is the same for all such observers. From these postulates he deduced:

1. An observer in one frame would find from his own measurements that lengths of objects in another frame are contracted by an amount given by the FitzGerald-Lorentz formula; this effect is reciprocal.
2. Similarly, each observer would find that time intervals between events in the other frame are dilated by a similar factor—i.e., the clocks in the other frame will appear to be running more slowly.
3. In addition to the elimination of any absolute space, there is no

absolute time; thus events that may be simultaneous for an observer in one frame will not in general be so for an observer in another frame. We cannot say that two events are really simultaneous unless they happen at the same place as well as the same time.

A well-known predicted consequence of the time-dilation effect is the "twin paradox" (sometimes called the "clock paradox"). If one twin goes off in a spaceship traveling at a speed close to that of light, then turns around and returns to earth, he should find that he is younger than the twin who stayed behind. This is difficult to believe; one's first reaction is, why isn't the situation symmetrical between the two twins? Without going into a detailed explanation it may be pointed out that the Earth is close to being an inertial frame of reference, whereas the spaceship clearly is not since it has to experience substantial accelerative forces to turn around. Thus the time-dilation effects are not the same for both twins.

4. In another paper published in 1905, Einstein showed that the observable mass of any object increases as it goes faster (relative to the observer), again by a factor similar to that for the FitzGerald-Lorentz contraction. This implies that it would take infinite energy to make such an object travel at the speed of light.

5. Closely connected with the mass-increase effect is Einstein's famous formula $E = mc^2$: mass and energy are no longer conserved but can be interconverted. In particular, nuclear reactions in which the total mass of the products is less than that of the reagents will release energy in the form of electromagnetic radiation, neutrinos, and excess kinetic energy of the product nuclei. Examples are the fission of heavy nuclei such as uranium or plutonium in the atom bomb or in a nuclear reactor and fusion of light nuclei such as hydrogen, as in the hydrogen bomb or in the interior of stars.

6. We can no longer treat space and time as independent entities; they are fused together into a four-dimensional "space-time manifold."

Each of these consequences is contrary to "common sense"; each has survived all possible experimental tests. The theory has been built into the core of all fundamental physical theories in the twentieth century. This does not mean that Newtonian physics is wrong, but rather that its region of validity is severely limited. On the other hand Maxwell's electromagnetic theory turns out to be completely correct (except for effects depending on the quantum nature of light), although the ether model that originally inspired it, like all ether models, has had to be abandoned.

Three historiographic or philosophical problems associated with

Einstein's invention of relativity theory might be mentioned. First, what was the role of the Michelson-Morley experiment? This has already been discussed in **8.5**. Second, was the Lorentz theory a reasonable alternative even after 1905, or was it justifiably discarded because of the unsatisfactory "ad hoc" character of the FitzGerald-Lorentz contraction hypothesis? Third, was the special theory of relativity really discovered by Lorentz and Poincare before Einstein? The latter suggestion is worth taking seriously only because it was advanced by E. T. Whittaker (1953); it seems to have been fairly completely refuted by Holton (1960) and others.

The second problem has been raised by Adolf Gruenbaum (1973), who argues that the FitzGerald-Lorentz contraction was not really an "ad hoc" hypothesis, if one defines "ad hoc" to mean not independently testable by any experiment other than the one it was cooked up to explain. Gruenbaum points out that the hypothesis predicts a finite fringe shift in the Kennedy-Thorndike experiment, which differs from the Michelson-Morley experiment in that the two arms of the interferometer apparatus are substantially different in length. The contraction hypothesis is falsified by the result of the Kennedy-Thorndike experiment unless one adds the additional hypothesis of time dilation; hence it is not ad hoc.

This argument illustrates the difference between the approach of historians and philosophers of science, for the Kennedy-Thorndike experiment was not done until 1932; Gruenbaum does not claim that anyone *before 1905* said that the contraction hypothesis was not ad hoc because it could be falsified by such an experiment. On the contrary, he admits that it was "psychologically ad hoc" in the sense that physicists *at that time* (circa 1900) *believed* that it was not independently testable. But logically it was. For the historian, the important point is that it was considered ad hoc in 1905, and this perception aided (and perhaps partly motivated) Einstein's theory.

Another philosopher of science, Elie Zahar, makes a similar claim on the basis of the Lakatos "methodology of scientific research programmes." (The British spelling, "programme," is used to indicate the special meaning of this term; see **11.5.**) He argues that the contraction hypothesis follows from a "molecular forces hypothesis" which Lorentz introduced independently of the Michelson-Morley experiment, hence should not have been counted as ad hoc. Lorentz's programme was "progressive" even after 1905, and Einstein's programme did not gain a decisive advantage until the successful explanation of the advance of Mercury's perihelion in 1915 (**8.7**). Zahar's proposal has been vigorously attacked by Arthur Miller (1974) for distorting the historical record of Lorentz's work (for references see **8.5**–*iii*).

BIBLIOGRAPHY

David Cassidy (1986), "Understanding the History of Special Relativity," *Historical Studies in the Physical and Biological Sciences* 16:177–95, reviews S. Goldberg (1984), *Understanding Relativity: Origin and Impact of a Scientific Revolution* (Boston: Birkhäuser); A. I. Miller (1981, cited below); L. Pyenson (1985), *The Young Einstein* (Boston: Hilger), and gives a bibliography of ninety-one works on this subject.

A. Pais (1982), *'Subtle Is the Lord': The Science and the Life of Albert Einstein* (New York: Oxford Univ. Press).

Themistocle M. Rassias and George M. Rassias, eds. (1982), *Selected Studies: Physics-Astrophysics, Mathematics, History of Science* (Amsterdam: North-Holland), contains papers by C. N. Yang, L. Debnath, L. Pyenson, E. Teller.

Harry Woolf, ed. (1980), *Some Strangeness in the Proportion* (Reading, Mass.: Addison-Wesley), includes G. Holton, "Einstein's Scientific Program in the Formative Years," 49–65; W. K. H. Panofsky, "Special Relativity Theory in Engineering," 94–105, with comments by E. M. Purcell, 106–8, and general discussion.

Gerald Holton (1960), "On the Origins of the Special Theory of Relativity," *American Journal of Physics* 28:627–36, reprinted in his *Thematic Origins of Scientific Thought* (Cambridge: Harvard Univ. Press, 1973).

Margaret C. Shields (1951), "Bibliography of the Writings of Albert Einstein," in *Albert Einstein Philosopher-Scientist,* ed. P. A. Schilpp (reprint, 1959, New York: Harper and Row), 689–760.

Topic *i:* 1905 Paper

Albert Einstein (1905), "Zur Elektrodynamik bewegter Körper," *Annalen der Physik,* series 4, 17:891–921. The most frequently used English translation, by W. Perrett and G. B. Jeffery (1923), may be found in *The Principle of Relativity* by Einstein et al. (London: Methuen), available in a paperback reprint (New York: Dover, n.d.) but is unreliable. I recommend the new translation in the Appendix of Arthur Miller's book (1981), cited below. In addition to Einstein's 1916 book *Relativity,* listed above under recommended readings, there is a recent paperback reprint of a short book containing translations of two lectures

(1920 and 1921), under the title *Sidelights on Relativity* (New York: Dover, 1983).

Nancy J. Nersessian (1984), "Aether/Or: The Creation of Scientific Concepts," *Studies in History and Philosophy of Science* 15:175–212; (1984), *Faraday to Einstein* (Dordrecht: Nijhoff), 121–59.

Arthur I. Miller (1984), *Imagery in Scientific Thought* (Boston: Birkhäuser), Chaps. 3, 5; (1981), *Albert Einstein's Special Theory of Relativity* (Reading, Mass.: Addison-Wesley), Chaps. 3–12 and Appendix. See also the essay review of his 1981 book by Robert M. Palter (1983), "History, Philosophy, and Physics in Einstein's Special Relativity Paper," *Annals of Science* 40:657–62. Miller's 1983 paper, listed above under recommended readings, is part of a symposium on the historiography of special relativity which includes also a paper by J. Earman et al. and a paper by K. F. Schaffner; see (1983), *PSA 1982,* vol. 2, ed. P. D. Asquith and T. Nickles (East Lansing, Mich.: Philosophy of Science Association), 403–28.

Dennis Dieks (1984), "The 'Reality' of the Lorentz Contraction," *Zeitschrift für allgemeine Wissenschafts theorie* 15:330–42.

P. W. Bridgman (1983), *A Sophisticate's Primer of Relativity,* 2d ed. (Middletown, Conn.: Wesleyan Univ. Press), with an introduction by Arthur Miller.

John Stachel and Roberto Torretti (1982), "Einstein's First Derivation of Mass-Energy Equivalence," *American Journal of Physics* 50:760–63.

Martin J. Klein (1982), "Some Turns of Phrase in Einstein's Early Papers," in *Physics as Natural Philosophy,* ed. A. Shimony and H. Feshbach (Cambridge: MIT Press), 364–75.

Enrico Bellone (1981), *La Relativita da Faraday a Einstein* (Turin, Italy: Editore Loescher).

E. G. Cullwick (1981), "Einstein and Special Relativity: Some Inconsistencies in His Electrodynamics," *British Journal for the Philosophy of Science* 32:167–76.

James T. Cushing (1981), "Electromagnetic Mass, Relativity, and the Kaufmann Experiments," *American Journal of Physics* 49:1133–49.

Adolf Gruenbaum (1973), *Philosophical Problems of Space and Time,* 2d ed. (Boston: Reidel). See A. I. Miller's review in *Isis* 66 (1975): 590–94, Gruenbaum's reply in *Isis* 68 (1977): 447–48, and Miller's rejoinder in *Isis* 68 (1977): 449–50.

Topic *ii:* Response of Scientific Community

Gerald Holton (1986), *The Advancement of Science, and Its Burdens: The Jefferson Lecture and Other Essays* (New York: Cambridge

Univ. Press), includes six essays on "Einstein and the Culture of Science."

Michel Biezunski (1985), "Popularization and Scientific Controversy: The Case of the Theory of Relativity in France," in *Expository Science,* ed. T. Shinn and R. Whitley (Boston: Reidel), 183–94.

Stanley Goldberg (1984), "Being Operational vs. Operationism: Bridgman on Relativity," *Rivista di Storia della Scienza* 1:333–54; (1976), "Max Planck's Philosophy of Nature and His Elaboration of the Special Theory of Relativity," *Historical Studies in the Physical Sciences* 7:125–60.

Gereon Wolters (1984), "Ernst Mach and the Theory of Relativity," *Philosophia Naturalis* 21:630–41, argues that statements attributed to Mach rejecting relativity are not authentic.

Lewis Pyenson (1982), "Relativity in Late Wilhelmian Germany: The Appeal to Preestablished Harmony between Mathematics and Physics," *Archive for History of Exact Sciences* 27:137–55; (1977), "Hermann Minkowski and Einstein's Special Theory of Relativity," ibid. 17:71–95.

A. Pais (1982), "How Einstein Got the Nobel Prize," *American Scientist* 70:358–65, concerns the question:"Why did the Nobel Committee for Physics wait so long before giving Einstein the prize, and why did they not award it for relativity?"

Lewis S. Feuer (1982), *Einstein and the Generations of Science,* 2d ed. (New Brunswick, N.J.: Transaction Books/Rutgers University).

Ryoichi Itagaki (1982), "Why did Mach Reject Einstein's Theory of Relativity?" *Historia Scientiarum* 22:81–95.

G. Battimelli (1981), "The Electromagnetic Mass of the Electron: A Case Study of a Non-Crucial Experiment," *Fundamenta Scientiae* 2:137–50, is on the experiments by W. Kaufmann (1906) and their failure to decide between the theories of M. Abraham, H. A. Lorentz, and Einstein.

Jeffrey Crelinsten (1980), "Physicists Receive Relativity: Revolution and Reaction," *Physics Teacher* 18:187–93.

Elie Zahar (1978), " 'Crucial' Experiments: A Case Study," in *Progress and Rationality in Science,* ed. G. Radnitzky and G. Andersson (Boston: Reidel), 71–97, concerns W. Kaufmann's 1905 electron experiment; (1977), "Mach, Einstein and the Rise of Modern Science," *British Journal for the Philosophy of Science* 28:195–213. See also the papers by Zahar and Miller cited in **8.5**-*iii.*

Topic *iii:* Later Applications and Tests

Mendel Sachs (1985), "On Einstein's Later View of the Twin Paradox," *Foundations of Physics* 15:977–80, claims that Einstein abandoned the view that there are material consequences such as asymmetrical aging implied by special relativity theory.

Arthur L. Robinson (1985), "Atomic Physics Tests Lorentz Invariance," *Science* 229:745–47, is a report on experiments by J. J. Bollinger et al. and by J. D. Prestage et al.

D. W. Allan, M. A. Weiss and N. Ashby (1985), "Around-the-World Relativistic Sagnac Experiment," *Science* 228:69–70, reports a test of the Sagnac effect, a special relativity effect due to the Earth's rotation.

Mark P. Haugan and C. M. Will (1985), "Modern Tests of Special Relativity," *Physics Today* 40, no. 5:69–76.

P. C. W. Davies (1977), *Space and Time in the Modern Universe* (New York: Cambridge Univ. Press), includes a detailed explanation of the twin paradox, 39–45.

Leslie Marder (1971), *Time and the Space-Traveller* (Philadelphia: Univ. of Pennsylvania Press), is a semihistorical review with extensive bibliography.

8.7 GENERAL THEORY OF RELATIVITY

i. Einstein's development of the general theory
ii. Tests of the theory: perihelion of Mercury, bending of light near sun, gravitational redshift
iii. Cosmology: theory of the expanding universe (before 1930)
iv. Einstein's later career; reputation and influence on world events; rejection of positivism

READINGS

McKenzie, *Major Achievements,* 308–13, 514–16, 530–33 (Einstein's 1933 lecture). Marks, *Science,* 26–75. Holton and Brush, *Introduction to Concepts and Theories,* 526–30. Schneer, *Evolution of Physical Science,* 312–14, 320–34. Olson, *Science as Metaphor,* 267–82 (cultural influence of relativity).

Solomon, *Structure of Space,* 141–62. J. Singh, *Great Ideas and Theories of Modern Cosmology* (1970), 123–57. Dickson, *Bowl of Night,* 110–31. Swenson, *Genesis of Relativity,* 188–221. Paul M. Clark, "Einstein: Philosophical Belief and Physical Theory" (cited in **8.5**), 47–57. Einstein and Infeld, *Evolution of Physics,* 209–45. Einstein, *Relativity: The Special and the General Theory* (1917), 69–137. F. Durham and

R. D. Purrington, *Frame of the Universe* (1983), 175–92. Pagels, *Cosmic Code,* 40–62.

Articles: Einstein, "On the Generalized Theory of Gravitation," *Scientific American,* Apr. 1950. Clifford M. Will, "Testing General Relativity: 20 Years of Progress," *Sky and Telescope* 66(1983): 294–99. John Stachel, "The Genesis of General Relativity," in *Einstein Symposion Berlin,* ed. H. Nelkowski et al. (1979, 1981), 428–42. N. T. Roseveare, "Leverrier to Einstein: A Review of the Mercury Problem," *Vistas in Astronomy* 23(1979):165–71. Timothy Ferris, "The Other Einstein," *Science 83* 4, no. 8(1983):35–41.

SYNOPSIS

The special theory of relativity can deal with only a limited range of phenomena, those in which gravitational forces are not involved. Moreover, it asserts the equivalence of frames of reference moving at constant velocity relative to each other, but not of those that are accelerated. The first limitation would have been of little concern to most physicists in 1900, since the theory of gravity was not then perceived as a major problem in physical theory. But when Einstein started to think about the second limitation he quickly found that the two had to be linked together. The result was his "principle of equivalence," proposed as early as 1907.

The principle of equivalence addresses a curious fact that had been well known but never properly understood before Einstein: the inertial mass of every object is the same as its gravitational mass. This was first illustrated by Galileo's discovery that all objects have the same gravitational acceleration at a given location, together with the interpretation of this discovery by Newton's laws. Since $a = F/m$, and F (the force of gravity) is proportional to the gravitational mass of the object, while m is its inertial mass, we see that if one object is twice as massive as another, it experiences twice the force but has twice the inertia, so the factors of 2 cancel out and leave the same acceleration.

Einstein argues that the equality of gravitational and inertial mass must mean that "the same quality of a body manifests itself according to circumstances as 'inertia' or as 'weight.' " Indeed, if one imagines an observer inside a box floating freely in space, which unknown to him is pulled by a uniform force, he will not be able to tell by experiments inside the box whether he is in a uniformly accelerated frame of reference or in a gravitational field. So the principle of equivalence asserts that acceleration is equivalent to a gravitational force. This implies that gravity is not a "real" force but rather a local property of space and time as described in a particular coordinate system.

In the same paper (1907) in which he proposed this principle, Einstein showed that a spectral line emitted by an atom in a strong gravitational field has, when observed at a place with lower gravitational potential, a greater wavelength. This is known as the "gravitational redshift." Its existence was not confirmed by observation until much later; the Pound-Rebka experiment (1959) is regarded as the best direct test.

In 1911 Einstein used his principle of equivalence to predict that light passing near the Sun would change its direction. The calculation was done with the Newtonian theory of gravity; in fact, J. Soldner in 1801 had attempted a similar calculation that would have given the same result, a deflection of about 0.83 seconds of arc, except that he made a numerical slip. Einstein urged astronomers to try to detect this effect, but (perhaps fortunately for the reputation of this theory) this was not done until after Einstein had developed his general theory from which a deflection twice as large was estimated.

During the years 1912–1914 Einstein worked with the Swiss mathematician Marcel Grossmann who helped him translate his physical ideas into mathematical form. This effort involved a major shift in Einstein's attitude toward the use of mathematics in physics, in particular a systematic exploitation of Riemannian geometry (**8.5-*i***) and the theory of differentiation of functions in Riemannian geometry developed by E. B. Christoffel (1869). A generalized "tensor calculus" applicable to curved spaces of any number of dimensions, worked out by G. Ricci-Curbastro and T. Levi-Civita at the end of the nineteenth century, proved useful. Einstein and Grossmann looked for a field equation in curved four-dimensional space that would be analogous to the Laplace-Poisson differential equation for the gravitational potential function in Newtonian theory. The components of the metric, g_{ij}, would specify the gravitational field and also the scale of time and distance intervals; they would have to be equated to something analogous to the mass density in the Laplace-Poisson equation.

In 1915 Einstein completed the development of his basic general theory by giving a set of field equations, constructed along the lines mentioned above. As it happened, the same equations were derived independently at the same time by the mathematician David Hilbert. Although Hilbert started to work on the problem only after Einstein had already formulated it and thus never claimed any priority as a discoverer of general relativity, his derivation of the final equations is in some respects more satisfactory than Einstein's. For example, he used a basic theorem just proved by Emmy Noether connecting the "action integral" (involving the Lagrangian function of classical mechanics) and the conservation of certain physical quantities.

Einstein showed that his equations predicted the bending of light by the sun, to the extent of about 1.7 seconds of arc (twice the value computed earlier from Newtonian gravitational theory). The successful test of this prediction by the British eclipse expedition of 1919 provided the first convincing evidence for the general theory; the two observational results were 1.98 ± 0.12 and 1.61 ± 0.30.

The advance of the perihelion of Mercury is another well-known test of general relativity; it is one that might be called an anomaly of the Newtonian theory since it was known in the nineteenth century that there was a discrepancy between the observations and the perihelion motion calculated from perturbations by the known planets. The residual effect was about 42.56 seconds per century; Einstein's calculation gave 43 seconds per century. His explanation of the effect depends on the idea that energy is associated with mass; even energy in "empty" space such as that associated with the sun's gravitational field has mass. The net mass of the sun as seen from outside the solar system is less than the sum of the masses of all the atoms in the sun (cf., the "binding energy" mass defect in a nucleus). As a planet gets close to the sun it experiences a greater "effective mass" of the sun and thus a deviation from Newtonian behavior if its orbit is eccentric.

These immediate successes of general relativity seem to have convinced Einstein that the true path to unlocking the secrets of the universe lay in the realm of abstract mathematical thought. He abandoned the positivistic-empiricist approach he had inherited from Ernst Mach and became what might be called a Platonic realist—at least so one might conclude from his oft-quoted 1933 lecture.

Within a few weeks after Einstein announced his field equations, Karl Schwarzschild discovered an exact solution of them for a spherically symmetric situation with a point mass at the origin of coordinates. The solution has a singularity at a certain distance from the mass; for a mass equal to that of the sun this distance is 2.5 km. The region inside this distance is now called a "black hole." Its interpretation goes back to John Michell (1784) and P. S. de Laplace (1798) who pointed out that if a star has a large enough ratio of mass to radius, the escape velocity from its surface would exceed the velocity of light and thus light could not escape from it.

In applying his theory to cosmology, Einstein wanted to avoid the well-known difficulty that a static distribution of attracting masses would not be stable but would collapse. For this and other reasons he added a term, later called the "cosmological constant," to his field equations; it produces a repulsive force proportional to the distance from the origin of coordinates. It depends on the radius of curvature of the space, which in turn depends on the total mass of the universe. His solution of the resulting equation, the "Einstein universe," has a uniform distribution of matter and is static.

In 1922 the Russian mathematician Aleksandr Friedmann developed nonstatic cosmologies based on Einstein's equations; these included both cyclic and expanding universes. During the 1920s the astronomical basis for the expanding universe was being developed by V. M. Slipher and E. P. Hubble (**13.1**). Einstein later abandoned his cosmological constant, but other cosmologists have continued to use it.

BIBLIOGRAPHY

A. Pais (1982), *'Subtle is the Lord': The Science and the Life of Albert Einstein* (New York: Oxford Univ. Press).

T. M. Rassias and G. M. Rassias, eds. (1982), *Selected Studies: Physics-Astrophysics* . . . (Amsterdam: North Holland), has papers by G. Birkhof, C. N. Yang, S. Bourne, L. Debnath, S. Weinberg.

Gerald Holton and Yehuda Elkana, eds. (1982), *Albert Einstein: Historical and Cultural Perspectives* (Princeton, N.J.: Princeton Univ. Press), is a collection of essays by G. Holton, A. I. Miller, P. G. Bergmann, P. A. M. Dirac, B. Hoffmann, L. R. Graham, R. Jakobson, E. H. Erikson, Y. Elkana, I. Berlin, F. Stern, and others.

Harry Woolf, ed. (1980), *Some Strangeness in the Proportion* (Reading, Mass.: Addison-Wesley), includes F. Gilbert, "Einstein's Europe," 13–27; H. Woolf, "Albert Einstein: Encounter with America," 28–37; I. I. Shapiro, "Experimental Challenges Posed by the General Theory of Relativity," 115–36, with comments by D. T. Wilkinson and others; S. W. Hawking, "Theoretical Advances in General Relativity," 145–52, with comments by W. G. Unruh and others; S.-S. Chern, "General Relativity and Differential Geometry," 271–80, with comments by T. Regge and others; other papers on cosmology cited in **13.3**; E. P. Wigner, "Thirty Years of Knowing Einstein," 461–68; H. Bethe et al., "Working with Einstein," 473–89.

Topic *i:* Einstein's Development of General Theory

Eva Isaksson (1985), "Der Finnische Physiker Gunnar Nordström und sein Beitrag zur Entstehung der allgemeinen Relativitätstheorie Albert Einsteins," *NTM* 22:29–52.

John Norton (1985), "What Was Einstein's Principle of Equivalence?" *Studies in History and Philosophy of Science* 16:203–46; (1985),

"Einstein's Struggle with General Variance," *Rivista di Storia della Scienze* 2:181–205; (1984), "How Einstein Found His Field Equations: 1912–1915," *Historical Studies in the Physical Sciences* 14:253–316.

E. Zahar (1979), "The Mathematical Origins of General Relativity and Unified Field Theories," in *Einstein Symposion Berlin . . .*, ed. H. Nelkowski et al. (New York: Springer-Verlag), 370–96.

J. Earman and C. Glymour (1978), "Einstein and Hilbert: Two Months in the History of General Relativity," *Archive for History of Exact Sciences* 19:291–308; (1978); "Lost in the Tensors: Einstein's Struggles with Covariance Principles, 1912–1916," *Studies in History and Philosophy of Science* 9:251–78.

Jon Dorling (1978), "Did Einstein Need General Relativity to Solve the Problem of Absolute Space? Or Had the Problem Already Been Solved by Special Relativity? *British Journal for the Philosophy of Science* 29:311–23.

Topic *ii:* Tests

Clifford M. Will (1986), *Was Einstein Right? Putting General Relativity to the Test* (New York: Basic Books).

Stuart L. Shapiro, Richard F. Stark, and Saul A. Teukolsky (1985), "The Search for Gravitational Waves," *American Scientist* 73:248–57.

Jeffrey Crelinsten (1983), "William Wallace Campbell and the 'Einstein Problem': An Observational Astronomer Confronts the Theory of Relativity," *Historical Studies in the Physical Sciences* 14:1–91.

Monroe G. Barnard (1983), "Star Displacement at Eclipse," *Physis* 25:577–84, discusses the report of Trumpler and Campbell on the 1922 eclipse.

P. W. Worden, Jr., and C. W. F. Everitt (1982), "Resource Letter GI-1: Gravity and Inertia," *American Journal of Physics* 50:494–500, is an annotated bibliography for physics teachers. Worden and Everitt also edited (1983), *Gravity and Inertia: Selected Reprints* (Stony Brook, N.Y.: American Association of Physics Teachers).

N. T. Roseveare (1982), *Mercury's Perihelion from Le Verrier to Einstein* (Oxford: Clarendon Press); (1979), "Leverrier to Einstein: A Review of the Mercury Problem," *Vistas in Astronomy* 23:165–71.

John Earman and Clark Glymour (1980), "The Gravitational Red Shift as a Test of General Relativity: History and Analysis," *Studies in History and Philosophy of Science* 11:175–214.

Norriss S. Hetherington (1980), "Sirius B and the Gravitational Redshift: An Historical Review," *Quarterly Journal of the Royal Astronomical Society* 21:246–52.

Donald F. Moyer (1979), "Revolution in Science: The 1919 Eclipse

Test of General Relativity," in *On the Path of Albert Einstein,* ed. A. Perlmutter and L. F. Scott (New York: Plenum Press), 55–101.

Topic *iii:* Cosmology

Jean Eisenstaedt (1986), "La Relativite Generale a l'Etiage: 1925–1955," *Archive for History of Exact Sciences* 35:115–85; (1982), "Histoire et Singularites de la Solution de Schwarschild (1915–1923), ibid. 27:157–98.

M. Heller and O. Godart (1981), "Origins of Relativitistic Cosmology," *Astronomy Quarterly* 4, no. 13:27–33.

John Archibald Wheeler (1981), "The Lesson of the Black Hole," *Proceedings of the American Philosophical Society* 125:25–37.

S. Chandrasekhar (1980), "The Role of General Relativity in Astronomy: Retrospect and Prospect," *Journal of Astrophysics and Astronomy* 1:33–45; (1980), "The General Theory of Relativity: The First Thirty Years," *Contemporary Physics* 21:429–49; (1979), "Einstein and General Relativity," *American Journal of Physics* 47:212–17.

Simon Schaffer (1979), "John Michell and Black Holes," *Journal for the History of Astronomy* 10:42–43.

Topic *iv:* Einstein's Later Career

Lewis Elton (1986), "Einstein, General Relativity, and the German Press," *Isis* 77:95–103.

Klaus Hentschel (1986), "Die Korrespondenz Einstein-Schlick: Zum Verhältnis der Physik zur Philosophie," *Annals of Science* 43:475–88.

Marshall Missner (1985), "Why Einstein Became Famous in America," *Social Studies of Science* 15:267–91.

Jamie Sayen (1985), *Einstein in America* (New York: Crown).

Ashley Montagu (1985), "Conversations with Einstein," *Science Digest* 93, no. 7:50–53, 75.

Richard Alan Schwartz (1983), "The F.B.I. and Dr. Einstein," *Nation* 237, no. 6:168–73.

J. Crelinsten (1980), "Einstein, Relativity, and the Press: The Myth of Incomprehensibility," *Physics Teacher* 18:115–22.

Elie Zahar (1980), "Einstein, Meyerson and the Role of Mathematics in Physical Discovery," *British Journal for the Philosophy of Science* 31:1–43.

9

ATOMIC STRUCTURE

9.1 THE ELECTRON

i. Faraday and electrolysis; basic unit of charge; Maxwell's views; G. J. Stoney's "electron"
ii. Cathode rays; particle vs. wave theory; J. J. Thomson's 1897 paper
iii. Quantization of charge: the Millikan oil-drop experiment; R. A. Millikan vs. F. Ehrenhaft

READINGS

Harold Sharlin, *The Convergent Century* (1966), 161–81. A. E. E. McKenzie, *Major Achievements* (1973), 276–85, 506–8. C. A. Russell and D. C. Goodman, "Atomism," Unit 5 in the Open University course AST 281, *Science and the Rise of Technology since 1800* (1973), 36–39 (on *ii*); Russell, "Early Electrochemistry," Unit 6, ibid., 59–75 (on *i*). H. P. Fyrth and M. Goldsmith, *Science, History and Technology,* Book 2, Part 2 (1969), 1–13 (background in late nineteenth-century European history), 14–19 (X rays, electron).

David L. Anderson, *Discoveries in Physics* (1973), 33–44 (on *i* and *ii*). Joan Solomon, *The Structure of Matter* (1974), 64–93. E. Segre, *From X-Rays to Quarks* (1980), 1–19 (includes overview of physics in 1895). Alex Keller, *The Infancy of Atomic Physics* (1983), Chaps. 1, 2, 3. R. B. Lindsay, *Early Concepts of Energy in Atomic Physics* (1979), 234–64 (includes Thomson's 1897 paper).

Articles: Marjorie Malley, "The Discovery of the Beta Particle," *American Journal of Physics* 39(1971):1454–60. B. A. Morrow, "On the Discovery of the Electron," *Journal of Chemical Education* 46(1969): 584–88.

One of the best textbooks for most of the material covered in Chaps. 9 and 11 is Victor Guillemin's *The Story of Quantum Mechanics* (1968). While it provides useful background material on the earlier history of ideas about electrons and light, it does not cover topics such as spectroscopy and radioactivity in enough detail or in appropriate chronological sequence for my purposes. I have not, therefore, given specific page references here, but recommend that you consider using it in any course dealing with quantum theory and elementary particles.

Conversely Cecil Schneer's *The Evolution of Physical Science* (1960) is suitable for a course going up to the twentieth century but devotes only one chapter to modern atomic physics (335–65). Both Guillemin and Schneer adopt the common but misleading view that Planck's introduction of quantum theory was motivated by the "ultraviolet catastrophe" of black-body radiation theory.

Similarly S. Mason gives only a very brief and somewhat inaccurate account of the history of quantum theory in his *History of the Sciences* (1962), 549–63, but this book does cover many other topics in the history of modern science and may therefore be useful.

SYNOPSIS

What we call the "discovery of the electron" is not a single historical event but rather a combination of experiments, debates, and theoretical speculations in the last part of the nineteenth century, culminating in the work of J. J. Thomson in 1897. Thomson, building on the work of his predecessors such as William Crookes and Jean Perrin, established the modern view that the so-called "cathode rays" are streams of discrete particles with negative electric charge and small but finite mass moving at speeds that are high but definitely less than that of light. The opposing theory was that cathode rays are some form of electromagnetic waves or ether disturbance. For about thirty years it was thought that Thomson had proved that electrons are particles rather than waves; thus we have the paradox that in 1906 he was awarded the Nobel Prize for proving the particle nature of the electron, and in 1937 his son George Thomson shared the Nobel Prize for proving the wave nature of the electron.

The discovery of the electron was in several ways a direct outgrowth of the work of Faraday. In addition to his research on the interaction of magnetic fields and electric currents, he discovered the quantitative law of electrochemical deposition in 1833; in publishing this research he introduced the modern terms "electrode," "cathode," "anode," "ion," "anion," "cation," and "electrolyte." Here he established the idea that a definite quantity of electricity is associated with each atom of matter. Finally it was Faraday who in 1838 studied electric discharges in a vacuum and discovered the "Faraday dark space" near the cathode.

Maxwell recognized that one could use Faraday's results on electrolysis to obtain a basic unit of electric charge by dividing the Faraday constant (total electric charge per mole of univalent atoms) by Avogadro's number but was reluctant to accept the idea that electricity is composed of atomic particles. G. J. Stoney was the first to cite a numerical value for the unit of electric charge found in this way in 1874 (published 1881), and in an 1894 article he established the use of the name "electron." Helmholtz supported the idea that electricity comes in atomic units in his Faraday lecture in London in 1881.

The study of vacuum discharges involved crucial developments in technology; only after the mid-nineteenth century was it possible to make observations at extremely low pressures where the most interesting phenomena of electronics occur. The invention of a new vacuum pump by J. H. W. Geissler in 1855, and its improvement by H. Sprengel in 1865, led directly to the flourishing of atomic physics as well as to the developments in electrical technology initiated by Thomas A. Edison (**8.2**). J. Pluecker used Geissler tubes to show that at low pressures the "Faraday dark space" becomes larger and there is an extended glow on the glass walls around the cathode. The glow was found to be affected by an external magnetic field.

In 1869 Pluecker's student J. W. Hittorf found that any solid body placed in front of the cathode cut off the glow on the walls as if it were blocking some kind of rays emanating from the cathode. By making an L-shaped tube, Hittorf established that these rays seemed to travel in straight lines from the cathode to the glass wall; the rays themselves are invisible but produce a phosphorescent glow on the glass when they strike it. Eugen Goldstein did extensive experiments on what he called "cathode rays." William Crookes extended this work; his best known experiment is the one that shows the shadow produced on the glass wall by a barrier shaped like a Maltese Cross.

The first published suggestion that cathode rays are composed of particles was by C. F. Varley in 1871. Crookes proposed that they are molecules that have picked up a negative electric charge from the cathode and then are repelled by it. That the rays are deflected by magnetic fields seems to support the idea they are negatively charged particles—but that they don't seem to be deflected by *electric* fields is puzzling. After H. Hertz succeeded in producing electromagnetic waves in 1888

(**8.4**), he and other German physicists argued that cathode rays are also electromagnetic waves, while most English physicists thought they must be some kind of particles.

J. J. Thomson (1856–1940) settled the debate about the nature of cathode rays in 1897. His work had three components. First, he improved Jean Perrin's method of collecting charge inside the vacuum tube, showing that an electric charge was collected only when a magnetic field was used to bend the rays into a path leading to the collector. (This avoided the objection that charged particles might be *produced* by the rays but were not identical with them.) Second, he showed that the rays *are* deflected by an electric field; Hertz had used his failure to observe such deflection as an argument that the rays are not charged particles. But Thomson explained Hertz's result by postulating that the discharge ionizes the gas inside the tube, and ions collect on the plates one is using to produce the field and neutralize the charge on the plates. If one uses a better vacuum pump to remove the gas ions from the neighborhood of the plates, this problem is avoided. Third, Thomson was able to obtain a reasonably good value for the charge/mass ratio of the electron by estimating its velocity in two ways, from the temperature rise of the charge collector and by balancing the magnetic and electric deflections of the beam.

Attempts were soon made in Thomson's laboratory (the Cavendish, at Cambridge University) to measure the electron charge directly. After several experiments by J. S. E. Townsend, C. T. R. Wilson, H. A. Wilson, and others, the definitive research was undertaken by R. A. Millikan at Chicago in 1906. He found a method to balance the electric and gravitational forces on a single oil drop. His experiments, concluded in 1914, showed that the electric charge always comes in integer multiples of a certain basic charge $e = 1.592 \times 10^{-19}$ coulomb. Millikan's conclusion was disputed by F. Ehrenhaft, whose own experiments yielded a range of noninteger multiples and fractions of this amount, but Millikan's conclusion was generally accepted.

A study by G. Holton (1977) analyzes that raw data of Millikan and Ehrenhaft and discusses the possible influence of pro- and anti-atomistic preconceptions on the interpretation of those data.

BIBLIOGRAPHY

John L. Heilbron and Bruce R. Wheaton (1981), *Literature on the History of Physics in the 20th Century* (Berkeley: Office for History of Science and Technology, University of California), 292–94.

L. Marton and C. Marton (1980), "Evolution of the Concept of the Elementary Charge," *Advances in Electronics and Electron Physics* 50:449–72.

David L. Anderson (1964), *The Discovery of the Electron* (Princeton, N.J.: Van Nostrand), is written for physics students but includes references to original sources.

Topic *i:* Faraday to Stoney

Jed Z. Buchwald (1985), *From Maxwell to Microphysics: Aspects of Electromagnetic Theory in the Last Quarter of the Nineteenth Century* (Chicago: Univ. of Chicago Press).

J. G. O'Hara (1975), "George Johnstone Stoney, F.R.S., and the Concept of the Electron," *Notes and Records of the Royal Society of London* 29:265–76.

Topic *ii:* Cathode Rays, Thomson

Isobel Falconer (1987), "Corpuscles, Electrons and Cathode Rays: J. J. Thomson and the 'Discovery of the Electron,'" *British Journal for the History of Science* 20:241–76.

Giora Hon (1987), "H. Hertz: 'The Electromagnetic Properties of the Cathode Rays Are Either *Nil* or Very Feeble.' (1883) A Case-Study of an Experimental Error," *Studies in History and Philosophy of Science* 18:367–82.

A. B. Arons (1982), "Phenomenology and Logical Reasoning in Introductory Physics Courses," *American Journal of Physics* 50:13–20, includes a discussion of Thomson's experiment.

A. N. Vyaltsev (1981), "The Discovery of Electron and Its Scientific and Technical Implications," *Acta Historiae Rerum Naturalium nec non Technicarum* 14:227–48.

Barbara M. Turpin (1980), "The Discovery of the Electron: The Evolution of a Scientific Concept 1800–1899," (Ph.D. diss., University of Notre Dame).

J. L. Heilbron (1976), "Thomson, Joseph John," *Dictionary of Scientific Biography* 13:362–72.

Topic *iii:* Millikan Oil Drop

William M. Fairbank, Jr. and Allan Franklin (1982), "Did Millikan Observe Fractional Charges on Oil Drops?" *American Journal of Physics* 50:394–97.

Harvey Fletcher (1982), "My Work with Millikan on the Oil-Drop Experiment," *Physics Today* 35, no. 6:43–47.

Allan Franklin (1981), "Millikan's Published and Unpublished Data on Oil Drops," *Historical Studies in the Physical Sciences* 1:185–201.

Gerald Holton (1977), "Electrons or Subelectrons? Millikan, Ehrenhaft and the Role of Preconceptions," in *History of Twentieth Century Physics,* ed. C. Weiner (New York: Academic Press), 266–89; (1977), "Subelectrons, Presuppositions, and the Millikan-Ehrenhaft Dispute," *Historical Studies in the Physical Sciences* 9:161–24.

9.2 RADIOACTIVITY AND NUCLEAR TRANSMUTATION, 1895–1930

i. Wilhelm Roentgen's discovery of X rays
ii. Henri Becquerel's discovery of radioactivity
iii. Life and work of the Curies; Marie Curie as a role model for women in science
iv. Life and work of Ernest Rutherford; nuclear transmutation hypothesis of Rutherford and Fredrick Soddy
v. Radioactive dating, age of the Earth, impact on geology
vi. The nuclear atom—Geiger-Marsden experiment
vii. Isotopes and atomic number; proton-electron model of the nucleus

READINGS

McKenzie, *Major Achievements,* 280, 286–98, 508–12. Gerald Holton and Stephen G. Brush, *Introduction to Concepts and Theories in Physical Science* (1973), 451–67. H. J. Fyrth and M. Goldsmith, *Science, History, and Technology,* Book 2, Part II (1969), 14–32.

Barbara L. Cline, *Men Who Made a New Physics* (1969), 9–31. Joan Solomon, *Structure of Matter* (1974), 95–103. Segre, *X-Rays to Quarks,* 19–60, 101–18. E. N. daC. Andrade, *Rutherford and the Nature of the Atom* (1964), 1–133, 144–70. T. Trenn, *Transmutation: Artificial and Natural,* 56–75. George L. Trigg, *Crucial Experiments in Modern Physics* (1975), 23–35 (on *iv*), 55–67 (on *vi*). Keller, *Infancy of Atomic Physics* (1983), Chaps. 5, 6.

Article: L. Badash, " 'Chance Favors the Prepared Mind': Henri Becquerel and the Discovery of Radioactivity," *Archives Internationales d'Histoire des Sciences* 18(1965):55–66.

SYNOPSIS

The beginning of "modern" physics is often defined as the discovery of X rays by Wilhelm Roentgen in 1895. He found that when certain substances are exposed to a beam of cathode rays, a new kind of radiation, having rather sensational properties, is produced. When X rays go through human flesh they produce a "picture" of the skeleton together with other incidental solid objects—bullets, swallowed pins, etc. Nevertheless, the idea that one kind of radiation could be transformed into another was not new to physics. Radioactivity, discovered by Henri Becquerel a few months later in 1896 in his attempt to find substances that emit X rays, was more revolutionary. Here was a substance—potassium-uranyl sulfate—spontaneously emitting radiation without any external stimulus and without apparently undergoing any change itself. Radioactivity was heralded in the popular press as an amazing new source of energy, which might even overthrow the principle of conservation of energy.

After Becquerel's initial discovery the pioneer researcher on radioactivity was Marie Sklodowska Curie (1867–1934). Born in Warsaw, Poland, she went to Paris in 1891 to study science, earned a degree in physics in 1893, and met the physicist Pierre Curie (eight years older, and already known for his work on piezoelectricity) in 1895. They collaborated on the isolation of radium, a highly radioactive element found in minute quantities in some uranium-bearing minerals. In 1902 they had a sample large and pure enough to determine the atomic weight, and the next year they were jointly awarded the Nobel Prize in Physics.

Marie Curie made radium available to the scientific community as a source of intense radioactivity. Other scientists with better facilities than the Curies had available in Paris (even after their Nobel Prize) quickly made a number of major discoveries and proposed bold new theories about the nature of radioactivity that the Curies were reluctant to accept. (See Malley's article on the possible influence of positivism in preventing them from recognizing transmutation.) Pierre was killed in a horse-cart accident in 1906. Marie continued her tedious experimental work, beginning to suffer the effects of radium exposure. Though a heroine of science to the public, she was still denied full recognition by the establishment in France; the Academy of Science refused to elect her to membership in 1911. In the same year a scandal broke out in the press about her alleged affair with the physicist Paul Langevin; the award to her of the Nobel Prize in Chemistry at this time may have been the result of sympathy among scientists in other countries rather than an objective assessment of her scientific achievements since the previous award.

In any case, after 1902 the leader in the new field was not Marie Curie but Ernest Rutherford (1871–1937), who came from New Zealand to work in J. J. Thomson's lab in 1895. He did much of his most important work as a professor at McGill University in Montreal from 1898 to 1907, when he was called to a chair at Manchester. Rutherford showed (1899) that uranium emits two different kinds of radiation, which he called "alpha" and "beta" rays—the latter are much more penetrating. A third kind, discovered by the French physicist Paul Villard in experiments with radium, is even more penetrating; Rutherford called these "gamma" rays. In 1903 Rutherford concluded that alpha rays are made up of doubly ionized helium atoms (thus having a positive charge) and beta rays are made up of electrons (negative charge).

Gamma rays and X rays are both forms of electromagnetic radiation, with wavelengths much shorter than visible light.

In 1902–1903 Rutherford and the British chemist Frederick Soddy proposed that radioactivity involves the actual *transmutation* of one element into another. Thus when uranium emits an alpha ray it becomes thorium + helium. The nineteenth-century doctrine that matter comes in a large number of qualitatively different kinds of *elements* had to be given up. Prout's hypothesis (**7.1**) could now be revived: all elements are compounds of hydrogen. In 1903 this was not yet completely acceptable because the atomic weights of the elements were known not to be integers.

Rutherford's conception of the radioactive decay process threw a new light on another nineteenth-century debate: randomness at the atomic level (**7.3**). Decay seemed to occur quite independently of any influences outside the atom and could be described only by assuming that there is a certain *probability* for an atom to decay in each small time interval. Starting around 1905, Egon von Schweidler and others constructed detailed statistical theories of radioactive processes in which no mention was made of any "causes" of decay; it was a purely random event. The success of such theories was one factor that led physicists to postulate randomness in other atomic theories in the twentieth century.

The geophysical significance of radioactivity was discovered fairly early. In 1903, Pierre Curie and Albert Laborde found that radium continually generates heat at a rate of about 1000 calories per gram per hour. It was immediately recognized by G. H. Darwin, J. Joly, and Rutherford that this heat, generated by radium and other radioactive minerals in the crust of the Earth, might well be sufficient to balance that lost by conduction and thus counteract the cooling of the Earth. Kelvin's estimate of the age of the Earth (**2.4**) would then be completely wrong, since he had assumed no sources of heat are present inside the Earth. Rutherford also showed how a minimum age of the Earth could be estimated from measurements of the amount of helium found in certain minerals; in 1905 he suggested values well over 100 million years. At about the same time R. J. Strutt (fourth Baron Rayleigh) had published an estimate of 2.4 billion years. While these estimates ignore the loss of helium from the rock and depend on other erroneous assumptions, there was no doubt that rocks had existed in a solid state on the Earth's surface for at least 20 times as long as Kelvin's original estimate of 100 million years for the "age of the Earth."

Kelvin's estimate of the age of the sun, based on the assumption that its heat come from gravitational contraction (**12.4**), was also wrong if the sun contains radium or other elements that can release energy by radioactive decay. Thus the atomic physicists rescued Darwin's theory of evolution from the earlier criticism that not enough time was available for the slow process of natural selection (**2.4-*ii***), and opened the way for new ideas about stellar evolution (**13.2**).

When he moved to Manchester, Rutherford inherited a German assistant Hans Geiger who quickly took an active part in his research on radioactivity and in 1913 invented the radiation detector well-known as the "Geiger counter." Together with a student, Ernest Marsden, Geiger carried out at Rutherford's suggestion a famous experiment on the scattering of alpha particles (1909) that was the basis for Rutherford's "nuclear model of the atom" published in 1911. Rutherford could estimate the nuclear charge from measurements of the scattering of alpha particles at various angles and concluded that it is approximately half of the atomic weight of the element. C. G. Barkla reached the same conclusion at about this time from the results of X-ray scattering experiments.

In 1913 a Dutch scientist, Antonius van der Broek, suggested that the nuclear charge is exactly equal to the "atomic number" in Mendeleeff's periodic table (**7.1-*iv***). This suggestion was shown to be correct by Rutherford's colleague Henry G. J. Moseley, on the basis of a systematic survey of X-ray spectra of the elements. He found that a certain line in the spectrum changes its position by a regular amount in going from one element to another whose atomic number differs by one. The conclusion was that the most important property of an element is not its atomic weight but its atomic number, the latter being equal to the positive charge on the nucleus and to the number of extranuclear electrons of the normal (neutral) atom.

Moseley was killed in action during World War I. He would probably have become one of the leading physicists of the twentieth century if he had lived ten years longer.

It was now possible to make sense out of radioactive transmutations, assuming that total nuclear charge is conserved in all decays. (When a beta ray is emitted one has to count its charge too.) Radioactive elements so short-lived that their atomic weights could not be determined directly could be assigned an atomic number by using the fact that they were produced by (or produced) a known element by emission of an alpha or beta ray. On this basis K. Fajans and F. Soddy independently assigned all the radioactive substances to places in the periodic table in 1913. Soddy decided that several substances previously considered to be different elements ended up in the same place in this table and could not be chemically distinguished from each other. Since they were therefore the same "element" but physically different, he introduced the term "isotope" (Greek, "same place") to describe them.

Since isotopes cannot be separated by chemical methods it is difficult to determine their individual atomic weights. The mass spectrograph, invented by F. W. Aston (1919) on the basis of earlier work by J. J. Thomson, does allow direct determination of atomic weights of isotopes, using an arrangement of electric and magnetic fields to separate them. In 1920 Aston announced that all atomic weights except that of hydrogen are integers when oxygen is taken as 16. The most serious discrepancy from the whole-number (Prout) hypothesis, chlorine's atomic weight of 35.46, was now resolved: there are 2 isotopes of C1 with atomic weights 35 and 37. (In nature 75 percent is 35, 25 percent is 37.) More accurate measurements soon showed that isotopic atomic weights are not quite integers, but the discrepancies can be attributed to the mass equivalent of the forces that hold together the parts of the nucleus. This interpretation could be verified by measuring the total kinetic energy of all the participants in a nuclear reaction, before and after.

The proton was discovered by J. J. Thomson in 1913, i.e., he showed that the "canal rays" found by Goldstein in cathode-ray discharges in 1886 are discrete particles with a definite charge/mass ratio. It had already been found by W. Wien (1898) that they have masses comparable to those of hydrogen atoms. The name "proton" was proposed by Rutherford in 1920.

Before the discovery of the neutron in 1932, it was generally assumed that the atomic nucleus consists of protons (number equal to the "mass number" A = integer closest to atomic weight) and electrons (number equal to A – Z, where Z = atomic number). See **10.1** for the modern theory of nuclear structure and definition of isotope.

BIBLIOGRAPHY

S. B. Sinclair (1985), "Crookes and Radioactivity: From Inorganic Evolution to Atomic Transmutation," *Ambix* 32:15–31.

Stephen G. Brush and Lanfranco Belloni (1983), *The History of Modern Physics: An International Bibliography* (New York: Garland), 205–7, 212–18.

J. L. Heilbron and Bruce R. Wheaton (1981), *Literature on the*

History of Physics in the 20th Century (Berkeley: Office for History of Science and Technology, University of California), 321–27, 339–57.

Walter Minder (1981), *Geschichte der Radioaktivität* (Berlin: Springer-Verlag).

Giuseppe Bruzzaniti (1980), *La Radioattività da Becquerel a Rutherford* (Turin, Italy: Loescher).

Henry A. Boorse and Lloyd Motz (1966), *The World of the Atom* (New York: Basic Books), includes extracts from the papers of Roentgen, Becquerel, the Curies, Rutherford, Villard, Soddy, Geiger and Marsden, Aston, van der Broek, and Moseley, with commentary.

Topic *i:* X rays

Giorgio Cosmacini (1984), *Röntgen* (Milan, Italy: Rizzoli).

Bruce R. Wheaton (1981), "Impulse X-Ray and Radiant Intensity: The Double Edge of Analogy," *Historical Studies in the Physical Sciences* 11:367–90.

Ronald A. Brown (1979), "X Rays and After," *Journal of Chemical Education* 56:191–93.

G. L'E Turner (1975), "Röntgen (Roentgen), Wilhelm Conrad," *Dictionary of Scientific Biography* 11:529–31.

Walter Thumm (1975), "Roentgen's Discovery of X rays," *Physics Teacher* 13:207–14.

Topic *ii:* Becquerel and Radioactivity

J. Van Brakel (1985), "The Possible Influence of the Discovery of Radio-active Decay on the Concept of Physical Probability," *Archive for History of Exact Sciences* 31:369–85.

Ann Prescott (1983), "Henri Becquerel, Discoverer of Radioactivity: The Family Background of a Great French Scientist," *Laurels* 54:37–47.

E. Amaldi (1979), "Radioactivity, A Pragmatic Pillar of Probabilistic Conceptions," in *Problems in the Foundations of Physics,* ed. G. Toraldo di Francia (New York: North-Holland), 1–28.

E. N. Jenkins (1979), *Radioactivity: A Science in Historical and Social Context* (New York: Crane, Russak).

Topic *iii:* The Curies

Marie Curie (1961), *Radioactive Substances* (English translation) (New York: Philosophical Library).

Marie Curie (1932), *Pierre Curie* (English translation) (New York: Macmillan), including autobiographical notes by Marie Curie.

Francoise Giroud (1986), *Marie Curie: A Life* (New York: Holmes and Meier).

Marjorie Malley (1979), "The Discovery of Atomic Transmutation: Scientific Styles and Philosophies in France and Britain," *Isis* 70:213–23, discusses why the Curies didn't discover it.

Lawrence Badash (1978), "Radium, Radioactivity, and the Popularity of Scientific Discovery," *Proceedings of the American Philosophical Society* 122:145–54.

Robert Reid (1974), *Marie Curie* (New York: Saturday Review Press).

Adrienne R. Weill (1971), "Curie, Marie (Maria Sklodowska)," *Dictionary of Scientific Biography* 3:497–503.

Alfred Romer, ed. (1970), *Radiochemistry and the Discovery of Isotopes* (New York: Dover), and (1964), *The Discovery of Radioactivity and Transmutation* (New York: Dover), contain extracts from original papers, with commentary.

Topic *iv:* Rutherford, Transmutation

Lawrence Badash (1987), "Ernest Rutherford and Theoretical Physics," in *Kelvin's Baltimore Lectures and Modern Theoretical Physics,* ed. R. Kargon and P. Achinstein (Cambridge: MIT Press), 349–73; (1979), *Radioactivity in America* (Baltimore: Johns Hopkins Univ. Press); (1975), "Rutherford, Ernest," *Dictionary of Scientific Biography* 12:25–36.

David Wilson (1983), *Rutherford, Simple Genius* (Cambridge: MIT Press).

Thaddeus J. Trenn (1981), *Transmutation* (Philadelphia: Heyden); (1977), *The Self-Splitting Atom: The History of the Rutherford-Soddy Collaboration* (London: Taylor and Francis).

Mario Bunge and W. R. Shea, eds. (1979), *Rutherford and Physics at the Turn of the Century* (New York: Science History), has papers by L. Badash, J. L. Heilbron, N. Feather, T. J. Trenn, S. L. Jaki, S. G. Brush, etc.

A. Pais (1977), "Radioactivity's Two Early Puzzles," *Reviews of Modern Physics* 49:925–38; reprinted in (1977), *A Festschrift for I. I. Rabi,* ed. L. Motz (New York: New York Academy of Sciences, 116–36, discusses the two puzzles: What is the energy source? What is the significance of half-life?

Topic *v:* Radiometric Dating, Age of Earth

Stephen G. Brush (1982), "Finding the Age of the Earth: By Physics or by Faith?" *Journal of Geological Education* 30:34–58, is a review of the development of radiometric dating methods and refutation of creationist "young earth" arguments.

Claude C. Albritton, Jr. (1980), *The Abyss of Time: Changing Conceptions of the Earth's Antiquity after the Sixteenth Century* (San Francisco: Freeman, Cooper), Chap. 16.

R. B. Lindsay (1979), *Early Concepts of Energy in Atomic Physics* (New York: Academic Press), 316–39, includes papers by Curie and Laborde and by Rutherford and Barnes on radioactive heat generation.

Henry Faul (1978), "A History of Geologic Time," *American Scientist* 66:159–65.

Topic *vi:* Nuclear Atom

Lawrence Badash (1983), "Nuclear Physics in Rutherford's Laboratory before the Discovery of the Neutron," *American Journal of Physics* 51:884–89.

T. J. Trenn (1974), "The Geiger-Marsden Scattering Results and Rutherford's Atom, July 1912 to July 1913: The Shifting Significance of Scientific Evidence," *Isis* 65:74–82.

Topic *vii:* Isotopes, Model of Nucleus

Roger H. Stuewer (1985), "Artificial Disintegration and the Cambridge-Vienna Controversy," in *Observation, Experiment, and Hypothesis in Modern Physical Science,* ed. P. Achinstein and O. Hannaway (Cambridge: MIT Press), 239–307.

George B. Kaufmann (1982), "The Atomic Weight of Lead of Radioactive Origin: A Confirmation of the Concept of Isotopy and the Group Displacement Laws," *Journal of Chemical Education* 59:3–8, 119–23.

Lawrence Badash (1979), "The Suicidal Success of Radiochemistry," *British Journal for the History of Science* 12:245–56, is on the work of K. Fajans, F. Soddy; isotopes, atomic weights.

J. W. Van Spronsen (1979), "Atomic Number before Moseley," *Journal of Chemical Education* 56:106.

J. L. Heilbron (1974), "Moseley, Henry Gwyn Jeffreys," *Dictionary of Scientific Biography* 9:542–45; (1974), *H. G. J. Moseley* (Berkeley: Univ. of California Press).

9.3 SPECTROSCOPY AND BLACK-BODY RADIATION

i. J. Fraunhofer lines in the solar spectrum
ii. Kirchhoff-Bunsen "spectrum analysis"

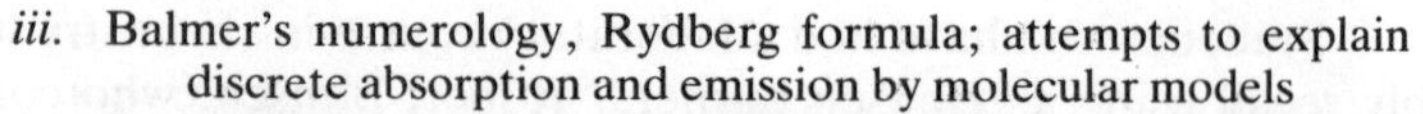

iii. Balmer's numerology, Rydberg formula; attempts to explain discrete absorption and emission by molecular models

iv. Stefan-Boltzmann law for radiation; relation to earlier work on cooling, heat conduction, thermodynamics, electromagnetic theory

v. The search for the frequency-distribution law of black-body radiation; Wien's laws

READINGS

Segre, *X-Rays to Quarks,* 61–68. Leo Klopfer, "Fraunhofer Lines" (32-page booklet, published by Science Research Associates, 1964, in the series "History of Science Cases"). Holton and Brush, *Introduction to Concepts and Theories,* 427–30, 470–78. Herbert Dingle (1963), "A Hundred Years of Spectroscopy," *British Journal for the History of Science* 1:199–216.

SYNOPSIS

The history of spectroscopy begins with observations of light from the Sun. In 1802 W. H. Wollaston noticed some dark lines in the solar spectrum. They were studied systematically by J. Fraunhofer (1787–1826), who invented a spectroscope (a glass prism combined with a theodolite, for measuring angles) and also developed the diffraction grating that enabled him to measure the wavelengths of spectral lines. Fraunhofer's other contributions to astronomy are noted in **12.1**.

The prehistory of quantum theory begins in 1859 with the discovery of the basic principles of spectrum analysis by G. R. Kirchhoff. The principle that there is a universal relation between emission and absorption of radiation by any substance, depending on frequency and temperature, led to Max Planck's distribution law for black-body radiation and the quantum hypothesis that explained that law in 1900. The principle that each chemical element has a characteristic spectrum of frequencies at which it can emit or absorb radiation led to the accumulation of large amounts of spectroscopic data and the search for numerical patterns in these data, culminating in Niels Bohr's atomic theory of 1913. Thus Kirchhoff's work is important in the "whiggish" history of modern physics when, as here, we reach back into the nineteenth century to find the roots of twentieth-century theories. It is also important in the contextual history of nineteenth-century physical science, where it can be seen on the one hand as the continuation of the earlier interest in connections between light and heat (**7.2-*i***), and on the other hand as a valuable technique for making new discoveries in astrophysics (**12.2**).

Students who have had a laboratory course in chemistry will probably recall using a "Bunsen burner." Robert Bunsen, who collaborated with Kirchhoff at Heidelberg in the 1850s, designed his burner to provide a hot flame with low luminosity; it could be used to heat various substances to a temperature at which they produced characteristic colors, without having that color obscured by the color of the gas flame itself. This provided a rudimentary kind of chemical analysis; Kirchhoff pointed out that the analysis could be made more precise by passing the light from the hot substance through a prism and examining its spectrum.

The idea that each substance produces a characteristic spectrum was certainly not new in 1859, but it had not led to a useful method of analysis because most spectra seemed to be excessively complicated; moreover, different substances seemed to have some of the brightest lines in common. William Swan showed in 1857 that extremely minute amounts of sodium produce a bright yellow color, so that the yellow line reported in many spectra is really due to sodium impurities.

Fraunhofer had noted the coincidence in wavelength of the bright yellow "R" line produced by gas lamps and the dark"D" line in the solar spectrum. Kirchhoff found in 1859 that dark lines could be produced in the laboratory by passing sunlight through a flame that would ordinarily produce a bright line by itself. Further work showed that any source emitting radiation at a particular wavelength could also absorb radiation at the same wavelength. Kirchhoff argued that the dark Fraunhofer lines in the solar spectrum must be due to absorption by substances in the Sun's atmosphere; and those substances could be identified as those that produced bright lines when heated in a terrestrial laboratory.

The principle that the absorption of radiation by a substance at any wavelength is proportional to its emission at the same wavelength had been proposed in 1858 by Balfour Stewart, who applied it to radiant heat. Kirchhoff gave it a more precise quantitative development but since Stewart had announced it first his friends claimed priority for him. Kirchhoff's formulation had the greater influence on later discussions of black-body radiation. In fact it was Kirchhoff who introduced (1860) the concept of a "black body" as one which absorbs all incident radiation, and postulated that for such a body there is a universal function of wavelength and temperature that describes its emissivity.

In 1885 J. J. Balmer, a Swiss schoolteacher, found a formula that fits the numerical values of the wavelengths of four known lines in the hydrogen spectrum. His approach was purely numerological and had no relation to any physical model of the atom. But it did "predict" an infinite sequence of wavelengths corresponding to assigning integer values to m in the ratio $km^2/(m^2 - 2^2)$, where $k = 3645.6 \times 10^{-7}$ mm. When wavelengths corresponding to other values of m were found, the formula gained credibility. In 1890 the Swedish physicist J. R. Rydberg showed that other spectral lines of hydrogen could be represented by a similar formula with the "2" replaced by another integer. He noted that the frequency of a spectral line (inversely proportional to its wavelength, since wavelength × frequency = speed of light) can be written in the form: frequency = $Rc(1/t_2^2 - 1/t_1^2)$, i.e., as the difference of two terms, each of which is the reciprocal of the square of an integer. This is now called the "Ritz combination principle." The constants are identified as: R = "Rydberg constant," c = speed of light.

The next major advance in radiation theory, the discovery of the Stefan-Boltzmann law, did not involve spectroscopy at all but emerged from investigations of the laws of heat transfer. In terms of Kirchhoff's universal emissivity function, one may identify the problem as determining the integral over all frequencies, as a function only of temperature. Historically, this problem was entangled with the problem of heat transfer by convection and conduction, since a radiating body is usually surrounded by a medium that can carry away some of the heat by molecular motion. When Newton proposed his "law of cooling" in 1701, he did not distinguish between different modes of heat transfer but considered hot iron in a uniform stream of cool air; here, convection would ordinarily be more important than radiation, and conduction would be negligible. Newton's law of cooling states that the rate of cooling is proportional to the difference between the temperature of a body and that of its surroundings.

During the eighteenth and nineteenth centuries attempts were made to improve Newton's law of cooling, both empirically and by trying to gain a theoretical understanding of the process of cooling. Until 1850 it was generally believed that fluids do not "conduct" heat but can transfer it only by convection, i.e., by actual motion of the fluid itself. ("Conduction" was at that time believed to involve radiation of heat from one atom to another.) In his kinetic theory of gases (**7.2-*v***), Maxwell proposed a "molecular convection" mechanism for conduction of heat through gases. Josef Stefan in Vienna was the first to make a thorough experimental test of Maxwell's theory of heat conduction, in particular of the prediction that the heat conduction coefficient, like the viscosity coefficient, is independent of density and increases with temperature. Previously it had simply been assumed that at low pressures heat is transferred mainly by radiation, but according to Maxwell's theory conduction is still important until the pressure is less than 1 mm of mercury. Taking this possibility into account, Stefan reanalyzed older data and found that the radiation component of heat transfer is best fitted by a term proportional to the fourth power of the absolute temperature.

In 1884 L. Boltzmann showed that Stefan's empirical T^4 law could be derived by combining Maxwell's formula for the pressure of electromagnetic radiation with the second law of thermodynamics. Emphasis was now placed on the amount of radiation energy that must be present in a space or "cavity" maintained at a particular temperature, rather than on the rate of cooling of a body.

In 1893 W. Wien extended Boltzmann's reasoning to calculate the effect of changing the volume of the cavity on the radiation of specified wavelengths. By using the Doppler-effect formula (**12.2**) he showed that the energy distribution must change with temperature in such a way that a wavelength L at temperature T maps into a wavelength L_0 at temperature T_0, where $LT = L_0T_0$. This is known as "Wien's displacement law." It implies that the function representing the distribution must have the form $F(L,T) = L^{-5}f(LT)$. This does not determine a particular distribution law but defines a set of possible laws.

In 1896 Wien proposed a specific form for the distribution law of black-body radiation: $F(L,T) = KL^{-5}e^{-b/LT}$. This is known as "Wien's distribution law" for black-body radiation.

BIBLIOGRAPHY

J. B. Hearnshaw (1986), *The Analysis of Starlight: One Hundred and Fifty Years of Astronomical Spectroscopy* (New York: Cambridge Univ. Press).

J. L. Heilbron and B. R. Wheaton (1981), *Literature on the History of Physics in the 20th Century* (Berkeley: Office for History of Science and Technology, University of California), 317–19.

William McGucken (1969), *Nineteenth-Century Spectroscopy* (Baltimore: Johns Hopkins Press).

Topic *i:* Fraunhofer Lines

Frank A. J. L. James (1985), "The Discovery of Line Spectra," *Ambix* 32:53–70; (1983), "The Debate on the Nature of the Absorption of Light, 1830–1835: A Core-Set Analysis," *History of Science* 21:335–68; (1983), "The Study of Spark Spectra, 1835–1859," *Ambix* 30:137–62; (1983), "The Conservation of Energy, Theories of Absorption and Resonating Molecules, 1851–1854: G. G. Stokes, J. A. Angstrom, and W. Thomson," *Notes and Records of the Royal Society of London* 38:79–107.

W. Gerlach and F. Fraunberger (1977), "Marginalien zur Frueh- und Vorgeschichte der Spektralanalyse," in *Prismata,* ed. Maeyama and Saltzer (Wiesbaden: Steiner), 133–50, contains notes on Kepler, Descartes, Marci, Grimaldi, Huygens, Voss, Newton, Fraunhofer, etc.

M. A. Sutton (1976), "Spectroscopy and the Chemists: A Neglected Opportunity?" *Ambix* 23:16–26.

Alfred Leitner (1975), "The Life and Work of Joseph Fraunhofer (1787–1826)," *American Journal of Physics* 43:59–68.

Reese V. Jenkins (1972), "Fraunhofer, Joseph," *Dictionary of Scientific Biography* 5:142–44.

Topic *ii:* Kirchhoff and Bunsen

Frank A. J. L. James (1985), "The Creation of a Victorian Myth: The Historiography of Spectroscopy," *History of Science* 23:1–24, argues, contrary to earlier writers, that the method of spectro-chemical analysis was not available before Bunsen and Kirchhoff. Their work should not be regarded as the "culmination of 40 years effort" since the earlier work was in physics, not chemistry. See the reply by M. A. Sutton (1986), "Spectroscopy, Historiography and Myth: The Victorians Vindicated," *History of Science* 24:425–32.

Daniel M. Siegel (1976), "Balfour Stewart and Gustav Kirchhoff: Two Independent Approaches to 'Kirchhoff's radiation law,' " *Isis* 67:565–60.

L. Rosenfeld (1973), "Kirchhoff, Gustav Robert," *Dictionary of Scientific Biography* 7:379–83.

Topic *iii:* Balmer and Rydberg; Molecular Models

Clifford L. Maier (1981), *The Role of Spectroscopy in the Acceptance of the Internally Structured Atom, 1860–1920* (1964, reprinted 1981, New York: Arno Press); (1975), "Rydberg, Johannes (Janne) Robert," *Dictionary of Scientific Biography* 12:42–45.

Paul Forman (1975), "Ritz, Walter," *Dictionary of Scientific Biography* 11:475–81.

Robert K. DeKosky (1973), "Spectroscopy and the Elements in the Late Nineteenth Century," *British Journal for the History of Science* 6:400–23.

J. MacLean (1972), "On Harmonic Ratios in Spectra," *Annals of Science* 28:121–37, is on precursors of Balmer (Mascart, Lecoq de Boisbaudran, Stoney, Maxwell, Soret); Schuster's conclusion was that looking for harmonic ratios is futile.

W. R. Hindmarsh (1967), *Atomic Spectra* (New York: Pergamon Press), includes reprints of papers by Balmer, Rydberg, etc.

Topic *iv:* Stefan-Boltzmann Law

Stephen G. Brush (1986), *The Kind of Motion We Call Heat* (1976, reprinted 1986, New York: North-Holland/American Elsevier), Chap. 13, "Heat Conduction and the Stefan-Boltzmann Law."

Morton L. Schagrin (1974), "Early Observations and Calculations on Light Pressure," *American Journal of Physics* 42:927–40.

Topic *v:* Frequency Law of Black-Body Radiation

James B. T. McCaughan (1980), "Jeans' Role in the Law of Black-Body Radiation," *Physics Education* 15:255–60.

Thomas Nickles (1979), "Scientific Problems and Constraints," *PSA 1978,* vol. 1, ed. P. D. Asquith and I. Hacking (East Lansing, Mich.: Philosophy of Science Association), 134–48.

Hans Kangro (1976), *Early History of Planck's Radiation Law* (New York: Crane, Russak); (1976), "Wien, Wilhelm Carl Werner Otto Fritz Franz," *Dictionary of Scientific Biography* 14:337–42.

9.4 MAX PLANCK AND THE ORIGIN OF QUANTUM THEORY

i. Life and work of Planck; his relations with Ostwald and Boltzmann

ii. The 1900 black-body radiation law and its theoretical interpretation; Lord Rayleigh and the myth of the "ultraviolet catastrophe"; Thomas S. Kuhn's claim that Planck did not introduce the "quantum discontinuity" in 1900

iii. Einstein's 1905 paper on the quantum; the photoelectric effect; Einstein's other papers on quantum theory (before 1913)

READINGS

McKenzie, *Major Achievements,* 314–18, 517–18. Russell Stannard and Noel G. Coley, "Introduction to Quantum Theory," Unit 7 in Open University course A381, *Science and Belief from Darwin to Einstein* (1981), 73–75 (on *iii*). Holton and Brush, *Introduction to Concepts and Theories,* 430–45.

Einstein and Infeld, *Evolution of Physics* (1967), 249–65. Segre, *From X-Rays to Quarks,* 66–77, 86–90. Solomon, *Structure of Matter,* 105–11. Cline, *Men Who Made . . . ,* 32–64. Heinz Pagels, *Cosmic Code* (1982), 25–31, 63–64. M. J. Klein, "The Beginnings of Quantum Theory," (1977), in *History of Twentieth Century Physics,* ed. C. Weiner,

1–39. George L. Trigg, *Crucial Experiments,* 4–20, 76–87. Lindsay, *Early Concepts of Energy,* 184–209 (includes Planck and Einstein papers).

Thomas S. Kuhn summarizes his controversial thesis (*ii*) in "Revisiting Planck," *Historical Studies in the Physical Sciences* 14(1984):231–52.

SYNOPSIS

The quantum hypothesis originated in Max Planck's search for a mathematical formula to describe the distribution of energy over frequencies in black-body radiation (**9.3-*v***). Despite the importance of quantum theory to modern science, the background of Planck's work in nineteenth-century physics has only recently been studied carefully by historians, and misconceptions persist in textbooks and popular expositions.

One of the best-known quotations about the nature of science is Max Planck's remark:

> An important scientific innovation rarely makes its way by gradually winning over and converting its opponents. . . . What does happen is that its opponents gradually die out, and that the growing generation is familiarized with the ideas from the beginning. [*Scientific Autobiography,* 1950, 33–34]

I suppose most people who read (or repeat) this quotation think Planck is referring to his quantum theory, but in fact he was talking about his struggle to convince scientists in the 1880s and 1890s that the second law of thermodynamics involves a principle of irreversibility (**7.3**) and that the flow of energy from hot to cold is *not* analogous to the flow of water from a high level to a low one, as Ostwald and the energetists claimed. He goes on to lament that his own efforts were fruitless, but the battle was eventually won because of advances from another direction: the statistical interpretation of entropy based on kinetic theory.

In his papers on physical chemistry in the 1880s Planck was hostile to atomistic theories, preferring to work with macroscopic thermodynamics. But as editor of Kirchhoff's lectures on heat he had to go through kinetic-theory derivations and had to defend one of them in a published exchange with Boltzmann. Thus we may assume that he was familiar with statistical calculations even though not an advocate of statistical mechanics before 1900.

Planck found that he could derive Wien's distribution law by assuming a simple relation between the entropy and energy of an idealized "oscillator" that absorbs and emits radiation. In particular, the reciprocal of the second derivative of entropy with respect to energy is proportional to the energy. But more accurate experiments on black-body radiation by Rubens and Kurlbaum indicated deviations from Wien's law at very low frequencies. Planck then tried adding a term quadratic in energy and found that the resulting distribution law gave a precise fit to the experimental data.

Having found the correct result by empirical methods, guided by thermodynamics, Planck then sought a plausible theoretical explanation. He introduced Boltzmann's statistical interpretation of entropy, $S = k \log W$, where W is the number of ways the atoms in a system can be arranged, corresponding to the thermodynamic state; W is thus proportional to the probability of that state. To compute W, he assumed that the total energy E is divided into P equal parts, and used a combinatorial formula for the number of ways of distributing P objects in N boxes (N = number of oscillators). He found that E/P is equal to a constant, h, multiplied by the frequency of the radiation.

Planck's very brief 1900 papers give little hint as to his motivation for introducing his assumption about energy or the physical meaning of that assumption. As it happened, Lord Rayleigh had published a paper on radiation a few months earlier in which he noted that if the equipartition theorem of statistical mechanics could be applied to the ether, then the distribution law for radiation would contain a factor proportional to the square of frequency. Rayleigh stated that this form would not apply for high frequencies, proposing instead to insert an exponential damping factor. But physicists such as Paul Ehrenfest later argued that strict application of classical physics would require just the frequency-squared factor, which when integrated over all frequencies would produce an infinite result. In other words, if there is equipartition of energy among all modes of vibration of the ether, then all the energy will be sucked into the higher frequency modes. This became known as the "ultraviolet catastrophe."

Rayleigh himself did not suggest any such difficulty in his 1900 paper, and there is no evidence that Planck was at all concerned about this problem; in any case Planck was not approaching the distribution law from this direction. The ultraviolet catastrophe cannot be regarded as an "anomaly" in classical physics that led to the introduction of the quantum hypothesis, but only as a retrospective justification of that hypothesis.

It is not clear whether Planck intended to propose a *physical* quantization of energy in 1900, or whether he merely introduced "energy elements" for mathematical convenience in doing combinatorial calculations. Kuhn (1978) has argued that the "quantum discontinuity" was first seriously proposed not by Planck but by Einstein and Ehrenfest, and only retrospectively associated with Planck's 1900 hypothesis. Although Kuhn's interpretation is not yet generally accepted by historians of physics, there is some evidence to support it (including Planck's 1920 Nobel lecture, reprinted in McKenzie's text) and none (as far as I know) that decisively refutes it. In any case, Kuhn has shown the need to study Planck's pre-1900 papers more carefully and to explore their relation to Boltzmann's work.

Einstein's 1905 paper on quantum theory is usually referred to as his paper on the photoelectric effect. He proposes to consider, as a "heuristic point of view," the hypothesis that light energy is distributed discontinuously in space, in "quanta" which can be absorbed or emitted only as units. If one assumes a thermal equilibrium of energy between these quanta and matter, one obtains Wien's rather than Planck's law. (The explanation for this discrepancy did not emerge until several years later.)

The photoelectric effect had been discovered by Hertz in 1887, but its quantitative laws were not yet known. Einstein proposed that a light quantum transfers all its energy to a single electron, which then emerges with an amount of kinetic energy equal, at most, to the energy of the light quantum, minus the work done in escaping from the metal. Thus the energy of the photoelectrons should increase linearly with the frequency of the incident light but should be independent of the intensity of the light. Quantitative verification was accomplished by Robert A. Millikan in 1916, after several years of work. Einstein received the Nobel Prize in 1921 for his theory of the photoelectric effect, and Millikan won it in 1923 for his experimental proof.

In 1907 Einstein applied the quantum hypothesis to the specific heats of solids, assuming that a vibrating atom would behave like one of Planck's oscillators. Assuming that all the atoms have the same frequency of vibration, he showed that the specific heat would go to zero at low temperatures but would level off at a constant value at high temperatures, in agreement with the Dulong-Petit law. The former conclusion agreed with Walther Nernst's new heat theorem or "third law of thermodynamics" published in 1906. After extensive experimental work, Nernst announced in 1911 that the specific heats of all substances, together with other thermodynamic functions, go to zero at absolute zero temperature and noted that this provides strong confirmation of the quantum theory.

Nernst also helped to clear up one of the anomalies in the classical kinetic theory of gases, using quantum theory: the specific heat ratio of diatomic gases. Lord Kelvin, in his famous lecture on "Clouds over the

19th Century Theories of Heat and Light," mentioned the apparent failure of the equipartition theorem to apply to internal motions of atoms and molecules. While this problem did not play an important role in the early development of quantum theory and certainly could not be called a reason for introducing it, it is a phenomenon that requires quantum theory for its explanation. The vibrations of atoms do not contribute to the specific heat of a diatomic or polyatomic molecule at room temperature because their fundamental frequency is so high that they rarely acquire even one quantum; hence the average energy in this degree of freedom is very small. Similarly, the motions of electrons do not contribute to the specific heats of molecules unless radiation is emitted or absorbed, which will not happen unless the temperature is high enough to excite electrons from their ground states. Finally, the rotation of a diatomic molecule around the interatomic axis may be ignored because the quantum of rotational energy (being inversely proportional to the moment of inertia) is much larger than kT at room temperature.

BIBLIOGRAPHY

See the bibliographies by Brush and Belloni and by Heilbron and Wheaton, cited in **9.2**, for additional references.

Jagdish Mehra and Helmut Rechenberg (1982), *The Historical Development of Quantum Theory*, vol. 1, *Quantum Theory of Planck, Einstein, Bohr, and Sommerfeld: Its Foundation and the Rise of Its Difficulties, 1900–1925* (New York: Springer-Verlag). See the review by Paul Forman (1983), *Science* 220:824–27 and the note by I. Prigogine (1983), *Science* 221:604, also the review by J. L. Heilbron (1985), *Isis* 76:388–93.

Daniela Longo and Nadia Robotti (1981), "Quantizzazione dell'Energia e Struttura Atomica: Alcuni Aspetti dei Primi Sviluppi delle Ipotesi Quantistiche," *Physis* 23:375–96.

D. R. Nilson (1976), "Bibliography on the History and Philosophy of Quantum Theory," in *Logic and Probability in Quantum Theory*, ed. P. Suppes (Boston: Reidel), 457–520.

D. ter Haar (1967), *The Old Quantum Theory* (New York: Pergamon Press), includes English translations of papers of Planck (1900), Einstein (1905).

Topic *i:* Planck's Life and Early Work

Max Planck (1958), *Physikalische Abhandlungen und Vortraege* (Braunschweig: Vieweg); (1949), *Scientific Autobiography and Other Papers* (New York: Philosophical Library).

J. L. Heilbron (1986), *The Dilemmas of an Upright Man; Max*

Planck as Spokesman for German Science (Berkeley: Univ. of California Press).

Stanley Goldberg (1976), "Max Planck's Philosophy of Nature and His Elaboration of the Special Theory of Relativity," *Historical Studies in the Physical Sciences* 7:125–60.

Hans Kangro (1975), "Planck, Max Karl Ernst Ludwig," *Dictionary of Scientific Biography* 11:7–17.

Topic *ii:* Planck's 1900 Black-Body Radiation Law

John Norton (1987), "The Logical Inconsistency of the Old Quantum Theory of Black Body Radiation," *Philosophy of Science* 54: 327–50.

D. de Casaubon (1985), "Le Rôle Heuristique des Mathématiques dans la Physique: Le Cas de Planck (1894–1900)," *Fundamenta Scientiae* 6:281–97.

Ulrich Hoyer (1980), "Von Boltzmann zu Planck," *Archive for History of Exact Sciences* 23:47–86.

Thomas S. Kuhn (1978), *Black-Body Theory and the Quantum Discontinuity, 1894–1912* (New York: Oxford Univ. Press), is reviewed by M. J. Klein, A. Shimony and T. J. Pinch (1979), "Paradigm Lost?" *Isis* 70:429–40; see also Peter Galison (1981), "Kuhn and the Quantum Controversy," *British Journal for the Philosophy of Science* 32:71–84; Joseph Agassi (1983), "The Structure of the Quantum Revolution," *Philosophy of the Social Sciences* 13:367–81.

Elizabeth Garber (1976), "Some Reactions to Planck's Law, 1900–1914," *Studies in History and Philosophy of Science* 7:89–126.

E. Schmutzer, ed. (1976), *75 Jahre Plancksches Wirkungsquantum. 50 Jahre Quantenmechanik* (Leipzig: Barth).

Hans Kangro (1976), *Early History of Planck's Radiation Law* (New York: Crane, Russak); (1972), *Planck's Original Papers in Quantum Physics* (London: Taylor and Francis).

Topic *iii:* Einstein's Quantum Theory

Bruce R. Wheaton (1981), "Impulse X-rays and Radiant Intensity: the Double Edge of Analogy," *Historical Studies in the Physical Sciences* 11:367–90, relates how ideas about X rays affected the reception of the quantum theory of radiation.

Joan L. Hawes (1981), "Matter for Illumination," *Physics Education* 16:178–85, concerns Planck's rejection of Einstein's quantum theory of light.

Martin J. Klein (1980), "No Firm Foundation: Einstein and the Early Quantum Theory" in *Some Strangeness in the Proportion,* ed. H.

Woolf (Reading, Mass.: Addison-Wesley), 161–85, is followed by remarks by T. S. Kuhn and others.

John Hendry (1980), "The Development of Attitudes to the Wave-Particle Duality of Light and Quantum Theory, 1900–1920," *Annals of Science* 37:59–79.

S. Bergia, P. Lugli and N. Zamboni (1979, 1980), "Zero-point Energy, Planck's Law and the Prehistory of Stochastic Electrodynamics," *Annales de la Fondation Louis de Broglie,* 4, no. 4:295–318; 5, no. 1:39–62, is on the papers of Einstein and Hopf (1910) and Einstein and Stern (1913).

A. Pais (1980), "Einstein on Particles, Fields, and the Quantum Theory," in *Some Strangeness in the Proportion,* ed. H. Woolf (Reading, Mass.: Addison-Wesley), 197–251, is followed by remarks by R. Jost and others; (1979), "Einstein and the Quantum Theory," *Reviews of Modern Physics* 51:861–914.

Max Jammer (1979), "Albert Einstein und das Quantenproblem," in *Einstein Symposion Berlin,* ed. H. Nelkowski et al. (New York: Springer), 146–67.

Arthur Miller (1976), "On Einstein, Light Quanta, Radiation and Relativity in 1905," *American Journal of Physics* 44:912–23.

9.5 THE BOHR ATOM

i. Atomic models before Bohr: Hantaro Nagoaka, J. J. Thomson, etc.; stability problems
ii. Life and work of Niels Bohr
iii. Bohr's 1913 paper
iv. Tests and extensions of Bohr's model; the "old quantum theory"; correspondence principle
v. Aufbau principle; explanation of periodic system, exclusion principle and spin

READINGS

McKenzie, *Major Achievements,* 318–20. Marks, *Science,* 315–20. Holton and Brush, *Introduction to Concepts and Theories,* 478–89.

Einstein and Infeld, *Evolution,* 265–70. J. L. Heilbron, "Lectures on the History of Atomic Physics 1900–1922," in *History of Twentieth Century Physics,* ed. C. Weiner, 40–108. Segre, *From X-Rays to Quarks,* 119–48. V. Guillemin, *The Story of Quantum Mechanics* (1968), 13–39,

55–60. Cline, *Men Who Made . . .*, 78–122. Pagels, *Cosmic Code,* 68–72. R. B. Lindsay, *Early Concepts of Energy,* 266–92 (Bohr, Franck, and Hertz). B. Hoffmann, *Strange Story of the Quantum* (1959), 43–69. J. L. Heilbron, "Bohr's First Theories of the Atom," *Physics Today* 38, no. 10 (1985):28–36.

SYNOPSIS

Bohr's earliest scientific work (1906) was a measurement of the surface tension of water, together with a theoretical analysis based on hydrodynamics; some of the ideas developed here were used in his liquid drop model of the nucleus in the 1930s. As part of his examination for the master's degree, Bohr was assigned to write about the application of the electron theory to the physical properties of metals; he used H. A. Lorentz's kinetic theory approach to this problem. According to J. L. Heilbron and T. S. Kuhn (1969), it was his calculations on the diamagnetism and paramagnetism of electrons in metals that led Bohr to consider the electronic structure of atoms. Paul Langevin had explained these properties by attributing magnetic moments to the molecules, but Bohr showed that this explanation was unsatisfactory if the electrons are bound inside atoms by forces obeying Newtonian mechanics. His 1911 dissertation was not available except in Danish until 1972, despite Bohr's own efforts to publish an English translation, and thus the relation between his work on the electron theory of metals and his atomic model has been obscured.

In 1912 Bohr visited Rutherford's group at Manchester. The nuclear model of the atom, recently proposed by Rutherford on the basis of the Geiger-Marsden scattering experiment(**9.2**-*vi*), provided the starting point for his theory of electron orbits. His work was influenced by R. Whiddington's 1911 cathode ray experiment (which suggested that a certain threshold energy is needed to knock an electron out of an atom), the Balmer and Rydberg formula for spectral lines (**9.3**-*iii*), which he learned from H. M. Hansen in Copenhagen, and J. W. Nicholson's atomic model.

Bohr's 1913 model for the hydrogen atom was constructed with the formulae long known for gravitational motion of a point mass around a heavy center of attraction, using electrostatic instead of gravitational force. Bohr postulated that the electron can move only in certain orbits, determined by the condition that angular momentum is a multiple of Planck's constant. A quantum of energy is absorbed or radiated only when the electron moves from one orbit to another. The Ritz combination principle for spectral line frequencies is directly connected with the assumption that the energy quantum depends on the difference between the energies of two orbits. Bohr succeeded in deriving from his model a formula for the Rydberg constant in terms of the electron's mass, charge, the speed of light, and Planck's constant.

An early test was provided by a series of lines observed in the spectrum of a star by E. C. Pickering in 1896, and in vacuum tubes containing a mixture of hydrogen and helium by A. Fowler in 1912. The lines were generally ascribed to hydrogen, but they did not fit Bohr's formula; he assumed that they were produced by once-ionized helium. This system would, like the neutral hydrogen atom, have only one electron, but the nuclear charge would be twice as great. With this assumption Bohr's formula did fit the Pickering-Fowler lines. A new experiment with a mixture of helium and chlorine by E. J. Evans (1913), confirmed Bohr's interpretation.

Further confirmation of Bohr's theory came from the experiment done by J. Frank and G. Hertz in 1914. While not originally intended as a test of the theory, it was later interpreted as evidence that electronic energy levels in atoms are quantized.

Several physicists then took up the problem of generalizing Bohr's theory to more complex systems. In 1915 Arnold Sommerfeld proposed that the stationary states of a periodic system should be determined by the condition that the integral around a complete cycle of $p_k dq_k$ is equal to an integral multiple of h, where p_k is the momentum corresponding to the coordinate q_k. Sommerfeld was able to treat three-dimensional electron orbits with relativistic corrections and thus to account for the fine structure of lines in the hydrogen and helium spectra. He introduced a new "azimuthal quantum number" k to characterize noncircular orbits.

Another advance was the "adiabatic principle" proposed by Paul Ehrenfest in 1913; this allows one to derive the stationary states of a system from another system whose states are known by infinitely slow changes of the relevant parameters. Ehrenfest's principle also provided insight into the physical significance of the postulates of Bohr and Sommerfeld.

In 1918 Bohr explicitly introduced his "correspondence principle" by which the frequencies for transitions between quantum states with large quantum numbers could be related to the classical frequencies for the system. The principle ensured that quantum theory results would reduce to classical results in the limit when Planck's constant goes to zero or the frequency goes to zero.

Much effort was devoted to explaining the effect of a magnetic field on spectral lines, first established by P. Zeeman in 1896. A third quantum number, the "magnetic quantum number" m, was introduced to characterize states that had the same energy in the absence of a magnetic field but split into two or more states when the field is present.

Bohr and others tried to work out a systematic procedure for distributing electrons among the possible quantum states, building up heavier elements in a way consistent with their known chemical properties. This was the "Aufbau" principle. The culmination of this line of work was Wolfgang Pauli's introduction of a fourth quantum number, and the association of this quantum number with an intrinsic "spin" of the electron by G. Uhlenbeck and S. Goudsmit in 1925. Pauli's "exclusion principle" states that no two electrons may be in the same quantum state, defined by the set of four quantum numbers.

BIBLIOGRAPHY

[Note: additional references may be found in the bibliographies of Brush and Belloni, and Heilbron and Wheaton, cited in **9.2.**]

Edward M. MacKinnon (1982), *Scientific Explanation and Atomic Physics* (Chicago: Univ. of Chicago Press), Chap. 5.

J. Mehra and H. Rechenberg (1982), *The Historical Development of Quantum Theory,* vol. 1 (New York: Springer-Verlag).

Sandro Petruccioli (1981), "Modello Meccanico e Regole di Corrispondenza nella Costruzione della Teoria Atomica,"*Physis* 23:555–79.

Richard Schlegel (1976), " 'Bohr and Einstein': A Course for Nonscience Students," *American Journal of Physics* 44:236–39.

Topic *i:* Pre-Bohr Atomic Models

S. B. Sinclair (1987), "J. J. Thomson and the Chemical Atom: From Ether Vortex to Atomic Decay," *Ambix* 34:89–116.

Chiyoko Fujisaki (1982), "P. Drude's Theory of Dispersion of Light and Atomic Model (1900–1913)," *Historia Scientiarum* 22:19–67.

Eri Yagi (1974), "Nagaoka, Hantaro," *Dictionary of Scientific Biography* 9:606–7; (1972), "The Development of Nagaoka's Saturnian Atomic Model II (1904–1905)," *Japanese Studies in the History of Science* 11:73–89 and earlier works cited therein.

Topic *ii:* Bohr's Life and Work

[Note: references on Bohr's philosophical views, the complementarity principle, etc. are given in **11.2**]

Niels Bohr (1972), *Collected Works,* general ed. L. Rosenfeld (New York: North-Holland), vol. 1, *Early Work (1905–1911),* ed. J. R.

Nielsen (1972), includes an English translation of Bohr's 1911 dissertation and related work; biographical sketch by Rosenfeld; commentaries on the dissertation by Nielsen.

A. P. French and P. J. Kennedy (1985), *Niels Bohr: A Centenary Volume* (Cambridge: MIT Press).

J. L. Heilbron et al. (1985), "Special Issue: Niels Bohr Centennial," *Physics Today* 38, no. 10.

Jørgen Kalckar et al. (1985), "Niels Bohr and the Infinitely Small," *Impact of Science on Society* 35:5–68.

F. Hund (1985), "Korrespondenz und Komplementarität: Bohrs Weg zur Atomdynamik," *Physikalische Blätter* 41:303–7, is part of a special issue on Bohr that also includes articles by C. F. von Weizsäcker and J. Bang.

W. Kuhn et al. (1985), *Praxis der Naturwissenschaften* 34, no. 7, is a special issue on Bohr.

Ulrich Röseberg (1985), *Niels Bohr: Leben und Werk eines Atomphysikers* (Stuttgart: Wissenschaftliche Verlagsgesellschaft mbH); (1985), *Niels Bohr: Bibliographie der Sekundärliteratur* (Berlin: Zentralinstitut für Philosophie der Akademie de Wissenschaften der DDR).

Peter Robertson (1979), *The Early Years: The Bohr Institute 1921–1930* (Copenhagen: Akademisk Forlag).

Leon Rosenfeld (1970), "Bohr, Niels," *Dictionary of Scientific Biography* 2:239–54.

Topic *iii:* Bohr's 1913 Paper

Niels Bohr (1981), *Collected Works,* vol. 2, *Work on Atomic Physics (1912–1917),* ed. Ulrich Hoyer (New York: North-Holland).

B. Carazza and G. P. Guidetti (1984), "Spettroscopia e Modelli Atomici Prima di Bohr," *Rendiconti del Seminario della Facolta di Scienze dell'Universita di Cagliari* 54, fasc. 1:73–86.

Girolamo Ramunni and Pierre Costabel (1981), *Les Conceptions Quantiques de 1911 à 1927* (Paris: Vrin).

J. L. Heilbron (1981), "Rutherford-Bohr Atom," *American Journal of Physics* 49:223–31; (1977), "J. J. Thomson and the Bohr Atom," *Physics Today* 30, no. 4:23–30.

J. L. Heilbron and T. S. Kuhn (1969), "The Genesis of the Bohr Atom," *Historical Studies in the Physical Sciences* 1:211–90.

Topic *iv:* Tests and Extensions of Bohr Model

Carsten Jensen (1984), "Two One-Electron Anomalies in the Old Quantum Theory," *Historical Studies in the Physical Sciences* 15:81–106.

Nadia Robotti (1983), "The Spectrum of Zeta Puppis and the Historical Evolution of Empirical Data," *Historical Studies in the Physical Sciences* 14:123–45.

Hans Radder (1982), "An Immanent Criticism of Lakatos' Account of the 'Degenerating Phase' of Bohr's Atomic Theory," *Zeitschrift für Allgemeine Wissenschaftstheorie* 13:99–109; (1982), "Between Bohr's Atomic Theory and Heisenberg's Matrix Mechanics: A Study of the Role of the Dutch Physicist H. A. Kramers," *Janus* 69:223–52.

Helge Kragh (1979), "Niels Bohr's Second Atomic Theory," *Historical Studies in the Physical Sciences* 10:123–86.

David C. Cassidy (1979), "Heisenberg's First Core Model of the Atom: The Formation of a Professional Style," *Historical Studies in the Physical Sciences* 10:187–224; (1978), "Heisenberg's First Paper," *Physics Today* 31, no. 7:23–28, is on a 1922 paper that attempted to explain the anomalous Zeeman effect.

Topic *v:* Aufbau Principle, Exclusion Principle, Spin

J. L. Heilbron (1983), "The Origins of the Exclusion Principle," *Historical Studies in the Physical Sciences* 13:261–310.

Karl von Meyenn (1980, 1981), "Paulis Weg zum Ausschliessungsprinzip," *Physikalische Blätter* 36:293–98, 37:13–19.

9.6 WAVE MECHANICS

i. Compton effect, particle nature of light
ii. Interpretation of spectra, development of matrix mechanics by Werner Heisenberg, Max Born, and Pascual Jordan
iii. Proposal of wave nature of matter: Louis de Broglie's thesis
iv. Erwin Schrödinger's route to wave mechanics
v. Reception of wave mechanics, its relation to matrix mechanics; electron diffraction and other evidence for wave nature of matter
vi. Paul Dirac's relativistic wave equation; prediction and discovery of the positron

READINGS

Marks, *Science,* 320–23. Holton and Brush, *Introduction,* 491–95. McKenzie, *Major Achievements,* 321–23. R. Stannard and N. G. Coley, "Introduction to Quantum Theory," Unit 7 in the Open University course A381, 76–99.

Einstein and Infeld, *Evolution of Physics,* 270–94. V. Guillemin, *Story of Quantum Mechanics,* 61–90. Segre, *From X-Rays to Quarks,* 149–74. Solomon, *Structure of Matter,* 105–27. Pagels, *Cosmic Code,* 72–84. Cline, *Men Who Made . . . ,* 123–59. Banesh Hoffmann, *Strange Story of the Quantum,* 72–139. G. L. Trigg, *Crucial Experiments,* 97–103 (on i), 105–33 (on v).

Articles: Bruce R. Wheaton, "Louis de Broglie and the origins of Wave Mechanics," *Physics Teacher* 22(1984):297–301. H. A. Medicus, "Fifty Years of Matter Waves," *Physics Today* 27, no. 2(1974):38–45. R. K. Gehrenbeck, "Electron Diffraction: Fifty Years Ago," *Physics Today,* 31, no. 1(1978):34–41, and exchange with R. Schlegel on "Who Discovered Matter Waves," ibid., no. 7(1978):9–13. P. A. M. Dirac, "The Evolution of the Physicist's Picture of Nature," *Scientific American* 208, no. 5(1963):45–53. Linda Wessels, "Erwin Schrödinger and the Descriptive Tradition," in *Springs of Scientific Creativity,* ed. R. Aris et al. (1983), 254–78.

SYNOPSIS

By 1920 Planck's quantum hypothesis had been generally accepted as a basis for atomic theory, and the photon concept had been given strong support by Millikan's experimental confirmation of Einstein's theory of the photoelectric effect. Yet, despite the early successes of Bohr's theory of the hydrogen atom, no one had found a satisfactory way to extend the theory to multielectron atoms and molecules without introducing arbitrary additional hypotheses tailored to fit each particular case. Moreover, one could hardly abandon the wave theory of light and return to a particle theory, since not only the properties such as interference and diffraction but also the formula for the energy of photon depended on wave properties.

Further evidence of the particle nature of light came in 1923 with the discovery by the American physicist Arthur Holly Compton that photons carry a definite momentum which they transfer in interactions with electrons. In the "Compton effect," X rays are scattered by individual electrons in such a way that the total momentum of the system photon + electron is conserved; the frequency-change of the scattered X ray and the momentum of the electron knocked out of the target can be calculated and checked experimentally.

The theory now known as "quantum mechanics" was discovered by Werner Heisenberg in 1925. It involved arrays of quantities that turned out to have the same mathematical properties as "matrices," introduced by Arthur Cayley in 1855. In the detailed development of his theory

Heisenberg was assisted by Max Born and Pascual Jordan. The calculational techniques of the theory were based on those previously derived for computing planetary motions in astronomy. But matrix mechanics still involved very complicated calculations for any but the simplest systems, and it abstained from providing any physical picture of what was going on in the atom.

An apparently quite different approach to atomic structure was proposed by Louis de Broglie in 1924. Just as photons behave like particles as well as waves, so electrons should behave like waves as well as like particles. In particular, the "de Broglie wavelength" of an electron is defined as equal to Planck's constant divided by its momentum. Quantization of allowed orbits in the Bohr atom then follows directly from the postulate that the circumference of the orbit must be equal to an integer number of wavelengths.

"Why Was It Schrödinger Who Developed de Broglie's Ideas?" is the title of an article by V. V. Raman and Paul Forman (1969). In other words, why didn't the physicists at Göttingen (Heisenberg and Born), Munich (Sommerfeld), or Copenhagen (Bohr) who were more deeply involved than Schrödinger in the theory of atomic structure and spectra, see the significance of de Broglie's matter waves? Raman and Forman have one answer to that question; other answers may be found in the articles listed below. In any case it was Einstein who first saw the merits of de Broglie's thesis and pointed out its significance in his 1925 paper on what is now called "Bose-Einstein statistics." This probably led Schrödinger to look into de Broglie's ideas and develop them further. Another stimulus to Schrödinger was Peter Debye's suggestion that he report on the de Broglie thesis at the physics colloquium in Zurich.

Schrödinger developed his theory by using the analogy, which the Irish mathematician W. R. Hamilton had pointed out a century earlier, between mechanics and geometrical optics. Wave propagation in optics is described by d'Alembert's equation; what is the corresponding generalization of particle mechanics to wave mechanics? The answer, using de Broglie's formula for the wavelength of a particle, is the Schrödinger equation—the most important equation in twentieth-century physical science. It states that a certain combination of spatial derivatives of a "wave function" added to a term representing the potential energy has the same effect as multiplying the wave function by a constant, E. For each possible solution ("eigenfunction") there is a corresponding value of E (the "eigenvalue"), which may be interpreted as the energy of the system when it is in the quantum state described by that wave function.

By using standard methods for solving partial differential equations, developed in the eighteenth and nineteenth centuries, Schrödinger could show that when the potential energy term is that for electrostatic attraction between an electron and a proton, the possible eigenvalues are just the energy levels found in Bohr's model of the hydrogen atom. Rather than introducing quantization as a postulate, Schrödinger deduced it as a consequence of the requirement that the wave function be single-valued. (This is a generalization of de Broglie's condition that an integral number of wave lengths must fit around the circumference of the orbit; but now there are no longer any definite orbits.)

Heisenberg immediately realized that wave mechanics threatened to replace his own matrix mechanics because it provided a picture (though a fuzzy one) of the behavior of the electron. It was more *anschaulich* (visualizable). In an attempt to demonstrate that his own theory was also *anschaulich,* Heisenberg wrote a paper that included the result for which he is now best known to the public, the "uncertainty principle" (or, preferably, "indeterminacy principle"—see **11.2**).

The wave nature of matter was directly confirmed by the electron-diffraction experiments done by C. J. Davisson and L. H. Germer in the United States and by G. P. Thomson in Scotland in 1927.

In 1926 Schrödinger also derived a "time-dependent" equation for the wave function. As he himself realized, it was not consistent with special relativity theory since the spatial coordinates entered as second derivatives while the time coordinate entered as a first derivative. According to Dirac (1963), Schrödinger had actually developed a relativistic form of the wave equation first, in which all coordinates appear as second derivatives, but did not publish it because it disagreed with observed spectral lines. (It is now known as the "Klein-Gordon" equation since it was first published by O. Klein and W. Gordon in 1926.) So Schrödinger presented only a nonrelativistic approximation; the disagreement with experiment was later cleared up when spin was taken into account. The moral of this story, according to Dirac, is "that it is more important to have beauty in one's equations than to have them fit experiment."

Schrödinger's own interpretation of the physical significance of his wave function was displaced by Born's statistical interpretation, though, like Einstein, Schrödinger refused to accept the "Copenhagen interpretation" of Bohr and Heisenberg (**11.2**). His influence on molecular biology, primarily through his book *What Is Life?,* is discussed in **4.7**.

Dirac discovered the correct relativistic wave equation for the electron in late 1927 by playing around with Pauli's spin matrices. He found that one could obtain a linear equation with the right properties if one replaced Pauli's 2×2 matrices with a new set of 4×4 matrices. The wave function now had to have four components. Dirac in essence *derived* the electron spin as a consequence of the mathematical requirements of relativistic invariance and linearity, rather than adding it as an arbitrary extra assumption as Pauli and Goudsmit and Uhlenbeck had done (**9.5**-*v*). But it turned out that two of the four components of the wave function corresponded to a particle similar to the electron but with positive charge. After some discussion about possible interpretations of this result, Dirac in 1931 suggested the existence of a new particle having the same mass and opposite charge to the electron. The next year Carl D. Anderson discovered tracks of positively charged electrons in cloud-chamber photographs of cosmic rays. These particles, which he called "positrons," were quickly recognized to be Dirac's predicted particle.

BIBLIOGRAPHY

Niels Bohr (1984), *Collected Works,* vol. 5, *The Emergence of Quantum Mechanics,* ed. K. Stolzenburg (Amsterdam: North-Holland), includes correspondence from the period 1924–1926.

Arthur I. Miller (1984), *Imagery in Scientific Thought: Creating 20th-Century Physics* (Cambridge: Birkhauser).

R. B. Lindsay (1983), *Energy in Atomic Physics, 1925–1960* (New York: Van Nostrand Reinhold), 1–43, 136–60, is an anthology of original sources.

Stephen G. Brush and Lanfranco Belloni (1983), *The History of Modern Physics: An International Bibliography* (New York: Garland), 171–76, 186–92.

Edward M. MacKinnon (1982), *Scientific Explanation and Atomic Physics* (Chicago: Univ. of Chicago Press), Chaps. 6–8.

David A. Woodrow (1982), "Atomic Theory, 1870–1930: A Case Study of Paradigm Change," Ph.D. diss., State University of New York at Stony Brook; see *Dissertation Abstracts International* 43(1982):1299-A.

Charlie D. Hurt (1982), "Identification of Important Literature in Quantum Mechanics," Ph.D. diss., University of Wisconsin; see *Dissertation Abstracts International* 42(1982):2916-A; (1981), "A Test of Differences in the Literature History of Four Historical Accounts of the Quantum Mechanics Problem," *Scientometrics* 3:457–66.

Elisabetta Donini (1982), *Il Caso dei Quanta: Dibattito in Fisica e Ambiente Storico, 1900–1927* (Milan: Clup-Clued).

Girolamo Ramunni and Pierre Costabel (1981), *Les Conceptions Quantiques de 1911 à 1927* (Paris: Vrin).

John L. Heilbron and Bruce R. Wheaton (1981), *Literature on the History of Physics in the 20th Century* (Berkeley: Office for History of Science and Technology, University of California), 366–69, 372–75; Heilbron (1968), "Quantum Historiography and the Archive for History of Quantum Physics," *History of Science* 7:90–111.

E. Weislinger (1978), *Eléments d'Histoire et d'Epistemologie de la Mécanique Quantique* (Paris: C.N.R.S., Centre de Documentation Sciences Humaines).

Gunther Ludwig (1968), *Wave Mechanics* (New York: Pergamon Press), includes translations of papers by de Broglie, Schrödinger, Heisenberg, Born, Jordan.

Topic *i:* Compton Effect

Roger H. Stuewer (1975), *The Compton Effect* (New York: Science History).

Topic *ii:* Matrix Mechanics, Heisenberg

Elisabeth Heisenberg (1984), *Inner Exile: Recollections of a Life with Werner Heisenberg* (Cambridge: Birkhauser).

John Hendry (1984), *The Creation of Quantum Mechanics and the Bohr-Pauli Dialogue* (Boston: Reidel). See the review by Mara Beller (1986), *Isis* 77:107–9.

Gerald Holton (1984), " 'Success Sanctifies the Means': Heisenberg, Oppenheimer, and the Transition to Modern Physics," in *Transformation and Tradition in the Sciences,* ed. E. Mendelsohn (New York: Cambridge Univ. Press), 155–73, is reprinted in his *The Advancement of Science and Its Burdens* (New York: Cambridge Univ. Press, 1986), 141–62.

David Cassidy and Martha Baker (1984), *Werner Heisenberg: A Bibliography of His Writings* (Berkeley: Office for History of Science and Technology, University of California).

Mara Beller (1983), "Matrix Theory before Schrödinger: Philosophy, Problems, Consequences," *Isis* 74:469–91.

Arthur I. Miller (1982), "Redefining *Anschaulichkeit,*" in *Physics as Natural Philosophy,* ed. A. Shimony and H. Feshbach (Cambridge: MIT Press), 376–41.

Jagdish Mehra and Helmut Rechenberg (1982), *The Historical Development of Quantum Theory,* vol. 2 (New York: Springer-Verlag).

Hans Radder (1982), "Between Bohr's Atomic Theory and Heisenberg's Matrix Mechanics: A Study of the Role of the Dutch Physicist H. A. Kramers," *Janus* 69:223–52.

Marshall Bowen and Joseph Coster (1980), "Born's Discovery of the Quantum-Mechanical Matrix Calculus," *American Journal of Physics* 48:491–92.

Edward MacKinnon (1977), "Heisenberg, Models, and the Rise of Matrix Mechanics,"*Historical Studies in the Physical Sciences* 8:137–88.

Felix Bloch (1976), "Heisenberg and the Early Days of Quantum Mechanics," *Physics Today* 29, no. 12:23–27.

B. L. Van der Waerden, ed. (1967), *Sources of Quantum Mechanics* (New York: Dover), includes translations of papers by H. A. Kramers, M. Born, Heisenberg, Jordan, Pauli, Dirac.

Topic *iii:* De Broglie

A. O. Barut et al., eds. (1984), *Quantum, Space and Time* (New York: Cambridge Univ. Press), includes articles by G. Lochak and others on de Broglie.

S. Diner et al., eds. (1984), *The Wave-Particle Dualism* (Boston: Reidel).

Jean-Pierre Vigier (1982), "Louis de Broglie: Physicist and Thinker," *Foundations of Physics* 12:923–30, is an introduction to a special issue honoring de Broglie's ninetieth birthday.

Edward MacKinnon (1976), "De Broglie's Thesis: A Critical Retrospective," *American Journal of Physics* 44:1047–55.

Topic *iv:* Schrödinger

Karl von Meÿenn (1984), "Gespensterfelder und Materiewellen: Schrödingers Hang zur Anschaulichkeit," *Physikalische Blätter* 40, no. 4:89–94.

Helge Kragh (1982–1983), "Erwin Schroedinger and the Wave Equation: The Crucial Phase," *Centaurus* 26:154–97.

Michelangelo De Maria and Francesco La Teana (1982), "I primi lavori di E. Schrödinger sulla meccanica ondulatoria e la nascita delle polemiche con la scuola di Göttingen-Copenhagen sull'interpretazione della meccanica quantistica," *Physis* 24:33–54.

Ulrich Hoyer (1981), "Wellenmechanik und Boltzmannsche Statistik," *Gesnerus* 38:347–49.

Patrick Suppes, ed. (1980), *Studies in the Foundations of Quantum Mechanics* (East Lansing, Mich.: Philosophy of Science Association), includes E. MacKinnon, "The Rise and Fall of the Schroedinger Interpretation," 1–51, and Linda Wessels, "The Intellectual Sources of Schroedinger's Interpretations," 59–76.

Linda Wessels (1979), "Schroedinger's Route to Wave Mechanics," *Studies in History and Philosophy of Science* 10:311–40.

Paul A. Hanle (1979), "The Schroedinger-Einstein Correspondence and the Sources of Wave Mechanics," *American Journal of Physics* 47:644–48.

V. V. Raman and Paul Forman (1969), "Why Was It Schrödinger Who Developed de Broglie's Ideas?" *Historical Studies in the Physical Sciences* 1:291–314.

Topic *v:* Reception of Wave Mechanics; Electron Diffraction

Bruce R. Wheaton (1983), *The Tiger and the Shark: Empirical Roots of Wave-Particle Dualism* (New York: Cambridge Univ. Press).

Karl von Meÿenn (1982), "Die Rezeption der Wellenmechanik und Schrödingers Reise nach Amerika im Winter 1926/27,"*Gesnerus* 39:261–77.

Yves Gingras (1981), "La Physique à McGill entre 1920 et 1940: La Reception de la Mecanique par une Communaute Scientifique Peripherique," *HSTC Bulletin: Journal of the History of Canadian Science, Technology, and Medicine* 5:15–39.

Katherine Sopka (1980), *Quantum Physics in America, 1920–1935* (New York: Arno Press).

Topic *vi:* Dirac, Relativistic Equation, Positron

Michelangelo de Maria and Arturo Russo (1985), "The Discovery of the Positron," *Rivista di Storia della Scienza* 2:237–86.

Asim O. Barut and Alwyn Van der Merwe (1983), "Paul Dirac on His Eightieth Birthday," *Foundations of Physics* 13:187–88, is an introduction to a special issue on Dirac.

Michelangelo de Maria and Francesco la Teana (1982), "Schrödinger's and Dirac's Unorthodoxy in Quantum Mechanics," *Fundamenta Scientiae* 3:129–48.

Helge Kragh (1981), "The Genesis of Dirac's Relativistic Theory of Electrons," *Archive for History of Exact Sciences* 24:31–67; (1979), *Methodology and Philosophy of Science in Paul Dirac's Physics* (Roskilde, Denmark: Roskilde Universitets Center).

Donald Franklin Moyer (1981), "Origins of Dirac's Electron, 1925–1928," "Evaluations of Dirac's Electron, 1928–1932," and "Vindications of Dirac's Electron, 1932–1934," *American Journal of Physics* 49:944–49, 1055–62, 1120–25.

I. V. Dorman (1981), "Die Theorie Diracs und die Entdeckung des Positrons in der kosmischen Strahlung," *NTM* 18, no. 1:50–57.

P. A. M. Dirac (1977), "Recollections of an Exciting Era," in *History of Twentieth Century Physics*, ed. C. Weiner (New York: Academic Press), 109–46; Joan Bromberg (1977), "Dirac's Quantum Electrodynamics and the Wave-Particle Equivalence," ibid., 147–57.

9.7 PROPERTIES OF MATTER

i. Interatomic forces and the chemical bond: has chemistry been reduced to physics?
ii. Low temperature phenomena: superfluidity and superconductivity
iii. Solid state physics: semiconductors and the transistor
iv. Phase transitions and critical point phenomena: solid-liquid and liquid-gas transitions; magnetic transitions; Lenz-Ising model; universality hypothesis

READINGS

Marks, *Science,* 453–60 (on *iii*). Stephen G. Brush, *Statistical Physics and the Atomic Theory of Matter* (1983), 145–258. K. Mendelssohn, *The Quest for Absolute Zero* (1977). G. L. Trigg, *Landmark Experiments in Twentieth Century Physics* (1975), chapters on superconductivity and liquid helium.

Articles: J. C. Slater, "Quantum Physics in America Between the Wars" and "Energy Bands in Solids," *Physics Today* 21, no. 1(1968): 43–51; no. 4(1968):61–71. T. H. Geballe, "The Golden Age of Solid-State Physics," *Physics Today* 34, no. 11(1981), 132–43. Linus Pauling, "G. N. Lewis and the Chemical Bond," *Journal of Chemical Education* 61(1984):201–3. C. A. Coulson, "The Influence of Wave Mechanics on Organic Chemistry," in *Wave Mechanics,* ed. W. C. Price et al. (1973), 255–71. L. H. Hoddeson and G. Bayn, "The Development of the Quantum Mechanical Electron Theory of Metals: 1900–28," *Proceedings of the Royal Society of London* A371(1980)8–23. W. H. Brattain, "Genesis of the Transistor," *Physics Teacher* 6(1980):109–14. Ernest Braun, "Science and Technological Innovation," *Physics Education* 14(1979):353–58 (on the development of microelectronics). Stuart Macdonald et al., "From Science to Technology: The Case of Semiconductors," *Bulletin of Science, Technology and Society* 1(1981):173–201. Robert Friedel, "Sic Transit Transistor," *American Heritage of Invention and Technology* 2, no. 1(1986):34–40. Robert G. Kooser and Lance Factor, *Cubes, Eights and Dots: A Student's Guide to the Octet Rule and Its History* (1983). Walter A. Harrison, "Fifty Years of Metal Theory," in *Felix Bloch and Twentieth-Century Physics,* ed. M. Chodorow et al. (1980), 39–55. L. H. Hoddeson and G. Baym, "The Development of the Quantum Mechanical Electron Theory of Metals: 1900–28," *Proceedings of the Royal Society of London* A371(1980):8–23.

Most of the above readings are "advanced" and are suitable only for students with some background in physics, though not necessarily a detailed knowledge of quantum theory. The topic is almost completely ignored in most general works on the history of modern science. An exception is Cecil Schneer's *Mind and Matter: Man's Changing Concepts of the Material World* (1969), which would be useful in a course oriented more toward chemistry and crystallography.

SYNOPSIS

In early theories of gases, various assumptions were made about the forces between atoms, ranging from "hard sphere" (repulsion at contact) to attractive forces decreasing gradually with distance. In the kinetic theories of Maxwell and Boltzmann, it was possible to deduce specific consequences (e.g., variation of viscosity coefficient with temperature) from the functional form of the force law, and thus it seemed possible to infer the force law from experimental measurements of gas properties. In practice this was not done successfully until the twentieth century when S. Chapman and D. Enskog solved the Maxwell-Boltzmann equations; J. E. Lennard-Jones was then able to use gas properties as one kind of evidence in estimating force laws, but this did not turn out to be decisive.

After 1900, when it was generally believed the atom was composed of electrically charged particles, several attempts were made to derive force laws from assumed charge distributions. For example, the force between two electric dipoles would vary inversely as the fourth power of their distance. But these estimates could not account for the forces that seemed to act between neutral atoms moving randomly between different orientations, and according to quantum mechanics most atoms and molecules have zero or very small permanent dipole moments.

In 1927 S. C. Wang estimated the interaction energy between two hydrogen atoms using an approximate solution of Schrödinger's wave equation, assuming that the electrons form instantaneous dipoles with the nuclei. He found that the energy levels of the two-atom system vary as the inverse sixth power of the distance between the two nuclei. A more accurate calculation was done by R. Eisenschitz and F. London. The attractive force varies inversely as the seventh power of the distance; London called this the "dispersion force" since it is due primarily to the outer electrons that are responsible for the dispersion of light. The standard explanation is that the instantaneous dipole of one atom induces a dipole moment in the other. The attraction itself will produce correlations in the motions of the electrons, thus enhancing this effect.

At short distances there will be additional forces between neutral atoms. One of these is the "exchange force" studied by Walter Heitler and London, which may lead to chemical bonding (see below) or to repulsion. J. C. Slater investigated the latter in the case of the interaction of two helium atoms. Using quantum mechanics, he found (1928) a repulsive force described by a potential that is an exponential function of the distance between the two nuclei.

Lennard-Jones adopted the attractive r^{-6} potential found by Wang and combined it with an r^{-m} repulsive potential. Using data on the interatomic distances and heats of sublimation of solids, he estimated values of m between 9 and 12, and eventually in the late 1930s settled on 12, primarily for mathematical convenience (e.g., in calculating virial coefficients). This became widely known and accepted as the "Lennard-Jones 6–12 potential." An alternative, based on Slater's exponential repulsion together with the same r^{-6} attraction, was called the "exp-6" potential; it was eventually found to be about as good. More recent work indicates that neither the 6–12, nor the exp-6, nor any simple function can be uniquely established as the "true" law of interatomic force.

Some of the earliest applications of quantum theory were to specific heats of solids and gases (**9.4**-*iii*). Whereas Einstein had assumed that all atoms in a solid have the same vibration frequency, Peter Debye in 1912 postulated a continuous spectrum of frequencies from zero up to a finite maximum value. His formula predicted that the specific heat goes to zero as the third power of the absolute temperature at very low temperatures, and this was confirmed by experiments.

Another early application of quantum theory was to chemical reactions and ionization equilibria at high temperatures, especially in stars. By analysis of stellar spectra it was possible to estimate chemical composition as well as temperatures. These topics are mentioned in **13.2**.

After the establishment of quantum mechanics it was found that a system of identical particles will have markedly different statistical properties, depending on whether the wave function remains the same or changes its sign when the coordinates of two particles are interchanged. These two possibilities are now called "Bose-Einstein statistics" and "Fermi-Dirac statistics" respectively.

The first kind of statistics was actually developed before Heisenberg and Schrödinger introduced their versions of quantum mechanics. The Indian physicist Satyendranath Bose suggested a new derivation of Planck's radiation law in 1924, based on counting the number of distributions of photons among cells in phase space, assuming that photons having the same energy are indistinguishable so that a permutation of two such photons does not produce a different arrangement. Einstein generalized this derivation to include particles having finite mass. He found that there is a maximum number of particles that can be placed in the excited energy levels of the system and suggested that if the total number of particles is increased beyond this value, the excess atoms must all go into the lowest quantum state.

The other kind of statistics, "Fermi-Dirac," is based on Pauli's "exclusion principle" (**9.5-v**). Enrico Fermi and (a few months later) P. A. M. Dirac in 1926 showed that a quantum theory of "ideal gases" (those in which no forces act between the particles) could be based on the exclusion principle. At low temperatures the particles will go into the lowest possible levels, and the pressure will be proportional to density$^{5/3}$ but independent of temperature to a first approximation. A peculiar feature of such a system is that it cannot respond to small changes in energy, because a particle cannot absorb energy except by jumping to a higher-energy state, and all the states just above it are already filled. Only the particles in the highest filled states—near the "Fermi level"—can absorb small amounts of energy by going into empty states just above them.

Astrophysicists soon recognized that the Fermi-Dirac theory, as applied to a gas of electrons, could be used to describe the state of matter in white dwarf stars (**13.3**). Pauli and Sommerfeld used the theory to construct a theory of electrons in metals. For example, Pauli was able to explain the relatively small paramagnetic susceptibility of metals. L. D. Landau estimated the diamagnetic susceptibility of metals in a similar way.

From Pauli's exclusion principle, using four quantum numbers to characterize possible electronic orbits in an atom, physicists could derive the "electron-shell" theory of the atom and give a rough account of chemical bonding and of the similarities of elements in the same column of the periodic table. Some of the outstanding problems to be solved by "quantum chemistry" were:

1. How can one explain "saturation"? If attractive forces hold atoms together to form molecules, why is there a limit on how many atoms can stick together (e.g., generally only two of the same kind)?
2. Stereochemistry—the three-dimensional structure of molecules, in particular the spatial directionality of bonds as in the "tetrahedral" carbon atom—needed to be explained.
3. One should be able to calculate the "bond length"—there seems to be a well-defined equilibrium distance between two atoms, which can be accurately determined by experiment.
4. Why do some atoms (e.g., helium) form no bonds, while others can form one or more? (These are the empirical rules of "valence.")

Soon after J. J. Thomson's discovery of the electron in 1897 (**9.1-*ii***), there were several attempts to develop theories of chemical bonds based on electrons. The most successful was that proposed by G. N. Lewis (1916) and Irving Langmuir (1919). They emphasized shared pairs of electrons and treated the atom as a static arrangement of charges. While the model as a whole was inconsistent with quantum theory, several specific features continued to be useful.

The key to the nature of the chemical bond is the phenomenon of quantum-mechanical "resonance," first described by Heisenberg in 1926–1927. Resonance is related to the requirement that the wave function must be symmetric or antisymmetric in the coordinates of identical particles. It may be thought of as a continual jumping back and forth or interchange of the electrons between two possible states. (In later papers this is called "exchange" while the word "resonance" is used to describe the idea that an electron or bond can be in different places.) In 1927 W. Heitler and F. London used this idea to obtain an approximate wave function for two interacting hydrogen atoms; they found that with a symmetric wave function there is an attractive force, while with an antisymmetric one there is a repulsive force. The complete wave function, including the electron spins, must be antisymmetric (since electrons obey Fermi-Dirac statistics) so the two hydrogen atoms can form a molecule if their electron spins are opposite but not if they are the same.

The Heitler-London approach to the theory of chemical bonds was rapidly developed by J. C. Slater and Linus Pauling. Slater proposed a simple general method for constructing many-electron wave functions that would automatically satisfy the Pauli exclusion principle by being antisymmetric on interchange of any two electrons, using the mathematical properties of determinants. Pauling introduced a "valence bond" method, picking out one electron in each of the two combining atoms and constructing a wave function representing a paired-electron bond between them. Pauling and Slater were able to explain the tetrahedral carbon structure in terms of a particular mixture of wave functions that has a lower energy than the original wave functions.

At about the same time R. S. Mulliken was developing an alternative technique based on what he called "molecular orbitals." Here the electron is not considered to be localized in a particular atom or two-atom bond but occupies a quantum state or "orbital" which is spread over the entire molecule.

The development of quantum chemistry has led to a general impression that all of chemistry has now been reduced, at least "in principle," to physics. All observable properties and reactions of atoms and molecules could be calculated if one could solve the Schrödinger equation for the system. But this means solving a partial differential equation in 3N variables for a system of N particles; for typical molecules of interest to chemists, N ranges from 20 to 200 (even if one considers only the electronic wave function, assuming that the nuclei are fixed at known positions).

Insofar as chemists believe that their subject is reducible to physics, the result has been rather damaging. If any chemical theory is regarded as only an approximation to some solution of the Schrödinger equation, one can always—if some of its predictions don't agree with experiment—go back and fiddle with some of the parameters. Chemical theories would thus lose the property of "falsifiability" that some philosophers and scientists, following Karl Popper, consider essential to science.

A further consequence of this belief in the reducibility of chemistry is the tendency to suppress history in teaching chemistry. Chemistry is defined as primarily an *experimental* science, having no theory of its own other than what it borrows from atomic physics. Since the history of science is usually written as a story of concepts and theories, the history of chemistry is likely to seem dull.

Recently some philosophers of science have challenged the idea that chemistry has been or can be reduced to physics. Thus the explanation of chemical phenomena does not follow directly from the Schrödinger equation unless one adds the Pauli exclusion principle, yet that principle itself was originally deduced at least in part from the periodic system of elements, i.e., from chemistry.

In addition to explaining certain known properties of metals not satisfactorily accounted for by classical physics, the new quantum statistics also elucidated two spectacular phenomena discovered in the early twentieth century, superconductivity and superfluidity. These are generally regarded as "macroscopic" quantum effects—phenomena whose very existence cannot be explained by classical physics. Both were discovered by the Dutch physicist H. Kamerlingh Onnes in 1911 (though the nature of superfluidity did not become clear until many years later).

Superconductivity is a thermodynamic equilibrium state that does not depend on the previous history of the substance; this was demonstrated in 1933 by W. Meissner and R. Ochsenfeld, who found that when a metal is cooled below its superconducting transition temperature the lines of magnetic field are completely pushed out of the metal. The brothers Fritz and Heinz London developed a macroscopic theory of superconductivity in 1935, based on a postulated set of equations relating electric currents and magnetic fields, but it took two more decades for a satisfactory microscopic explanation to be developed. It was suggested by H.

Froehlich in 1950 that superconductivity is due to a special kind of interaction between electrons in a metal, in which they absorb and emit "phonons." These are quantized vibrations of the lattice formed by the atomic ions. The phonon interaction has the effect of producing an attractive force between electrons that have nearly the same energy, and this leads to new many electron states lying below the normal ground state of the system. A quantitative theory of superconductivity was worked out on this basis by J. Bardeen, L. Cooper and J. Schrieffer in 1957. It may need substantial revision in order to explain the remarkable room-temperature superconductors recently discovered.

The recognition that liquid helium undergoes a definite transition to a peculiar "superfluid" state also came first in the 1930s, primarily as a result of research by W. H. Keesom and several collaborators. Paul Ehrenfest suggested that this be called the "lambda" transition because of the similarity between the specific heat versus temperature curve and the shape of the Greek letter lambda. In 1938 F. London proposed that the transition is an example of the condensation of a Bose-Einstein gas of particles, discovered theoretically by Einstein in 1925. The term "superfluidity" was introduced by P. L. Kapitza to describe the ability of helium, at a temperature below the lambda transition, to flow through a thin capillary tube as if it had practically no viscosity. L. Tisza showed how superfluidity might be explained by an extension of London's Bose-Einstein condensation model and also predicted the phenomenon later known as "second sound" (propagation of temperature waves).

A radically different theory of superfluid helium was proposed in 1941 by L. D. Landau, based on a macroscopic "quantized hydrodynamics" rather than on Bose-Einstein statistics. Landau postulated two kinds of collective excitations, "phonons" and "rotons." During the next two decades there was a competition between the theories of Landau and of London and Tisza; Landau succeeded in describing and predicting the low-temperature properties more successfully, while the nature of the lambda transition seemed to be better described by a modification of the Bose-Einstein condensation. N. N. Bogoliubov and R. P. Feynman proposed new theoretical approaches that reconciled the viewpoint of Landau with that of London and Tisza.

The modern history of solid-state physics provides a good example of an interaction between science and technology that has had a major social impact. The theory of semiconductors was one of the early applications of the quantum mechanics to electrons in metals, initiated by A. H. Wilson in 1931. The Bell Telephone Laboratory gained a reputation for supporting fundamental research with the Davisson-Germer discovery of electron diffraction in 1927 (**9.6**-*v*) and was able to attract bright young physicists like Walter H. Brattain and William Shockley in the 1930s. The need to improve the crystal detector for microwaves and radar stimulated materials research, especially during World War II and afterwards. Bell Labs established a semiconductor research group in 1945, and John Bardeen joined it in the fall of that year. The first point-contact transistor was made in December 1947, and the discovery was announced in July 1948. Bardeen, Shockley, and Brattain received the Nobel Prize for their work in 1956. The subsequent use of the transistor in communications links up with the recent history of radio and television (**8.4**), and its use in computers is related to the growth of cognitive science, the successor to behaviorism (**6.3**-*iv*).

Shockley is now better known to the U.S. public through his involvement in the race-IQ controversy (**6.8**) than as coinventor of the transistor. But it was presumably his status as a Nobel Laureate that forced the press to take notice of his views on other issues.

One of the triumphs of the kinetic theory of gases (**7.2**-*v*) was J. D. van der Waals's theory of the continuity of gaseous and liquid states (1873). Van der Waals's work, though very approximate in nature, suggested that it might be possible to derive from a single atomic model not only the properties of all three states of matter but the nature of the transitions between them. In particular, the van der Waals "equation of state," relating pressure, volume, and temperature with two parameters representing molecular size and the strength of the intermolecular attractive force, displays the "critical point" phenomenon: at temperatures above the critical point, there is no longer a sharp distinction between liquid and gas.

A rather different kind of phenomenon, the loss of permanent magnetization at high temperature, was not recognized as a "phase transition" similar to those between gases, liquids, and solids until near the end of the nineteenth century. Pierre Curie in 1895 defined and distinguished three types of magnetism: ferro-, para-, and dia-, and investigated their variation with temperature. He proposed an analogy between magnets and fluids, noting the similarity between the temperature variation of the intensity of magnetization and the density of a fluid. A detailed atomic theory of magnetism was proposed by W. Lenz and E. Ising in the 1920s; in the Lenz-Ising model, each atom has two possible states which may be called "spin up" (spin axis parallel to an external magnetic field) or "spin down" (antiparallel).

Theories of order-disorder transitions in alloys and of solubility properties of liquid mixtures were found to be mathematically equivalent to the Lenz-Ising model, which gradually became the focus for intensive research. In 1942 Lars Onsager found an exact solution for a special case (two dimensions, zero magnetic field), exhibiting an ordered state below the critical temperature and a logarithmic discontinuity in the specific heat.

An alternative theory of the gas-liquid transition, based on the kinetic theory of gases, was developed by J. E. Mayer and Sally R. Harrison starting in 1936. While their theory was ultimately unsuccessful, it did make interesting predictions about critical phenomena that stimulated more accurate experimental work, and the new data proved to be essential in testing later theories.

Since the time of Maxwell, kinetic theorists had treated molecular motion statistically on the grounds that even if all the positions and velocities of the particles were known at an instant of time, it would be impossible to use this information to compute the history of the system because a macroscopic system contains such a large number of particles. But in the 1950s, the development of high-speed, large-memory computers at weapons-development laboratories such as Livermore and Los Alamos made it possible to follow the actual motions and collisions of simple systems consisting of a few tens or hundreds of particles. In 1957 B. Alder and T. Wainwright at Livermore discovered in this way that a system of "hard spheres" with no attractive forces can condense into an ordered solidlike state at high densities. It appears from their results that freezing is primarily due to repulsive forces that cause the molecules to fall into an ordered pattern at high densities, whereas condensation (gas to liquid) is primarily due to attractive forces that hold the molecules together.

During the 1960s and 1970s theorists were able to deduce the detailed shape of the curves representing physical properties very close to the critical point, using the same model (Lenz-Ising model or one of its modifications) to explain phase transitions in different kinds of physical systems. Kenneth Wilson received the 1982 Nobel Prize in Physics for his application of the "renormalization group" technique, borrowed from quantum theory, to this problem.

The "universality hypothesis" that underlies and is supported by much of this work asserts the atomic nature of all phase transitions—the way in which long-range order suddenly appears in a system as a result of short-range forces—is essentially the same, whether the transition is gas condensation, spontaneous magnetization, order-disorder in alloys, phase separation in liquid mixtures, or the superfluidity of liquid helium. In each case a macroscopic property, such as being gaseous or solid, has been "reduced to" (explained in terms of) a microscopic property (intermolecular forces), thus accomplishing one of the major goals of scientific research over the past three centuries.

BIBLIOGRAPHY

J. L. Heilbron and B. R. Wheaton (1981), *Literature on the History of Physics in the 20th Century* (Berkeley: Office for History of Science and Technology, University of California), 277–85, 383–91, 405.

S. G. Brush and L. Belloni (1983), *The History of Modern Physics: An International Bibliography* (New York: Garland), 114–20, 129–42.

John C. Slater (1975), *Solid-State and Molecular Theory: A Scientific Biography* (New York: Wiley/Interscience), is a personal history of the development of quantum mechanics and its application to atomic, molecular, and solid-state physics.

Topic *i:* Chemical Bond

Judith Goodstein (1984), "Atoms, Molecules, and Linus Pauling," *Social Research* 51:691–708.

Edvige Schettino (1984), "Il Modello Ring-Electron di A. L. Parson," *Physis* 26:361–71.

Arturo Russo (1982), "Mulliken e Pauling: Le Due Vie della Chimica-Fisica in America," *Testi e Contesti: Quaderni di Scienze, Storia e Societa* 6:37–59.

Anthony V. Stranges (1982), *Electrons and Valence: Development of the Theory, 1900–1925* (College Station: Texas A and M Univ. Press).

Yuko Abe (1981), "Pauling's Revolutionary role in the Development of Quantum Chemistry," *Historia Scientiarum* 20:107–24.

R. E. Kohler (1975), "G. N. Lewis's Views on Bond Theory 1900–1916," *British Journal for the History of Science* 8:233–39.

Norman H. Nachtrieb (1975), "Interview with Robert S. Mulliken," *Journal of Chemical Education* 52:560–64.

Robert J. Paradowski (1972), "The Structural Chemistry of Linus Pauling," Ph.D. diss., University of Wisconsin.

J. H. Van Vleck (1970), "Spin: The Great Indicator of Valence Behavior," *Pure and Applied Chemistry* 24:235–55.

Topic *ii:* Low Temperature Physics

Theodore Christidis, Yorgos Goudaroulis, and Maria Mikou (1987), "The Heuristic Role of Mathematics in the Initial Development of Superconductivity Theory," *Archive for History of Exact Sciences* 37:183–91.

Per F. Dahl (1986), "Superconductivity after World War I and

Circumstances Surrounding the Discovery of a State B = 0," *Historical Studies in the Physical and Biological Sciences* 16:1–58; (1984), "Kamerlingh Onnes and the Discovery of Superconductivity: The Leyden Years, 1911–1914," *Historical Studies in the Physical Sciences* 15:1–37.

Kostas Gavroglou and Yorgos Goudaroulis (1986), "Some Methodological and Historical Considerations in Low Temperature Physics II: The Case of Superfluidity," *Annals of Science* 43:137–46; (1985), "From the History of Low Temperature Physics: Prejudicial Attitudes That Hindered the Initial Development of Superconductivity Theory," *Archive for History of Exact Sciences* 32:377–83; (1984), "Some Methodological and Historical Considerations in Low Temperature Physics I: The Case of Superconductivity 1911–1957," *Annals of Science* 41:135–49.

R. B. Lindsay (1983), *Energy in Atomic Physics, 1925–1960* (New York: Van Nostrand Reinhold), 52–97.

Robert B. Hallock (1982), "Resource Letter SH-1: Superfluid Helium," *American Journal of Physics* 50:202–12, is an annotated bibliography of works suitable for students and teachers. See also Hallock, ed. (1983), *Superfluid Helium: Selected Reprints* (Stony Brook, N.Y.: American Association of Physics Teachers).

David Pines (1981), "Elementary Excitations in Quantum Liquids," *Physics Today* 34, no. 11:106–31, is a survey of developments since 1931.

J. K. Hulm, E. Kunzler, and B. T. Matthias (1981), "The Road to Superconducting Materials," *Physics Today* 34, no. 1:34–43.

P. J. Ford (1981), "Towards the Absolute Zero: The Early History of Low Temperatures," *South African Journal of Science* 77:244–48.

Anna Livanova (1980), *Landau* (New York: Pergamon Press).

M. Delbrueck (1980), "Was Bose-Einstein Statistics Arrived at by Serendipity?" *Journal of Chemical Education* 57:467–70.

E. N. Hiebert (1978), "Chemical Thermodynamics and the Third Law: 1884–1914," *Proceedings of the XV International Congress of History of Science, 1977* (Edinburgh: Edinburgh Univ. Press), 305–13.

R. P. Hudson (1977), "A Century of Cryogenics," *Journal of the Washington Academy of Sciences* 67:119–30.

Zygmunt M. Galasiewicz (1971), *Helium 4* (New York: Pergamon Press), includes reprints of original papers by H. Kamerlingh Onnes and J. Boks, W. H. Keesom et al., P. Kapitza, L. D. Landau, and others.

Topic *iii:* Solid State

Michael Eckert (1987), "Propaganda in Science: Sommerfeld and the Spread of the Electron Theory of Metals," *Historical Studies in the Physical and Biological Sciences* 17:191–233.

Lillian Hoddeson, Gordon Baym, and Michael Eckert (1987), "The Development of the Quantum-Mechanical Theory of Metals, 1928–1933," *Reviews of Modern Physics* 59:287–327.

S. T. Keith and Paul K. Hoch (1986), "Formation of a Research School: Theoretical Solid State Physics at Bristol 1930–54," *British Journal for the History of Science* 19:19–44.

T. R. Reid (1985), "The Chip," *Science 85,* 6, no. 1:32–41.

Paul Hoch (1984), "Held in Check by Years of Tyranny," *Times Higher Education Supplement,* 20 Jan. 1984, 14–5, explains how the Russians fell behind in solid-state physics and thus failed to invent the transistor; (1983), "A Key Concept from the Electron Theory of Metals: History of the Fermi Surface 1933–60," *Contemporary Physics* 24:3–23.

Krzysztof Szymborski (1984), "The Physics of Imperfect Crystals: A Social History," *Historical Studies in the Physical Sciences* 14:317–55.

Frank Herman (1984), "Elephants and Mahouts: Early Days in Semiconductor History," *Physics Today* 37, no. 6:56–63.

R. B. Lindsay, ed. (1983), *Energy in Atomic Physics, 1925–1960* (New York: Van Nostrand Reinhold), 98–111, 324–55; (1979), *Early Concepts of Energy in Atomic Physics* (New York: Academic Press), 352–96. Both are anthologies of original sources.

S. Millman, ed. (1983), *A History of Engineering and Science in the Bell System: Physical Sciences (1925–1980)* (Indianapolis: AT&T Bell Laboratories).

Ernest Braun and Stuart MacDonald (1982), *Revolution in Miniature: The History and Impact of Semiconductor Electronics,* 2d ed. (New York: Cambridge Univ. Press).

Joop Schopman (1982), "The History of Semiconductor Electronics: A Kuhnian Story?" *Zeitschrift für allgemeine Wissenschaftstheorie* 12:297–302; (1982), "The Dutch Contribution to Barrier-Layer Semiconductors in the Pre-Germanium Era," *Janus* 69:1–28.

Lillian Hoddeson (1981), "The Discovery of the Point-Contact Transistor," *Historical Studies in the Physical Sciences* 12:41–76; (1980, published 1982), "The Entry of Quantum Theory of Solids into the Bell Telephone Laboratories, 1925–1940: A Case-Study of the Industrial Application of Fundamental Science," *Minerva* 18:422–47; (1977), "The Roots of Solid-State Research at Bell Labs," *Physics Today* 30, no. 3:23–30.

Lillian Hoddeson and G. Baym (1980), "The Development of the Quantum Mechanical Electron Theory of Metals: 1900–1928," *Proceedings of the Royal Society of London* A 371:8–23, and other papers in this issue.

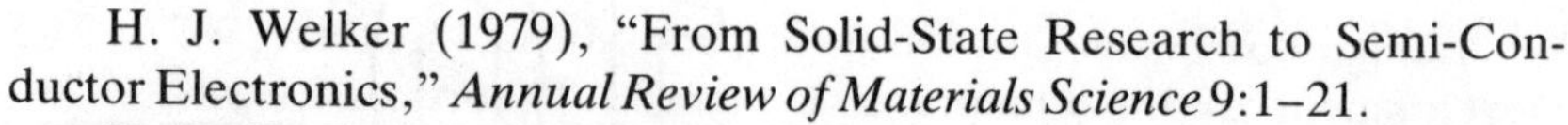

H. J. Welker (1979), "From Solid-State Research to Semi-Conductor Electronics," *Annual Review of Materials Science* 9:1–21.

H. B. G. Casimir (1977), "Development of Solid-State Physics," in *History of Twentieth Century Physics,* ed. C. Weiner (New York: Academic Press), 158–69.

Topic *iv:* Phase Transitions

J. H. Hildebrand (1981), "A History of Solution Theory," *Annual Review of Physical Chemistry* 32:1–23.

K. G. Wilson (1979), "Problems in Physics with Many Scales of Length," *Scientific American* 241, no. 2:158–79.

J. M. H. Levelt Sengers (1979), "Liquidons and Gasons: Controversies about the Continuity of States," *Physica* 98a:363–402; (1976), "Critical Exponents at the Turn of the Century," *Physica* 82a:319–51.

G. E. Uhlenbeck (1978), "Some Historical and Critical Remarks about the Theory of Phase-Transitions," in *The Ta-You Festschrift: Science of Matter,* ed. S. Fujita (New York: Gordon and Breach), 99–107.

H. Eugene Stanley (1973), *Cooperative Phenomena near Phase Transitions: A Bibliography with Selected Readings* (Cambridge: MIT Press).

10

THE EXPLOSION OF PHYSICS

10.1 CONSTRUCTING THE NUCLEUS

i. Discovery of the neutron and of deuterium
ii. Invention (and later discovery) of the neutrino; Fermi's theory of beta decay
iii. Heisenberg's model of the nucleus
iv. Early accelerators: E. O. Lawrence
v. Bethe-Weizsaecker theory of nucleosynthesis and energy generation in stars

READINGS

E. Segre, *From X-Rays to Quarks* (1980), 175–99. Thaddeus Trenn, *Transmutation: Natural and Artificial* (1981), 75–80. David Anderson, *Discoveries in Physics* (1973), Chap. 4 (on *ii*). *Project Physics Course* (1978), Unit on "The Nucleus," 49–81, 95–103. Charles Weiner, "1932: Moving into the New Physics," *Physics Today* 25, no. 5(1972):40–49. E. N. da C. Andrade, *Rutherford and the Nature of the Atom* (1964), 168–213. G. L. Trigg, *Landmark Experiments in Twentieth Century Physics* (1975), 191–208 (neutrino).

SYNOPSIS

Prout's hypothesis (**7.1**-*iii*) that all elements are compounds of hydrogen was revived in the twentieth century when it appeared that the deviations of atomic weights from integer values could be attributed to (1) mixing of isotopes of an element with different atomic weights, and (2) deviation of isotopic weights from integers because of the mass-energy conversion involved in forces between parts of the nucleus. By the 1920s it was known that a hydrogen atom consists of a positively charged particle, the proton, which contains most of the mass, and a negatively charged particle, the electron, with mass about 1/1836 that of the proton. If the mass number A is defined as the integer closest to the atomic weight of an isotope, then one might assume that the isotope has a nucleus consisting of A protons. The nuclear charge, Z, is generally only about one-half of A. Since it was known that electrons can be emitted in radioactive decay ("beta decay"), it seemed reasonable to suppose that the nucleus *contains* electrons, namely A – Z electrons, which partly neutralize the positive charge of the protons and leave a net charge of +Z.

For a few years in the 1920s it was thought that the nucleus actually does consist of protons and electrons. But detailed calculations based on quantum mechanics showed that this assumption was inconsistent with other known properties of the nucleus. Rutherford predicted the existence of the "neutron," a neutral particle with about the same mass as the proton, in 1920. In 1932 James Chadwick identified the neutron as a particle formed when beryllium is bombarded with alpha particles and confirmed that its mass is close to that of the proton.

Heisenberg then suggested that the nuclei of all atoms (except hydrogen) contain protons and neutrons rather than protons and electrons. Thus a nucleus of charge Z and mass number A is composed of Z protons and A – Z neutrons.

The year 1932 saw not only the discovery of the neutron but also of the positron, predicted by Dirac's relativistic quantum mechanics (**9.6**), and of deuterium. Harold Urey discovered this form of hydrogen with atomic weight 2, named from the Greek work *deuteros,* "second." Its nucleus, the "deuteron," contains a neutron and a proton. It is called an "isotope" of hydrogen because, having the same atomic number, it occupies the same place in the periodic table (**9.2**-*vii*). According to Heisenberg's theory of nuclear structure, different isotopes of an element have the same number of protons in the nucleus but different numbers of neutrons.

By this time fairly accurate atomic weights were known for the stable isotopes of all the elements. By comparing them with the masses of the proton and neutron one could calculate a *curve of binding energy,* showing how much energy would be released or absorbed in nuclear transmutations. The curve, drawn as a function of mass number (A), is U-shaped, starting with a high value for the light elements, going to a wide valley with the lowest point corresponding to iron, then slowly rising. Energy is released by any reaction that goes down the curve (just like rolling downhill) but energy must be supplied to go up. In general, if the density and temperature are high enough to allow nuclear reactions to occur, light elements will fuse to form medium-weight elements, and heavy elements will split (fission) to form medium-weight elements. Iron is the ultimate equilibrium state (though it is not stable against much greater increases in density).

To understand the rapid development of the physics of elementary particles and the atomic nucleus in the 1930s, one must go back to the development of quantum mechanics in the previous decade (**9.6**). One of the earliest applications of quantum mechanics to nuclear physics was made in 1928 by George Gamow, R. W. Gurney, and E. U. Condon. They pointed out that if a particle is held inside a nucleus by attractive forces, one would not ordinarily expect it to escape unless it can somehow acquire enough speed so that its kinetic energy exceeds the potential energy of those forces. One can imagine the particle sitting at the bottom of a volcano; it would take some kind of violent upheaval to kick it up to the top and then out. However, according to quantum mechanics, the particle has a small but finite chance of finding itself outside the volcano even if it did not get up to the top; in effect it has tunneled through the side. This theory gave a very effective explanation of radioactive decay, a process that seems to be completely random and involves the escape of a particle that doesn't have enough energy to overcome the forces holding it in the nucleus. (If it *did* have enough energy, it would be hard to explain why it doesn't escape right away instead of waiting thousands or millions of years.)

Other physicists, especially R. d'E. Atkinson and F. Houtermans, saw that tunneling could work in the other direction too; in particular, two protons that ordinarily repel each other would have a small but finite probability of getting close enough together for the short-range nuclear forces to fuse them. This opened up the possibility of explaining how elements are built up from hydrogen and how stars get enough energy to keep shining for billions of years.

It was generally assumed that mass-energy is conserved in nuclear reactions; whenever there is a change in mass, the difference shows up as kinetic energy in accordance with Einstein's formula $E = mc^2$. The assumption was confirmed by Cockcroft and Walton in 1932, using their accelerator to bombard lithium atoms with protons to produce the reaction

$$_3Li^7 + {}_1H^1 = 2{}_2He^4$$

They measured the kinetic energies of the nuclei taking part in the reaction and found that the mass-energy equation balances within experimental error.

But late in the 1920s it was found that mass-energy conservation does not seem to be valid for beta-decay reactions. The emitted electron has on the average only about one-third of the energy it should have, and the rest of the energy seems to disappear. To avoid violating energy conservation, Wolfgang Pauli postulated that another particle is emitted and carries off the missing energy. Enrico Fermi worked out a detailed theory of beta decay based on this proposed new particle, which he called the neutrino (Italian for "small neutral one"). The neutrino was assumed to have zero mass and to travel at the speed of light, but it is different from a photon. Its interaction with matter is so extremely weak that is can pass all the way through the Earth without being deflected. It was not possible to detect neutrinos experimentally for more than twenty years after their existence was postulated. Yet physicists had so much faith in the law of conservation of mass-energy that they preferred to believe in an apparently unobservable particle rather than abandon the law. The neutrino was eventually detected in 1956 by Frederick Reines and Clyde Cowan using a nuclear reactor to produce an extremely dense beam of neutrinos.

The usual beta-decay reaction amounts to changing a neutron into a proton and an electron, and comparison of the masses of the particles involved shows that this reaction goes with a significant liberation of energy. But it is possible to reverse the reaction, squeezing together a proton and an electron to form a neutron. This "inverse beta-decay reaction" can occur only under conditions of very high pressure or temperature, but it does play an important role in stars.

Two versions of inverse beta decay were studied in the 1930s. The first was originally proposed in 1934 by two astronomers, Walter Baade and Fritz Zwicky: if a star collapses to high enough density, nearly all of its protons and electrons will become neutrons and we will have a "neutron star." This idea was not taken seriously until about thirty years later (**13.2**).

The second version was studied by George Gamow and Edward Teller: the collision of two protons at high enough speeds can form a deuteron plus a positron. In 1938, H. A. Bethe and C. L. Critchfield showed that this reaction could start a chain of nuclear reactions leading to the formation of helium in stars, generating enough energy to account for a large part of that emitted by the sun. C. F. von Weizsaecker also suggested about this time that sequences of nuclear reactions starting from hydrogen and using other light elements such as lithium and carbon could build up the elements and provide sufficient energy to keep the stars shining. Although Weizsaecker was actually the first to present a comprehensive qualitative discussion of the various fusion reactions, he did not succeed in establishing by detailed calculations just which reactions would be the most likely to take place.

By 1938 the time was right for a synthesis of ideas about stellar evolution and nuclear reactions. The astronomical background for this synthesis is presented in **13.2**, but you may wish to insert it at this point in the discussion of nuclear physics. Briefly, the astronomers were ready to assume that hydrogen is the primordial material, and so the most likely way to start the sequence of nuclear reactions is by fusing protons together to form deuterium, whether in stars or in the Big Bang. The probability that this reaction will occur when any two protons collide is so small that it had never been observed in the laboratory [ask your colleague in the physics department whether it has yet been observed] but theoretical physicists were so confident of the validity of the beta-decay theory that they seemed quite confident in the calculation indicating that the reaction will eventually occur at high densities and high temperatures.

The observed rate of energy generation by stars such as the sun is considerably greater than one would expect if hydrogen fusion provided the only source of energy. If one may assume that heavier elements are already present in a star (perhaps seeded by explosions of a previous generation of stars), a number of other reactions become possible. Bethe's scheme, for which he received the Nobel Prize in Physics (1967), is the "carbon cycle" of nuclear reactions. A carbon nucleus ($_6C^{12}$) absorbs a proton, turning into nitrogen ($_7N^{13}$); the resulting isotope of nitrogen is unstable since it has more protons than neutrons, so it expels a positron and turns back into a heavier carbon isotope ($_6C^{13}$). (A neutrino is also produced in this reaction, which is another version of inverse beta decay.) This carbon isotope absorbs yet another proton, forming a heavier isotope of nitrogen ($_7N^{14}$) that is stable and can absorb yet another proton, turning into an oxygen isotope ($_8O^{15}$). This oxygen isotope is unstable (more protons than neutrons), so it expels a positron and a neutrino, decaying back to nitrogen ($_7N^{15}$). The nitrogen absorbs a proton and emits an alpha particle, leaving the original ($_6C^{12}$) carbon nucleus from which we started. So the net result is to convert four protons into a helium nucleus ($_2He^4$); one might say that the carbon nucleus acts as a "catalyst" to facilitate the process, remaining unchanged itself in the long run.

Bethe showed that the carbon cycle would produce about the right amount of energy to explain the present rate of radiation from the sun;

subsequent work using more accurate data on nuclear reactions indicated that the carbon cycle should be the main source of energy for stars brighter than the sun, while the proton-proton reaction should be the main source for stars dimmer than the sun. The sun itself could use either or both; this question does not seem to be completely settled yet, especially in view of the disagreement of the observed and predicted fluxes of neutrinos from the sun.

Bethe did not, in his 1939 article on stellar energy, explain the origin of the carbon atoms that seemed to be needed as catalysts for his reactions. He concluded that no elements heavier than helium can be built up in "ordinary" stars; the heavier elements found in stars must exist already when the star is formed. Further developments of this topic are discussed in **13.2**.

BIBLIOGRAPHY

W. H. Brock (1985), *From Protyle to Proton: William Prout and the Nature of Matter, 1785–1985* (Boston: Hilger), Chap. 9.

Roger H. Stuewer (1984), "Nuclear Physicists in a New World. The Emigres of the 1930s in America," *Berichte zur Wissenschaftsgeschichte* 7:23–40; Stuewer, ed. (1979), *Nuclear Physics in Retrospect: Proceedings of a Symposium on the 1930s* (Minneapolis: Univ. of Minnesota Press), includes H. A. Bethe, "The Happy Thirties"; E. M. McMillan, "Early History of Particle Accelerators"; E. P. Wigner, "The Neutron: The Impact of Its Discovery and Its Uses"; R. Peierls, "The Development of Our Ideas on the Nuclear Forces"; and a long paper by J. A. Wheeler, "Some Men and Moments in the History of Nuclear Physics."

E. Amaldi (1984), "From the Discovery of the Neutron to the Discovery of Nuclear Fission," *Physics Reports* 111:1–331.

Stephen G. Brush and Lanfranco Belloni (1983), *The History of Modern Physics: An International Bibliography* (New York: Garland), 226–35.

R. Bruce Lindsay (1983), *Energy in Atomic Physics, 1925–1960* (Stroudsburg, Pa.: Hutchinson Ross), 161–63, 183–91, 251–78, is an anthology of original sources.

(1982), *Colloque International sur l'Histoire de la Physique des Particules* (Paris: Editions de Physique), contains several papers on cosmic ray discoveries, research done with accelerators, beta decay and the neutrino, etc.

J. L. Heilbron and Bruce R. Wheaton (1981), *Literature on the History of Physics in the 20th Century* (Berkeley: Office for History of Science and Technology, University of California), 347–51.

Glenn T. Seaborg (1979), "The Periodic Table: Tortuous Path to Man-Made Elements," *Chemical and Engineering News* 57, no. 16:46–52.

Topic *i:* Neutron, Deuterium

John Hendry, ed. (1984), *Cambridge Physics in the Thirties* (Philadelphia: Heyden), includes articles by N. Feather, J. Chadwick, and P. Dee on the discovery of the neutron.

Ferdinand G. Brickwedde (1982), "Harold Urey and the Discovery of Deuterium," *Physics Today* 35, no. 9:34–39.

J. Byrne, J. M. Pendlebury, and K. F. Smith (1982), "Fifty Years of Neutrons," *Physics Bulletin* 33:285–88.

B. M. Kedrov, ed. (1979), *Das Neutron. Eine Artikelsammlung* (Berlin: Akademie-Verlag).

Edoardo Amaldi (1977), "Personal Notes on Neutron Work in Rome in the 30s and Post-War European Collaboration in High-Energy Physics," in *History of Twentieth Century Physics,* ed. C. Weiner (New York: Academic Press), 294–351.

Topic *ii:* Neutrino

Ettore Fiorini, ed. (1982), *Neutrino Physics and Astrophysics* (New York: Academic Press), has historical articles by R. Peierls, F. Reines, and B. Pontecorvo.

Laurie M. Brown (1978), "The Idea of the Neutrino," *Physics Today* 31, no. 9:23–28.

Leon Lederman (1970), "Resource Letter Neu-1. History of the Neutrino," *American Journal of Physics* 38:129–36.

F. L. Wilson (1968), "Fermi's Theory of Beta Decay," *American Journal of Physics* 36:1150–60, includes a complete English translation of Fermi's paper originally published in 1934.

Topic *iii:* Heisenberg's Nucleus

Arthur I. Miller (1985), "Werner Heisenberg and the Beginning of Nuclear Physics," *Physics Today* 38, no. 11:60–68.

William R. Shea, ed. (1983), *Otto Hahn and the Rise of Nuclear Physics* (Boston: Reidel), includes articles by Shea, R. H. Stuewer, and R. H. Kargon on nuclear physics before the discovery of fission.

Joan Bromberg (1971), "The Impact of the Neutron: Bohr and Heisenberg," *Historical Studies in the Physical Sciences* 3:307–41; (1971), "Heisenberg's Papers on Nuclear Structure," *Actes du XIIe Congres International d'Histoire des Sciences, Paris, 1968* 5:13–15.

Topic *iv:* Accelerators

Edwin M. McMillan (1984), "A History of the Synchrotron," *Physics Today* 37, no. 2:31–4.

Guy Hartcup and T. E. Allibone (1984), *Cockcroft and the Atom* (Philadelphia: Heyden).

John Hendry, ed. (1984), *Cambridge Physics in the Thirties* (Philadelphia: Heyden), has articles by E. T. S. Walton and J. Cockcroft.

J. L. Heilbron, Robert W. Seidel, and Bruce R. Wheaton (1981), "Lawrence and His Laboratory: Nuclear Science at Berkeley," *LBL News Magazine* 6, no. 3, 106 pp.

M. Stanley Livingston (1980), "Early History of Particle Accelerators," *Advances in Electronics and Electron Physics* 50:1–88; Livingston, ed. (1966), *The Development of High-Energy Particle Accelerators* (New York: Dover), contains reprints of original papers.

Topic *v:* Nucleosynthesis

Karl Hufbauer (1981), "Astronomers Take Up the Stellar-Energy Problem," *Historical Studies in the Physical Sciences* 11:277–303.

Jeremy Bernstein (1981), *Prophet of Energy: Hans Bethe* (New York: Dutton).

10.2 SMASHING THE NUCLEUS

i. Discovery of nuclear fission: experiments of Frederic and Irene Joliot-Curie, Enrico Fermi, and others; Otto Hahn, Fritz Strassmann, Lise Meitner, Otto Frisch

ii. Organization of the Manhattan Project, first chain reaction, development of atomic bomb, role of J. Robert Oppenheimer

iii. The decision to drop the bomb on Japan, and its consequences. Was it necessary to destroy Hiroshima and Nagasaki? Could physicists in Germany or Japan have developed the bomb, and did they try?

READINGS

T. Trenn, *Transmutation,* 80–92. Segre, *From X-Rays to Quarks,* 200–222 (mainly on Fermi). D. L. Anderson, *Discoveries in Physics* (1973), 47–60. H. G. Graetzer and D. L. Anderson, *The Discovery of Nuclear Fission* (1971), 1–113. D. Kevles, *The Physicists* (1978), 342–48. Joan Solomon, *The Structure of Matter* (1973), 128–46. Margaret

Gowing and Lorna Arnold, *The Atomic Bomb* (1979), about 25 pages of text plus extracts from documents. Jack Dennis, ed., *The Nuclear Almanac: Confronting the Atom* (1984), articles by A. K. Smith, "Manhattan Project: The Atomic Bomb," 21–42, and photo essay on "Hiroshima and Nagasaki," 43–51.

Articles: O. R. Frisch and J. A. Wheeler, "The Discovery of Fission," *Physics Today* 20, no. 11(1967):43–52. Richard Feynman, "Los Alamos from Below," *Science 84,* 5, no. 10(1984):63–69. Alan P. Lightman, "To Cleave an Atom," *Science 84,* 5, no. 9(1984):103–8. George B. Kauffman, "The Lesson of Hiroshima and Nagasaki, Part I: The Past," *Journal of College Science Teaching* 14(1985):463–70.

The Center for History of Physics has produced a packet of audio tapes, slides, illustrated scripts, and teacher's guide on "The Discovery of Fission" in its *Moments of Discovery* series.

SYNOPSIS

In 1934 Irene Curie and her husband Frederic Joliot (they both took the name Joliot-Curie) found that when certain light elements (boron, magnesium, aluminum) were bombarded with alpha particles from polonium, positrons as well as protons and neutrons were emitted, and the source continued to emit positrons after the alpha source was removed. It appeared that an initially stable nucleus had been changed into a radioactive one. Enrico Fermi and his colleagues in Italy then undertook a systematic study of nuclear reactions induced by neutrons, looking for new elements that might be produced in this way. In particular, Fermi thought that by bombarding uranium, the heaviest known element, an even heavier ("transuranium") element might be produced. In fact he did produce a radioactive element but in such a small amount that he could not determine what it was.

Ida Noddack, a German chemist, suggested that Fermi's experiments did not produce a transuranium element but might have caused the nucleus to break into several large fragments; but this suggestion was ignored.

The discovery of fission was accomplished by a group in Berlin, after Fermi's group in Rome and the Joliot-Curies in Paris had failed to achieve it. The Berlin group originally consisted of Otto Hahn, Lise Meitner, and Fritz Strassmann; it was Meitner, a physicist, who recognized the need for working with chemists—Hahn was an organic and nuclear chemist, Strassmann was an analytical and physical chemist. (The Rome and Paris groups were almost exclusively composed of physicists.) The political/social situation in Berlin kept the three members of the team together: Meitner had Jewish ancestors, and Strassmann refused to join the Nazi party, hence they had few opportunities aside from working with Hahn.

In 1938 Hahn and Strassmann showed that one of the substances Fermi thought might be a transuranium element was actually an isotope of barium (atomic number 56); they also found that isotopes of lanthanum (57), strontium (38), yttrium (39), krypton (26), and xenon (54) were produced. They were reluctant to accept what now seems the obvious conclusion, that the nucleus of uranium (92) had split into two small nuclei of unequal size. This conclusion was soon proposed by Meitner and Frisch, who introduced the term "fission" and noted that enormous amounts of energy could be released by the transformation of a small amount of the mass of the uranium nucleus.

Scientists in several countries immediately started to study nuclear fission, and it was quickly realized that the fission reaction produces neutrons that could initiate other fission reactions, resulting in a self-sustaining "chain reaction." Niels Bohr and John Wheeler developed a theory of fission that could explain why some nuclei undergo fission only when hit by a very fast neutron, while others split when hit by a slow neutron.

Leo Szilard, while living in England in 1935, recognized the possibility of weapons based on fission, thought this research should be controlled, and applied for a patent on the chain reaction. After hearing of the discovery of fission in 1939 he did an experiment that showed more than one neutron is liberated in the disintegration of uranium, hence a chain reaction is possible. Fermi confirmed this independently. Szilard and other emigres in the United States tried to convince their colleagues in England, France, and America not to publish further results on fission, but Joliot-Curie refused to comply (he learned that some of the American work had already been mentioned in the press), and the attempt at voluntary secrecy had to be temporarily abandoned.

In September 1939 the British magazine *Discovery* carried a piece by its editor, C. P. Snow, which sums up remarkably well the knowledge and attitudes of physicists at that time:

> Some physicists think that, within a few months, science will have produced for military use an explosive a million times more violent than dynamite. It is no secret; laboratories in the United States, Germany, France and England have been working on it feverishly since the Spring. . . . The principle is fairly simple: . . . a slow neutron knocks a uranium nucleus into two approximately equal pieces, and two or more *faster* neutrons are discharged at the same time. These faster neutrons go on to disintegrate other uranium nuclei,

> and the process is self-accelerating . . . [the weapon] must be made, if it really is a physical possibility. If it is not made in America this year, it may be next year in Germany. There is no ethical problem; if the invention is not prevented by physical laws, it will certainly be carried out somewhere in the world.

Leo Szilard and other European scientists who had emigrated to the United States persuaded the American government to sponsor a secret project to develop the atomic bomb. It is well known that Einstein was persuaded to sign a letter to President Franklin D. Roosevelt endorsing this proposal; but while this did lead to a small-scale project in the United States, the prospects for actually producing a weapon in time for use in the war seemed so remote that little effort was devoted to the project until news of progress in Britain was received. Indeed, nuclear energy was considered to have such low military priority that emigre physicists in the United States and Britain were left free to work on it while being excluded (because of their "alien" status) from participating in projects such as radar.

Two weeks after the outbreak of World War II, the British government received a report of a speech by Adolf Hitler in which he threatened the use of a "weapon which is not yet known." This later turned out to be a bad translation—he was really talking about the German Air Force—but it worried one British official, Lord Hankey, who recalled that several years earlier Rutherford had told him that experiments on nuclear transformation might someday be of great importance to the defense of Britain. He asked James Chadwick, discoverer of the neutron (**10.1**), to look into the question. It turned out that two physicists familiar with nuclear theory, Otto Frisch and Rudolph Peierls, were thinking about fission in Birmingham, and their calculations launched the atomic bomb project, first in Britain and subsequently in the United States.

Frisch and Peierls at first came to the conclusion that a bomb was impossible (i.e., it would have to weigh several tons) because only the 235 isotope, which makes up a very small part of natural uranium, is subject to fission. Bohr had already suspected that U^{235} is responsible for fission; Frisch and Peierls applied the Bohr-Wheeler fission theory and concluded that U^{235} would have a very high probability of fission by neutrons of any energy. If it were possible to obtain this isotope in nearly pure form, a fission weapon would probably work. The energy liberated by a 5-kilogram bomb would be equivalent to several thousand tons of dynamite.

The British government then appointed a committee chaired by G. P. Thomson, known as the MAUD committee (named after Maud, the governess of Niels Bohr's children). The committee was so secret that Frisch and Peierls, as foreigners, were excluded from it, as were most of the other scientists whose expertise was needed; so another "technical" committee had to be created to allow them to be consulted. The problem of obtaining uranium enriched in the 235 isotope was studied by Francis Simon; by the end of 1940 he had completed the design for a gaseous-diffusion separation plant that theoretically could produce 1 kilogram per day with 99 percent purity.

By this time two French physicists who had come to Cambridge, Hans Halban and Leo Kowarski, had shown experimentally that a divergent chain reaction maintained by slow neutrons could be produced in a mixture of uranium oxide and heavy water. This reaction could be used to produce element 94 from U^{238}, following the process suggested by Louis Turner at Princeton ($n + U^{238} = 94^{239}$), and confirmed by E. McMillan and P. Abelson in May 1940 (and published in *Physical Review*). On reading the McMillan-Abelson paper, E. Bretscher and N. Feather in England argued from nuclear theory that element 94 could fission more efficiently than U^{235} so the required critical mass would be lower. Two other American papers published in the spring of 1940 by A. Nier's group confirmed the prediction from Bohr-Wheeler theory that U^{235} is responsible for fission by slow neutrons.

The MAUD committee reported that a U^{235} bomb could be produced in two years. An American physicist, C. C. Lauritsen, heard about this report and informed Vannevar Bush (director of the U.S. Office of Scientific Research and Development) in summer 1941. Bush urged President Roosevelt to accelerate the American work, which up to that time had been stalled because there seemed little chance of success. Roosevelt appointed a committee, known as "S-1," including Bush, J. B. Conant (president of Harvard and chairman of the National Defense Research Committee), and Henry L. Stimson (Secretary of War). In 1942 Brig. Gen. Leslie R. Groves of the Army Corps of Engineers became officer-in-charge of the "Manhattan Project," which was set up to develop the atom bomb. Stimson was responsible for overseeing the project and was Roosevelt's adviser on atomic energy but had little direct involvement with it until the spring of 1945; Bush, Conant, and Groves were the most important administrators.

Problems soon developed between the British and Americans. Conant, in 1942, urged that the United States should no longer share information with the British since atomic energy work in the United States was now based almost entirely on American research; he wanted the United States to have a monopoly after the war, even at the cost of some delay in developing the bomb, according to Martin Sherwin (*A*

World Destroyed). Roosevelt accepted this suggestion, but Winston Churchill protested strongly. At a meeting in Quebec in August 1943 the policy of complete interchange of information was restored. The British then terminated their own project and sent as many of their scientists as possible to help the American project produce the bomb quickly and also so that the British would acquire knowledge and experience that would be useful after the war.

Producing the Material for the Bomb

In order to make a fission bomb, it was necessary either to separate the 235 isotope of uranium or to use the 238 isotope to produce the new element 94, called plutonium. Four different methods were proposed to separate 235:

1. Gaseous diffusion: convert uranium into a gaseous compound and allow it to flow through a barrier with very small holes. In thermal equilibrium the lighter isotope will have greater average speed (**7.2**) and thus will move more quickly through the barrier, as shown by Lord Rayleigh in 1896. Unfortunately the only suitable gaseous compound of uranium is a highly corrosive hexafluoride; fluorine has only one isotope, whereas other elements that could produce a gaseous compound with uranium are themselves mixtures of isotopes.
2. Thermal diffusion: according to the kinetic theory of gases developed by S. Chapman and D. Enskog in 1916–1917, a mixture subjected to a temperature gradient will tend to separate into components, in a way that depends on the details of the interatomic force law, as well as on the temperature and original concentration of the mixture. H. Clusius and G. Dickel had used thermal diffusion to separate isotopes in 1938; they found that the heavier isotopes usually go to the cooler side.
3. Electromagnetic separation, as in the mass spectroscope: since the forces operate on individual atoms rather than statistically on large numbers, it has a very small yield but can produce very pure material.
4. Centrifuge: the differential force of gravity pushes the heavier isotope to the outside of a spinning container.

In addition to these methods of isotope separation, one needs a *chemical* process if one is going to use plutonium, since it has to be extracted from the uranium in which it is created.

Until 1945 no one was certain which if any of these processes would be able to produce enough fissionable material to make a bomb. The centrifuge process was eliminated fairly early, although it was developed for isotope separation for other purposes. In November 1942 General

Groves appointed a committee chaired by Warren K. Lewis, a chemical engineering professor at MIT, to recommend which of the other methods should be pursued. The Lewis committee visited a gaseous diffusion project at Columbia University, Fermi's pile (to produce plutonium) at Chicago, and E. O. Lawrence's electromagnetic separation project (called the "Calutron") at Berkeley. Thermal diffusion was not at that time considered a serious contender, perhaps because it was not well understood theoretically.

Fermi's group rushed to complete their pile and succeeded in obtaining a controlled chain reaction on 2 December, with a member of the Lewis committee in attendance. This historic event took place at the University of Chicago, under the football field. (A. H. Compton authorized this experiment himself since he was afraid President Robert Maynard Hutchins, if asked, would not allow it.)

The Lewis committee recommended continuing the pile project and also constructing a gaseous diffusion plant, thus keeping open the option of developing either a uranium or a plutonium bomb or both. They didn't think Lawrence's Calutron would produce enough fissionable material to be militarily significant but noted that it could immediately produce a few grams of 235 isotope that could be used for physical measurements. Conant insisted on revising the report to give more support to the Calutron. Roosevelt approved the revised report on 28 December 1942, and construction of a gaseous diffusion plant began early in 1943 at Oak Ridge, Tennessee, while work on the Calutron was also supported.

The plutonium-producing pile did not stem directly from Fermi's experiment of December 1942, but from related research in Chicago. Thomas V. Moore and Miles C. Leverett designed the pile; Glenn T. Seaborg and others worked on chemical separation of plutonium; Charles M. Cooper supervised the implementation. General Groves chose a site at Hanford, Washington, for the production plant.

By spring 1944 it appeared that both the gaseous diffusion process and the electromagnetic process were running into formidable obstacles and could not produce enough U^{235} to make a bomb by mid-1945. In the meantime Philip Abelson had been working on a thermal diffusion process with liquid uranium hexafluoride at the Naval Research Laboratory in Anacostia, just outside Washington, D.C. He introduced the liquid into the annular space between concentric vertical pipes, heating the outer surface and cooling the inner one; the lighter isotope would tend to concentrate near the hot wall and then rise by convection to the top.

There had been little communication between the army's Manhattan Project and the navy until Oppenheimer learned about Abelson's progress and informed General Groves in April 1944. Groves decided to build a full-scale thermal diffusion plant at Oak Ridge next to the gaseous diffusion plant so it could use the same source of steam. An electromagnetic separation plant was also built at Oak Ridge. Eventually all three processes were used to enrich the uranium in the 235 isotope in a sequence of operations: first thermal diffusion, then gaseous diffusion, finally electromagnetic separation. This system started to produce material in April 1945. The first sample of plutonium from Hanford was available two months earlier.

Designing the Bomb

In Fall 1942 J. R. Oppenheimer and others suggested setting up a new laboratory to design the atomic bomb. To preserve secrecy it should bring scientists together at a remote location. A site at Los Alamos, New Mexico, was selected at the first staff assembled there in March 1943, with Oppenheimer as director. There were four divisions: (1) theoretical, headed by Hans Bethe; (2) experimental, headed by Robert Bacher; (3) chemistry and metallurgy, headed by Joseph W. Kennedy, with Cyril S. Smith in charge of metallurgy; (4) ordnance, headed by Navy Captain William S. Parsons.

The Los Alamos group considered two possible methods for assembling a critical mass of fissionable material. The simpler was a "gun" that would fire one subcritical mass into another. The more complex was "implosion," suggested by Seth Neddermeyer: surrounding a subcritical mass with high explosives, deployed in such a way as to produce a converging shock wave that would compress the material into a small volume; this would be equivalent to exceeding the critical mass. George B. Kistiakowski, an expert on explosives, was brought in to develop this method.

It was discovered that in addition to the Pu^{239}, which could be caused to fission by neutrons, another isotope Pu^{240} was produced in the pile, and this isotope could undergo spontaneous fission. Calculations indicated that because of this effect, the neutron background would be too high for the gun method to work on plutonium; only implosion could be used.

To be sure that at least one workable bomb could be produced, the Los Alamos group developed two kinds of bombs: a gun weapon, using U^{235}, and an implosion weapon, using plutonium. Since they were not sure the implosion method would work, it was thought necessary to test it first, and this was done at Alamogordo, New Mexico, on 16 July 1945.

The bomb dropped on Hiroshima on 6 August 1945 was a uranium weapon using the gun method, called "Little Boy"; the bomb dropped on Nagasaki on 9 August 1945 was a plutonium weapon using the implosion method, called "Fat Man."

The Decision to Use the Bomb

We now come to the topic that should have the greatest interest for any class of students, even if they have trouble grasping the technical details summarized above. How did the physicists involved in the development of the atomic bomb justify putting such a terrible weapon at the disposal of the politicians, and to what extent did they try to influence the actual use of this weapon in World War II and afterwards? Conversely, how did the atomic bomb influence the outcome of the war and the history of international relations after the war?

Niels Bohr escaped to England from Nazi-occupied Denmark in September 1943. When he learned about the Manhattan Project, he became concerned about the possibility of a postwar arms race. He proposed to Churchill and Roosevelt that the Soviets be invited to participate in plans for international control of atomic energy *before* the bomb was used in war; i.e., they should be informed of the existence of the project though not of the details about the bomb. Bohr underestimated the political difficulty of getting his proposals accepted, though several elements of his plan were actually adopted by the United States in its later proposals to the Soviet Union and to the United Nations Atomic Energy Commission.

After initially being encouraged by an indirect approach to Roosevelt through Felix Frankfurter, Bohr met with Churchill in May 1944 but Churchill bruskly rejected his plan. Possibly Churchill felt that the plan threatened his hopes of exploiting atomic energy after the war, or perhaps he was suspicious that Bohr would leak crucial information to the Soviets. Bohr returned to the United States and met with Roosevelt in September 1944; though more friendly than Churchill, Roosevelt also rejected Bohr's proposal.

In November 1944 a scientific intelligence mission, "Alsos," discovered that the German atomic bomb program was far behind the American; thus there was no longer a strong basis for concern that the Germans might be able to use the bomb. "Gen. Groves may have deliberately kept this news from the physicists at Los Alamos in order not to dampen their enthusiasm" (S. Goudsmit, quoted in Sherwin, *A World Destroyed,* 134).

Robert Jungk, in his book *Brighter Than a Thousand Suns,* claimed that Heisenberg and other German physicists intentionally pushed the German atomic bomb project into a dead end because they did not want Hitler to have the bomb. Jungk's account has been strongly criticized by several physicists. There was also a Japanese nuclear project, which did not come close to developing a workable bomb. If one is going to discuss hypothetical alternative histories (what if the United States had refused to drop the bomb?) or assign moral credit and discredit (were the German and Japanese physicists more virtuous because they didn't develop a bomb?) these facts need to be taken into account.

Before the defeat of Germany, policymakers assumed that the atomic bomb would actually be used in the war; the question of refraining from its use was apparently not even discussed. Instead, the issue was postwar international control; Roosevelt and Churchill wanted to maintain an Anglo-American monopoly to support their diplomatic position against the Soviet Union, according to Sherwin. (Other historians claim that Roosevelt was more naive about post-war Soviet intentions.)

Could a bomb that had not been used serve any diplomatic or military purpose? In January 1944 Szilard had written to Bush that postwar international control of atomic energy could not be achieved unless bombs had "actually been used in the war and the fact of their destructive power has deeply penetrated the mind of the public." But in June 1945 he took the opposite position, while those advocating use of the bomb based their case on his 1944 argument. The bomb was ready too late for use against the Germans, and no one thought the Japanese could have developed their own atomic bomb. If the Japanese could somehow be convinced that the Americans had such a weapon, they might surrender to the threat rather than the actual use of it. A group of physicists at Chicago, chaired by James Franck, argued that the bomb's power should be demonstrated first at an uninhabited site where it could be observed by the Japanese leaders. They feared that a surprise attack would start an arms race and damage the chances for international control of atomic weapons. But Arthur Compton, Fermi, and Oppenheimer opposed this plan and favored direct military use. A preannounced demonstration might fizzle, or the Japanese might place American prisoners at the site.

President Roosevelt died on 12 April 1945 and Harry S Truman became president. Apparently Truman had not been informed about the atomic bomb project and Roosevelt's plans for its use. According to Gar Alperovitz (*Atomic Diplomacy*), Truman knew or should have known that it was not really necessary to use the bomb to end the war, but he wanted to exploit it as part of a diplomatic offensive designed to reduce Soviet influence in world affairs, and this plan played a major role in his decision to use the bomb against Japan. At the Yalta Conference in February 1945 the Soviets had agreed to enter the war against Japan two or three months after Germany was defeated, in return for restoration of property they had lost to Japan in 1904. But Truman and his advisers feared that the Soviets would gain too much influence in the Far East if they entered the war and wanted to use the atomic bomb to defeat Japan without any help from the Russians. Moreover, by avoiding a full-scale invasion of Japan, the United States could leave large numbers of troops in Europe where they would strengthen our position during postwar negotiations. According to Alperovitz, Stimson tried to speed up the bomb development so it would be essentially completed before the meeting scheduled with Stalin at Potsdam in summer 1945, and even postponed it to 15 July (over Churchill's objections) to make it more likely that Truman would be able use the bomb to negotiate from strength.

Recall that the plutonium bomb was tested at Alamogordo on 16 July; this was the day after Truman arrived at Potsdam. According to Stimson's account Truman was given great confidence in presenting his demands to Stalin by the report on the success of the test, and this was the "psychological turning point of the conference." For example, Truman refused to recognize Russian puppet governments in Balkan countries unless they were reorganized along democratic lines. On 24 July Truman mentioned to Stalin that he had a new weapon at his disposal though he didn't mention the words "nuclear" or "atomic." Stalin didn't seem especially surprised, either because he didn't recognize the significance of the remark or because he already knew about the American atomic bomb project from spies. But Stalin did tell his aides to push Igor Kurchatov, the director of Soviet atomic energy research, to speed up his efforts. The Soviets did develop their own bomb much sooner than the U.S. political leaders expected (despite fairly accurate forecasts by scientists that it would take them only three to five years).

Martin J. Sherwin (*A World Destroyed*) concludes, contrary to Alperovitz and William A. Williams (*The Tragedy of American Diplomacy*), that the primary reason for Truman's decision to use the bomb against Japan was to win the war with minimum loss of life. Herbert Feis (*The Atomic Bomb and the End of World War II*) rejects Alperovitz's claim that Truman wanted the Soviets to stay out of the war against Japan until after it had been ended by the atomic bomb, and accepts Truman's own statement that the "most urgent" reason for going to Potsdam "was to get from Stalin a personal affirmation of Russia's entry into the war against Japan, a matter which our military chiefs were most anxious to clinch" (*Year of Decisions,* 411). Influencing the Soviets may have been a secondary consideration in the decision to use the bomb, especially for advisers like Stimson. Other reasons were: policymakers feared the wrath of Congress if they didn't use a weapon on which so much money had been spent (James Byrnes asked, "how would you get

Congress to appropriate money for atomic energy research if you do not show results for the money which has been spent already?"); and, on the other hand, the world might be convinced of the need to abolish war if it saw the terrible results of this weapon.

There is still considerable disagreement about whether the Japanese would have surrendered without an invasion if the bomb had not been used. The U.S. Strategic Bombing Survey concluded that Japan would almost certainly have surrendered by the end of 1945. But it also found that the Japanese had more than 5000 planes ready for kamikaze (suicide) use, and plenty of pilots, so they certainly had the capability of inflicting enormous loss of life on an invading force if they chose to hold out.

Another controversy revolves around the U.S. insistence on "unconditional surrender"; had the United States explicitly promised to preserve the Japanese emperor on his throne, the Japanese leaders who wanted to end the war before August 1945 might have prevailed.

Hiroshima was bombed on 6 August 1945. About 100,000 people were killed immediately, and tens of thousands more were left dying of radiation poisoning. This may be compared with the March raids on Tokyo that killed 83,000 and injured 40,000—that was the most destructive "conventional" air raid in history.

The original schedule called for the second bomb to be dropped on Nagasaki on 11 August but was advanced two days because bad weather was expected. The result was another controversial action: many historians argue that after Hiroshima the Japanese surrender was inevitable, but they simply did not have time enough to respond and reach a decision before 9 August; perhaps the loss of life at Nagasaki was unnecessary. In fact the Japanese surrendered on 14 August.

The Soviet Union had declared war on Japan 8 August, and the next day the Red Army crossed the Manchurian border. Although the bomb had not prevented the Soviets from entering Manchuria, it did alter the balance of power in the Far East; for example, Stalin accepted unilateral American occupation of Japan.

BIBLIOGRAPHY

R. Rhodes (1987), *The Making of the Atomic Bomb* (New York: Simon and Schuster).

Alwyn McKay (1984), *The Making of the Atomic Age* (New York: Oxford Univ. Press).

For references to earlier works see the bibliographies by Brush and Belloni and Heilbron and Weart cited in **10.2**.

Topic *i:* Fission

Lawrence Badash, Elizabeth Hodes, and Adolph Tiddens (1986), "Nuclear Fission: Reaction to the Discovery in 1939," *Proceedings of the American Philosophical Society* 130:196–231.

Roger Stuewer (1985), "Bringing the News of Fission to America," *Physics Today* 38, no. 10:48–56; Stuewer, ed. (1979), *Nuclear Physics in Retrospect* (Minneapolis: Univ. of Minnesota Press), includes papers by E. Segre on "Nuclear Physics in Rome" and by O. R. Frisch on "Experimental Work with Nuclei: Hamburg, London, Copenhagen."

W. Gerlach and D. Hahn (1984), *Otto Hahn 1879–1968* (Stuttgart: Wissenschaftliche Verlagsgesellschaft).

E. Amaldi (1984), "Neutron Work in Rome in 1934–36 and the Discovery of Uranium Fission," *Rivista di Storia della Scienze* 1:1–24; (1984), "From the Discovery of the Neutron to the Discovery of Nuclear Fission," *Physics Reports* 111:1–331; (1977), "Personal Notes on Neutron Work in Rome in the 30s and Post-War European Collaboration in High-Energy Physics," in *History of Twentieth Century Physics,* ed. C. Weiner (New York: Academic Press), 294–351.

W. R. Shea, ed. (1983), *Otto Hahn and the Rise of Nuclear Physics* (Boston: Reidel), includes articles by S. Weart, F. Krafft, and others.

Patricia E. Rife (1983), "Lise Meitner: The Life and Times of a Jewish Woman Physicist," Ph.D. diss., Union for Experimenting Colleges and Universities; see *Dissertation Abstracts International* 45(1984):922-A.

R. B. Lindsay (1983), *Energy in Atomic Physics 1925–1960* (Stroudsburg, Pa.: Hutchinson Ross), 192–99, 222–50.

Werner Stolz (1983), *Otto Hahn, Lise Meitner* (Leipzig: Teubner).

E. Segre (1981), "Fifty Years Up and Down a Strenuous and Scenic Trail," *Annual Review of Nuclear and Particle Science* 31:1–18, has personal recollections of his work with E. Fermi and others.

Spencer Weart (1979), *Scientists in Power* (Cambridge: Harvard Univ. Press), concerns the French scientists working on nuclear fission in the 1930s and 1940s and their interactions with society.

Kurt Starke (1979), "The Detours Leading to the Discovery of Nuclear Fission," *Journal of Chemical Education* 56:771–75.

Topic *ii:* Chain Reaction, Developing the Bomb

Malcolm C. MacPherson (1986), *Time Bomb: Fermi, Heisenberg, and the Race for the Atomic Bomb* (New York: Dutton).

Vincent C. Jones (1986), *Manhattan: The Army and the Atomic Bomb* (Washington: Government Printing Office).

Rudolf Peierls (1985), *Birds of Passage* (Princeton, N.J.: Princeton

Univ. Press), is an autobiography of a physicist involved in development of the atomic bomb.

Martin J. Sherwin (1985), "How Well They Meant," *Bulletin of the Atomic Scientists* 41, no. 7(1985):9–15.

Ferenc Morton Szasz (1984), *The Day the Sun Rose Twice: The Story of the Trinity Site Nuclear Explosion, July 16, 1945* (Albuquerque, N.M.: Univ. of New Mexico Press).

Gregg Herken (1983), "Mad about the Bomb," *Harper's* 267, no. 1603:48–55. "The inventors of nuclear weapons gather for the fortieth anniversary of guilt and pride."

Victor F. Weisskopf (1983), "Los Alamos Anniversary: 'We Meant so Well,' " *Bulletin of the Atomic Scientists* 39, no. 7:24–26.

David Hawkins, Edith C. Truslow, and R. C. Smith (1983), *Project Y: The Los Alamos Story* (Los Angeles: Tomash).

Albert Wattenberg (1982), "December 2, 1942: The Event and the People," *Bulletin of the Atomic Scientists* 38, no. 10:22–32, recalls the first self-sustaining nuclear chain reaction.

Lawrence Badash et al., eds. (1980), *Reminiscences of Los Alamos, 1943–1945* (Boston: Reidel).

Ronald W. Clark (1980), *The Greatest Power on Earth* (New York: Harper and Row).

E. O. Wollan (1980), "The Other Record of the First Nuclear Reactor Start-up," *American Journal of Physics* 48:979–80.

Spencer Weart (1977), "Secrecy, Simultaneous Discovery, and the Theory of Nuclear Reactors," *American Journal of Physics* 45:1049–60; (1976), "Scientists with a Secret," *Physics Today* 29, no. 2:23–30.

A. G. Brown and C. B. MacDonald, eds. (1977), *The Secret History of the Atomic Bomb* (New York: Dial Press), includes reprints of previously classified or unpublished documents.

Topic *iii:* Dropping the Bomb; Projects in Germany, Japan

Lawrence Badash (1987), "Werner Heisenberg and the German Atomic Bomb," *Physics and Society* 16, no. 1:10–12.

Len Ackland and Steven McGuire, eds. (1986), *Assessing the Nuclear Age* (Chicago: Educational Foundation for Nuclear Science), includes articles by M. J. Sherwin, J. Rotblat, V. F. Weisskopf, R. R. Wilson, R. L. Messer, and S. R. Weart.

Dan Kurzman (1985), *Day of the Bomb: Countdown to Hiroshima* (New York: McGraw-Hill).

Peter Wyden (1985), "The Sudden Dawn," *Washingtonian* 20, no. 10:8–95, 182–89; see also his review of books by Szasz and Williams and Cantelon (1985), in *Isis* 76:286–88.

Martin J. Sherwin (1975), *A World Destroyed: The Atomic Bomb and the Grand Alliance* (New York: Knopf), includes a discussion of Roosevelt's views on the use of the bomb.

David Irving (1968), *The German Atomic Bomb: The History of Nuclear Research in Nazi Germany* (New York: Simon and Schuster).

Herbert Feis (1966), *The Atomic Bomb and the End of World War II* (Princeton, N.J.: Princeton Univ. Press).

10.3 SCIENCE AND POLITICS AFTER THE BOMB

i. The impact of nuclear weapons on geopolitics
ii. The "scientists' movement" (1945–1946) leading to civilian (?) control of atomic energy; Pugwash and the peace movement, campaigns for nuclear disarmament, opposition to nuclear power; test bans
iii. The trial of J. Robert Oppenheimer; McCarthyism, anticommunism, Teller's push for H-bomb; involvement of scientists in politics (apart from nuclear-related issues); atomic spies (K. Fuchs, the Rosenbergs)

READINGS

John Marks, *Science and the Making of the Modern World* (1983), 384–87, 441–52. H. J. Fyrth and M. Goldsmith, *Science, History and Technology,* Book 2, Part III (1969), 1–57.

Kevles, *The Physicists,* 349–66. Joan Solomon, *The Structure of Matter,* 147–57. Jack Dennis, ed., *The Nuclear Almanac,* articles by H. F. York and G. A. Greb, "The Superbomb," 53–66, and J. Lamperti, "Government and the Atom," 67–79.

Articles: Herbert F. York, "The Debate over the Hydrogen Bomb," *Scientific American* 233, no. 4(1975):106–13. G. A Greb and W. Heckrotte, "The Long History: The Test Ban Debate," *Bulletin of the Atomic Scientists* 39, no. 7(1983):36–42. Giorgio de Santillana, "Galileo and Oppenheimer," *Reporter* 17, no. 11(1957):10–18. Murray Kempton, "The Ambivalence of Robert Oppenheimer," *Esquire* 100, no. 6(1983):236–48. Spencer R. Weart, "The Heyday of Myth and Cliche," *Bulletin of the Atomic Scientists* 41, no. 7(1985):38–43.

SYNOPSIS

The scientists involved in developing the atomic bomb had begun to think and argue about the problems of controlling it long before Hiroshima, but politicians and the public were suddenly faced with these issues in August 1945. In October 1945 an administration bill was introduced by Congressman Andrew J. May (Kentucky) and Senator Edwin C. Johnson (Colorado) to vest authority over nuclear research in an Atomic Energy Commission appointed by the president. The May-Johnson bill seemed to give military needs first priority and continued a high degree of secrecy. Despite support for this bill from V. Bush, J. B. Conant, E. O. Lawrence, E. Fermi, and A. Compton (Oppenheimer vacillated), other atomic scientists opposed it and lobbied Congress and the administration favoring a civilian-controlled program. They started a new journal, *Bulletin of the Atomic Scientists,* to discuss such issues; its cover featured a clock with hands close to midnight, calling attention to impending doom. (The position of the hands has been adjusted at various times since to reflect changes in the international situation.) They joined forces with sympathetic officials like Don K. Price in the Bureau of the Budget and James R. Newman in the Office of War Mobilization and Reconversion to defeat the May-Johnson bill. Senator Brien McMahon (Connecticut) introduced a bill to establish a civilian-controlled Atomic Energy Commission (AEC). President Truman decided to support McMahon's bill, and it was passed in July 1946.

The scientists wanted to continue the high level of funding for basic research established during World War II but remove it from the control of a few large corporations that benefitted from patents, as well as from military control. Bush, in his book *Science: The Endless Frontier* (1945), proposed a program that would be insulated from political control. After several failures, a bill to establish a National Science Foundation was finally passed and signed in March 1950. But much research was still supported by the Office of Naval Research and the Public Health Service.

Brookhaven National Laboratory was opened on Long Island in 1947; it was operated for the Atomic Energy Commission by a consortium of nine universities and provided nuclear research facilities in the Northeast comparable to those already available at Argonne (in Illinois), Oak Ridge (Tennessee), and Berkeley.

During the first postwar decade United States graduate schools produced Ph.D.'s in physics three times as fast as in the prewar period—more than 500 in 1955. But participation of women and Catholics was very low, while Jews and Orientals made up a disproportionately high fraction (details in Kevles's book, *The Physicists*).

Scientists attempted to influence U.S. government policy with varying degrees of success. During World War II Bush had been the scientific adviser to President Roosevelt, but the postwar dissolution of the Office of Scientific Research and Development had eliminated any direct formal source of scientific advice. This gap was partly filled by the

AEC General Advisory Committee (GAC) when the AEC was created in 1946. When the GAC's advice on the hydrogen bomb was overruled in 1950 (see below), its influence declined. But the influence of other scientists, e.g., Edward Teller, increased. Teller's critics claimed that he represented only a minority of scientists. One response to Sputnik (1957) was the appointment of James Killian, president of MIT, as President Eisenhower's special assistant for science and technology. This began a revival of formal advisory mechanisms, e.g., the President's Science Advisory Committee (PSAC); but PSAC did not survive the animosity between scientists and the administration created by the Vietnam War.

The atomic bomb did not change the structure of international diplomacy [according to Michael Mandelbaum, *The Nuclear Question* (1979)], it was just incorporated into it since countries would not give up their sovereignty. But it did convince many people that the chief purpose of a military establishment should now be to avert wars rather than win them. An American plan for international control of atomic energy was presented by Bernard Baruch to the United Nations in June 1946. The Soviets rejected the Baruch Plan, calling instead for a complete ban on the bomb (which only the United States possessed at that time).

In proposing the Baruch and other plans the United States insisted on enforcement mechanisms but these always seemed to involve some intrusion on sovereignty that the Soviets refused to accept. Thus, while it was difficult to conceal weapons tests in the atmosphere, underground tests might be conducted in such a way that they would be almost indistinguishable from earthquakes at a distance. At a conference in Geneva in 1958, American scientists recommended setting up a network of control posts, including some on Soviet territory, to monitor seismic signals to detect possible weapons tests; the Soviets demanded a veto power and a much smaller number of posts than the Americans considered necessary.

In September 1949 the United States detected evidence for an atomic bomb explosion, which had taken place in the Soviet Union on 29 August. The Soviet project, directed by Igor Kurchatov, had been started in 1943, then put aside and revived in July 1945 (perhaps as a result of Truman's remark to Stalin at the Potsdam conference, see **10.2**). Thus it took them about four years, compared to the three years and seven months from the time the U.S. project was initiated in December 1941 to the Alamogordo test in July 1945.

Americans perceived this event as the latest in a series of Russian threats, following their 1948 takeover of Czechslovakia and erection of the Wall dividing East and West Berlin, and the 1949 takeover of mainland China by the Communists. In considering an appropriate response, the possibility of a crash program to develop a much more powerful bomb, based on the fusion of hydrogen, immediately arose. The hydrogen bomb, known as the "Super," had been discussed by physicists ever since Bethe's theory of thermonuclear reactions in stars (**10.1**), and Edward Teller had been strongly urging its development at Los Alamos. The first public discussion of the hydrogen bomb appeared in Hans Thirring's *Die Geschichte der Atombombe* (1946), but it was still supposed to be secret in the United States. Thirring estimated that a hydrogen-helium or hydrogen-lithium bomb triggered by a uranium bomb would have as much as 1000 times the power of a simple uranium bomb.

J. Robert Oppenheimer, chairman of the GAC, opposed the development of the Super, and persuaded the GAC to adopt a report to this effect in October 1949. The GAC report actually supported one of the two thermonuclear programs in progress at Los Alamos, the "booster" that would improve the efficiency of fission bombs by incorporating a small thermonuclear reaction. The reaction is ignited by the fission bomb and produces neutrons that increase the fission yield. Moreover, the GAC favored the production of more and better nuclear weapons in the absence of any effective arms-control agreement with the Soviets. Thus the GAC was not dominated by "doves," and according to Herbert York (*The Advisors* 1976) their assessment of the situation was basically sound.

Oppenheimer and the GAC opposed the Super for two primary reasons: (1) it was too big and murderous, (2) it was not needed to ensure American security. Secondary factors were: (3) it would compete for resources needed elsewhere, (4) the existing design was awkward and not certain to work, (5) there were promising alternative approaches to the design of bombs larger than those in the stockpile but smaller than the Super, (6) there was a possibility of using the special danger to mankind as a basis for reopening arms-control talks.

Teller continued to lobby for the Super, and he had influential friends in the administration. In January 1950 Truman decided to go ahead with the hydrogen bomb. A month later it was revealed that Klaus Fuchs had been passing secret information about the bomb to the Russians; this news encouraged the United States to go ahead with the hydrogen bomb. (Fuchs had also been giving information on the hydrogen bomb to the British; according to R. W. Clark, *The Greatest Power on Earth,* 266, the British got more information about the hydrogen bomb from Fuchs than they had from the Americans.)

In June 1950 communist North Korea invaded South Korea, and the United States, under United Nations auspices, soon entered the war on the side of South Korea. This situation encouraged the U.S. government to revise its defense policy, shifting toward the development of small tactical atomic weapons that could be used in a "limited" war—as advocated by Oppenheimer. This meant that Oppenheimer and his supporters were on the side of the army, against the air force, which wanted to maintain the priority of strategic bombing. This may account for some of the subsequent hostility of the air force toward Oppenheimer.

In December 1950 Teller and S. Ulam conceived a new idea that made the Super practical. While the idea is still officially classified top secret, according to Howard Morland (1979) its essence is to use a container to reflect the radiation energy produced by a fission trigger, concentrating it on the thermonuclear fuel. In any case, the scientists to whom the idea was revealed (including Oppenheimer) immediately agreed that it was a major breakthrough.

The booster principle was tested in the "Item" shot at Eniwetok in May 1951. Another test designed to "observe phenomena expected to be associated with the Super," the "George" shot in the same month, was successful although the contribution of thermonuclear reactions to the total energy release was small (York, *Advisors*). The Teller-Ulam idea was tested in the "Mike" shot on 1 November 1952; the device had a yield equivalent to 10 million tons of TNT (10 megatons) but was not a deliverable weapon since it used liquid deuterium, which had to be kept under refrigeration. (Theoretical calculations were more reliable with pure D than with LiD.) Finally on 1 March 1954 came the "Bravo" test of a deliverable Super using LiD; its yield was 15 megatons, the largest of any American test.

In the meantime the Soviets had also begun to develop a hydrogen bomb. Andrei Sakharov is known as the "father of the Soviet H-Bomb"; we don't know how accurate this is, but in any case he was elected to the Soviet Academy of Sciences in 1950 at the early age of 29. According to York, the Soviets did not "win" the race of the H bomb as is sometimes said; their August 1953 test (called "Joe 4" by the Americans) was of a device no more powerful than the largest previous U.S. fission explosion, and only 3 to 5 percent as powerful as the first U.S. thermonuclear explosion that had taken place nearly a year before, "Mike." Moreover it was not a Super using the Teller-Ulam idea; their first Super-like device came only in November 1955, three years after the United States had experimentally verified the basic ideas.

York argues that if the United States had not tested the Super in 1952, the Soviets would probably not have developed their own Super as early as they did; the U.S. test informed them that a hydrogen bomb of very large yield was possible, and their analysis of its fallout may have given them a clue to how it worked. He thinks that if the Soviet test had come first it would not have taken the United States very long to catch up, with the help of the fast computers (needed for optimal weapon design) available after 1953.

In December 1953 the AEC suspended Oppenheimer's security clearance, accusing him of associating with Communists and opposing the development of the hydrogen bomb. Most of the charges of Communists association went back to 1943 or earlier and had been taken into account when his clearance was reviewed; as Director of the Los Alamos laboratory, Oppenheimer's services were considered so vital to the war effort that the government decided to put aside any doubts about his loyalty. But in 1953 those services were not longer needed—he was only a consultant to the AEC for a few days a year, though he exerted considerable influence through its advisory committees. In the 1954 hearings on the Oppenheimer case, damaging testimony by Edward Teller seemed to play a crucial role; also, it appears that most of the charges against Oppenheimer related to the hydrogen bomb were based on FBI interviews with Teller. The AEC voted not to restore Oppenheimer's security clearance. Teller became a pariah in the scientific community, losing many of his best friends; he became identified with the right-wing "hawks" in subsequent debates on policy.

Thermonuclear Geopolitics

According to Robert Gilpin (*American Scientists and Nuclear Weapons Policy* 1962), American atomic scientists can be categorized in three major groups or schools with respect to their views on nuclear weapons policy after 1946:

1. The "control school," Linus Pauling, Philip Morison, Harlow Shapley, argued that the U.S.-USSR hostility (the "cold war") was a result of the danger of the atomic bomb, and that the first priority should be control of weapons leading to disarmament; they attacked the U.S. position as "inflexible"; they assumed that one could solve problems in international relations, as in science, by reason without the use of physical force; what is needed is education and a solution agreeable to all sides.

2. The "finite containment" school, which comprised the majority of politically active scientists, including Hans Bethe, Oppenheimer, Teller, Arthur Compton, I. Rabi, Conant, and Harold Urey, argued that the cold war originated in the aggressive policies of the USSR; the only way to contain Soviet aggression was the threat of the American atomic bomb. They wanted the United States to draw closer to Europe, to help reconstruct it. Teller and Oppenheimer at first agreed that scientists

must help rebuild U.S. military strength; and, after the Soviet rejection of the Baruch plan, efforts toward control and disarmament had to be abandoned. This school advocated limiting the nuclear arms race by international agreement at some *finite* point prior to settlement of political differences between the United States and the USSR, while *containing* Soviet aggression. They (except Teller) recommended against developing the hydrogen bomb because it would destroy hope for control; it would make the United States and other countries totally reliant on nuclear weapons, and they would not agree to give them up.

3. The "infinite containment" school, led by Teller when he broke with the finite school, argued that a feasible control system is possible in a completely open world; political differences must be resolved before the arms race can be ended. Modern science and technology will always produce surprises, which can circumvent any control system if there is secrecy. In a sense this school was extending Bohr's thesis of the need for an open world; Teller insisted that each nation should have the right to inspect all the facilities of any other nation, and he has always advocated (at least in public) reducing U.S. secrecy as much as possible. But, he argued, since a democracy is already partly open, it is at a disadvantage relative to a totalitarian country that preserves complete secrecy; the latter can gain lead time in the development of new weapons by secret violation of a control agreement. If the Soviets develop a hydrogen bomb and the United States does not, they will have an advantage and fission weapons can no longer deter them.

In 1953 the Eisenhower administration proclaimed the policy of deterrence: the threat to use the bomb to support "containment" of the USSR, as "massive retaliation" for a wide range of aggressive acts. For Republicans this policy was attractive because it seemed cheaper than maintaining a large standing army. But when the Soviets had their own atomic weapons, the U.S. strategy had to be revised to achieve "survivability" of its retaliatory force after a first strike by the Soviets.

In 1951 a report by Albert Wohlstetter at the RAND corporation (one of the "think tanks" set up to provide advice and policy research for the government) argued that overseas air bases could not be protected against a Soviet strike, hence all planes should be based in the United States. This recommendation was adopted by the U.S. Air Force. Also, as pointed out by William Kaufmann in 1956, the United States was unlikely to use the bomb in retaliation except for very major provocations. Thus it did not really deter Soviet aggression in areas that did not seriously affect our interests. So if the United States wanted to respond to Soviet advances in Korea and Indochina, it had to use conventional forces after all. The nuclear deterrent was not enough because it would not actually be used in such wars.

The development of the Polaris submarine, which could launch nuclear missiles from under water, appeared to provide an invulnerable retaliatory force and thus solved for the time being *part* of the deterrence problem (to deter a first strike against the United States).

In 1957 Henry Kissinger, a professor of government at Harvard, published *Nuclear Weapons and Foreign Policy* in which he argued that nuclear warfare could remain limited and that limited nuclear war was an appropriate way to defend Western Europe against the USSR. But "war games" experiments cast doubt on this theory—it seemed that several million Europeans would be killed and there would be some chance of escalation into global war.

A new factor was introduced with the Bravo hydrogen bomb test in March 1954. Radioactive rain fell on a small Japanese tuna trawler, "Lucky Dragon Number 5." One fisherman died and others developed radiation sickness. Fish contaminated with radioactivity reached the Japanese market. There was a worldwide clamor about the dangers of radioactive fallout from nuclear tests. Lewis Strauss, chairman of the AEC, claimed that radioactivity from tests would "decrease rapidly after the tests until the radiation level has returned approximately to the normal background" (AEC press release, 31 March 1954, quoted by Gilpin). But many scientists criticized the AEC for underestimating the effects of fallout, especially the possible genetic damage.

In 1949 D. E. Uphoff and C. Stern had found that the genetic effect of radiation is proportional to the dosage—there is no "threshold" below which there is no damage at all. Biologists argued that the effects are cumulative so even a small amount of testing would affect future generations by producing mutations.

Another danger is the isotope strontium 90 (half-life, twenty-eight years) produced in a nuclear explosion. Strontium is chemically similar to calcium, and is therefore absorbed by plants eaten by cattle, which produce milk drunk by children whose growing bones have a strong affinity for calcium. Since bone marrow produces white blood corpuscles, the result may be leukemia.

After the death of Joseph Stalin (1953), Soviet leaders recognized the need to reduce tensions (to "thaw" the cold war), perhaps by accepting steps toward disarmament and control, and cessation of nuclear tests. Having acquired their own hydrogen bomb, the Soviets had the capacity to destroy the United States. In any case, it was believed by many U.S. scientists that "nuclear parity" would make the Soviets more reasonable.

There was also a shift in the nature of the arms race after 1953: the emphasis was on the means of delivery (e.g., the intercontinental ballistic missile, ICBM) rather than on simply increasing the power of the bomb itself. This shift reduced the need for testing huge weapons.

The "control" school became more active beginning in 1955. In the 1956 presidential election, Adlai Stevenson proposed a ban on nuclear tests; this may have helped to defeat him, but it publicized the issue, and pressures for a ban increased in 1957. The infinite containment school (Teller, Willard Libby, L. Strauss) defended Eisenhower's antiban position; Teller argued that the shortening of average life span due to fallout was only comparable to that due to smoking. He also suggested that a "clean" bomb could be made if further research were allowed; this was understood to mean a fusion-only bomb since the radioactivity was thought to come only from the fission trigger. In 1960 Freeman Dyson began to discuss what later became known as the "neutron bomb."

Teller and David Griggs proposed underground tests of nuclear weapons in 1956; such tests would be suitable for small tactical weapons and would not produce any fallout. The first such test, "Rainier," was held in fall 1957. The AEC at first announced that Rainier was detected only within 250 miles; this claim was protested by several scientists, and the AEC finally admitted that it had been detected as far as 2300 miles away. This created the impression that the AEC had tried to falsify its initial report to show that underground tests couldn't be detected outside the borders of the country and thus a test ban couldn't be enforced.

Both the United States and the USSR began to fear the instability resulting from acquisition of nuclear weapons by many other countries; by 1960 France had become the fourth nuclear power after the United States, the USSR, and the United Kingdom.

In January 1958 Linus Pauling presented to the United Nations a petition signed by 9,000 scientists, calling for a ban on nuclear tests as a "first step" toward eventual disarmament. On 31 March the Soviets announced that they would discontinue all nuclear tests forever provided other nations ceased also. This was just after they had completed a major test series and just before an announced American series. The United States response was to propose a conference of experts from the United States and the USSR to be held in Geneva in July to discuss the possibility of enforcing a test ban.

The conference recommended a network of control posts with instruments to monitor seismic signals. The scientists assumed that an underground nuclear explosion would produce a "positive first motion" in all directions since the initial impulse would be outwards from the point of the explosion, whereas most earthquakes produce "negative first motion" in some directions since they begin with a shearing displacement. Thus only those events with positive first motion in all directions would need on-site inspection. But the Soviets demanded a veto power over the commission operating the control posts, so they could prevent any violation of their sovereignty. Thus the conference could not agree on an absolute right of each country to investigate suspicious events in the other.

The United States agreed to stop its own tests in October 1958, even though no firm agreement on enforcement of a ban had been reached. But Teller argued that it would be possible to disguise an underground explosion in a large cavern by "decoupling" it; at his suggestion, A. Latter made a theoretical study of decoupling and found that a 300-kiloton explosion could be made to look seismically like a 1-kiloton explosion. The Soviets rejected this idea as absurd, and accused the Americans of simply trying to sabotage the Geneva proposals. But Latter's calculation was confirmed by the "Cowboy" test in the spring of 1960, and the argument about whether small nuclear underground tests can be detected continues up to the present day.

In September 1961 the USSR started testing again with a series of 50 atmospheric explosions, one of them 57 megatons. In response, President Kennedy decided that the United States should also resume its own tests. In August 1962 the United States proposed a ban on atmospheric tests only; this avoided the problem of inspection since satellites could now detect atmospheric tests quite reliably. In October 1962 came the Cuban missile crisis, which seems to have produced a less aggressive Soviet policy afterwards. In June 1963 Kennedy called for a moratorium on polemics with the Soviets, argued that conflict is not inevitable, and invited the Soviets to join in a search for peace. This speech was favorably received by the Soviet leaders; it came at a time when the Chinese Communists were about to break with the Soviets. The Soviets wanted to discourage both the Chinese and the Germans from developing nuclear weapons.

The final version of the Test Ban Treaty was signed by the United States, the United Kingdom, and the USSR on 5 August 1963 in Moscow. Ultimately over 100 nations signed. Kennedy promised the Joint Chiefs of Staff that they could continue underground testing, which was not prohibited by the treaty. The limited treaty of 1963 was followed by several other agreements between the United States and the USSR, e.g., restrictions on deployment of antiballistic missiles (ABMs). But these agreements have not eliminated the fear of catastrophic nuclear war.

Recently several scientists have argued that the effects of nuclear war would be much greater than previously thought: "Subfreezing temperatures, low light levels, and high doses of ionizing and ultraviolet radiation extending for many months after a large-scale nuclear war could destroy the biological support systems of civilization, at least in the Northern Hemisphere. . . . Postwar survivors would face starvation as well as freezing conditions in the dark and be exposed to near-lethal doses of radiation . . . the extinction of a large fraction of the Earth's animals, plants, and microorganisms seems possible. The population size of *Homo sapiens* conceivably could be reduced to prehistoric levels or below, and extinction of the human species itself cannot be excluded" [Ehrlich et al., *Science* 222(1983):1293].

BIBLIOGRAPHY

Len Ackland and Steven McClure, eds. (1986), *Assessing the Nuclear Age* (Chicago: Educational Foundation for Nuclear Science), contains articles on the arms race by H. E. DeWitt, David Holloway et al.

George B. Kauffman (1985), "The Lesson of Hiroshima and Nagasaki, Part II: The Present and Future," *Journal of College Science Teaching* 15:14–23.

Robert C. Williams and Philip L. Cantelon, eds. (1984), *The American Atom: A Documentary History of Nuclear Policies from the Discovery of Fission to the Present, 1939–1984* (Philadelphia: Univ. of Pennsylvania Press).

Peter Wyden (1984), *Day One: Before Hiroshima and After* (New York: Simon and Schuster).

Karl Hufbauer (1984), *Nuclear Age Bibliography: Historical Studies Published since 1/1/75* (twenty-two-page mimeographed report available from author at History Department, University of California, Irvine, Calif. 92717).

Margaret M. Gowing (1983), "The History of Science, Politics and Political Economy," in *Information Sources in the History of Science and Medicine,* ed. P. Corsi and P. Weindling (Boston: Butterworth Scientific), 99–115, is mostly on the atomic bomb and postwar science policy.

Betty Reardon, John Anthony Scott, and Samuel Totten, eds. (1983), "Nuclear Weapons: Concepts, Issues, Controversies," *Social Education* 47:473–522, includes Dan B. Fleming, "Nuclear War: What Do High School History Textbooks Tell Us?" and J. A. Scott, "John Hersey's *Hiroshima:* A Teaching Tool."

Fred Kaplan (1983), *The Wizards of Armageddon* (New York: Simon and Schuster), is a history of nuclear weapons policy from the late 1940s to 1980.

Joseph Ellis (1983), "Apocalypse and American Culture: The Threat of Nuclear War," *Newsletter, Intellectual History Group,* no. 5:3–5, is a syllabus for a course at Mt. Holyoke College.

Dietrich Schroeer (1982), "Resource Letter PNAR-1: Physics and the Nuclear Arms Race," *American Journal of Physics* 50:786–95.

Topic *i:* Nuclear Geopolitics

Peter Wyden (1985), "The Sudden Dawn," *Washingtonian* 20, no. 10:88–95, 182–89.

Gar Alperovitz (1985), *Atomic Diplomacy,* rev. ed. (New York: Penguin Books).

Gregg Herken (1981), *The Winning Weapon: The Atomic Bomb in the Cold War, 1945–1950* (New York: Knopf).

Ronald W. Clark (1981), *The Greatest Power on Earth: The International Race for Nuclear Supremacy* (New York: Harper and Row).

Michael Mandelbaum (1979), *The Nuclear Question: The United States and Nuclear Weapons, 1946–1976* (New York: Cambridge Univ. Press).

Martin J. Sherwin (1975), *A World Destroyed: The Atomic Bomb and the Grand Alliance* (New York: Knopf).

Topic *ii:* Scientists and the Control of Nuclear Energy

Milton S. Katz (1986), *Ban the Bomb: A History of SANE, the Committee for a Sane Nuclear Policy, 1957–1985* (Westport, Conn.: Greenwood).

Howard Ball (1986), *Justice Downwind: America's Atomic Testing Program in the 1950s* (New York: Oxford Univ. Press).

A. Costandina (1983), "A-Bombs in the Backyard: Southern Nevada Adapts to the Nuclear Age, 1951–1963," *Quarterly of the Nevada Historical Society* 26:235–54.

Glenn Seaborg (1981), *Kennedy, Krushchev, and the Test Ban* (Berkeley: Univ. of California Press).

Carolyn Kopp (1979), "The Origins of the American Scientific Debate over Fallout Hazards," *Social Studies of Science* 9:403–22.

Robert A. Divine (1978), *Blowing in the Wind: The Nuclear Test Ban Debate, 1954–1960* (New York: Oxford Univ. Press).

Alice K. Smith (1965), *A Peril and a Hope: The Scientists' Movement in America, 1945–47* (Chicago: Univ. of Chicago Press).

Robert Gilpin (1962), *American Scientists and Nuclear Weapons Policy* (Princeton, N.J.: Princeton Univ. Press).

Topic *iii:* Oppenheimer; Political Scientists; H-Bomb

Robert C. Williams (1987), *Klaus Fuchs, Atom Spy* (Cambridge: Harvard Univ. Press).

Frederick Warner et al. (1985), *The Environmental Consequences of Nuclear War* (New York: Wiley), concludes,"nuclear winter may be the least of our worries . . . starvation seems more likely," as summarized by R. J. Smith in *Science* 229:1245–46.

Barton J. Bernstein (1985), "The Oppenheimer Conspiracy," *Discover* 6, no. 3:22–32; (1982), "In the Matter of J. Robert Oppenheimer," *Historical Studies in the Physical Sciences* 12:195–252.

Walter Sullivan (1984), "Stanislaw Ulam, Theorist on Hydrogen Bomb," *New York Times,* 15 May, B8 (obituary).

Gian-Carlo Rota et al., eds. (1984), *Uncommon Sense: J. Robert Oppenheimer* (Cambridge: Birkhauser), includes his 1951 essay "The Consequences of Action," 41–54.

Hans A. Bethe (1982), "Comments on the History of the H-Bomb," *Los Alamos Science* 3, no. 3:43–53.

James W. Kunetka (1982), *Oppenheimer: The Years of Risk* (Englewood Cliffs, N.J.: Prentice-Hall).

Peter Goodchild (1981), *J. Robert Oppenheimer: Shatterer of Worlds* (Boston: Houghton Mifflin).

Jeremy Bernstein (1981), *Prophet of Energy: Hans Bethe* (New York: Dutton).

Howard Morland (1979), "The H-Bomb Secret," *The Progressive* 43, no. 11:14–23.

Roger M. Anders (1978), "The Rosenberg Case Revisited: The Greenglass Testimony and the Protection of Atomic Secrets," *American Historical Review* 83:388–400.

Michael E. Parrish (1977), "Cold War Justice: The Supreme Court and the Rosenbergs," *American Historical Review* 82:805–42.

Herbert F. York (1976), *The Advisors: Oppenheimer, Teller and the Superbomb* (San Francisco: Freeman).

S. Ulam (1976), *Adventures of a Mathematician* (New York: Scribner's), Chaps. 8 and 11 recall his work at Los Alamos and the development of the Super.

Edward Teller (1955), "The Work of Many People," *Science* 121:267–74, is on the development of the hydrogen bomb.

10.4 ELEMENTARY PARTICLES

i. Quantum electrodynamics and field theory; difficulties in dealing with strong interactions; retreat to phenomenology; dispersion relations, S-matrix

ii. H. Yukawa's meson theory; proliferation of "elementary" particles and attempts to classify and understand them; nonconservation of parity

READINGS

E. Segre, *From X-Rays to Quarks,* 235–63. V. Guillemin, *Story of Quantum Mechanics* (1968), 139–191, 198–217. D. L. Anderson, *Discoveries in Physics* (1973), 63–90 (on the neutrino). Haven Whiteside, *Elementary Particles* (1971) (Project Physics Course, suppl. unit). Trigg, *Landmark Experiments,* 265–81. H. A. Boorse and L. Motz, *World of the Atom* (1966), 1431–1581 (extracts from original sources). Heinz Pagels, *The Cosmic Code* (1982), Part II, Chaps. 1–9.

Articles: Laurie M. Brown and Lillian Hoddeson, "The Birth of Elementary-Particle Physics," *Physics Today* 35, no. 4(1982):36–43. James W. Cronin and Margaret S. Greenwood, "CP Symmetry Violation," *Physics Today* 35, no. 7(1982):38–44. George Gale, "Forces and Particles: Concepts Again in Conflict," *Journal of College Science Teaching* 3(1973):29–35. Hugh Taylor, "From 'The Lynxes' to Stanford's 'LinAc,'" *American Scientist* 54(1966):333–44. Steven Weinberg, "The Search for Unity: Notes for a History of Quantum Field Theory," *Daedalus* 106(1977):17–35. Jeremy Bernstein, "Pauli's Puzzle," *Science Digest* 94, no. 8(1986):40–43, 84–85.

SYNOPSIS

This section takes up the history of particle physics starting from the end of WWII, with a brief look back at the beginnings of quantum field theory in the late 1920s and the Yukawa meson theory of nuclear forces proposed in 1935. It ends with the discovery of parity nonconservation, signalling the beginning of a new era; the subject of particles and fundamental interactions since 1957 lies outside the scope of my course.

A major theme is the fundamental uncertainty about whether one should describe nature in terms of elementary *particles,* including the assumption that forces between particles are due to the "virtual exchange" of other particles, or in terms of one or a small number of basic *forces* on which the particles appear as somewhat transient quanta or epiphenomena. This theme has obvious connections with earlier periods in the history of physics, such as the attempt of "dynamists" to replace atoms as substantial objects by forces, in the early nineteenth century (**7.2**).

Aside from gravity, which seemed to be negligible on the atomic scale, the only fundamental forces known in 1926 were electricity and magnetism. Both were described in a unified way by Maxwell's electromagnetic theory (**8.3**); in fact, since magnetic poles had never been found to exist separately, it was possible to describe magnetism in terms of circulating electric currents as Ampère had suggested (**7.2**). With the development of quantum field theory, theorists began to think of electromagnetic forces in terms of the creation, propagation, and absorption of photons by charged particles, rather than as instantaneous action at a distance. That does not necessarily mean they adopted a Cartesian view—that the world consists only of pieces of matter acting solely by contact forces—though that view did become briefly popular after WWII. Rather than reducing forces to the exchange of particles, theorists could also say that they were reducing particles to manifestations of force fields; and the latter viewpoint has recently become popular because of the difficulty of regarding any particular set of particles as "elementary."

Dirac's relativistic version of the Schrödinger wave equation led to the prediction of a new particle, the antielectron or "positron," subsequently discovered by Anderson in 1932 (**9.6**). Since the proton is described by a similar equation, it should also have an antiparticle. The antiproton was not discovered until 1955, when it was literally manufactured in the Bevatron at Berkeley by a group headed by Emilio Segre and Owen Chamberlain. All that was needed was to provide enough kinetic energy in a system of two interacting particles to supply the rest-mass energy of a proton-antiproton pair. (Whiteside's book, cited above, gives a simple calculation of this energy.) Perhaps the U.S. taxpayers should have shared the Nobel Prize for this discovery.

With this confirmation of the theory, it was established that elementary particles generally come in pairs, though some particles like the photon may be identical to their antiparticle. It also became clear that our part of the universe contains a remarkable asymmetry that does not seem to be a logical consequence of the laws of nature: it is composed only of particles, and antiparticles (which theoretically have an equal right to exist and can be produced in the laboratory) are so rare that when they appear they are quickly annihilated by combination with the corresponding particle. This is a puzzle for current theories of cosmology (**13.3**).

Although the discovery of antiparticles might be regarded as a proliferation relative to the set of particles previously known, one could also regard it as a reduction in the total number of possible particles, since two particles are to be considered as merely different states or versions of the same particle. (The electron and positron count as only one fundamental particle rather than two.) This kind of reduction through the recognition of similarities has been a major theme in the development of elementary particle theory. For example, when Heisenberg proposed his theory of nuclear structure (**10.1**), he suggested that the neutron and proton could be regarded as two states of the same particle, having different values of a certain quantum number that was later called "isotopic spin" by analogy with "ordinary" spin. (Electron spin, which was itself a rather abstract concept related only indirectly to mechanical spin in the classical sense when Goudsmit and Uhlenbeck introduced it in 1925 [**9.5**], was now considered ordinary.)

In spite of the fact that he regarded the neutron as an independent fundamental particle *not* composed of a proton and an electron, Heisenberg proposed to describe the *force* between a proton and a neutron in terms of the exchange of an electron between two protons. (In a significant aside, he remarked that, strictly speaking, the particle to be exchanged might better be considered an electron without spin, obeying Bose statistics.) The proton-neutron system would thus be similar to the He_2^+ ion as treated in the Heitler-London theory (**9.7**). He also assumed a weak attractive force between two neutrons and a pure Coulomb repulsion between two protons. Later he assumed also a short-range repulsion between both protons and neutrons.

Heisenberg's assumptions about the forces between protons and neutrons were rejected in the late 1930s as a result of scattering experiments. It was decided that the "strong" interactions are the same for any pair of nucleons whether neutron or proton; the electrical interaction between protons is simply added to the strong interaction. This became known as the assumption of "charge independence" of nuclear forces and helped to define a third basic kind of force in nature.

Fermi's theory of beta decay (**10.1**) introduced another new force, now called the "weak interaction."

Forces between charged particles were described in the "quantum field theory" developed around 1930 in terms of the continual emission and absorption of "virtual" photons. (The law of mass-energy conservation prevents a particle at rest from emitting a "real" photon, since the particle cannot have less than its rest-mass energy. If this statement is true in the frame of reference in which the particle is at rest, it must be true in any other frame.) Moreover, the same theory could be extended to describe other particles such as electrons and protons as quanta of various fields; thus, as Weinberg puts it (1977), "particles were reduced in status to mere epiphenomena . . . the essential reality is a set of fields . . . all else is derived as a consequence of the quantum dynamics of these fields."

In 1935 Hideki Yukawa published a theory that attributed the strong nuclear force to exchange of a particle. This was clearly intended as a modification of Heisenberg's 1932 exchange theory, following up Heisenberg's suspicion that the particle exchanged is not really an electron but a particle with no spin obeying Bose statistics. That nuclear forces have a very short range (about 10^{-15} meter, the size of the nucleus) implied that the particle should have a mass equal to about 200 times that of the electron.

Yukawa's prediction became known to Western physicists in 1937, about the same time as the discovery of the muon by C. D. Anderson and S. H. Neddermeyer. It was believed for several years that the new particle was in fact the one postulated by Yukawa (informally known as the "yukon") and as a result Yukawa's theory attracted much attention.

Yukawa had originally proposed the meson-exchange force only for the proton-neutron interaction and had assumed that the proton-proton force was nothing more than Coulomb repulsion. However, proton-proton scattering experiments by M. A. Tuve, L. R. Hafstad, and N. Heydenberg in 1936 indicated that there is a strong short-range force in this case, similar to that between the proton and neutron. Consequently N. Kemmer in 1938 proposed that the force between two protons is produced by the exchange of *neutral* mesons and the np, nn, and pp interactions all have the same strength apart from Coulomb forces.

When research in elementary-particle physics was resumed after WWII, it was found that the muon could not be the carrier of nuclear forces postulated by Yukawa. An experiment by M. Conversi, E. Pancini, and O. Piccioni (1947) showed that the interaction between muons and atomic nuclei is very weak. Several theorists argued that any forces transmitted by such particles would correspondingly be extremely weak.

About the same time C. F. Powell and G. Occhialini discovered two new charged particles (one +, one −) which do interact strongly with nuclei; these were later named the pions (+ and −) and were identified as Yukawa's particles. The neutral pion postulated by Kemmer was somewhat more difficult to detect and was not identified until 1950.

The year 1947, which saw the discovery of the long-awaited Yukawa particle, also marked the beginning of the final phase in the history of quantum electrodynamics. Theorists had previously worried about the "self-energy" of the electron in a radiation field, due to emission and absorption of virtual photons in 1930. J. R. Oppenheimer tried to calculate this energy as a shift in atomic energy levels but found that it came out infinite. It was suggested by V. F. Weisskopf and others in the mid-1930s that the infinite correction could be eliminated by "renormalization," i.e., by redefining the observed energy levels or the electron's mass so that the observed value already includes the correction. However, it was not known whether this could be done in a consistent way for all energy levels.

In 1947 Willis Lamb and Robert Retherford reported a very accurate determination of the difference between two excited states of the hydrogen atom, using microwave apparatus developed during WWII. According to Dirac's electron theory these two states should have the same energy. Hans Bethe immediately showed how the observed value of the level difference could be approximately obtained by mass renormalization. Sin-itiro Tomonaga had already (1943) developed a general method for performing such calculations in a way consistent with relativity theory but had not yet carried out detailed calculations in this particular case. During the years 1947–1949 the calculations were done independently by Julian Schwinger, Richard Feynman, and Tomonaga and his colleagues. Freeman Dyson proved the equivalence of the three methods and gave a unified exposition of the results that helped to establish quantum electrodynamics as the most complete and accurate theory in physics.

The calculations done by Schwinger, Tomonaga, and Feynman in 1947–1949 could have been done at least ten years earlier since, as Steven Weinberg had pointed out, they were based not on a new physical theory but "simply the old quantum field theory of Heisenberg, Pauli, Fermi, Oppenheimer, Furry, and Weisskopf, but cast in a form far more convenient for calculation, and equipped with a more realistic definition of physical parameters like masses and charges." Quantum field theory had not been taken seriously before, partly because Dirac's 1928 theory had successfully accounted for the hydrogen atom spectrum without taking account of self-energy effects, and partly because "the appearance of infinities discredited quantum field theory altogether in many physicists' minds." But, Weinberg suggests, the "deepest reason is a psychological difficulty. . . . There is a huge apparent distance between the equations that theorists play with at their desks, and the practical reality of atomic spectra and collision processes." Until the Lamb-Retherford experiment, there was not enough motivation to do the work needed to extract the consequences of the theory.

Without getting into the details of quantum electrodynamics it should also be pointed out that there is a philosophical difference between the Tomonaga-Schwinger approach and the Feynman method. The former is a field theory; its adherents "believed in electric and magnetic fields as physically meaningful quantities. . . . On the other hand Feynman . . . did not believe that fields exist except as some kind of classical limit. . . . The Feynman theory was a pure particle theory" (Dyson 1965). Feynman's theory was more attractive to the younger generation educated after WWII, but did not appeal to Bohr when it was presented at a meeting in 1948. When Dyson recounted the reception of Feynman's theory, he considered it "ironic" that now (i.e., 1965) Feynman needed no defender—his diagram method for doing calculations had been widely adopted, whereas the field theory approach was obsolescent. But twelve years after that a leader of the next generation of theoretical physicists, Steven Weinberg, was proclaiming the triumphal revival of field theory and writing its history as a success story.

A third major discovery in 1947 (in addition to the pion and the Lamb-Retherford energy shift) was the detection of the K meson, or kaon, by G. D. Rochester and C. C. Butler in England. It was soon established that two new particles are produced in collisions of pions with protons: K, which has a mass greater than the pion but less than the nucleon, and lambda, which is heavier than the nucleon.

The new particles decay quickly, K to two pions and lambda to a pion and a proton, but not as quickly as they should in view of their copious production in pion-proton interactions. Thus the "interaction time" for the production process—the time needed for a pion to travel a distance equal to the range of nuclear forces—is about 10^{-23} seconds, whereas the decay time (mean lifetime) for the lambda is much longer, about 10^{-10} seconds.

To explain the discrepancy between the rapid production and the slow decay of the K and lambda particles, A. Pais and Y. Nambu in 1952 proposed that they are produced only in pairs by strong interactions but decay separately in weak interactions. In 1953 Murray Gell-Mann and Kazuhiko Nishijima suggested that they have a property called "strangeness," which is conserved in strong but not in weak interactions. The "reason" why a K must be produced together with a lambda is that K has strangeness +1 while lambda has strangeness −1. Since pions and protons both have zero strangeness, the total strangeness is unchanged by the reaction in which a pion and a proton yield K and lambda. This explanation, which might seem to be nothing more than a fancy way of reformulating the original puzzle, was subsequently justified by its success in helping to systematize the description of many other reactions involving "strange" particles. (Cf., Clausius's apparently ad hoc introduction of an unknown molecular diameter in his kinetic theory, subsequently justified by Maxwell's use of it—see **7.2**).

Another problem involving the K meson was the "theta-tau puzzle," which emerged around 1954. It was found that the K meson could decay to either two or three pions; these were called the theta and tau modes of decay, respectively. One might think that theta and tau were actually two different particles, but they seemed to have exactly the same masses and lifetimes. The puzzle arises from the fact that the "parity" of the pi meson is *odd.* Parity is the behavior of the wave function on inversion of the coordinates (substitute $-r$ for r). If this operation leaves the wave function unchanged, the parity is even; if it changes the sign, the parity is odd. For a system of several particles the parity is the product of the parities of the individual particles. Before 1956 it was generally believed that parity is conserved in all reactions; thus theta and tau would have opposite parities and could not correspond to the same particle.

In 1956 T. D. Lee and C. N. Yang analyzed the evidence for parity conservation and concluded that there was actually no direct evidence for its validity in weak interactions such as beta (electron) decay. They suggested that the law could be directly tested by observing the emission of electrons from nuclei aligned in a particular direction by a magnetic field. Ordinarily one would expect that electrons would be equally likely to come out in all directions, but their theoretical analysis showed that if parity is not conserved, the distribution would not be spherically symmetric.

Chien-Shiung Wu (Columbia University) performed the experiment with the cobalt-60 isotope, together with a group of physicists (E. Ambler, R. W. Hayward, D. D. Hoppes, and R. P. Hudson) at the National Bureau of Standards. Similar experiments were done about the same time by R. L. Garwin, L. M. Lederman, and M. Weinrich, and by J. L. Friedman and V. L. Telegdi. The result was an unsymmetrical distribution, showing that parity is not conserved. This opened the way to a new class of elementary particle theories, especially for the mysterious neutrino.

Yang and Lee received the Nobel Prize for proposing their hypothesis, but the experimentalists who confirmed it did not. (A possible reason is that it would have been unfair to honor only one of the several physicists mentioned above who participated in the experiments. As a matter of policy the Nobel Prize is not split among more than three people.) One consequence is that Wu, one of the most eminent living women physicists, remains unknown to the general public and to many girls who might be inspired by this "role model." Even worse is the fact that some authors who should know better, like C. P. Snow, refer to the "Lee and Yang experiment"! [*The Two Cultures and the Scientific Revolution* (1959), 16–17.]

One consequence of the discovery of parity violation was a new theory of the neutrino. It was proposed that there are two kinds of neutrinos, one a "left-handed" neutrino whose spin points opposite to its direction of motion, the other (now called the antineutrino) whose spin points in the same direction as its motion. This "two-component" neutrino theory was verified in experiments by Raymond Davis, who showed that antineutrinos do not change ${}_{17}Cl^{37}$ to ${}_{18}A^{37}$, a reaction that is theoretically possible for neutrinos but not for antineutrinos.

During the 1950s and 1960s the United States and other governments provided large sums of money to construct ever more powerful accelerators for research in particle physics. New observational apparatus, such as the "bubble chamber" invented by Donald Glaser in 1953, made it possible to obtain large numbers of tracks from the particles generated by these accelerators, and computerized techniques made it possible to analyze the tracks as fast as they were produced. The result was the discovery of many new particles, so many that it was difficult to consider them all "elementary." Thus there was an obvious need for new theories to explain why these particles should exist with the particular properties they have. It was assumed that a satisfactory explanation would be based on a much smaller number of truly elementary particles, with the others being described as "excited states" or antiparticles of the former, or as quanta carrying the forces between them.

In constructing such theories one could look back at three earlier theories as successful examples to follow: (1) the Heisenberg-Schrödinger quantum mechanics (1925–1926); (2) Dirac's theory of the electron (1928); (3) Yukawa's theory of the meson (1935).

(1) Quantum mechanics had shown how a spectrum of energy levels, which previously had to be postulated somewhat arbitrarily, could be derived from simple assumptions; it was assumed that the masses of families of particles such as the "leptons" (electron, neutrino, muon . . .) and the "baryon" (proton, neutron, lambda, sigma, . . .) could similarly be derived. Quantum mechanics also included a procedure, based on the mathematical theory of symmetry groups, for classifying and predicting the possible quantum states of a many-particle system even when the actual energy levels could not be derived exactly.

(2) Dirac's relativistic wave equation had predicted the existence of the positron as the antiparticle of the electron. The success of this prediction, and the successful manufacture of the predicted antiproton, led

theorists to assume that all charged particles and even some neutral particles have antiparticles. Moreover, Dirac's theory demonstrated that the incorporation of the principle of relativity could produce a fundamental modification, not merely a "correction term" important only at high speeds. Thus all fundamental theories should satisfy the requirements of relativity theory from the beginning.

(3) That the pion has the properties postulated by Yukawa as the carrier of the strong interaction, just as the photon is the carrier of electromagnetic interactions, suggested that there should be a carrier of weak interactions (the "intermediate boson" or weakon, W) and also for gravity (the hypothetical graviton).

Despite the success of quantum field theory in dealing with electromagnetic interactions—quantum electrodynamics as perfected by Schwinger, Feynman, and Tomonaga—it could not be assumed that field theory would be able to deal with strong nuclear forces. In the former case, calculations could be done by constructing a power series in powers of a parameter equal to about 1/137, whereas in the latter case the corresponding parameter would be greater than 1. The apparent failure of field theory led to a temporary retreat from the program of explaining the motions of particles in terms of a postulated force acting at short distances. Instead, theorists tried to develop a more empirical approach based on Heisenberg's S (for "scattering") matrix. This matrix simply characterizes the relation between what goes into an interaction and what comes out, without relying on detailed assumptions about how the particles interact. This approach is sometimes referred to as "dispersion relations." A later, more radical version developed by G. Chew eliminated the assumption that any particles are more fundamental than others and attempted to explain each in terms of the others by a "bootstrap" procedure.

BIBLIOGRAPHY

Abraham Pais (1986), *Inward Bound: Of Matter and Forces in the Physical World* (New York: Oxford Univ. Press).

Laurie M. Brown and Lillian Hoddeson, eds. (1983), *The Birth of Particle Physics* (New York: Cambridge Univ. Press), covers from 1930 to 1952.

J. T. Cushing (1983), "Models, High-Energy Theoretical Physics and Realism," *PSA 1982,* 2:31–56 and following papers by M. L. G. Redhead and P. Teclar.

S. G. Brush and L. Belloni (1983), *The History of Modern Physics: An International Bibliography* (New York: Garland), 204, 240–50.

(1982), *Colloque International sur l'Histoire de la Physique des Particules* (Paris: Editions de Physique), includes several personal recollections of research on cosmic rays, weak interactions, and the introduction of new ideas in the 1950s.

J. L. Heilbron and B. R. Wheaton (1981), *Literature on the History of Physics in the 20th Century* (Berkeley: Office for History of Science and Technology, University of California), 358–65, 377–78.

Topic *i:* QED, Field Theory

R. P. Feynman (1966), "The Development of the Space-Time View of Quantum Electrodynamics," *Physics Today* 19, no. 8:31–34.

Julian Schwinger (1982), "Quantum Electrodynamics: An Individual View," in *Colloque International* . . . (cited above), 409–24; Schwinger, ed. (1958), *Selected Papers on Quantum Electrodynamics* (New York: Dover), contains reprints of thirty-four papers by Dirac, Weisskopf, Lamb and Retherford, Schwinger, Tomonaga, Feynman, etc.

Silvan S. Schweber (1986), "Shelter Island, Pocono, and Oldstone: The Emergence of American Quantum Electrodynamics after World War II," *Osiris,* ser. 2, 2:265–302; (1986), "Feynman and the Visualization of Space-Time Processes," *Reviews of Modern Physics* 58:449–508.

Olivier Darrigol (1986), "The Origin of Quantized Matter Waves," *Historical Studies in the Physical and Biological Sciences* 16:197–253; (1984), "La Genése du Concept de Champ Quantique," *Annales de Physique,* ser. 15, 9:433–501.

Gloria B. Lubkin (1983), "Seeds of History Sown on Shelter Island," *Physics Today* 36, no. 9:67–69, 123, reports on a reunion of most of the participants in the 1947 conference on quantum electrodynamics.

Marcello Cini (1982), "Cultural Traditions and Environmental Factors in the Development of Quantum Electrodynamics (1925–1933), *Fundamenta Scientiae* 3:229–53; (1980), "The History and Ideology of Dispersion Relations: The Pattern of Internal and External Factors in a Paradigm Shift," *Fundamenta Scientiae* 1:157–72, explains how the popularity of this theoretical technique for interpreting elementary particle phenomena in the late 1950s is related to the role of the American physics community which was markedly different from that of the European physics community in 1925 when the relations were first proposed. The papers by M. L. Goldberger (1955), G. F. Chew (1958), and others are discussed.

David C. Cassidy (1981), "Cosmic Ray Showers, High Energy Physics, and Quantum Field Theories: Programmatic Interactions in the 1930s," *Historical Studies in the Physical Sciences* 12:1–39.

Yehudah Freundlich (1980), "Theory Evaluation and the Bootstrap

Hypothesis," *Studies in History and Philosophy of Science* 11:267–77, focuses on the rise and fall of the hypothesis proposed by G. Chew.

Joan Bromberg (1977), "Dirac's Quantum Electrodynamics and the Wave-particle Equivalence," in *History of Twentieth Century Physics,* ed. C. Weiner (New York: Academic Press), 147–57.

W. E. Lamb (1977), "Some History of the Hydrogen Fine Structure Experiment," in *A Festschrift for I. I. Rabi,* ed. Lloyd Motz (New York: New York Academy of Sciences), 82–86.

George Gale (1974), "Chew's Monadology," *Journal of the History of Ideas* 35:339–48.

Freeman J. Dyson (1965), "Old and New Fashions in Field Theory," *Physics Today* 18, no. 6:21–24.

Topic *ii:* Particles and Parity

Yuval Ne'eman and Yoram Kirsh (1986), *The Particle Hunters* (New York: Cambridge Univ. Press).

John Krige and Dominique Pestre (1986), "The Choice of CERN's First Large Bubble Chambers for the Proton Synchrotron (1957–1958)," *Historical Studies in the Physical Sciences* 16:255–79.

James T. Cushing (1986), "The Importance of Heisenberg's S-Matrix Program for the Theoretical High-Energy Physics of the 1950s," *Centaurus* 29:110–49.

K. Gavroglu (1985), "Popper's Tetradic Scheme, Progressive Research Programs and the Case of Parity Violation in Elementary Particle Physics," *Zeitschrift für allgemeine Wissenschaftstheorie* 16:261–86.

Joseph L. Spradley (1985), "Particle Physics in Prewar Japan," *American Scientist* 73:563–69; (1985), "Yukawa and the Birth of Meson Theory," *Physics Teacher* 23:283–89.

R. B. Lindsay (1983), *Energy in Atomic Physics 1925–1960* (Stroudsburg, Pa.: Hutchinson Ross), reprints papers on accelerators.

Paul Forman (1982), "The Fall of Parity," *Physics Teacher* 20, no. 5:281–88, is based on an exhibit at the Smithsonian.

I. V. Dorman (1982), "History of the Discovery of Elementary Particles in Cosmic Rays," *Acta Historiae Rerum Naturalium necnon Technicarum,* special issue 18:369–406.

Visapriya Mukherji and Sudhaneu Kumar Roy (1982), "Particle Physics since 1930: A History of Evolving Notions of Nature's Simplicity and Uniformity," *American Journal of Physics* 50:1100–03.

Helge Kragh (1981), "The Concept of the Monopole: A Historical and Analytical Case-Study," *Studies in History and Philosophy of Science* 12:141–72, is mostly on Dirac's 1931 theory; also discusses the "false 1975 discovery" by P. B. Price et al.

Andrew Pickering (1981), "Constraints on Controversy: The Case of the Magnetic Monopole," *Social Studies of Science* 11:63–93.

R. R. Wilson (1981), "US Particle Accelerators at Age 50," *Physics Today* 34, no. 11:86–103.

M. Jacob, ed. (1981), *CERN—25 Years of Physics* (New York: North-Holland), includes articles on experiments done at CERN and areas of particle physics in which CERN had a leading role.

Alan Franklin (1981), "Justification of a 'Crucial' Experiment: Parity Nonconservation," *American Journal of Physics* 49:109–12; (1981), "What Makes a 'Good' Experiment?" *British Journal for the Philosophy of Science* 32:367–79, has examples from twentieth-century particle physics.

Edward J. Barboni (1977), "Functional Differentiation and Technological Specialization in a Specialty in High Energy Physics: The Case of Weak Interactions of Elementary Particles," Ph.D. diss., Cornell University; see *Dissertation Abstracts International* 38(1978):4381-A.

Daniel Sullivan et al. (1977), "The State of a Science: Indicators of the Speciality of Weak Interactions," *Social Studies of Science* 7:167–200, is a statistical study of the literature, 1950–1972.

N. P. Chang, ed. (1977), "Five Decades of Weak Interactions," *Annals of the New York Academy of Sciences* vol. 294, 102 pp., is a symposium in honor of R. E. Marshak, with papers by Y. Ne'eman, C. S. Wu, H. A. Bethe, and others.

K. Gavroglu (1976), "Research Guiding Principles in Modern Physics: Case Studies in Elementary Particle Physics,"*Zeitschrift für Allgemeine Wissenschaftstheorie* 7:223–48.

Yuval Ne'eman (1974), "Concrete versus Abstract Theoretical Models," in *The Interaction between Science and Philosophy,* ed. Y. Elkana (Atlantic Highlands, N.J.: Humanities Press), 1–25, comments on the history of particle theory, including Marxist influences.

V. Mukherji (1974), "A History of the Meson Theory of Nuclear Forces from 1935 to 1952," *Archive for History of Exact Sciences* 13:27–102.

M. Stanley Livingston, ed. (1966), *The Development of High-Energy Accelerators* (New York: Dover), includes reprints of papers by D. Bohm and L. Foldy (1946–1947), M. S. Livingston et al. (1950), E. D. Courant et al. (1952), N. Christofilos (1956), etc.

11

PHILOSOPHICAL AND SOCIAL PERSPECTIVES

11.1 PHILOSOPHY OF SCIENCE BEFORE 1914

i. The Newtonian method and its interpretation in the eighteenth and early nineteenth centuries
ii. Kant and his influence on philosophy of science
iii. John Herschel, William Whewell, J. S. Mill, and others, on scientific method
iv. Auguste Comte and positivism
v. Mach, Pierre Duhem, Poincaré, and the critique of mechanism

READINGS

A. E. E. McKenzie, *Major Achievements of Science* (1973), 43–44, 122–31, 239–45, 338–41, 379, 399–405, 424–35, 492–98, 527–30. John Marks, *Science* (1983), 201–2 (on *iv*). John Losee, *A Historical Introduction to the Philosophy of Science* (1980), 80–170. Philipp Frank, *Modern Science and Its Philosophy* (1941), Chaps. 1–3.

SYNOPSIS

In my course I do not present a systematic account of the philosophy of science as such, but refer only to those philosophers and scientists whose views (or supposed views) on scientific method and epistemology have exerted some influence on the development of science itself. Thus Newton's claim that he did not make hypotheses—"hypotheses non fingo"—but rather proceeded by induction from observed facts, was sometimes used in the nineteenth century to denigrate proposed new theories, such as Thomas Young's wave theory of light (**7.2**) or John

Herapath's kinetic theory of gases. Recent historical scholarship, beginning with Alexandre Koyre's remark that *fingo* means "feign," not "frame," has shown that Newton did indeed make many hypotheses. His slogan was apparently directed against those, like Descartes, who used hypotheses that were implausible or even inconsistent with their own basic principles.

The revival of the wave theory of light and other challenges to Newtonian theories in nineteenth-century physics were associated with the rise of the "hypothetico-deductive method." This meant that scientists had greater freedom to invent imaginative hypotheses that could not be derived from phenomena by induction but could ultimately be tested by deducing their consequences. According to L. Laudan, ether theories of the late eighteenth and early nineteenth centuries played an important role in this development, culminating in the unexpected discovery of the "Poisson bright spot" predicted from the wave theory. Herschel and Whewell argued that a theory that successfully predicts previously unknown phenomena is entitled to support, whereas Mill complained that logically it should make no difference whether the experiment was done before or after the theoretical deduction. Indeed, the stress on prediction does seem a step toward recognizing the importance of psychological or sociological factors in the choice of scientific theories.

Nevertheless, one of the most publicized philosophies of science in the nineteenth and early twentieth centuries, "positivism," urged that hypotheses were of little epistemological value, regardless of their methodological effectiveness. Auguste Comte, who gave the movement its name in his *Cours de Philosophie Positive* (1830–1842), aspired to a stage of intellectual development in which, going beyond the theological and metaphysical stages,

> the mind has given over the vain search after absolute notions, the origin and destination of the universe, and the causes of phenomena, and applies itself to the study of their laws—that is, their invariable relations of succession and resemblance. [From H. Martineau's English translation of 1855, 26]

Comte was impressed by Fourier's theory of heat conduction, which accurately described the phenomena of heat without (apparently) relying on any specific assumption about the nature of heat itself. Thus the entire development of the kinetic-molecular theory of heat, which was one of the major achievements of nineteenth-century physical science (**7.2**), was an illegitimate excursion beyond Comte's posted limits of scientific knowledge. (Comte did not, however, reject the atomic structure of matter.)

A recurring theme in Comte's *Cours* is the impossibility of knowing anything about the universe beyond our own solar system. When he wrote the early volumes of this work, astronomers did not yet know the distances of any stars. He asserted that we cannot assume the law of gravity applies to the stars, ignoring the fact that analysis of double-star motions had already shown that they conform to Kepler's laws. He considered the physical nature of heavenly bodies even more inaccessible than the dynamics of their motion and stated that we will never be able to study the temperature or chemical composition of the stars. Likewise, the limits of cosmogony are set at the stage when the fluid solar nebula was rotating on its axis; Comte was willing to discuss what happened after that stage (and even fancied that he could provide a new confirmation of Laplace's nebular hypothesis) but admonished his readers not to speculate about earlier events and chastised Descartes for doing so.

Comte was equally positive that no answers could ever be obtained to other questions that intrigued nineteenth-century scientists. Physicists should stop speculating about "agents" of the phenomena of heat, light, electricity, and magnetism, i.e., ethers and imponderable fluids—not because they are nonexistent, but because their existence can neither be proved nor disproved. He doubted that it made any sense to talk about heat apart from a luminous body; in any case it would never be possible to reduce the phenomena of light to any other. Moreover, no one should even try to explain the colors of bodies in terms of "supposed derangements of the molecules." Finally, Comte excluded theoretical chemistry from science:

> Every attempt to refer chemical questions to mathematical doctrines must be considered, now and always, profoundly irrational, as being contrary to the nature of the phenomena. [Martineau's translation, 256]

James Clerk Maxwell, commenting on the work of the American scientist J. Willard Gibbs, which successfully transgressed this limit, surmised:

> The warning which Comte addressed to his disciples, not to apply dynamical or physical ideas to chemical phenomena, may be taken, like several other warnings of his, as an indication of the direction in which science was threatening to advance. [1876; see *Philosophical Magazine* 16(1908):818–24]

Indeed, the history of modern physical science shows that all of Comte's negative claims have been falsified—not in the sense that we have established the *true* answers to his forbidden questions, but rather that we have accepted the questions as legitimate ones for science to consider and have found reasonably satisfactory working hypotheses about the nature of gravity, heat, light, the molecular reasons for the color of bodies, and the mathematical laws of chemical structure and reactions.

With such a disastrous record of failure, one might wonder why Comte's positivism deserves any serious consideration by historians and philosophers of science. For the historian, negative influences on scientific progress call for investigation; in fact, a recurrent question is: what caused the decline of French theoretical physics in the nineteenth century? Or, if we wish to avoid the pitfalls of the "Whig interpretation of the history of science," we should ask why French scientists rejected certain research problems and methods without assuming that they were foolish to do so merely because those problems and methods now seem to be in the mainstream of modern physical science. Can we blame the positivists for failing to anticipate the discoveries of Maxwell and Einstein?

For the philosopher of science it may be interesting to note that Comte, like Herschel and Whewell, stressed the importance of successful prediction as the goal of science:

> All science has prevision for its end: an axiom which separates science from erudition, which relates the events of the past, without any regard to the future. [Martineau's translation, 135]

But it is this very distinction (or, as Karl Popper puts it, line of demarcation) between science that can predict the future and "erudition" or some other intellectual activity that can only deal with the past, that has been used to stigmatize evolutionary biology and Freudian psychoanalysis as nonscientific (**4.7, 5.5**). Elsewhere, however, as Laudan (1971, 1981) points out, Comte admits that predictions don't have to be restricted to future events but may include events that have already happened but are previously unknown to us, such as eclipses in antiquity.

The term "positivism" has come to acquire another meaning, often found in popular writings on science: the claim that there is no valid knowledge other than scientific knowledge and that the methods of the natural sciences should be extended to all other domains of inquiry. The proposition that one can explain, predict, and control such phenomena as the evolution of species and human behavior, using a "mechanistic" approach, is repulsive to many people. When they proclaim limits of scientific knowledge, they may characterize their standpoint as a *rejection* of positivism. This is the "computers can never think" dogma; rather than being based on confidence that the limits will never be violated, it seems to be inspired by a fear that they will be. (Notice how the definition of "think" keeps changing in order to ensure that computers can't think!)

Much writing on what we now call the philosophy of science was produced toward the end of the nineteenth century in connection with the perceived breakdown of the mechanical worldview and debates about the reality of atoms. Some of this writing was by scientists like Ernst Mach who claimed that they were not doing philosophy at all but simply trying to clarify their ideas and eliminate metaphysics from science. They also took a considerable interest in the history of science (for example, Pierre Duhem uncovered many important medieval scientific works), though there is always the suspicion that their interpretations of history are part of their advocacy of views about how modern science should be conducted.

Mach became known for his claim that the purpose of science is not to find out what the world is really like, but only to provide the most efficient way of organizing our knowledge of phenomena—"economy of thought." He criticized those who attributed reality to atoms, though he did not deny that the atomic hypothesis may sometimes be useful. It is sometimes said that Mach finally accepted atoms when he was shown the scintillations produced on a screen by alpha particles, but this "conversion experience" lacks reliable documentation and is contradicted by other evidence. Mach's position was that reality should be attributed only to what can be directly perceived by the senses; and since no one had actually seen an atom, it must be considered only hypothetical.

Mach was also skeptical about the reality of absolute space and time, and his critique of Newton on this point seems to have encouraged the young Einstein to think about possible alternatives. For several years the logical positivists, who saw themselves as Mach's disciples, claimed that Einstein's theory of relativity vindicated their views on science. Einstein's friend and biographer Philipp Frank, one of those responsible for promoting this view of Einstein, was at least honest enough to record his own surprise when he learned in 1929 that Einstein was opposed to positivism. Recent historical work, especially by Gerald Holton, has shown that Einstein turned away from Mach's empirical philosophy toward a realist view of nature. In any case, Einstein's attack on the Copenhagen interpretation of quantum mechanics put him squarely in the antipositivist camp (**11.2**).

There is some reason to believe that positivism, insofar as scientists actually took it seriously, was injurious to the progress of science. John Herivel has attributed the decline in French theoretical physics partly to the dominance of the positivist tradition, articulated by Comte and exemplified by Fourier. Marjorie Malley suggests that the positivist tradition still dominating French science in 1900 prevented the Curies from recognizing that radioactivity involved transmutation of one element to another. John Blackmore (1983) has listed several Austrian scientists who were led astray by Mach's views. Poincaré, guided or limited by his "conventionalist" philosophy of science, failed to discover the theory of relativity although he had worked out much of the mathematical basis for it.

One further remark should be made here: Pierre Duhem, whose philosophical views led him to undervalue the discoveries of Galileo and Maxwell, did advance the philosophy of science by his criticism of the concept of "crucial experiment." When Leon Foucault measured the velocity of light in water and showed that it is greater than the velocity in air, this was widely considered as proof of the wave theory of light, since the Newtonian particle theory predicted the opposite result. But Duhem pointed out that any experiment can test only a combination of hypotheses and assumptions, and if one obtains a negative result, one does not know which of them is false. Thus a modified particle theory might account for the observed facts. Duhem was clever enough to publish this in 1905, just before Einstein revived the particle theory of light. The philosophical point is now known as the "Duhem-Quine thesis."

BIBLIOGRAPHY

John Neu, ed. (1980, 1985), *Isis Cumulative Bibliography 1966–1975,* vol. 1, *Personalities and Institutions;* vol. 2, *Subjects, Periods and Civilizations* (London: Mansell), has numerous entries in vol. 1 for Kant, Herschel, J. S. Mill, Whewell, Comte, Mach, Duhem, Poincaré, and in vol. 2 for "Philosophy and Methodology of Science" (sections Ak and 9 Ak).

Larry Laudan (1981), *Science and Hypothesis: Historical Essays on Scientific Methodology* (Boston: Reidel), has chapters on the Newtonians, the "epistemology of light," Comte, Whewell, Mach, etc.; (1969), "Theories of Scientific Method from Plato to Mach: A Bibliographic Review," *History of Science* 7:1–63.

Ralph M. Blake, Curt J. Ducasse, and Edward H. Madden (1960), *Theories of Scientific Method: The Renaissance through the Nineteenth Century* (Seattle: Univ. of Washington Press).

Topic *i:* Newtonian Method

Ronald Laymon (1983), "Newton's Demonstration of Universal Gravitation and Philosophical Theories of Confirmation," in *Testing*

Scientific Theories, ed. John Earman (Minneapolis: Univ. of Minnesota Press), 179–99.

Ryan D. Tweney (1980), "Isaac Newton's Two Uses of Hypothetical Reasoning: Dual Influences on the History of Psychology," *Storia e Critica della Psicologia* 1:235–49.

Peter Beer, ed. (1978), "Newton and the Enlightenment: Proceedings of an International Symposium held at Cagliari, Italy, on 3–5 October 1977," *Vistas in Astronomy* 22:367–557.

Imre Lakatos (1978), "Newton's Effect on Scientific Standards," in his *Philosophical Papers,* vol. 1, ed. J. Worrall and G. Currie (New York: Cambridge Univ. Press), 193–222.

Peter Wallis and Ruth Wallis (1977), *Newton and Newtoniana 1672–1975: A Bibliography* (Folkestone, Kent, Eng.: Dawson).

R. E. Butts and J. W. Davis, eds. (1970), *The Methodological Heritage of Newton* (Toronto: Univ. of Toronto Press).

Topic *ii:* Kant

Immanuel Kant (1985), *Philosophy of Material Nature,* ed. and trans. by J. W. Ellington (Indianapolis: Hackett), contains translations of Kant's *Prolegomena to Any Future Metaphysics That Will Be Able to Come Forward as Science* and *Metaphysical Foundations of Natural Science.*

Allen W. Wood (1984), *Self and Nature in Kant's Philosophy* (Ithaca, N.Y.: Cornell Univ. Press), includes Philip Kitcher, "Kant's Philosophy of Science."

Robert E. Butts (1984), *Kant and the Double Government Methodology: Supersensibility and Method in Kant's Philosophy of Science* (Dordrecht: Reidel).

Clark Zumbach (1984), *The Transcendent Science: Kant's Conception of Biological Methodology* (The Hague: Nijhoff).

Timothy Lenoir (1982), *The Strategy of Life* (Boston: Reidel); (1981), "The Goettingen School and the Development of Transcendental Naturphilosophie in the Romantic Era," *Studies in History of Biology* 5:111–205; (1980), "Kant, Blumenbach, and Vital Materialism in German Biology," *Isis* 17:77–108.

William R. Shea, ed. (1983), *Nature Mathematized* (Dordrecht: Reidel), includes articles on Kant by P. M. Harman and Kathleen Okruhlik.

Kwang-Sae Lee (1981), "Kant on Empirical Concepts, Empirical Laws, and Scientific Theories," *Kant-Studien* 72:398–414.

Gordon G. Brittan, Jr. (1978), *Kant's Theory of Science* (Princeton, N.J.: Princeton Univ. Press).

David H. Galaty (1974), "The Philosophical Basis of Mid-19th Century German Reductionism," *Journal of History of Medicine* 29:295–316.

Topic *iii:* Herschel, Whewell, Mill

William Whewell (1984), *Selected Writings on the History of Science,* ed. and with an introduction by Yehuda Elkana (Chicago: Univ. of Chicago Press); (1967), *The Philosophy of the Inductive Sciences Founded upon Their History* (1840, 1847), reprinted with a new introduction by John Herivel (New York: Johnson Reprint).

John Herschel (1966), *A Preliminary Discourse on the Study of Natural Philosophy* (1830), reprinted with a new introduction by Michael Partridge (New York: Johnson Reprint).

John Losee (1983), "Whewell and Mill on the Relation between Philosophy of Science and History of Science," *Studies in History and Philosophy of Science* 14:113–26.

Paolo Casini (1981), "Herschel, Whewell, Stuart Mill e 'l'Analogia della Nature,' " *Rivista di Filosofia* 72:372–91.

Richard Yeo (1979), "William Whewell, Natural Theology and the Philosophy of Science in Mid-19th-Century Britain,"*Annals of Science* 36:493–516.

Susan Faye Cannon (1978), *Science in Culture: The Early Victorian Period* (New York: Science History); (1976),"The Whewell-Darwin Controversy," *Journal of the Geological Society of London* 132:377–84.

Michael Ruse (1975), "Darwin's Debt to Philosophy: An Examination of the Influence of the Philosophical Ideas of John F. W. Herschel and William Whewell on the Development of Charles Darwin's Theory of Evolution," *Studies in History and Philosophy of Science* 6:159–81. See the comment by Paul Thagard (1977), "Darwin and Whewell," ibid. 8:353–56 and Ruse's reply (1978), ibid. 9:323–31.

Topic *iv:* Comte

Stephen G. Brush (1983), "Negativism Sesquicentennial," in *The Limits of Lawfulness,* ed. N. Rescher (Lanham, Md.: Univ. Press of America), 3–22.

Larry Laudan (1981), *Science and Hypothesis* (Boston: Reidel), Chap. 9 is a revised version of his 1971 article on Comte.

Marjorie Malley (1979), "The Discovery of Atomic Transmutation: Scientific Styles and Philosophies in France and Britain," *Isis* 70:213–23.

Barbara Haines (1978), "The Inter-Relations between Social, Biological, and Medical Thought, 1750–1850: Saint-Simon and Comte," *British Journal for the History of Science* 11:19–35.

John Herivel (1966), "Aspects of French Theoretical Physics in the Nineteenth Century," *British Journal for the History of Science* 3:109–32.

Topic *v*: Mach, Duhem, Poincaré

Pierre Duhem (1962), *The Aim and Structure of Physical Theory* (New York: Atheneum), trans of French ed. of 1914 (1st ed., 1905).

Ernst Mach (1976), *Knowledge and Error: Sketches on the Psychology of Inquiry,* trans. by Paul Foulkes and T. J. McCormack (Boston: Reidel); (1943), *Popular Scientific Lectures* (La Salle, Ill.: Open Court), includes English translation of his 1872 article on economy of thought.

John T. Blackmore and K. Hentschel, eds. (1985), *Ernst Mach als Aussenseiter: Machs Briefwechsel über Philosophie und Relativitätstheorie mit Persönlichkeiten seiner Zeit* (Vienna: Braumüller).

Henri Poincaré (1963), *Mathematics and Science: Last Essays* (New York: Dover); (1958), *The Value of Science* (New York: Dover); (1952), *Science and Hypothesis* (New York: Dover); (1952), *Science and Method* (New York: Dover). These are all translations of collections of essays first published in French around 1900–1910.

Klaus Hentschel (1985), "On Feyerabend's Version of 'Mach's Theory of Research and Its Relation to Einstein,'" *Studies in History and Philosophy of Science* 16:387–94, comments on the papers of Zahar and Feyerabend cited below.

Katherine Arens (1985), "Mach's 'Psychology of Investigation,'" *Journal of the History of the Behavioral Sciences* 21:151–68.

Zeljko Loparic (1984), "Problem-Solving and Theory Structure in Mach," *Studies in History and Philosophy of Science* 15:23–49.

Gereon Wolters (1984), "Ernst Mach and the Theory of Relativity," *Philosophia Naturalis* 21:630–41.

Paul K. Feyerabend (1984), "Mach's Theory of Research and Its Relation to Einstein," *Studies in History and Philosophy of Science* 15:1–22.

Stanley L. Jaki (1984), *Uneasy Genius: The Life and Work of Pierre Duhem* (The Hague: Nijhoff).

John Blackmore (1983), review of F. Stadler (1982), *Vom Positivismus . . .* (Vienna: Löcker), *Isis* 74:621–22; (1972), *Ernst Mach* (Berkeley: Univ. of California Press).

J. Giedymin (1982), *Science and Convention: Essays on Henri Poincaré's Philosophy of Science and the Conventionalist Tradition* (New York: Pergamon Press).

Ryoichi Itagaki (1982), "Why Did Mach Reject Einstein's Theory of Relativity?" *Historia Scientiarum* 22:81–95.

Wolfram W. Swoboda (1982), "Physics, Physiology, and Psychophysics: The Origins of Ernst Mach's Empiriocriticism," *Rivista di Filosofia* 73:234–74.

Miodrag Cekic (1981), "Mach's Phenomenalism and Its Consequences in Physics," *International Philosophical Quarterly* 21:249–59.

John Nyhof (1981), "Instrumentalism and Beyond" (Ph.D. diss., University of Otago, Dunedin, N.Z.), examines the historical relations between kinetic theory and positivism.

E. Zahar (1977), "Mach, Einstein and the Rise of Modern Science," *British Journal for the Philosophy of Science* 28:195–213; Paul Feyerabend (1980), "Zahar on Mach, Einstein, and Modern Science," ibid. 31:273–82; Zahar (1981), "Second Thoughts about Machian Positivism: A Reply to Feyerabend," ibid. 32:267–76. See also references on the atomic debates, **7.4.**

11.2 PHILOSOPHY OF QUANTUM MECHANICS

i. Decline of mechanism and determinism in physical science before 1925
ii. Heisenberg's indeterminacy principle and Max Born's statistical interpretation of the wave function
iii. Niels Bohr: the complementarity principle and the Copenhagen interpretation of quantum mechanics
iv. The Bohr-Einstein debate; Einstein's objections to the Copenhagen interpretation; the Einstein-Podolsky-Rosen paradox; Schrödinger's cat paradox
v. Hidden variables theories proposed by David Bohm and others
iv. Bell's theorem and the recent experimental tests of hidden variables vs. quantum mechanics; the Aspect experiment
vii. The anthropic principle and Wheeler's hypothesis; breakdown of the barrier between subject and object
viii. Relation of quantum physics to Eastern mysticism and parapsychology

READINGS

G. Holton and S. Brush, *Introduction to Concepts and Theories in Physical Science* (1985, 2d ed.), 495–500. McKenzie, *Major Achievements of Science,* 323–26, 518–21. Marks, *Science,* 272–74 (on *iv*). Russell Stannard and Noel G. Coley, "Introduction to Quantum Theory," Unit 7 in the Open University course A381, *Science and Belief from Darwin to Einstein* (1981), 99–104 (on Heisenberg principle); Coley and

Stannard, "Quantum Theory: The Bohr-Einstein Debate," Unit 8, 107–35; optional readings, extracts from Einstein, Heisenberg, etc., in the Coley and Hall anthology, *Darwin to Einstein.*

Albert Einstein and Leopold Infeld, *Evolution of Physics* (1938), 280–97. V. Guillemin, *Story of Quantum Mechanics* (1968), 91–102, 218–63. Heinz Pagels, *The Cosmic Code,* Part I, Chaps. 4–13. Barbara L. Cline, *Men Who Made a New Physics* (1969), 160–79, 193–212.

Articles: Abner Shimony, "The Reality of the Quantum World," *Scientific American* (Jan. 1988):46–53; "Metaphysical Problems in the Foundations of Quantum Mechanics," *International Philosophical Quarterly* 18(1978):3–17. Trevor J. Pinch, "The Hidden Variables Controversy in Quantum Physics," *Physics Education* 14(1979):48–52. Arthur L. Robinson, "Loophole Closed in Quantum Mechanics Test," *Science* 219(1983):40–41 [on the experiment by Aspect et al.]. Halley D. Sanchez, "Shimony, the Dilemma of Quantum Mechanics, and the History of Philosophy," *Dialogos* 45(1985):79–92. Jeremy Bernstein, "Quantum Reality," *American Scholar* 54(1984–1985):7–14.

Popular articles that explain the current state of affairs are: N. David Mermin, "Bringing Home the Atomic World: Quantum Mysteries for Anybody," *American Journal of Physics* 49(1981):940–43 (see also *Journal of Philosophy,* 1981); "Is the Moon There When Nobody Looks? Reality and the Quantum Theory," *Physics Today* 38, no. 4(1985):38–47. Martin Gardner, "Quantum Weirdness," *Discover* 3, no. 10(1982):69–76. Douglas R. Hofstadter, "Metamagical Themes. Pitfalls of the Uncertainty Principle and Paradoxes of Quantum Mechanics," *Scientific American* 245, no. 1(1981):18–29. Ann Finkbeiner, "A Universe in Our Own Image," *Sky and Telescope* 68(1984):106–10 (on *vii*). D. E. Thomson, "Going Bohr's Way in Physics," *Science News* 129(11 Jan. 1986):26–27.

For a course in the history of modern physics, in which a more extended treatment of this topic is appropriate, I would recommend the anthology by Stephen Toulmin, *Physical Reality* (1970), described at the beginning of the Bibliography for this section. Richard Schlegel has described a course presenting the ideas of relativity and quantum mechanics to nonscience students, featuring the philosophical issues in the Bohr-Einstein debate, in *American Journal of Physics* 44(1976):236–39.

SYNOPSIS

The surprising discoveries and strange theories that entered physics after 1895 (**8.6**, **8.7**, **9.4**, **9.5**) broke down the Newtonian "mechanistic" view of the world but did not clearly point to any comprehensible alternative. As noted in **7.4**, nineteenth-century attempts to reconcile the principle of irreversibility with Newtonian mechanics led some physicists to abandon determinism and suggest that atoms behave randomly. Early in the twentieth century, physicists were puzzled by two phenomena that could best be described as random processes: radioactivity and Brownian movement. In the first, there seemed to be no reason for an atom to explode at a particular time, yet the properties of a radioactive substance could be accurately described by assuming a certain *probability* of radioactive decay for each atom in a small time interval. In Brownian movement, one could directly observe the (apparently) random motion of a microscopic particle; the theory of Einstein and Marian von Smoluchowski described this motion quantitatively in terms of statistical fluctuations at the atomic level. Thus it seems to me that indeterminism did not, as often assumed, emerge suddenly in the 1920s as a result of quantum mechanics, but that physicists had been prepared to accept a probabilistic interpretation of quantum mechanics by several decades of discussion of the behavior of atoms, partly stimulated by the problem of irreversibility in thermodynamics.

Another aspect of the interpretation of quantum mechanics that emerged after 1926—the so-called "Copenhagen interpretation"—is a rejection of philosophical realism in favor of the view that one should not ascribe reality to entities that cannot be directly observed; in some cases reality is even denied to entities that one does not happen to observe on a particular occasion even though they could have been observed. In other words, reality is ascribed only to one's own perception, not to the external world. This view also has historical roots, both in the positivism of nineteenth-century writers such as Comte and Mach, and in earlier "idealist" philosophies. Whether the recurrence of philosophical idealism, or the more pragmatic version known as "instrumentalism," in the twentieth century has any direct connection with (or influence on) the views of physicists is of course open to dispute; and whether quantum mechanics legitimizes a mystical or supernatural view of the world is even more controversial.

I will begin by summarizing the views of the most important physicists with respect to the interpretation of quantum mechanics, starting with those who might be called "realists."

Einstein was influenced by the positivist philosophy of Ernst Mach in his earlier work (**8.5**) but later adopted a realist view. This shift was concurrent with his increasing uneasiness about randomness as a fundamental property of atoms (in spite of the fact that his own early work on quantum theory gave a strong impetus to this idea), leading to his famous declaration that God does not play dice. It was, however, surprising to those who considered him a positivist on the basis of his 1905 relativity paper.

Max Planck, though a skeptic about atomistic hypotheses before 1900, attacked Ernst Mach from a realist standpoint in 1909 and later became a defender of determinism.

Erwin Schrödinger, carrying on the tradition of Boltzmann, employed statistical methods in his early papers and appeared to be a supporter of indeterminism but sought a realist interpretation of his wave function in opposition to Born and Heisenberg; later he supported Einstein by proposing the "cat paradox" against Bohr's interpretation of quantum mechanics.

Louis de Broglie initially preferred a realist interpretation of wave mechanics but did not actively defend it until after 1950, when he supported a hidden variable theory.

This roster of realists among the founders of quantum theory looks at first quite impressive, yet their philosophical views on their own theories seem to have carried remarkably little weight.

On the other side, Niels Bohr was the leading spokesman for the new movement in physics, and thus it acquired the name "Copenhagen interpretation" (CI). His main contribution, the "complementarity principle," was announced in 1927: phenomena may have two apparently incompatible properties, such as wave and particle aspects. Neither by itself exhausts the nature of the phenomena. The experimental situation determines which aspect will be displayed; entities such as electrons cannot be said to have definite properties apart from our observation of them.

Werner Heisenberg provided the most potent ammunition for the claim that theories must deal with relations between observable quantities, since his matrix mechanics—the *first* successful version of quantum mechanics (though its scope was quite limited) was constructed on just that principle. Subsequently he proposed that physical variables such as position and momentum cannot be determined precisely, but there is a reciprocal relation such that the more accurately one quantity is determined, the less accurately the other one can be determined. The product of the average errors of the two quantities is greater than or equal to a number proportional to Planck's constant.

Heisenberg's indeterminacy principle is sometimes said to predict the "uncertainty" in our knowledge of a property resulting from the uncontrollable disturbance due to simultaneous measurement of another property—as if an atomic property really has a definite value before we try to measure it. But that was not what Heisenberg meant; as he remarked, "it is possible to ask whether there is still concealed behind the statistical universe of perception a 'true' universe in which the law of causality would be valid. But such speculation seems to us to be without value and meaningless, for physics must confine itself to the description of the relationship between perceptions" (1927, quoted from M. Jammer, *Philosophy of Quantum Mechanics,* 1974, 76).

Heisenberg consistently rejected what he called "materialism"—the belief in "an objectively real world whose smallest parts exist objectively in the same sense as stones or trees exist, independently of whether or not we observe them" (*Physics and Philosophy* 1958, 129). But in his later years he returned to the Platonic worldview that had impressed him before 1920; the ultimate reality is neither our sensations nor substantial atoms, but mathematical forms. At the same time Heisenberg explicitly identified the wave function with Aristotle's *potentia*—that is, an entity that indicates the possibility that an event may occur—it has some degree of reality but not as much as the event itself. In making a measurement, we convert the potential to the actual. (This is called by other physicists "reduction of the wave packet.")

Max Born received the Nobel Prize in 1954 for his statistical interpretation of wave mechanics, published in 1926. The term "statistical" is ambiguous, for Born's view is not the same as the recent "statistical interpretation of quantum mechanics" which assumes, following Einstein, that the wave function describes only the statistical properties of a large collection of systems. Born's interpretation is that the wave function (or rather its absolute square) describes the probability distribution of a single particle or system of several interacting particles. The statistical interpretation of the wave function does not mean that the particle *has* a well-defined position and momentum, and the wave function only represents our knowledge of it. Rather than talk about the probability distribution of a particular property of the particle, one should really talk about the probability that a particular measurement will give a particular result; the property inferred from that result is not necessarily consistent with the property that would be inferred from the result of a different measurement on the same system.

In 1932 John von Neumann provided what seemed to be a solid bulwark protecting the CI against the claim that determinism could be recovered by going to a deeper level of "hidden variables." He argued that it is impossible to find a theory based on such variables that would agree with the observable consequences of quantum mechanics. Thus, as long as predictions from quantum mechanics agree with experiment, the theory must remain fundamentally indeterministic.

Von Neumann's proof of his antihidden variable theorem was occasionally criticized but generally accepted by physicists through the 1950s.

Finally, in 1964, J. S. Bell discovered that the proof of the theorem depended on the assumption that expectation values are additive—i.e., that the average value of a sum of observed quantities is equal to the sum of the averages of the individual quantities. This is true in quantum mechanics but need not be true in some other theory. But Bell's analysis, while encouraging the development of hidden variable theories, also revealed a new way in which they could be submitted to direct experimental test, as will be noted below.

The debate between Einstein and Bohr, pursued intermittently over a period of more than ten years, is comparable in historic significance to the Newton-Leibniz debate (the Clark-Leibniz correspondence) at the beginning of the eighteenth century. Unlike Newton and Leibniz, who were bitter enemies and communicated only through third parties, Bohr and Einstein remained personal friends throughout this period. There is one striking similarity in the two debates: in each, the man who made the more substantial contributions to physics (Newton, Einstein) was attacking a philosophical position that seemed to be supported by his own work and which in fact triumphed (at least temporarily). Newton was attacking, and Leibniz defending, the "clockwork universe" concept (**1.4**-*iii*) that subsequently became known as the "Newtonian world machine." Einstein was attacking, and Bohr defending, the quantum indeterminism and instrumentalism that others based on Einstein's theories of photons and relativity.

After considerable discussion of thought experiments intended to refute the CI, Einstein was forced to admit that the indeterminacy principle could not be violated by any conceivable experiment, so the quantum limitations on the *measurement* of physical properties could not be circumvented. But he still rejected Bohr's claim that quantum mechanics tells us all we can expect to know about nature. Bohr succeeded in persuading physicists that the CI is logically self-consistent and accounts for the experimental facts, but only at the cost of abandoning the idea that a particle has definite properties independent of our observation. For Einstein this was clearly unsatisfactory and so he concentrated on showing that quantum mechanics is an "incomplete" theory that fails to account for every element of reality.

Einstein's objection to the CI was cogently expressed in a paper by Einstein, Boris Podolsky, and Nathan Rosen (1935), now known as the "EPR paradox." It has had a major influence on subsequent discussions of the foundations of quantum mechanics and inspired some of the recent experiments. EPR postulated that "if, without in any way disturbing a system, we can predict with certainty the value of a physical quantity, then there exists an element of physical reality corresponding to this physical quantity." They considered two systems (which might be simply two electrons) that interact for a short time and then remain completely separated so that neither one can possibly influence the other. (It is *assumed* that such a complete separation is possible.) According to quantum mechanics (QM) the wave function for the combined system at any later time can be expressed in terms of the various possible wave functions for the separate systems in such a way that a particular wave function for system I is associated with a particular wave function for system II. In the simplest case, if the experiment is designed so that the total spin or angular momentum is zero, and each system has spin ½, then a wave function which assigns spin +½ along the x axis to I will be associated with a function that assigns spin −½ to II.

EPR then pointed out that a measurement of a property *a* for system I singles out a particular wave function of that system corresponding to a definite numerical value of that property (e.g., the momentum) and this will be associated with a particular wave function for II. As a result, one can determine a property of II without influencing it in any way merely by making a measurement on I. But suppose we had chosen to measure a different property *b* of I, e.g., its position. Then we would have been able to determine that property for II. Since we *could* have determined definite values of either *a* of *b* for II without perturbing that system, then II must actually *have* definite values (using the above-quoted definition of reality), yet Heisenberg's principle forbids QM from specifying both values simultaneously. Hence QM does not give a complete description of reality.

In his reply Bohr rejected the EPR definition of reality, arguing that one cannot talk about the reality of II apart from the experiment on I; even though there is no direct interaction by which the measurement on I might influence the state of II, they are bound together by the very fact that one it trying to get information about II by observing I. Thus the EPR argument merely illustrates a characteristic feature of QM, and in particular shows how it involves a new conception of physical reality—a conception which, Bohr claimed, had already been introduced by Einstein's general theory of relativity.

Thus Bohr and Einstein agreed that their disagreement was essentially philosophical. Einstein had maintained the realist position but had stripped it of all reference to unobservable quantities, in deference to the acknowledged success of QM; he insisted only that if a quantity can be measured with certainty, then it must be real. Bohr chose to defend an extreme instrumentalist position: A quantity is not real just because it *can* be measured, it is also necessary that it *is* measured. Or rather, reality cannot be attributed to the quantity itself in any case, but only to the measurement of the quantity.

Schrödinger now entered the debate on the side of Einstein, arguing that the EPR paradox indicates a serious deficiency of QM. He introduced an example that captures the spirit of Einstein's own critique better than the EPR paper: a cat is placed in a chamber with a radioactive sample and a Geiger counter connected to an electrical device that will kill the cat with probability ½ in a certain time interval. At the end of the experiment the cat is represented by a wave function that is a combination of the function corresponding to a living cat and one for a dead cat. According to the CI the cat is neither alive nor dead but in some intermediate state—until we open the chamber and look at it. At the instant when we *observe* the cat, its wave function collapses to either one or the other of its two components, and the cat becomes either alive or dead.

It is important to realize that the position defended by Einstein and Schrödinger is not so much determinism as common sense realism: the cat must be really alive or dead before someone looks at it, and QM is incomplete if it cannot account for this fact. Most of us still find it hard to believe that the world has no real existence apart from ourselves, and even those who claim to accept the CI often refuse to admit that this is what it implies.

Among the numerous attempts to develop a realist interpretation of QM, three alternatives to the CI attracted some attention among physicists in the 1950s and 1960s:

1. The "statistical interpretation," suggested (though not worked out in any detail) by Einstein, claims that QM applies only to "ensembles" (collections) of many systems; like classical statistical mechanics it is compatible with deterministic behavior in a single system. While claiming the incompleteness of QM, proponents of the statistical interpretation did not usually attempt to go beyond it by advancing a deterministic theory of individual particles. In the USSR it provided an ideologically acceptable alternative to the CI for a brief period in the 1950s; as one might expect, the subjectivist flavor of the latter aroused the suspicions of dogmatic Marxists. Yet there has been remarkably little difference between the dominant interpretation of QM in the USSR and the West during the last several decades.

2. Theories of "hidden variables" try to go beyond QM, but encountered difficulty in remaining consistent with the empirically validated predictions of QM while adding any new physical content. Even after it

was recognized that von Neumann's theorem does not exclude all such theories, the assumptions needed to achieve consistency with QM seemed distasteful. According to Bell's results, they must be "non-local"—i.e., "there must be a mechanism whereby the setting of one measuring device can influence the reading of another instrument, however remote. Moreover, the signal involved must propagate instantaneously," which would violate relativity theory (Bell 1964, 199). One might say that this is just what happens in the EPR paradox if one assumes that the measurement on system I determines the state of II, but QM does not seem to involve any "mechanism" or signal for the transmission of this determination and thus does not obviously contradict relativity.

As a result of Bell's work, attention turned to hidden variable theories that made predictions different from those of QM, but for situations that had not yet been experimentally investigated.

3. The "many-worlds" interpretation proposed by Hugh Everett in 1957 rejects the usual assumption that a system that is in a mixture of all possible states before an observation "collapses" into just one of them ("reduction of the wave packet"). Instead, all these states continue to evolve in separate universes, corresponding to different possible outcomes of the observation. Since the state of the "observer" is considered part of the state of the universe, each observer interacts only with his or her own universe and unaware of the others. Thus the cat paradox is resolved by asserting that the cat remains alive in half of the universes and dies in the other half.

During the past fifteen years there have been a number of experiments designed to test a special kind of correlation, analogous to the EPR effect, which is predicted by QM but violates an inequality that must be satisfied by all *local* hidden-variables theories. ("Local" here means that a measurement of one system cannot instantaneously affect another system widely separated from it.) The first inequality of this kind was derived by J. S. Bell in 1964; others, more susceptible to practical experiments, were obtained subsequently by J. F. Clauser, M. A. Horne, A. Shimony, and R. A. Holt. The experiments involve an "atomic cascade" in which an excited state of an atom (e.g., mercury) decays by emission of two photons. QM predicts, for example, that the correlation of polarization directions of photons observed at certain angles will have a definite numerical value such as $(\frac{1}{4})^{1/2}$ in a situation where all local hidden variables theories require a value less than or equal to ¼ for the same quantity.

It should be noted that it is not the *deterministic* character of hidden variables theories that is incompatible with QM; any local nondeterministic theory is also incompatible. As pointed out by Clauser and Shimony, "it is the objectivity of the associated systems and their locality which produces the incompatibility. Thus, the whole realistic philosophy is in question."

Most experiments gave results clearly in favor of QM; the most decisive is that conducted recently by Alain Aspect. Using the description of the EPR paradox above, we can take property *a* to be the spin of a photon, which is quantized to be +1 or −1 along the axis we choose to measure it. In the Aspect experiment it is impossible for any signal, even one travelling at the speed of light, to inform photon II that we have determined the spin of photon I to be +1 in a particular direction; yet the measurement of the spin of photon II reveals just the correlation with the spin of I that is required by QM. In other words, the spins cannot be said to have real values before we did the experiment. The consequence is: "either one must totally abandon the realistic philosophy of most working scientists, or dramatically revise our concept of time and space" (ibid. 1881). Physical systems cannot be said to have definite properties independent of our observations; "perhaps an unheard tree falling in the forest makes no sound after all."

Few physicists have seriously attempted to work out the implications of QM for our view of the world. One of these is John Wheeler, who summarizes his position with the assertion: "no elementary phenomenon—whether now or in the earliest days of the universe—is a phenomenon until it is an observed phenomenon," where "observed" is to be interpreted in the sense of "registered" by a physical measuring instrument. The Copenhagen doctrine that reality cannot be attributed to the external world, only to our observation of it, is interpreted by Wheeler to mean that the *past* has no existence except as it is recorded in the present. This would imply that the early history of the universe comes into existence only with the development of intelligent life. To implement this view he describes a "delayed-choice" experiment by which an observer can confer reality on events in the past by observation in the present. Indeed, the creation of the universe itself may depend on the existence of the observers that will evolve from that creation.

Wheeler links his proposal with the "anthropic principle" of Brandon Carter, founded on Robert Dicke's remark that the age and mass of the universe and the value of the gravitational coupling constant must have the order of magnitude they actually do in order to account for the existence of life—"it is well known that carbon is required to make physicists" but carbon can be produced only under certain physical conditions [*Nature* 192(1961):440–41]. In 1968 or 1970 Carter suggested that "what we can expect to observe must be restricted by the conditions necessary for our presence as observers," and these conditions imply quantitative limits on the values of various physical constants, including the rate of expansion of the universe (*Confrontation of Cosmological Theories with Observational Data,* ed. M. S. Longair, 1974, 291–98).

Carter, in the spirit of the "many-worlds" interpretation of QM, proposed that one could consider an ensemble of universes having all possible combinations of values of these constants, then simply note that we happen to live in one of the subsets for which the evolution of intelligent life is possible. This might be called a "weak" anthropic principle. Wheeler's stronger version is that "the universe, through some mysterious coupling of future with past, required the future observer to empower past genesis" (1977 "Genesis and Observership," 21). The universe cannot really exist until it has evolved someone to observe it, according to the CI. Wheeler notes that his thesis was vaguely anticipated by the nature-philosopher F. W. J. Schelling; indeed it is consistent with the doctrine, prevalent up to the nineteenth century, that the universe has been designed to provide a home for man—except that man himself, or rather his potential existence, now assumes the divine role of designer.

As one might expect, Wheeler's conclusions have been exploited by proponents of mysticism and parapsychology who wish to use the prestige of modern physics to support their own view of the world. Fritjof Capra and others have shown that there are numerous verbal similarities between the usual formulations of contemporary physical theories and of the CI and selected texts of Eastern mysticism. The apparent instantaneous (or "superluminal," faster than the speed of light) propagation of an effect produced by human observation in the EPR experiment could be explained by telekinesis or some similar "mind-over-matter" influence, though there is no compelling reason to adopt any particular explanation of this kind. When a respectable physicist asserts, quite seriously, that Schrödinger's cat can by an act of will reduce the probability of the radioactive decay that would cause its own death [O. Costa de Beauregard, *Foundations of Physics* 6(1976):539–59], should we conclude that science has now explained why a cat has nine lives?

Wheeler himself has strongly protested attempts to confer scientific respectability on parapsychology through such alleged links with QM, and has called on the American Association of the Advancement of Science (AAAS) to rescind the recognition granted earlier to that subject in a "decade of permissiveness." (See M. Gardner, *New York Review of Books,* 17 May 1979.) This brings us to what I call "Wheeler's dilemma": how can one maintain a strong version of the CI, in which the observer is inextricably entangled with that which is observed, while at the same time denying that our consciousness affects that which we are conscious of and thus accepting the possibility of telekinesis and other psychic effects? For Wheeler himself there is no dilemma at all—one simply has to recognize "the clear distinction between (1) the strange but well verified and repeatable features of quantum mechanics and (2) the pseudo-scientific, non-repeatable and non-verified so-called extra sensory perception" (*New York Review of Books,* 27 Sept. 1979, 68). But Wheeler's own views are likely to strike a nonphysicist as being just as bizarre as those of the parapsychologists he deplores. Indeed, no one has yet formulated a consistent world view that incorporates the CI of QM while excluding what most scientists would call pseudosciences—astrology, parapsychology, creationism, Velikovsky's theories, and thousands of other cults and doctrines.

BIBLIOGRAPHY

P. Lahti and P. Mittelstaedt, eds. (1986), *Foundations of Modern Physics* (Philadelphia: Taylor and Francis), contains articles on the EPR thought experiment and recent developments in the interpretation of quantum mechanics.

Ulrich Röseberg (1984), *Szenarium einer Revolution: Nichtrelativistische Quantenmechanik und philosophische Widerspruchsproblematik* (Berlin: Akademie-Verlag).

Arthur I. Miller (1984), *Imagery in Scientific Thought* (Cambridge: Birkhauser).

Howard Stein (1983), "On the Present State of the Philosophy of Quantum Mechanics," in *PSA 1982,* ed. P. D. Asquith and T. Nickles (East Lansing, Mich.: Philosophy of Science Association), 563–81.

John Archibald Wheeler and Wojciech Hubert Zurek, eds. (1983), *Quantum Theory and Measurement* (Princeton, N.J.: Princeton Univ. Press), is an anthology of forty-nine papers on the interpretation of quantum mechanics.

S. G. Brush and L. Belloni (1983), *History of Modern Physics: An International Bibliography* (New York: Garland), 193–200.

M. L. Dalla Chiara and P. A. Metelli (1982), "Philosophy of Quantum Mechanics," in *Contemporary Philosophy: A New Survey,* vol. 2, *Philosophy of Science,* ed. Guttorm Fløistad, ed. (The Hague: Nijhoff), 219–47.

Edward M. MacKinnon (1982), *Scientific Explanation and Atomic Physics* (Chicago: Univ. of Chicago Press).

J. L. Heilbron and B. R. Wheaton (1981), *Literature on the History of Physics in the 20th Century* (Berkeley: Office for History of Science and Technology, University of California), 436–39.

George Arkell Riggan (1982), "Quantum Physics and Freedom in a Whiteheadian Perspective," *Zygon* 17:255–65.

S. G. Brush (1980), "The Chimerical Cat: Philosophy of Quantum Mechanics in Historical Perspective," *Social Studies of Science* 10:393–447.

Max Jammer (1974), *The Philosophy of Quantum Mechanics* (New York: Wiley/Interscience).

B. S. DeWitt and R. N. Graham (1971), "Resource Letter IQM-1 on the Interpretation of Quantum Mechanics," *American Journal of Physics* 39:724–38, is an annotated bibliography.

Stephen Toulmin, ed. (1970), *Physical Reality* (New York: Harper and Row), is an anthology of papers including the Mach-Planck debate of 1909–1910, philosophical essays by T. P. Nunn and Moritz Schlick, the EPR paper and Bohr's reply, N. R. Hanson on the CI, P. K. Feyerabend on David Bohm's philosophy of nature, and a 1961 essay by Bohm. Most of this material would be understandable by students with some background in physics and an interest in philosophical problems.

Topic *i:* Decline of Mechanism before 1925

Milic Capek (1961), *The Philosophical Impact of Contemporary Physics* (Princeton, N.J.: Van Nostrand).

For additional references see **7.4.**

Topic *ii:* Heisenberg and Born

Werner Heisenberg (1977), "Remarks on the Origin of the Relations of Uncertainty," in *The Uncertainty Principle and Foundations of Quantum Mechanics,* ed. W. C. Price and S. S. Chissick (New York: Wiley), 3–6; (1977), *Physics and Beyond* (New York: Harper and Row); (1958), *Physics and Philosophy* (New York: Harper and Row); (1958), *The Physicist's Conception of Nature* (New York: Harcourt Brace).

[Max Born, ed.] (1971), *The Born-Einstein Letters* (New York: Walker), includes some of Einstein's frequently quoted statements about quantum indeterminism.

Max Born (1964), "The Statistical Interpretation of Quantum Mechanics," in *Nobel Lectures: Physics, 1942–1962* (New York: Elsevier), 256–67; (1949), *Natural Philosophy of Cause and Chance* (Oxford: Clarendon Press).

Mara Beller (1985), "Pascual Jordan's Influence on the Discovery of Heisenberg's Indeterminacy Principle," *Archive for History of Exact Sciences* 33:337–49; (1983), "Matrix Theory before Schroedinger: Philosophy, Problems, Consequences," *Isis* 74:469–91, argues that matrix mechanics was originally a deterministic theory and that Heisenberg and Born adopted indeterminism in response to wave mechanics.

J. Van Brakel (1985), "The Possible Influence of the Discovery of Radio-Active Decay on the Concept of Physical Probability," *Archive for History of Exact Sciences* 31:369–85.

S. G. Brush (1983), *Statistical Physics and the Atomic Theory of Matter* (Princeton, N.J.: Princeton Univ Press), Chap. 2.

Arthur Miller (1982), "Redefining *Anschaulichkeit,*" in *Physics as Natural Philosophy,* ed. A. Shimony and H. Feshbach (Cambridge: MIT Press), 376–411.

A. Pais (1982), "Max Born's Statistical Interpretation of Quantum Mechanics," *Science* 218:1193–98.

Paul A. Hanle (1979), "Indeterminacy before Heisenberg: The Case of Franz Exner and Erwin Schroedinger," *Historical Studies in the Physical Sciences* 10:225–69.

Topic *iii:* Bohr, Copenhagen Interpretation

Niels Bohr (1987), *Philosophical Writings,* Vol. I: *Atomic Theory and the Description of Nature;* Vol. II: *Essays 1932–1957 on Atomic Physics and Human Knowledge;* Vol. III: *Essays 1958–1962 on Atomic Physics and Human Knowledge* (Woodbridge, Conn.: Ox Bow Press); (1985), *Collected Works,* vol. 6, *Foundations of Quantum Physics (1926–1932),* ed. Jørgen Kalckar (New York: Elsevier); (1934), *Atomic Theory and the Description of Nature* (London: Cambridge Univ. Press), includes "The Quantum Postulate and the Recent Development of Atomic Theory," based on the 1927 Como lecture in which Bohr presented his interpretation.

Henry J. Folse (1985), *The Philosophy of Niels Bohr* (New York: Elsevier).

F. Hund (1985), "Korrespondenz und Komplementarität: Bohrs Weg zur Atomdynamik," *Physikalische Blätter* 41:303–7.

Hans-George Schöpf (1983), "Das Bohrsche Komplementaritätskonzept im historischen Kontext der physikalischen Ideen," in *Beiträge zur Komplementarität,* ed. W. Buchheim (Berlin: Akademie-Verlag), 10–21.

John Honner (1982), "Niels Bohr and the Mysticism of Nature," *Zygon* 17:243–53; (1982), "The Transcendental Philosophy of Niels Bohr," *Studies in History and Philosophy of Science* 13:1–29.

Leon Rosenfeld (1979), *Selected Papers (Boston Studies in the Philosophy of Science* 21), is a reprint of several papers on Bohr's philosophy and the Copenhagen interpretation.

Jan Faye (1979), "The Influence of Harald Høffding's Philosophy on Niels Bohr's Interpretation of Quantum Mechanics," *Danish Yearbook of Philosophy* 16:37–72.

David Favrholdt (1976), "Niels Bohr and Danish Philosophy," *Danish Yearbook of Philosophy* 13:206–20.

Topic *iv:* Bohr-Einstein Debate, EPR, Cat Paradox

Niels Bohr (1949), "Discussion with Einstein on Epistemological Problems in Atomic Physics," in *Albert Einstein Philosopher-Scientist,* ed. P. A. Schilpp (New York: Library of Living Philosophers), 199–241.

Albert Einstein (1936), "Physics and Reality," *Journal of the Franklin Institute* 221:349–82; Einstein, B. Podolsky, and N. Rosen (1935), "Can Quantum-Mechanical Description of Reality Be Considered Complete?" *Physical Review* 47:777–80; reply by N. Bohr, same title, ibid. 48:696–702. Both reprinted in S. Toulmin, ed. (1970), *Physical Reality* (New York: Harper and Row).

Karl von Meyenn (1987), "Pauli, Schrödinger und die Streit um die Deutung der Quantentheorie," *Gesnerus* 44:99–124.

Arthur Fine (1986), *The Shaky Game: Einstein, Realism, and the Quantum Theory* (Chicago: Univ. of Chicago Press); (1984), "What Is Einstein's Statistical Interpretation, or, Is It Einstein for Whom Bell's Theorem Tolls?" *Topoi* 3:23–36; (1984), "Einstein's Realism," in *Science and Reality,* ed. J. T. Cushing et al. (Notre Dame, Ind.: Univ. of Notre Dame Press), 106–33; (1981), "Einstein's Critique of Quantum Theory: The Roots and Significance of EPR," in *After Einstein,* ed. P. Barker and C. G. Shugart (Memphis: Memphis State Univ. Press), 147–58.

Don Howard (1985), "Einstein on Locality and Separability," *Studies in History and Philosophy of Science* 16:177–201.

Pekka Lahti and Peter Mittelstaedt, eds. (1985), *Symposium on the Foundations of Modern Physics: 50 Years of the Einstein-Podolsky-Rosen Gedankenexperiment* (Philadelphia: Taylor and Francis).

Abner Shimony (1983), "Reflections on the Philosophy of Bohr, Heisenberg, and Schrödinger," in *Physics, Philosophy, and Psychoanalysis,* ed. R. S. Cohen and L. Laudan (Boston: Reidel), 209–21.

Michelangelo De Maria and Francesco La Teana (1982), "I Primi Lavori di E. Schrödinger sulla Meccanica Ondulatoria e la nascita delle Polemiche con la Scuola di Göttingen-Copenhagen sull-Interpretazione della Meccanica Quantistica," *Physis* 24:33–55.

Gino Tarozzi (1981), "Realisme d'Einstein et Mecanique Quantique: Un Cas de Contradiction entre une Theorie Physique et une

Hypothèse Philosophique clairement définie," *Revue de Synthèse* 102:125–58.

Linda Wessels (1981), "The 'EPR' Argument: A Post-Mortem," *Philosophical Studies* 40:3–30.

Olivier Costa de Beauregard (1980), "The 1927 Einstein and 1935 EPR Paradox," *Physis* 22:211–42, explains the relation of EPR to Loschmidt's reversibility paradox [**7.3**] and Born's principle of adding partial amplitudes rather than probabilities.

Mario Bunge (1979), "The Einstein-Bohr Debate over Quantum Mechanics: Who Was Right about What?" in *Einstein Symposion Berlin,* ed. H. Nelkowski et al. (Berlin: Springer-Verlag), 204–19.

Nathan Rosen (1979), "Can Quantum-Mechanical Description of Physical Reality Be Considered Complete?" in *Albert Einstein: His Influence on Physics, Philosophy, and Politics,* ed. P. C. Aichelburg and R. U. Sexl (Braunschweig: Vieweg), 57–67, is a review of the EPR paper by its third author, discussing Bohr's reply to it.

Topic *v:* Hidden Variables

Trevor Pinch (1977), "What Does a Proof Do if It Does Not Prove? A Study of the Social Conditions and Metaphysical Divisions Leading to David Bohm and John von Neumann Failing to Communicate in Quantum Physics," in *The Social Production of Scientific Knowledge,* ed. E. Mendelsohn et al. (Boston: Reidel), 171–215.

F. J. Belinfante (1973), *Survey of Hidden Variables Theories* (New York: Pergamon Press).

Topic *vi:* Bell's Theorem, Recent Experiments

J. S. Bell (1964), "On the Einstein-Rosen-Podolsky Paradox," *Physics* 1:195–200.

Alain Aspect, Jean Dalibard, and Gerard Roger (1982), "Experimental Test of Bell's Inequalities Using Time-Varying Analyzers," *Physical Review Letters* 49:1804–7.

L. E. Ballentine (1987), "Resource Letter IQM-2: Foundations of Quantum Mechanics Since the Bell Inequalities," *American Journal of Physics* 55:785–96.

Arthur L. Robinson (1986), "Testing Superposition in Quantum Mechanics," *Science* 231:1370–72, reports a meeting at which several recent experiments were discussed; (1986), "Demonstrating Single Photon Interference," *Science* 231:671–72, reports an experiment by P. Grangier, G. Roger, and A. Aspect showing that a single photon can manifest both wave and particle natures.

C. O. Alley, O. G. Jakubowicz and W. C. Wickes (1986), "Results of the Delayed-Random-Choice Quantum Mechanics Experiment with Light Quanta," *Proceedings of the Second International Symposium on Foundations of Quantum Mechanics, Tokyo, 1986,* 36–52, also news item, "The Photon's Split Personality," *Science 86* 7, no. 5:4–5, reports the results of the "delayed choice experiment" proposed by John Wheeler (1978), cited below under *vii.*

A. Aspect et al. (1985), "Reality and the Quantum Theory," *Physics Today* 38, no. 11:9–15, 136–42, is a collection of letters to the editor discussing F. Mermin's article "Is the Moon There . . . ?" cited above under READINGS.

[J. M. Martinis et al.] (1985), "Quantum Cats," *Scientific American* 253, no. 6:80, reports an experiment on tunneling at a Josephson junction showing a macroscopic quantum effect analogous to the Schrödinger cat paradox.

Bill Harvey (1981), "The Effects of Social Context on the Process of Scientific Investigation: Experimental Tests of Quantum Mechanics," in *The Social Process of Scientific Investigation, Yearbook of the Sociology of the Sciences* 4:139–63; (1981), "Plausibility and the Evaluation of Knowledge: A Case-Study of Experimental Quantum Mechanics," *Social Studies of Science* 11:95–130, is on the "hidden variables" experiments done in the 1970s, and how they were interpreted by physicists (work of J. S. Bell, S. J. Freedman and J. F. Clauser, R. A. Holt and F. Pipkin).

Bernard d'Espagnat (1979), "Quantum Theory and Reality," *Scientific American* 241, no. 9:158–81.

J. F. Clauser and A. Shimony (1978), "Bell's Theorem: Experimental Tests and Implications," *Reports on Progress in Physics* 41:1881–1927.

Wolfgang Buchel (1978), "Der Bellsche Beweis: Eine Fallstudie," *Zeitschrift für Allgemeine Wissenschaftstheorie* 8:221–36, argues that contrary to Kuhn and Lakatos, a "purely metaphysical" thesis became falsifiable and was falsified by a "crucial" experiment, so that followers of an older paradigm changed their opinion on rational grounds.

Topic *vii:* Anthropic Principle, Wheeler

John Barrow and Frank Tipler, *The Anthropic Cosmological Principle* (New York: Oxford Univ. Press); Barrow (1981), "The Lore of Large Numbers: Some Historical Background to the Anthropic Principle," *Quarterly Journal of the Royal Astronomical Society* 2:388–420.

Quentin Smith (1985), "The Anthropic Principle and Many-Worlds Cosmology," *Australasian Journal of Philosophy* 63:336–48.

Evelyn Fox Keller (1985), "Cognitive Repression in Contemporary

Physics," in her *Reflections on Gender and Science* (New Haven: Yale Univ. Press), 139–49, argues that discussions about the meaning of QM remain stymied because of failure to formulate a cognitive paradigm adequate to the theory. Conventional interpretations are inadequate because they retain a tenet of classical physics: either objectivity or knowability of nature. "Cognitive repression" means knowledge has been acquired but not yet assimilated.

John Maddox (1983), "Can Observation Prevent Decay?" *Nature* 306:111; M. Bunge and A. J. Kalnay, "Solution to Two Paradoxes in the Quantum Theory of Unstable Systems" and "Real Successive Measurements on Unstable Quantum Systems Take Nonvanishing Time Intervals and Do Not Prevent Them from Decaying," *Nuovo Cimento* 77B:1–9, 10–18.

George Gale (1981), "The Anthropic Principle," *Scientific American* 245, no. 6:154–71.

John A. Wheeler (1979), "Frontiers of Time" in *Problems in the Foundations of Physics,* ed. N. Toraldo di Francia (Amsterdam: North-Holland), 395–497; (1978), "The 'Past' and the 'Delayed-Choice' Double-Slit Experiment," in *Mathematical Foundations of Quantum Theory,* ed. A. R. Marlow (New York: Academic Press), 9–48, describes an experiment later performed by C. Alley (see above); (1977), "Genesis and Observership," in *Proceedings of the 5th International Congress on Logic, Methodology and Philosophy of Science,* ed. R. E. Butts and J. Hintikka (Boston: Reidel), Part 2, 3–33; (1977), "Include the Observer in the Wave Function?" in *Quantum Mechanics, a Half Century Later,* ed. J. L. Lopes and M. Paty (Dordrecht: Reidel), 1–18.

Topic *viii:* Mysticism and Parapsychology

Ken Wilber (1984), *Quantum Questions: Mystical Writings of the World's Great Physicists* (Boulder, Colo.: Shambhala/New Science Library; New York: Random House).

Sal P. Restivo (1983), *The Social Relations of Physics, Mysticism and Mathematics* (Boston: Reidel); (1982, 1978), "Parallels and Paradoxes in Modern Physics and Eastern Mysticism," *Social Studies of Science* 12:37–71, 8:143–81, includes a bibliographical essay on the relations between physics and mysticism.

H. M. Collins and T. J. Pinch (1982), *Frames of Meaning* (London: Routledge and Kegan Paul), Chap. 4.

David Harrison (1979), "What You See Is What You Get!" *American Journal of Physics* 47:576–82; (1979), "Teaching the Tao of Physics," ibid., 779–83.

G. Zukav (1979), *The Dancing Wu Li Masters* (New York: Morrow).

11.3 MATHEMATICAL KNOWLEDGE

i. Attempts to establish firm foundations for mathematics in the nineteenth century; "rigor" in the calculus; David Hilbert's program

ii. Debates about mathematics and its problems, 1900–1930; set theory and the axiom of choice; continuum hypothesis; Bertrand Russell and Alfred North Whitehead; logicism, intuitionism, formalism

iii. Goedel's theorem and the elusiveness of mathematical certainty; implications for the mechanistic view of nature and mind

iv. The nature of mathematics and its role in science in the twentieth century

READINGS

Morris Kline, *Mathematics: The Loss of Certainty* (1980), 172–354. Carl B. Boyer, *A History of Mathematics* (1968), 649–76. Rudy Rucker, "Master of the Incomplete," *Science 82,* Oct., 56–60 (on *iii*), or expanded version in his *Infinity and the Mind* (1982), 157–88. Philip J. Davis and Reuben Hersh, *The Mathematical Experience* (1981), 136–40 (on formalization), 217–36 (non-Euclidean geometry and non-Cantorian set theory), 318–44 (foundations), 345–59 (Lakatos). G. T. Kneebone, *Mathematical Logic* (1963), 311–56.

Articles: Freeman J. Dyson, "Is Real Mathematics of Any use to Physics?" in *Changing Views of the Physical World 1954–1979,* ed. G. K. White (1980), 1–8. Gregory H. Moore, "Beyond First-Order Logic: The Historical Interplay between Mathematical Logic and Axiomatic Set Theory," *History and Philosophy of Logic* 1(1980):95–137, is for advanced students. Eric T. Bell, "Mathematics Up-To-Date," *Current History* (April 1934):54–60.

Some students may have already read Douglas Hofstadter's book *Goedel, Escher, Bach* (1979) or some of his columns published under the title "Metamagical Themas" in *Scientific American,* dealing with self-referential statements. This is one angle that might be used to get students interested in what would otherwise seem to be a rather esoteric subject; another is the claim that Goedel's theorem proves the impossibility of accurately representing the human mind as a machine, thus impinging on the domain of cognitive science.

This section continues the history of selected aspects of mathematics from **7.5**; if that section was not used, one should at least assign some of the readings listed there.

SYNOPSIS

Mathematical knowledge has often been regarded as more certain and reliable than the knowledge we obtain about the world from the natural and social sciences. It is thought to be completely objective, unaffected by the personal biases and social influences that may distort scientific research. But the preeminence of mathematical knowledge has been undermined over the last century by the recognition that there is no necessary connection between a particular mathematical result and the physical world, followed by the demonstration that it is impossible to establish even the self-consistency of mathematics as an independent system of axioms and theorems. Perhaps the final blow is the development of the "sociology of mathematics."

By the middle of the nineteenth century many mathematicians had realized that they could no longer justify their results by appealing to their success in describing the physical world. It was thought necessary to find consistent axioms and rigorous proofs. The discovery of non-Euclidean geometries provided part of the impetus for this.

Augustin-Louis Cauchy was a leader in the movement to rigorize the calculus. He based this program on the limit concept, as suggested earlier by J. Wallis, J. Gregory, and J. L. d'Alembert. Cauchy proposed definitions of continuity and tests of convergence of series but did not use them consistently in his own work.

Karl Weierstrass, starting in the 1860s, freed analysis from dependence on motion and other intuitive physical or geometrical ideas. This began the "arithmetization of analysis"—e.g., he showed that continuity does not imply differentiability by presenting (1872) an example of a function that is continuous for all x but nowhere differentiable. He proposed a definition of irrational numbers on the basis of rational numbers. But then mathematicians such as Richard Dedekind and Giuseppe Peano in the 1880s recognized the need for an axiomatic approach to *rational* numbers as well. Similarly, the consistency of non-Euclidean geometries was proved on the assumption that Euclidean geometry is consistent, but then it was recognized that Euclidean geometry (which was previously accepted without question because it was thought to represent physical truth) had not been proved consistent.

Foundation work in mathematics relied on logic, which was formalized by George Boole in the 1840s and 1850s. He proposed to express the laws of reasoning in symbolic form. This was extended by Augustus de Morgan (1847), C. S. Peirce (1870s and 1880s), and Gottlob Frege (1880s and 1890s).

At the Second International Congress of Mathematics in Paris

(1900), Henri Poincaré claimed that absolute rigor had at last been attained in the foundations of analysis. But David Hilbert presented a list of twenty-three unsolved problems, one of which was to prove that the science of arithmetic is consistent. This was soon recognized to present serious difficulties because contradictions began to turn up in the theory of infinite sets. How big is the set of all sets? Georg Cantor asked in 1895. On one hand its number should be the largest that can exist, but on the other hand the set of all subsets of a given set must have larger number than the set itself, so there is a contradiction. It could be avoided by simply asserting that one may not consider the set of all sets.

Other paradoxes were discussed in the decades after 1900, e.g., Bertrand Russell's "barber paradox." A village barber advertised that he doesn't shave any people in the village who shave themselves, but he does shave all those who don't shave themselves. Should he shave himself? If he does, he doesn't by the first part of the statement; if he doesn't, he does by the second part.

In examining critically the assumptions used in proofs, Ernst Zermelo pointed out in 1904 that Cantor implicitly used an "axiom of choice"—the assumption that given any collection of sets, finite or infinite, one can select an object from each set and form a new set. Other mathematicians objected that unless a definite law specified which element was chosen, no real choice had been made so no new set was formed. For example, if I have 100 pairs of shoes, I can choose the left shoe from each pair and create a new set of left shoes; but if I have 100 pairs of socks, there is no definite rule for doing so since the two socks in each pair are identical.

Around 1900 those concerned with the foundations of mathematics seemed to be divided into two camps, the logicists and the intuitionists. The logicist program was developed independently by Frege and Russell and is best known from the work by Russell and Alfred North Whitehead, *Principia Mathematica* (1910–1913). To avoid the contradictions mentioned above, Russell and Whitehead introduced a "theory of types." As explained by Kline (*Mathematics: The Loss of Certainty,* 221), "Individual objects are of type 0, sets of objects are of type 1, sets of sets are of type 2, etc. . . . Every assertion must be of higher type than what it asserts of some lower type . . . . Thus, if one says *a* belongs to *b, b* must be of higher type than *a.* Also, one cannot speak of a set belonging to itself." A statement such as "all rules have exceptions" is no longer paradoxical since it cannot be interpreted as applying to itself.

The theory of types resolved one kind of problem only to create another. It was no longer as easy to carry out certain proofs, for example those involving least upper bounds or irrational numbers, both of which are defined in a way that makes them a higher type than other numbers in the proof. To avoid such difficulties Russell and Whitehead introduced an "axiom of reducibility," which asserted that propositions of any type are "equivalent" to those of type 1; but this was not really a satisfactory procedure. The use of the axiom of choice, the axiom of reducibility, and other dubious axioms cast doubt on the program of founding mathematics on logic. Nor was it clear that "nonarithmetic" fields like geometry could be deduced from logic.

Intuitionism, the opposite of logicism, came from the writings of Leopold Kronecker who argued that one should simply accept the integers rather than try to derive them from other assumptions. Kronecker is known for the dictum, "God Himself made the whole numbers—everything else is the work of men." He rejected irrational numbers that could not actually be constructed (even if they could be "proved to exist") and infinite sets, and severely criticized the work of Weierstrass and Cantor.

L. E. J. Brouwer developed a systematic intuitionist philosophy starting in 1907. He asserted that mathematics should be based on self-evident primitive notions. Principles such as the law of the excluded middle, developed originally in connection with finite sets, cannot legitimately be applied to infinite sets. "Thus if one proves that not all integers of an infinite set of whole numbers are even, the conclusion that there exists at least one integer which is odd was denied by Brouwer, because this argument applies the law of excluded middle to infinite sets" (Kline, 237). When this law is abandoned, many important theorems are thrown into doubt, and a new class of "undecidable propositions"—those that can neither be proved nor disproved—is created. Other propositions had to be painfully reconstructed by the intuitionists in order to have any significant mathematics left at all.

There may be some similarity between intuitionism in mathematics and certain schools of positivist or idealist approaches to physics. Tito Tonietti (1982) has suggested that the cultural milieu in Germany after World War I may have fostered intuitionism in the same way that (according to Paul Forman) this milieu fostered a causality in quantum physics (**11.4**).

Hilbert rejected both the logicist and intuitionist programs, preferring a "formalist" approach that he began to develop in the 1920s. He adopted the logical axioms of Russell and Whitehead but added others with more specifically mathematical content. He was willing to operate in the world of symbols without worrying too much about what the symbols represent. He constructed a metamathematical "proof theory" with the aim of establishing the absolute consistency of formal mathematical systems.

A fourth school, competing with logicism, intuitionism, and formalism, relies on set theory as founded by Dedekind and Cantor. Zermelo proposed in 1908 a system of axioms for set theory (including the axiom of choice); it was revised in 1922 by Abraham Fraenkel and subsequently became known as the Zermelo-Fraenkel system. Like the Russell-Whitehead system it avoids the paradoxes of self-inclusive sets by forbidding them; it does not address the problems of pure logic. The Zermelo-Fraenkel system has not been proved consistent, but it has not yet encountered any disastrous inconsistencies.

A group of mathematicians writing collectively under the pseudonym "N. Bourbaki" worked out the set theory program starting in 1936, applying it in some detail to all areas of mathematics. Bourbaki's influence spread as far as American elementary school classrooms in the 1960s, where it inspired the controversial "new math" curriculum.

The major foundational problem for all four schools was to establish the *consistency* of a mathematical system; if a system contains inconsistencies, then the correctness of any particular result is suspect. Another problem is *completeness:* it should be possible to prove that any proposition within the domain of a theory is either true or false. By 1930 several systems had been proved to be either consistent or complete, or—in a few very simple cases—both. There seemed good reason for optimism that such results could be extended to all systems, and Hilbert urged this as a high-priority goal of mathematical research.

But in 1931 Kurt Goedel proved "that the consistency of any mathematical system that is extensive enough to embrace even the arithmetic of whole numbers cannot be established by the logical principles adopted by the several foundational schools, the logicists, the formalists, and the set-theorists" (Kline, 261). (The intuitionists survived Goedel's onslaught somewhat better.) Moreover, he showed that any such system is incomplete; there is at least one true statement in its domain that cannot be proved or disproved from the axioms of the system. An example of an unprovable statement is: "This sentence is unprovable." It is either true or unprovable. (Note that statements that are unprovable from the axioms of the system may be provable by bringing in arguments from outside the system, but simply *adding* these to the system does not make it complete in the mathematical sense.)

Goedel's theorems were immediately accepted and extended by other mathematicians, though they were sometimes called "unpleasant" since they seemed to destroy the reasonable hope of "a complete redemption of classical mathematics by the formal axiomatic"—one must either abandon classical mathematics or accept a hypothesis not logically justifiable [A. A. Fraenkel, *Scripta Mathematica* 13(1947):17–36]. They also seemed to have implications for science, especially those disciplines that still aspired to establish mechanistic and deterministic theories to explain and predict all phenomena. As Karl Popper pointed out [*Conjectures and Refutations* (1962), 269–70], Rudolph Carnap's thesis of one universal language for a unified science, which became a major goal of the logical positivists, was refuted even before it was published and was further demolished by a result of Alfred Tarski a few years later; even though Carnap was one of the first philosophers to recognize the importance of Goedel's results, he and other members of the Vienna Circle failed to see how it devastated their own program (**11.5**).

Jacob Bronowski (who should be familiar to users of this *Guide* through his *Ascent of Man* series) argued in the 1960s that Goedel's theorem has serious consequences for the natural and behavioral sciences. Together with Tarski's theorem, it shows that "there cannot be a universal description of nature in a single, closed, consistent language . . . the unwritten aim that the physical sciences have set themselves since Isaac Newton's time cannot be attained. The laws of nature cannot be formulated as an axiomatic, deductive, formal and unambiguous system which is also complete. . . ." Insofar as one expects all other sciences to be reducible to physics, the same applies to them too (*American Scientist,* 1966). Thus Goedel's theorems produced a "crisis of mechanism" that reinforced but was independent of the crisis produced by quantum mechanics. Moreover, according to J. R. Lucas and others, it showed that minds cannot be explained as machines. This contention has provoked considerable discussion among philosophers; a critical review with many references has recently been published by Judson Webb (1983).

An example of some of the extreme conclusions drawn from Goedel's theorems is found in a paper by J. S. Kafka [*Archives of General Psychiatry* 25(1971):232–39]. Contrary to the "double-bind theory" of G. Bateson et al., that schizophrenia is caused by exposure to paradoxical situations, Kafka argues that the *lack* of childhood exposure to ambiguity and paradox leads to schizophrenia in the adult; parental fear and intolerance of ambiguity prevent the child from learning how to cope with the paradoxical aspects of reality. This argument is bolstered by appeal to Goedel's proof of the inevitability of paradoxes in nontrivial systems; the mathematician's theorem seems to be a metaphor for the moral that the child cannot forever live within a neatly ordered world in which all problems have solutions but must be prepared to break out of each system by bringing in new facts and ideas. At no stage is everything self-consistent, but growth requires tolerance of ambiguity and willingness to consider new, unproven concepts.

If a statement is unprovable but is needed in order to derive important results, one can postulate it as a new axiom. However, this means that there are one or more alternative systems in which the axiom is denied; recall the case of Euclid's fifth postulate. When it was found that this postulate could not be derived from the others, the result was the creation of non-Euclidean geometries and the realization that Euclidean geometry does not necessarily tell the truth about the physical world.

A similar situation arose in 1963 when Paul Cohen proved that the axiom of choice and the continuum hypothesis are independent of the other axioms in the Zermelo-Fraenkel system. (The continuum hypothesis states that there is no transfinite cardinal number between aleph-zero and c; see **7.5**.) These axioms are unprovable statements; others have been discovered since 1963. Mathematicians are thereby presented with a variety of choices in the future development of their subject, and the discipline seems in danger of fragmenting into warring camps with each advocating its own set of axioms yet unable to give compelling reasons for doing so. What will remain of the old idea that mathematics gives reliable knowledge—at least about its own abstract concepts, if not about the real world?

There is still a fundamental disagreement about the existence of the subject matter of mathematics: does it exist in the world or only in our minds? According to Reuben Hersh (1979), most mathematicians are Platonists on weekdays and formalists on Sundays: while *doing* mathematics they think they are dealing with objective reality, but when challenged to give an objective account of this reality they find it easier to pretend that it doesn't exist after all. Only a handful of mathematicians—including Goedel and R. Thom—explicitly defend Platonic realism. (This is similar to the situation in modern physics where physicists behave as if they thought particles and fields exist independently of their measurements even though officially almost all of them subscribe to the Copenhagen interpretation; see **11.2**).

Mathematicians have also realized in the twentieth century that there are no absolute standards for the "rigor" of a proof. As Kline notes (315), "A proof is accepted if it obtains the endorsement of the leading specialists of the time or employs the principles that are fashionable at the moment." It follows (historically, not logically) that mathematics is susceptible to sociological analysis, not merely in the Mertonian style but in the European "sociology of knowledge" tradition

(**11.6**). An example of such analysis is Imre Lakatos's discussion of the changing attitudes toward L. Euler's polyhedron theorem, recently elaborated by David Bloor (1978) with the help of the anthropological "grid/group" typology of Mary Douglas. Mathematical knowledge, like scientific knowledge in general, is now regarded by some scholars as the result of "social construction" rather than of objective discoveries about a world independent of ourselves.

BIBLIOGRAPHY

David Abbott, ed. (1986), *The Biographical Dictionary of Scientists: Mathematicians* (New York: Bedrick).

Joseph W. Dauben (1985), *The History of Mathematics from Antiquity to the Present: a Selective Bibliography* (New York: Garland).

Douglas M. Campbell and John C. Higgins, eds. (1984), *Mathematics: People, Problems, Results* (Belmont, Calif.: Wadsworth International).

S. A. Jayawardene (1983), "Mathematical Sciences," in *Information Sources in the History of Science and Medicine,* ed. P. Corsi and P. Weindling (Boston: Butterworth Scientific), 259–84.

Robert Blanche (1973), "Axiomatization," *Dictionary of the History of Ideas,* ed. P. P. Wiener (New York: Scribner), 1:162–72.

Morris Kline (1972), *Mathematical Thought from Ancient to Modern Times* (New York: Oxford Univ. Press).

Howard Delong (1970), *A Profile of Mathematical Logic* (Reading, Mass.: Addison-Wesley), Chaps. 1, 2, 5.

Topic *i:* Foundations, Up to 1900

Michael Detlefsen (1986), *Hilbert's Program: An Essay on Mathematical Instrumentalism* (Boston: Reidel).

Judith V. Grabiner (1981), *The Origins of Cauchy's Rigorous Calculus* (Cambridge: MIT Press).

I. Grattan-Guinness, ed. (1980), *From the Calculus to Set Theory, 1630–1910* (London: Duckworth), has articles by the editor and by T. Hawkins, J. W. Dauben, R. Bunn.

Carl B. Boyer (1959), *The History of the Calculus,* reprint of 1939 ed. (New York: Dover), Chap. 7.

Topic *ii:* Debates, 1900–1930

Jean van Heijenoort (1967), *From Frege to Goedel: A Source Book in Mathematical Logic, 1879–1931* (Cambridge: Harvard Univ. Press), includes writings of Hilbert (1904, 1925, 1927), Russell (1901, 1908), Zermelo (1904, 1908), Fraenkel (1922), Brouwer (1923, 1927), J. von Neumann (1923, 1925), Goedel (1930, 1931) and others.

Paul Benacerraf and Hilary Putnam, eds. (1964), *Philosophy of Mathematics: Selected Readings* (Englewood Cliffs, N.J.: Prentice-Hall), includes a symposium on the foundations of mathematics (Carnap on logicism, Heyting on intuitionism, v. Neumann on formalism, Brouwer, Hilbert, etc.)

Carl J. Posy (1984), "Kant's Mathematical Realism," *Monist* 67:115–34.

Gregory H. Moore (1983), "Lebesgue's Measure Problem and Zermelo's Axiom of Choice: The Mathematical Effects of a Philosophical Dispute," *Annals of the New York Academy of Sciences* 412:129–54; (1982), *Zermelo's Axiom of Choice: Its Origins, Development, and Influence* (New York: Springer-Verlag).

Tito Tonietti (1982), "A Research Proposal to Study the Formalist and Intuitionist Mathematicians of the Weimar Republic," *Historia Mathematica* 9:61–64.

Peter Eggenberger (1977), "The Philosophical Background of L. E. J. Brouwer's Intuitionistic Mathematics" (Ph.D. diss., University of California, Berkeley).

B. Van Rootselaar (1970), "Brouwer, Luitzen Egbertus Jan," *Dictionary of Scientific Biography* 2:512–14.

Topic *iii:* Goedel's Theorem

J. van Heijenoort, ed. (1970), *Fregel and Goedel: Two Fundamental Texts in Mathematical Logic* (Cambridge: Harvard Univ. Press), includes translations of Goedel's 1930–1931 papers.

Alfred Tarski (1956), *Logic, Semantics, Metamathematics: Papers from 1923 to 1938* (Oxford: Clarendon Press), includes translation of "The Concept of Truth in Formalized Languages" (1931, published later with revisions).

Rudy Rucker (1982), *Infinity and the Mind* (Cambridge: Birkhauser), Chap. 4, "Robots and Souls" and Excursion II, "Goedel's Incompleteness Theorems."

Gina Kolata (1982), "Does Goedel's Theorem Matter to Mathematics?" *Science* 218:779–80. "The recent discovery [by J. Paris] of two natural but undecidable statements indicates that Goedel's theorem is more than just a logician's trick."

Gerald E. Lenz (1980), "Kurt Goedel: Mathematician and Logician," *Mathematics Teacher* 73:612–14.

I. Grattan-Guinness (1979), "In Memoriam Kurt Goedel: His 1931

Correspondence with Zermelo on His Incompletability Theorem," *Historia Mathematica* 6:294–304.

Michael Detlefsen (1976), "The Importance of Goedel's Second Incompleteness Theorem for the Foundations of Mathematics" (Ph.D. diss., Johns Hopkins University, 1976). See *Dissertation Abstracts International* 37A: 374.

Ernest Nagel and James R. Newman (1959), *Goedel's Proof* (London: Routledge and Kegan Paul), explains the background of the problem and provides a version of the proof which can be followed, with some effort, by a reader with only an elementary knowledge of mathematics. See also their article with the same title in *Scientific American,* June 1956.

Implications for mechanism:

Judson Webb (1983), "Goedel's Theorem and Church's Thesis: A Prologue to Mechanism," in *Language, Logic and Method,* ed. R. S. Cohen and M. W. Wartofsky, vol. 31 of *Boston Studies in the Philosophy of Science* (Boston: Reidel), 309–53, argues, contrary to Lucas and others, that Goedel's proof *supports* mechanism. Extensive bibliography.

J. Bronowski (1978), *The Origins of Knowledge and Imagination* (New Haven: Yale Univ. Press), Chap. 4; (1966),"The Logic of the Mind," *American Scientist* 54:1–14.

William H. Desmonde (1971), "Goedel, Non-Deterministic Systems, and Hermetic Automata," *International Philosophical Quarterly* 11:49–74, discusses the claim that Goedel's theorem refutes mechanism; proposes an alternative model and applies it to salvationistic theologies such as that of Teilhard de Chardin; also discusses "the Hermetic tradition in which the scientist and technologist are seen as outgrowths of the magician-alchemist who assists in the redemption of the cosmos."

J. R. Lucas (1970), *The Freedom of the Will* (New York: Oxford Univ. Press); (1961), "Minds, Machines, and Goedel," *Philosophy* 36:112–27, reprinted (1964) in *Minds and Machines,* ed. A. R. Anderson (Englewood Cliffs, N.J.: Prentice-Hall), 43–59, argues that Goedel's theorem refutes mechanism, i.e., it shows that minds cannot be explained as machines. See also the discussion in *Monist* 51(1967):9–33, 52(1968):145–58.

Topic *iv:* Nature and Role of Mathematics; Education; Sociology

Morris Kline (1985), *Mathematics and the Search for Knowledge* (New York: Oxford Univ. Press), 141–245.

W. Baldamus (1984), "Epistemology and Mathematics," in *Society and Knowledge: Contemporary Perspectives in the Sociology of Knowl-*

edge (New Brunswick, N.J.: Transaction Books), 349–64, discusses whether there could be a sociology of mathematical knowledge.

Philip Kitcher (1983), *The Nature of Mathematical Knowledge* (New York: Oxford Univ. Press).

Randall Collins and Sal Restivo (1983), "Robber Barons and Politicians in Mathematics: A Conflict Model of Science," *Canadian Journal of Sociology* 8:199–227.

Sal Restivo (1981), "Mathematics and the Limits of the Sociology of Knowledge," *Social Science Information* 20:679–701.

Raymond L. Wilder (1981), *Mathematics as a Cultural System* (New York: Pergamon Press).

Penelope Maddy (1980), "Perception and Mathematical Intuition," *Philosophical Review* 89:163–96, is on Goedel's set theoretic realism.

Reuben Hersh (1979), "Some Proposals for Reviving the Philosophy of Mathematics," *Advances in Mathematics* 31:31–50.

David Bloor (1978), "Polyhedra and the Abominations of Leviticus," *British Journal for the History of Science* 11:247–72; see also 12(1979):71–81 for a reply by Worrall and response by Bloor. Bloor argues that Lakatos himself advocates a radical position of the flexibility of mathematical concepts and proofs, contrary to the views of his editors who have tried to "correct" him.

Imre Lakatos (1976), *Proofs and Refutations: The Logic of Mathematical Discovery,* ed. J. Worrall and E. Zahar (New York: Cambridge Univ. Press).

Herbert Mehrtens (1976), "T. S. Kuhn's Theories and Mathematics: A Discussion Paper on the 'New Historiography' of Mathematics," *Historia Mathematica* 3:297–320.

Felix E. Browder (1976), "Does Pure Mathematics Have a Relation to the Sciences?" *American Scientist* 64:542–49.

Garrett Birkhoff et al. (1975), "Foundations of Mathematics," *Historia Mathematica* 2:503–33, has papers by E. Bishop, N. Kopell, G. Stolzenberg, G. Sacks, H. Putnam, and discussion by participants in a workshop on the Evolution of Modern Mathematics.

11.4 PHYSICS AND TWENTIETH-CENTURY CULTURE

i. The impact of Einstein and his theory of relativity; comparable trends in art, literature, and music

ii. Cultural influences on physical theory: Paul Forman's thesis about causality and Weimar culture—Does physics follow the *Zeitgeist?*

iii. Quantum mechanics, randomness, and entropy
iv. The two cultures debate and the relations between sciences and humanities in recent years

READINGS

Richard Olson, ed., *Science as Metaphor* (1971), Chaps. 9, 10, 267–312 (T. J. Craven on "Art and Relativity," H. I. Rogers on "Charles Beard, the 'New Physics' and Historical Relativity," H. Margenau on "Quantum Mechanics, Free Will and Determinism," J. Lukacs on "Quantum Mechanics and the End of Scientism," Olson on the contemporary revolt against science).

For *i:* L. Pearce Williams, ed., *Relativity Theory* (1968), 129–57 (extracts from *N.Y. Times,* H. W. Carr, H. Elliott, T. J. Craven, P. C. Squires, and J. Ortega y Gasset). Edward A. Purcell, Jr., *The Crisis of Democratic Theory* (1973), Chap. 4 "Non-Euclideanism: Logic and Metaphor," 47–73. Gerald Holton and Yehuda Elkana, eds., *Albert Einstein: Historical and Cultural Perspectives* (1982), introduction by Holton "Einstein and the Shaping of Our Imagination," vii-xxxii, and Loren R. Graham, "The Reception of Einstein's Ideas: Two Examples from Contrasting Political Cultures," 107–36. Holton's introduction is reprinted in his *The Advancement of Science* (1982), 105–22. An expanded version of Graham's article is in his book *Between Science and Values* (1981), 1–158, which includes views on the humanistic/social implications of modern physics presented by Einstein, Bohr, Eddington, Fock, Heisenberg, and Bergson.

For *ii:* John Fauvel, "Physics and Society," Unit 9 in the Open University course A381, *Science and Belief, Darwin to Einstein* (1981), 139–52; extracts from papers by Forman and Hendry in the Chant-Fauvel anthology, *Darwin to Einstein* (1980); extract from Spengler's *Decline of the West* in N. G. Coley-M. D. Hall anthology, *Darwin to Einstein* (1980).

For *iii:* Erwin Schrödinger, *Science and Humanism* (1951), is a 68-page essay including his views on indeterminism, causality, and the alleged breakdown of the subject-object barrier. Alfred Bork, "Randomness and the Twentieth Century," *Antioch Review* 27(1967):40–61; Henry Eulau, "From Utopia to Probability: Liberalism and Recent Science," ibid. 26(1966):5–16; Jeremy Campbell, *Grammatical Man* (1982).

For *iv:* C. P. Snow, "The Two Cultures and the Scientific Revolution"—the original 1959 lecture is a small booklet of about 50 pages; it was reprinted in 1964 with a 50-page supplement, as *The Two Cultures and a Second Look.* The best-known response is a lecture by F. R. Leavis, *Two Cultures? The Significance of C. P. Snow* (1962), reprinted (1963) with a new preface for the American reader and an essay on Snow's lecture by Michael Yudkin (New York: Pantheon Books). While the "Two Cultures" controversy is sometimes called the "Snow-Leavis debate," I find Leavis's style so ill-mannered and his criticisms of Snow so irrelevant to the main issue that it hardly seems fair to the humanists to let Leavis be their standard-bearer. Lionel Trilling's response is somewhat more effective: "Science, Literature and Culture," in *The Scientist versus the Humanist,* ed. George Levine and Owen Thomas (1963). A more balanced presentation, published at the same time as Snow's lecture, may be found in H. M. Jones, *One Great Society* (1959), chapters titled "What Are the Humanities" and "The Humanities as Information," 3–40. For general background I suggest R. J. Bieniek, "Evolution of the Two Cultures Controversy," *American Journal of Physics* 49(1981):417–23.

SYNOPSIS

After Einstein's theory of the electrodynamics of moving bodies (1905) was given the name "theory of relativity" by Planck (1907) and Einstein reluctantly accepted this name (he would have preferred "invariant theory"—see Holton's "Shaping" article cited above), it became customary to blame it for propagating the vague idea that "everything is relative." Twentieth-century Europeans and Americans seem to have lost their confidence in absolute moral standards; to what extent is this due to Einstein's theories? Cubist art abandoned representational realism in favor of abstract geometrical forms; was Guillaume Apollinaire correct in attributing this to relativity and non-Euclidean geometry? Is Lawrence Durrell's *Alexandria Quartet* an attempt to portray the four-dimensional space-time continuum? Is Arnold Schönberg's *Tonreihe,* which treats each of the twelve notes in the octave on an equal basis, inspired by Einstein's postulate that every inertial coordinate system is as good as any other?

More likely there is no question of a direct influence on culture of Einstein's theory, published in 1905 but not accepted by most of the scientific community until the following decade. If we see Picasso and the cubists abandoning perspective, Schönberg setting aside tonality, and Franz Boas insisting that each culture must be judged on its terms rather than against our own (**3.4**)—all within the same short period of time—we must conclude either that this is a meaningless coincidence or that all were responding to the same *Zeitgeist.* Fifteen years later, when the public became aware that Einstein had done some amazing things with "non-Euclidean geometry" involving the bending of light and the curvature of space, the idea got about that one formal system, even though internally self-consistent and apparently plausible, could just as well be replaced by another one. At that point scientific naturalism became a threat to traditional moral codes; cultural relativism and pragmatism gave birth to "functionalism," the doctrine that scientists need only describe how things work in a particular society without worrying about how things *should* work.

The *Zeitgeist* of the early twentieth century has been called "neo-realism." It differed from mid-nineteenth-century realism in emphasizing mathematical structure over physical content—relativity and cubism are the best known examples. Paul Souriau "crystallized a broad current of thought" when he exalted the beauty of geometrical forms (Ragon 1976). The IQ rapidly became a quantitative, though one-dimensional, measure of the human mind (**6.4, 6.5**). In many of the new theories and discoveries of the years around 1905, the concept of *time* changed from an absolute, uniformly flowing, independent variable to a contingent, discrete variable dependent on random events. More precisely, the transition was from an evolutionary worldview infected by physical dissipation and biological degeneration, to a stochastic worldview shaken and possibly rejuvenated by collisions and catastrophes (Brush 1979).

Randomness as a component of the twentieth-century worldview is often attributed to Heisenberg's indeterminacy principle, combined more recently with the thermodynamic concept of *entropy* representing *disorder* as explained by Boltzmann (**7.3**). Paul Forman and Lewis Feuer have turned around the causal relation, arguing that the indeterminacy principle itself symbolizes a response of mathematicians and physicists to the cultural movements and social currents of the 1920s. Their descriptions of the period certainly demonstrate a striking similarity between the movements in physics and the cultural environment but do not (in my opinion) prove that the former is the result of the latter. As the physicist Gustav Mie said in a 1925 speech quoted by Forman, "Even physics . . . is led into paths which run perfectly parallel to the paths of the intellectual movement in other areas." Nevertheless, Forman's thesis has been enthusiastically welcomed by a new school of sociologists who aim to interpret all scientific discoveries as the product of "social construction" (**11.6**).

The "two cultures" thesis of C. P. Snow implies on the other hand separation and antagonism between the sciences and the humanistic culture, rather than mutual influence or sharing of similar tendencies. Some American observers have claimed that Snow's remarks are relevant only to the British scene, and they can certainly be seen in the context of a science vs. humanities debate going back to T. H. Huxley and Matthew Arnold. But the ideas and the rhetoric—complaints about disregard for humanistic values on the part of scientists, and complaints about ignorance of science on the part of almost everyone else—continue to surface in the United States, most recently in the 1983 controversy about the "rising tide of mediocrity" in education (that phrase, along with much of the other striking language in the report of the Commission on Excellence, is apparently due to physics-historian Gerald Holton).

BIBLIOGRAPHY

The quarterly *Publication of the Society of Literature and Science* (started 1985) contains news of current activities in field. Contact Lance Schachterle, SLS Project Center, Worcester Polytechnic Institute, 100 Institute Road, Worcester, MA 01609 for subscription information.

Erwin Hiebert (1986), "Modern Physics and Christian Faith," in *God and Nature,* ed. D. C. Lindberg and R. L. Numbers (Berkeley: Univ. of California Press), 424–47.

Alan J. Friedman and Carol C. Donley (1985), *Einstein as Myth and Muse* (New York: Cambridge Univ. Press), includes sections on the impact of quantum theory as well as relativity on modern culture.

S. G. Brush and L. Belloni (1983), *The History of Modern Physics: An International Bibliography* (New York: Garland), 265–69.

Lewis Feuer (1982), *Einstein and the Generations of Science* (New Brunswick, N.J.: Transaction Books).

J. L. Heilbron and B. R. Wheaton (1981), *Literature on the History of Physics in the 20th Century* (Berkeley: Office for History of Science and Technology, University of California), 414–31.

Alan J. Friedman (1973), "Physics and Literature in this Century: A New Course," *Physics Education* 8:305–8.

John M. Bailey (1971), "Physics and Everything: A Bibliography," *American Journal of Physics* 39:1347–52, is a list of 203 items on relations of physics with other fields, for use by students in writing research papers.

John A. Richardson (1971), *Modern Art and Scientific Thought* (Urbana: Univ. of Illinois Press).

Stanley L. Jaki (1966), *The Relevance of Physics* (Chicago: Univ. of Chicago Press).

Topic *i:* Einstein, Relativity

Emile Meyerson (1985), *The Relativistic Deduction: Epistemological Implications of the Theory of Relativity, with a Review by Albert Einstein* (Boston: Reidel).

Linda Dalrymple Henderson (1983), *The Fourth Dimension and Non-Euclidean Geometry in Modern Art* (Princeton, N.J.: Princeton Univ. Press).

Stephen Kern (1983), *The Culture of Time and Space 1880–1918* (Cambridge: Harvard Univ. Press).

Iain Paul (1982), *Science, Theology and Einstein* (New York: Oxford Univ. Press).

Stephen G. Brush (1979), "Scientific Revolutionaries of 1905: Einstein, Rutherford, Chamberlin, Wilson, Stevens, Binet, Freud," in *Rutherford and Physics at the Turn of the Century,* ed. M. Bunge and W. R. Shea (New York: Science History), 140–71.

Mendel Sachs (1978), "On the Philosophy of General Relativity Theory and Ideas of Eastern and Western Cultures,"in *The Ta-You Festschrift,* ed. S. Fujita (New York: Gordon and Breach), 9–24, suggests that whereas the empiricist view is anthropocentric the realist (general relativity) view implies unification of man with all of nature.

William J. Scheick (1978), "The Fourth Dimension in Wells's Novels of the 1920s," *Criticism* 20:167–90.

Ernestine Schlant (1978), "Hermann Broch and Modern Physics," *Germanic Review* 53:69–75.

Michel Ragon (1976), "Art, Science and Technology," *Cultures* 3, no. 3:105–46.

Francine Ringold (1976), "The Metaphysics of Yoknapatawpha County: 'Airy Space and Scope for Your Delirium,' " *Hartford Studies in Literature* 8:23–40, considers the influence of theories of Einstein and Bergson on Faulkner's *Absalom, Absalom!*

Paul M. Laporte (1966), "Cubism and Relativity (with a Letter from Albert Einstein)," *Art Journal* 25:246–48.

Topic *ii:* Cultural Influence on Theory; Forman Thesis

Paul Forman(1984), "*Kausalität, Anschaulichkeit,* and *Individualität,* or How Cultural Values Prescribed the Character and the Lessons Ascribed to Quantum Mechanics," in *Society and Knowledge,* ed. N. Stehr and V. Meja (New Brunswick, N.J.: Transaction Books), 333–47; (1971) "Weimar Culture, Causality and Quantum Theory, 1918–1927: Adaptation by German Physicists and Mathematicians to a Hostile Cultural Environment," *Historical Studies in the Physical Sciences* 3:1–115.

P. Kraft and P. Kroes (1984), "Adaptation of Scientific Knowledge to an Intellectual Environment. Paul Forman's 'Weimar Culture, Causality, and Quantum Theory, 1918–1927': Analysis and Criticism," *Centaurus* 27:76–79.

Hans Radder (1983), "Kramers and the Forman Theses," *History of Science* 21:165–82.

Bill Harvey (1981), "The Effect of Social Context on the Process of Scientific Investigation: Experimental Tests of Quantum Mechanics," *Yearbook of the Sociology of the Sciences* 5:139–63.

John Hendry (1980), "Weimar Culture and Quantum Causality," *History of Science* 18:155–80.

S. G. Brush (1980), "The Chimerical Cat: Philosophy of Quantum Mechanics in Historical Perspective," *Social Studies of Science* 10:393–447.

Topic *iii:* Quantum Mechanics, Randomness, Entropy

Robert John Russell (1984), "Entropy and Evil," *Zygon* 19:449–68.

John W. Haas, Jr. (1983), "Complementarity and Christian Thought: An Assessment," *Journal of the American Scientific Affiliation* 35:145–51, 203–9.

Robert Nadeau (1981), *Readings from the New Book on Nature: Physics and Metaphysics in the Modern Novel* (Amherst: Univ. of Massachusetts Press).

Virginia P. Williams (1981), "Surrealism, Quantum Philosophy, and World War I" (Ph.D. diss., Duke University).

Jeremy Rifkin (1980), *Entropy: A New World View* (New York: Viking).

Steven T. Ryan (1979), "Faulkner and Quantum Mechanics," *Western Humanities Review* 33:329–39.

Richard Schlegel (1979), "Quantum Physics and the Divine Postulate," *Zygon* 14:163–85, argues that quantum mechanics reinforces a line of current religious thought (H. Wieman, G. Riggan).

Allen E. Hye (1978), "Bertolt Brecht and Atomic Physics," *Science/Technology and The Humanities* 1:157–70.

Daniel Simberloff (1978), "Entropy, Information, and Life: Biophysics in the Novels of Thomas Pynchon," *Perspectives in Biology and Medicine* 21:617–25.

D. Stanley Tarbell (1978), "Perfectibility vs. Entropy in Recent Thought," *Science/Technology and the Humanities* 1:103–13.

Florence A. Falk (1977), "Physics and the Theatre: Richard Foreman's *Particle Theory,*" *Educational Theatre Journal* 29:395–404.

Topic *iv:* Two Cultures

C. P. Snow (1960), *The Two Cultures and the Scientific Revolution* (New York: Cambridge Univ. Press).

Ronald J. Bieniek (1981), "Evolution of the Two Cultures Controversy," *American Journal of Physics* 49:417–24.

Alan J. Friedman (1979), "Contemporary American Physics Fiction," *American Journal of Physics* 47:392–95.

11.5 PHILOSOPHY OF SCIENCE IN THE TWENTIETH CENTURY

i. Logical positivism and the Vienna Circle; Bridgman's operationalism; distinction between observation and theoretical languages; reduction

ii. Methodologies for accepting and rejecting theories—Karl Popper: falsifiability and the problem of demarcation between science and pseudoscience; Imre Lakatos: "methodology of scientific research programmes"; rational reconstruction of the history of science

iii. Paradigms and scientific revolutions: T. S. Kuhn's theory; reaction of philosophers of science; relativism and incommensurability; views of Feyerabend

iv. Thematic analysis: Holton's theory that certain thematic ideas recur throughout the history of science despite major changes in the content of scientific theories and experiments

v. Historical realism: attempts to unite the history and philosophy of science; evolutionary models of Toulmin and others; Dudley Shapere's theory of domains; relations between scientific fields; Laudan's focus on problem solving.

READINGS

Marks, *Science,* 357–66. Paul Wood, "Philosophy of Science in Relation to History of Science," in *Information Sources in the History of Science and Medicine,* ed. P. Corsi and P. Weindling (1983), 116–33. Losee, *Historical Introduction* (1980), 170–218; this should be supplemented by a more explicit presentation of logical positivism, either P. Frank's personal account in *Modern Science and Its Philosophy,* 30–61, or H. Feigl's systematic survey "Positivism in the Twentieth Century" in *Dictionary of the History of Ideas* (1973), 3:545–51. P. K. Feyerabend, *Philosophical Papers* (1981), 1:3–16; 2:1–33. G. Holton, *Introduction to Concepts and Theories in Physical Science,* 2d. ed., rev. by S. G. Brush (1973), 173–221.

The anthology ed. by Ian Hacking, *Scientific Revolutions* (1981), includes substantial extracts from writings of Kuhn, Shapere, Putnam, Popper, Lakatos, Laudan, and Feyerabend. Another useful collection is *Introductory Readings in the Philosophy of Science,* ed. E. D. Klemke et al. (1980), which has selections by P. Thagard, Popper, Hanson, Kuhn, T. Roszak, and others.

Articles: Michael Pollak, "From Methodological Prescription to Socio-Historical Description: The Changing Metascientific Discourse," *Fundamenta Scientiae* 4(1983):1–27; Stephen Toulmin, "Are the Principles of Logical Empiricism Relevant to the Actual Work of Science?" *Scientific American* 214(1966):129–33. Gerald Holton, "On the Role of Themata in Scientific Thought," *Science* 188:323–38, with following comment by R. Merton. I. Lakatos, "History of Science and Its Rational Reconstructions," in *Method and Appraisal in the Physical Sciences,* ed. C. Howson (1976), 1–39.

A special issue on "Teaching Philosophy of Science" was published by *Teaching Philosophy* 2, no. 2(1979).

As noted in **11.1**, I don't try to give a survey of the philosophy of science in my own course but discuss only its interaction with the history of science. For the earlier period this means: how did philosophical writings influence science? For the modern period, it means: do philosophical writings provide useful models for the development of science? Thus, one may take each of the five kinds of philosophies listed above and treat it as a hypothesis or theory about the history of science, to be tested against historical evidence, independently of its place in the history of the philosophy of science or its status among modern philosophers. The advantage of this approach is that hypotheses of the type *i, ii* or *iii* that may be considered out of date or refuted by philosophers of science can still provide useful idealized models which direct attention to certain features of science that might be overlooked in a more realistic descriptive approach of the type mentioned under *iv* and *v*. ("Realistic" as used here does not refer to philosophical realism but to historical accuracy.)

SYNOPSIS

Logical positivism, later called "logical empiricism" by its advocates (also called "logical reconstructionism" by Losee, and in a later version the "received view" by Suppe), is now generally considered obsolete. Yet some of its premises may capture the goals of the scientist even though they fail to be realized in the actual practice of science. One example is the "neutral observation language"; another is "reduction." Scientists would like to believe that they can describe their experimental results in a manner independent of the theory being tested. Similarly, while it is fashionable in many circles to reject "reductionism," a large part of the development of theoretical physics has been devoted to the attempt to reduce macroscopic phenomena to microscopic theories. Indeed many scientists still regard the reductions achieved by kinetic theory and statistical mechanics as some of the outstanding achievements of modern science, and the reduction of thermodynamics to statistical mechanics is often presented (somewhat misleadingly) by philosophers as a paradigm for reduction in general.

Karl Popper's doctrine, that science progresses by generating conjectures and then trying to refute them, has been widely accepted; his view that a hypothesis must be testable and thus *falsifiable* is occasionally cited by scientists and may even have influenced their choice of hypotheses in some cases. As I have noted elsewhere in this *Guide,* his insistence that a theory must make predictions about the results of future experiments has fostered misconceptions about the nature of science and reinforced existing prejudices against the legitimacy of theories dealing with the distant past (**4.7**-*iii*), as well as creating well-deserved skepticism about theories that are too easily modified to evade refutation by experiments (**5.5**-*i*).

A curious aspect of the development of the history of science within the past twenty years has been the anomalous role of Thomas S. Kuhn, or rather of the thesis presented in his book *The Structure of Scientific Revolutions* (*SSR*) (1962, 1970). I assume that the reader of this *Guide* will already have read *SSR* and formed some opinion about it, but may not be aware that Kuhn himself has revised his "paradigm" concept (see his "Second Thoughts on Paradigms" cited below) and does not use it at all in his more recent work. Most historians of science reject or ignore the Kuhnian thesis, for reasons articulated by Nathan Reingold in his article "Through Paradigm-Land. . . ." On the other hand it has had an enormous impact on the philosophy of science and on the social sciences, and the term "paradigm" has entered the language in the Kuhnian sense, displacing the older linguistic meaning. Philosophers of science devoted considerable effort to demolishing Kuhn's thesis but at the same time recognized that their own doctrines needed to be radically revised in order to give a plausible account of the way scientists actually behave, as an alternative to Kuhn's description. One result has been the "methodology of scientific research programmes" developed by Imre Lakatos and his followers in England. In the United States, philosophers of science such as Stephen Toulmin, Dudley Shapere, and Larry Laudan have proposed other theories of the growth of science. A common feature of these accounts is the attempt to find a "rational reconstruction" of the history of science—to preserve the assumption that scientists adopt and change their theories for good reasons, amenable to philosophical discussion. At the same time the penchant of some philosophers for the invention of formal axiomatic systems has found expression in the "structuralist" view of theories proposed by Joseph Sneed and Wolfgang Stegmueller, an approach that is said to be in harmony with Kuhn's ideas [see Kuhn's article in *Erkenntnis* 10(1976):179–99].

It is ironic that while historians of science such as L. P. Williams disdain Kuhn's approach as being not truly historical [*History of Science* 18(1980):68–74], those in other fields of history are much more enthusiastic about it [D. A. Hollinger, *American Historical Review* 78(1973):370–93]. On the other hand the trend toward sociological interpretations of the history of science, while perhaps stimulated by Kuhn's emphasis on the role of the scientific community in the establishment of paradigms, seems to have gone beyond his position to explore the ultrarelativist views Kuhn disclaimed when they were attributed to him (**11.6**-*iii*).

My recommendation is to present logical positivism, falsificationism (in both the Popper and Lakatos versions), and the paradigm-revolutionary view of Kuhn as three distinct models for the growth of science, each of which captures some element of reality but perhaps pushes it too far by neglecting other aspects. A fourth alternative can then be selected, according to personal preference, from the views of Holton, Toulmin, Shapere, Laudan, and other contemporary writers; each offers a somewhat realistic approach, avoiding extreme statements that can be directly refuted, while providing insight into a number of historical cases.

BIBLIOGRAPHY

John Neu, ed. (1985), *ISIS Cumulative Bibliography 1966–1975*, vol. 2, *Subjects, Periods and Civilizations* (London: Mansell), 554–59.

Richard J. Blackwell, ed. (1983), *A Bibliography of the Philosophy of Science, 1945–1981* (Westport, Conn.: Greenwood).

Guttorm Fløistad, ed. (1982), *Contemporary Philosophy: A New Survey*, vol. 2, *Philosophy of Science* (Boston: Nijhoff).

Robert J. Richards (1981), "Natural Selection and Other Models in the Historiography of Science," in *Scientific Inquiry and the Social Sciences*, ed. M. B. Brewer and B. E. Collins (San Francisco: Jossey-Bass), 37–76.

Ilya Prigogine et al. (1980), "Philosophy of Science: Contemporary Issues," *Revue International de Philosophie* 34:3–292.

Peter D. Asquith and Henry E. Kyburg, Jr., eds. (1979), *Current Research in Philosophy of Science* (East Lansing, Mich.: Philosophy of Science Association), includes articles on relation of philosophy of science to history of science (L. Laudan, E. McMullin, T. S. Kuhn) and other disciplines.

Frederick Suppe (1974), "The Search for Philosophic Understanding of Scientific Theories," in *The Structure of Scientific Theories,* ed. F. Suppe (Urbana: Univ. of Illinois Press), 1–241, and (1977), "Afterward—1976" in ibid., 2d ed.

Topic *i:* Logical Positivism

Percy W. Bridgman (1927), *The Logic of Modern Physics* (New York: Macmillan), is the original presentation of operationism.

A. J. Ayer, ed. (1966), *Logical Positivism* (1959, New York: Macmillan/Free Press), includes editor's introduction and extracts from writings of Russell, M. Schlick, R. Carnap, C. G. Hempel, H. Hahn, O. Neurath, A. J. Ayer, C. L. Stevenson, F. D. Ramsey, G. Ryle, F. Waissmann; extensive bibliography.

Ernest Nagel (1961), *The Structure of Science* (New York: Harcourt), is a widely used exposition of the modern version of logical positivism.

Harold J. Allen (1980), "P. W. Bridgman and B. F. Skinner on Private Experience," *Behaviorism* 8:15–29, discusses the influence of Bridgman on Skinner (cf., **6.3**).

F. Suppe and P. D. Asquith, eds. (1977), *PSA 1976,* vol. 2 (East Lansing, Mich.: Philosophy of Science Association), includes J. Alberto Coffa, "Carnap's *Sprachanschauung Circa 1932,*" 205–42; Lindley Darden, "The Heritage from Logical Positivisim," 242–58; and S. G. Brush, "Statistical Mechanics and the Philosophy of Science: Some Historical Notes," 551–84, on reduction and other philosophical theses illustrated by kinetic theory.

Peter Achinstein and Stephen F. Barker, eds. (1969), *The Legacy of Logical Positivism* (Baltimore: Johns Hopkins Univ. Press), includes H. Feigl, "The Origin and Spirit of Logical Positivism"; S. Toulmin, "From Logical Analysis to Conceptual History"; N. R. Hanson, "Logical Positivism and the Interpretation of Scientific Theories"; Mary Hesse, "Positivism and the Logic of Scientific Theories," etc.

Topic *ii:* Popper, Lakatos (Demarcationism, Falsificationism)

Karl Popper (1983), *Realism and the Aim of Science* (Totowa, N.J.: Rowman and Littlefield); (1962), *Conjectures and Refutations* (New

York: Basic Books); (1934/1959), *The Logic of Scientific Discovery* (London: Hutchinson).

Imre Lakatos (1978), *Philosophical Papers* (New York: Cambridge Univ. Press).

Steven Yearley (1985), "Imputing Intentionality: Popper, Demarcation and Darwin, Freud and Marx," *Studies in History and Philosophy of Science* 16:337–50.

Barry Gholson and Peter Barker (1985), "Kuhn, Lakatos, and Laudan: Applications in the History of Physics and Psychology," *American Psychologist* 40:755–69, conclude that the accounts of Lakatos and Laudan are more accurate than "popularized Kuhnian versions" applied to the competition of behaviorism and cognitive psychology in learning theory since 1930, and (rather superficially) to the Newton/Maxwell/Einstein tradition in physics.

Larry Laudan (1983), "The Demise of the Demarcation Problem," in *Physics, Philosophy and Psychoanalysis,* ed. R. S. Cohen and L. Laudan (Boston: Reidel), 111–27.

Elie G. Zahar (1982), "The Popper-Lakatos Controversy," *Fundamenta Scientiae* 3:21–54.

Robert S. Cohen et al., eds. (1976), *Essays in Memory of Imre Lakatos* (Boston: Reidel).

Colin Howson, Ed. (1976), *Method and Appraisal in the Physical Sciences* (New York: Cambridge Univ. Press), is the best introduction to the Lakatos "methodology" and its application to the "rational reconstruction" of history of science, with a critique by Paul Feyerabend.

M. A. B. Deakin (1976), "On Urbach's Analysis of the 'IQ Debate,' " *British Journal for the Philosophy of Science* 27:60–65.

P. A. Schilpp, ed. (1974), *The Philosophy of Karl Popper* (La Salle, Ill.: Open Court), includes an autobiography, 3–181, and essays by H. Putnam, Lakatos, T. Settle, H. Margenau, P. Suppes, A. Gruenbaum, T. S. Kuhn, etc.

Imre Lakatos and Alan Musgrave, eds. (1970), *Criticism and the Growth of Knowledge* (New York: Cambridge Univ. Press), are proceedings of a symposium on the theories of Popper, Kuhn, and Lakatos; includes Lakatos, "Falsification and the Methodology of Scientific Research Programmes," 91–195.

Topic *iii:* Kuhn; Paradigms and Revolutions; Feyerabend

Thomas S. Kuhn (1983), "Reflections on Receiving the John Desmond Bernal Award," *4S Review* 1, no. 4:26–30, comments on the relation between *SSR* and the sociology of scientific knowledge (see **11.6-*iii***), and complains that the term "interest" has been limited to socioeconomic ones, excluding "the special cognitive interests inculcated by scientific training"; (1983), "Commensurability, Comparability, Communicability," in *PSA 1982,* ed. P. D. Asquith and T. Nickles (East Lansing, Mich.: Philosophy of Science Association), 669–89, with following comments by P. Kitcher and M. Hesse; (1979), "History of Science," in *Current Research in Philosophy of Science,* ed. P. D. Asquith and H. E. Kyburg, Jr., (East Lansing, Mich.: Philosophy of Science Association), 121–28; (1977), *The Essential Tension* (Chicago: Univ. of Chicago Press), see especially "Second Thoughts on Paradigms," 293–319; (1976), "Theory-Change as Structure-Change: Comments on the Sneed Formalism," *Erkenntnis* 10:179–99; (1962/1970), *The Structure of Scientific Revolutions* (Chicago: Univ. of Chicago Press).

Paul K. Feyerabend (1981), *Philosophical Papers* (New York: Cambridge Univ. Press); (1975), *Against Method* (Atlantic Highlands, N.J.: Humanities Press); (1965), "Problems of Empiricism," in *Beyond the Edge of Certainty,* ed. R. G. Colodny (Englewood Cliffs, N.J.: Prentice-Hall), 145–260; (1970), " . . . Part II," in *The Nature and Function of Scientific Theories,* ed. R. G. Colodny (Pittsburgh: Univ. of Pittsburgh Press), 275–353.

Daniel Goldman Cedarbaum (1983), "Paradigms," *Studies in History and Philosophy of Science* 14:173–213, traces the concept back to G. C. Lichtenberg and L. Wittgenstein, and discusses the influence of W. V. Quine on Kuhn's views.

John D. Heyl (1981–1982), "Kuhn, Rostow, and Palmer: The Problem of Purposeful Change in the 60s," *Historian* 44:299–313.

J. Hintikka et al., eds. (1981), *Proceedings of the 1978 Pisa Conference on the History and Philosophy of Science* (Boston: Reidel), includes discussions of the ideas of Kuhn, Sneed, Stegmueller, and others on the relations between history and philosophy of science and several applications of philosophical analysis to historical cases.

Gary Gutting, ed. (1980), *Paradigms and Revolutions: Applications and Appraisals of Thomas Kuhn's Philosophy of Science* (Notre Dame, Ind.: Univ. of Notre Dame Press), is an anthology of articles by Shapere, Musgrave, MacIntyre, Stegmueller, King, Hollinger, Laudan, Greene, etc.

Nathan Reingold (1980), "Through Paradigm-Land to a Normal History of Science," *Social Studies of Science* 10:475–96, expresses the viewpoint of a leading historian of science.

J. C. Sheldon (1980), "A Cybernetic Model of Physical Science Professions: The Causes of Periodic Normal and Revolutionary Science between 1000 and 1870 A.D.," *Scientometrics* 2, no. 2:147–67.

Douglas Lee Eckberg and Lester Hill, Jr. (1979), "The Paradigm Concept and Sociology: A Critical Review," *American Sociological Review* 4:925–37.

Robert K. Merton (1977), "The Sociology of Science: An Episodic Memoir," in *The Sociology of Science in Europe,* ed. R. K. Merton and J. Gaston (Carbondale: Southern Illinois Univ. Press), 3–141, includes an account and interpretation of Kuhn's career, 76–109.

I. Lakatos and A. E. Musgrave, eds. (1970), *Criticism and the Growth of Knowledge* (New York: Cambridge Univ. Press), includes Kuhn, "Logic of Discovery or Psychology of Research?" and "Reflections on My Critics"; Margaret Masterman's "The Nature of a Paradigm," 59–89, is a frequently cited analysis of the multiple meanings Kuhn gave to paradigms in *SSR.*

Topic *iv:* Thematic Analysis (Holton)

Gerald Holton (1988), *Thematic Origins of Scientific Thought,* rev. ed. (Cambridge: Harvard Univ. Press); (1986), *The Advancement of Science, and Its Burdens* (New York: Cambridge Univ. Press); (1978), *The Scientific Imagination* (New York: Cambridge Univ. Press), Chaps. 1–4; (1964), "Stil und Verwirklichung in der Physik," *Eranos Jahrbuch* 33:319–63; (1962), "Ueber die Hypothesen, welche der Naturwissenschaft zugrunde liegen," ibid. 31:351–425.

John Losee (1987), *Philosophy of Science and Historical Enquiry* (New York: Oxford Univ. Press), Chap. 8.

Angèle Kremer-Marietti (1987), "Thematic Analysis," In *Encyclopedia of Library and Information Science,* ed. Allen Kent, vol. 41, Suppl. 16 (New York: Dekker), 332–39; (1984), "Holton, Gerald," *Dictionnaire* des Philosophes, ed. D. Huisman (Paris: Presses Universitaire de France), 1:1247–51.

Stephen Toulmin (1982), "The Intellectual Authority and the Social Context of the Scientific Enterprise: Holton, Rescher and Lakatos," *Minerva* 18:652–67.

I. C. Jarvie (1980), "On the History of Science," *Queen's Quarterly* [Canada] 87:65–68, is a comparison of views of Holton and Toulmin.

Robert K. Merton (1975), "Thematic Analysis in Science: Notes on Holton's Concept," *Science* 188:335–38.

Topic *v:* Historical Realism (Toulmin, Shapere, Laudan, etc.)

Dudley Shapere (1986), "External and Internal Factors in the Development of Science," *Science and Technology Studies* 4, no. 1:1–9, and other papers cited therein; comments by P. T. Carroll and S. Turner and reply by Shapere, ibid. 10–23; (1984), *Reason and the Search for Knowl-*

edge (Boston: Reidel); (1974), "Scientific Theories and Their Domains," in *The Structure of Scientific Theories,* ed. F. Suppe (Urbana: Univ. of Illinois Press), 518–65; (1969), "Notes toward a Post-Positivistic Interpretation of Science," in *The Legacy of Logical Positivism,* ed. P. Achinstein and S. F. Barker (Baltimore: Johns Hopkins Univ. Press), 115–60.

Larry Laudan (1984), *Science and Values: The Aims of Science and Their Role in Scientific Debate* (Berkeley: Univ. of California Press); (1981), *Science and Hypothesis: Historical Essays on Scientific Methodology* (Boston: Reidel); (1979), "Historical Methodologies: An Overview and Manifesto," in *Current Research in Philosophy of Science,* ed. P. D. Asquith and H. E. Kyburg (East Lansing, Mich.: Philosophy of Science Association), 40–54; (1977), *Progess and Its Problems* (Berkeley: Univ. of California Press).

Stephen Toulmin (1982), *The Return to Cosmology* (Berkeley: Univ. of California Press); (1981), "The Emergence of Post-Modern Science," *Great Ideas Today,* 69–114; (1977), "From Form to Function: Philosophy and History of Science in the 1950s, and Now," *Daedalus* 106:143–62; (1972), *Human Understanding,* vol. I (Princeton, N.J.: Princeton Univ. Press); (1966), "The Evolutionary Development of Natural Science," *American Scientist* 55:456–71.

Thomas Nickles, ed. (1980), *Scientific Discovery: Case Studies* (Boston: Reidel).

Lindley Darden and Nancy Maull (1977), "Interfield Theories," *Philosophy of Science* 4:43–64; Lindley Darden (1979), "Discoveries and the Emergence of New Fields of Science," in *PSA 1978,* ed. P. D. Asquith and I. Hacking (East Lansing, Mich.: Philosophy of Science Association) 1:149–60.

Imre Lakatos (1976), "Understanding Toulmin," *Minerva* 14:126–43, concerns his relation to Wittgenstein.

11.6 THE SOCIOLOGY OF SCIENCE

i. Social history of science, influenced by Marx and Max Weber; Merton's study of the rise of science in seventeenth-century England; J. D. Bernal and the social function of science; Soviet sociology of science

ii. The social system of contemporary science; Robert Merton and the American school; studies of the reward system, Matthew effect, simultaneous discovery; norms and their violation: fraud in science

iii. Sociology of knowledge and other European traditions; the

Edinburgh group; David Bloor, Henry Collins, "strong programme," and relativism in sociology of science

iv. Anthropological approaches; Mary Douglas and grid-group analysis; laboratory observation

v. Scientometrics: quantitative studies of science; citation analysis

READINGS

I do not treat this as a separate topic but briefly discuss the major approaches at the beginning of my course and then present some of the case studies mentioned below in connection with the corresponding scientific subjects. A good survey is in Marks, *Science,* 203–8, 480–88; see also L. J. Jordanova, "The Social Sciences and History of Science and Medicine," in *Information Sources in the History of Science and Medicine,* ed. P. Corsi and P. Weindling (1983), 81–96.

For *i:* Joseph Ben-David, *The Scientist's Role in Society* (1984), Chap. 1, "The Sociology of Science," 1–20; Chap. 6, "The Rise and Decline of the French Scientific Center . . . ," 88–107; Chap. 7, "German Scientific Hegemony and the Emergence of Organized Science," 108–38; "The Professionalization of Research in the United States," 139–68; Chap. 9, "Conclusion," 169–85. J. D. Bernal, *The Social Function of Science* (1939/1967), 1–34, 408–16.

Robert Merton, "Puritanism, Pietism, and Science," *Sociological Review* 28(1936):1–30 and "Science and the Economy of Seventeenth-Century England," *Science and Society* 3(1939):3–27, reprinted in his *Social Theory and Social Structure* (1957), 574–606, 607–27, and in Bernard Barber and Walter Hirsch, eds., *The Sociology of Science* (1962), 33–66, 67–88. A. Rupert Hall, "Merton Revisited . . . ," *History of Science* 2(1963):1–16.

For *ii:* Merton's major contributions to the subject are reprinted in his *The Sociology of Science* (1973), including an introductory essay by Norman Storer on the development of Merton's theories, and the frequently cited papers on behavior patterns of scientists, multiple discoveries, and the Matthew effect. The articles on "Science and the Social Order" and "Priorities in Scientific Discovery . . ." are also reprinted in the Barber-Hirsch book cited above. It is probably easier for students to read Merton's own papers than articles by others about his ideas.

For *iii:* H. M. Collins, "The Sociology of Scientific Knowledge: Studies of Contemporary Science," *Annual Review of Sociology* 9(1983):265–85. David Bloor, *Knowledge and Social Imagery* (1976), Chap. 1, "The Strong Programme in the Sociology of Knowledge," 1–19. Collins has edited a collection of studies using this approach, "Knowledge and Controversy," published as a special issue of *Social Studies of Science* 11, no. 1 (1981); his introduction includes an extensive bibliography.

For *iv:* Mary Douglas, ed., *Essays in the Sociology of Perception* (1982), editor's "Introduction to Grid/Group Analysis," 1–8, and "Introduction" to section on history and history of ideas, 115–19; articles in this book by D. Bloor, "Polyhedra . . . ," 191–218, and M. Rudwick, "Cognitive Styles in Geology," 219–41.

For *v:* Derek J. de Solla Price, *Little Science Big Science* (1963), 1–91. Henry W. Menard, *Science: Growth and Change* (1971). Arnold Thackray, "Measurement in the Historiography of Science," 11–24, and other articles in Y. Elkana et al., eds., *Toward a Metric of Science* (1978) (see Bibliography).

SYNOPSIS

According to Marx and Engels, science and technology are strongly influenced by the economic needs and power relations in a society. In the 1930s several British scientists undertook Marxist analyses of science to promote its application to social problems; the permanent legacy of this movement is J. D. Bernal's remarkable collection of writings interpreting the history of science. Among the historic examples of scientific advances said to be strongly influenced by the economic needs of society is thermodynamics; even non-Marxist historians generally admit that attempts to improve the efficiency of the steam engine, a prime mover of the Industrial Revolution in Europe, were at least partly responsible for fundamental research in this area of physics.

During the same period in the United States, Robert Merton was following the inspiration of Max Weber and R. H. Tawney in explaining the rise of science in seventeenth-century England as a phenomenon assisted by the Protestant or more specifically Puritan ethic.

Merton went on to become one of the most influential sociologists in the United States in the 1940s and 1950s; he founded the "American school" of the sociology of science. The Mertonian approach stresses science as a social system supported by a set of norms governing proper behavior, and a reward system that enforces those norms. Merton's best-known papers discuss the fact that scientists claim to be uninterested in "priority" for their discoveries while in fact fighting vigorously for it, and the fact that credit for a coauthored discovery is generally given to the author who subsequently becomes more famous (Matthew effect).

While many of Merton's students do not seem to share his interests in the history of science (see e.g., Stephen Cole's explicit rejection of historical evidence in his 1970 paper), the sociological approach to the development of science has been taken up by others. An interesting example is Joseph Ben-David's study showing how the overcentralization of French medical science in the nineteenth century contributed to its decline relative to Germany, where competition between several academic centers helped new specialties to become established [*American Sociological Review* 25(1960):828–43].

Around 1970 another movement known as "sociology of scientific knowledge" arose, partly in reaction to the refusal of the Mertonians to consider social aspects of the *content* of scientific theories. Most of its advocates were British and were familiar with the European tradition of "sociology of knowledge" developed by K. Mannheim; there was also some inspiration from T. S. Kuhn's *Structure of Scientific Revolutions.* Barry Barnes, David Bloor, and H. M. Collins argue for a "symmetrical" or "relativist" approach: whereas other scholars had assumed that a sociological explanation is appropriate only for irrational or wrong ideas, they claimed that *all* knowledge is "socially constructed." Bloor was especially interested in applying this claim to mathematics, generally supposed to be more objective than other sciences; if one can show that the truth of a theorem depends on the worldview or personality of the mathematician, then it should not be hard to establish a relativist conception of truth in physics and biology.

This view, also known as the "strong programme" in the sociology of science, is of course subject to criticism on the grounds of self-reference (cf., the logical paradoxes mentioned in **11.3**); this and other criticisms were forcefully presented by the philosopher of science Larry Laudan in 1980 (see the published version of the Laudan-Bloor debate and related papers in *Philosophy of the Social Sciences,* June 1981).

Whereas Collins, Pinch, Barnes, Bloor, Pickering, Harvey, and others engaged in the sociology of scientific knowledge attempt to become knowledgeable observers of (and sometimes even participants in) the scientific community concerned with a particular problem, another group takes the approach of the anthropologist who can only observe the behavior of another culture without really understanding it. This is the "laboratory life" approach exemplified by the book of that title by Bruno Latour and Steve Woolgar (1979). June Goodfield's *An Imagined World* (1981) is somewhat more accessible as an introduction to the work and ideas of a scientist, closely observed. On a more theoretical level, a few scholars have recently tried to apply the "grid-group" classification of worldviews proposed by the anthropologist Mary Douglas.

"Scientometrics," the quantitative study of science (especially its literature), is a subject inspired by the work of Derek Price; Eugene

Garfield developed the "science citation index" and its computerized data base, which has made possible new kinds of research on the connections between scientists. These studies have obvious applications to the sociology of contemporary science and to science policy research; a group at the National Science Foundation is actively engaged in compiling and analyzing "science indicators" in order to assess the state of United States compared to world science. But these methods have not yet been fully appreciated or exploited by historians of science. The recent project to extend the citation index back to 1920 for physics (and possibly for related sciences such as astronomy and chemistry) should be of considerable value for understanding the development of twentieth-century science.

Presumably those with leftist leanings would favor the claim by Jose Ortega y Gasset (*Revolt of the Masses,* 1930) that science has progressed by adding up the contributions of large numbers of researchers of modest ability who pave the way for the more noteworthy advances of the genius. This "Ortega hypothesis" has been tested and refuted by Jonathan and Stephen Cole (1972), using the 1965 *Science Citation Index.*

Sociological studies of bias in science have debunked some widely held suspicions. In *Fair Science* (1979), Jonathan Cole concluded that the universalistic norms of science are generally satisfied in practice, and in particular that "the measurable amount of sex-based discrimination against women scientists is small" (86). A recent study by Stephen and Jonathan Cole and Gary Simon (1981) concluded that there is "no evidence of systematic bias in the selection of NSF reviewers" and that "getting a research grant depends to a significant extent on chance." All of these conclusions, based on statistical analysis, have of course been challenged by those whose experience and intuition suggest otherwise.

BIBLIOGRAPHY

John Ziman (1985), *An Introduction to Science Studies: The Philosophical and Social Aspects of Science* (New York: Cambridge Univ. Press).

James Robert Brown, ed. (1984), *Scientific Reality: The Sociological Turn* (Boston: Reidel), includes L. Laudan's critique and D. Bloor's defense of the "strong programme"; other essays by G. Gutting, B. Barnes, E. McMullin, I. Jarvie, A. Lugg, J. Gaston, and J. M. Nicholas.

Michael Mulkay (1981), "Sociology of Science in the West," *Current Sociology* 28, no. 3:1–184, is a 116-page survey dealing with the emergence of the specialty, patterns of scientific growth, social inequality, social norms and evaluative repertoires, social construction of scientific knowledge, and the political dimension; annotated bibliography of 342 items, mostly since 1970, with "a strong bias in favour of items written in English." The same issue contains an article by Vojin Milic, "The Science of Science in European Socialist Countries," 185–342.

Jerry Gaston, ed. (1978), *Sociology of Science* (San Francisco: Jossey-Bass) (also published as a special issue of *Sociological Inquiry,* vol. 48). Although the emphasis is on American work, all five approaches are discussed or at least mentioned in one of these articles. Includes articles on sexism and racism, theory choice, relations between history and sociology of science, quantitative studies, and comparison of United States and British traditions.

Jonathan R. Cole and H. Zuckerman (1975), "The Emergence of a Scientific Specialty: The Self-Exemplifying Case of the Sociology of Science," in *The Idea of Social Structure,* ed. L. Coser (New York: Harcourt Brace Jovanovich), 139–74, includes an analysis of the most-cited authors in various time periods.

Roger Hahn (1975), "New Directions in the Social History of Science," *Physis* 17:205–18.

Topic *i:* Social History of Science

Robert K. Merton (1970), *Science, Technology, and Society in Seventeenth Century England* (New York: Harper and Row), is a reprint of the 1936 essay with a new introduction by Merton.

J. D. Bernal (1972), *The Extension of Man: A History of Physics before the Quantum* (Cambridge: MIT Press); (1971), *Science in History,* 4 vols. (Cambridge: MIT Press); (1953/1969), *Science and Industry in the Nineteenth Century* (Bloomington: Indiana Univ. Press); (1939/1967), *The Social Function of Science* (Cambridge: MIT Press), the 1967 reprint includes a new preface, "After Twenty-Five Years."

Loren R. Graham (1985), "The Socio-Political Roots of Boris Hessen: Soviet Marxism and the History of Science," *Social Studies of Science* 5:705–22, explains the context of Hessen's famous 1931 paper on Newton: it was intended to help protect Einstein's relativity theory, at that time under attack in the USSR by Marxist ideologists.

Steven Shapin (1982), "History of Science and Its Sociological Reconstructions," *History of Science* 20:157–211; (1980), "A Course in the Social History of Science," *Social Studies of Science* 10:231–58, has suggested readings on various topics, mainly seventeenth to nineteenth century, including psychology and social reform; heredity in medicine; science and woman's role; scientific naturalism and the reaction against it.

Jerome Ravetz (1981), "Bernal's Marxist Vision of History," *Isis*

72:393–402, followed by R. S. Westfall, "Reflections on Ravetz's Essay," ibid. 402–5.

Robert Fox and Georg Weisz, eds. (1980), *The Organization of Science and Technology in France 1808–1914* (New York: Cambridge Univ. Press).

Barry Barnes and Steven Shapin, eds. (1979), *Natural Order: Historical Studies of Scientific Culture* (Beverly Hills, Calif.: Sage), includes articles on theories of the Earth, social Darwinism, reception of non-Euclidean geometry, physics and psychics, the biometry-Mendelism controversy, heredity vs. environment debate.

Steven Shapin and Arnold Thackray (1978), "Prosopography as a Research Tool in History of Science: The British Scientific Community 1700–1900," *History of Science* 12:1–28.

Robert E. Filner (1978), "J. D. Bernal and the Social Function of Science," *Humanities Perspectives on Technology* 7:7–10; (1976), "The Roots of Political Activism in British Science," *Bulletin of the Atomic Scientists* 32, no. 1:25–29, covers the 1930s and 1940s.

Linda L. Lubrano (1976), *Soviet Sociology of Science* (Columbia, Ohio: American Association for the Advancement of Slavic Studies), is a comprehensive review of the literature.

Robert Young (1973), "The Historiographic and Ideological Contexts of the Nineteenth-Century Debate on Man's Place in Nature," in *Changing Perspectives in the History of Science,* ed. M. Teich and R. Young (Boston: Reidel), 344–438, is a partly autobiographical Marxist-oriented account, critique of "internalist" historians of science; argues that Kuhn has dissociated himself from the radical implications of his own theory.

Topic *ii:* Social System of Contemporary Science

Robert K. Merton (1973), *The Sociology of Science* (Chicago: Univ. of Chicago Press).

Richard Whitley (1985), *The Intellectual and Social Organization of the Sciences* (New York: Oxford Univ. Press); (1972), "Black Boxism and the Sociology of Science: A Discussion of the Major Developments in the Field," in *The Sociology of Science,* ed. Paul Halmos (New York: Humanities Press), 61–92, argues that Kuhn "legitimated sociologists' revolt against Merton," especially in Britain. Inside the "black box" is the content of science, ignored by the Mertonians.

David A. Hollinger (1983), "The Defense of Democracy and Robert K. Merton's Formulation of the Scientific Ethos," *Knowledge and Society* 4:1–15.

Robert K. Merton and Jerry Gaston, eds. (1977), *The Sociology of*

Science in Europe (Carbondale: Southern Illinois Univ. Press), includes Merton's "episodic memoir" on the history of sociology of science, the abortive attempt to develop an archive for historical sociology of science in connection with the *Dictionary of Scientific Biography,* remarks on the citation index, and on T. S. Kuhn's career; chapters on sociology of science in West Germany and Austria, Poland, Britain, France, Italy, the USSR, and Scandinavia.

Joseph Ben-David and Teresa A. Sullivan (1975), "Sociology of Science," *Annual Review of Sociology* 1:203–22.

Fraud and cheating:

C. J. List (1985), "Scientific Fraud: Social Deviance or the Failure of Virtue?" *Science, Technology and Human Values* 10, no. 4:27–36, proposes an evolutionary explanation of why fraud may occur in science, and why women scientists (according to List) rarely cheat.

William Broad and Nicholas Wade (1982), *Betrayers of the Truth* (New York: Simon and Schuster), argue that the contemporary scientific system fails to provide protection against fraud; a rather sensationalized account, generalizing from cases the authors have investigated themselves; critical reviews by scientists have gone too far in the other direction in refusing to acknowledge that a problem exists.

David L. Hull (1978), "Altruism in Science: A Sociobiological Model of Cooperative Behaviour among Scientists," *Animal Behaviour* 26:685–97.

Harriet Zuckerman (1977), "Deviant Behavior and Social Control in Science," in *Deviance and Social Change,* ed. Edward Sagarin (Beverly Hills, Calif.: Sage), 87–138.

Stephen Cole (1970), "Professional Standing and the Reception of Scientific Discoveries," American Journal of Sociology 76:286–306, argues that historical evidence is of little value because the social structure of science has changed so much in recent years.

Topic *iii:* Sociology of Knowledge, Strong Programme

David Bloor (1976), *Knowledge and Social Imagery* (Boston: Routledge and Kegan Paul); see below under Laudan (1981) for discussion.

Jonathan Harwood (1986), "Ludwik Fleck and the Sociology of Knowledge," *Social Studies of Science* 16:173–87, is an essay review of recent editions of Fleck's work and works about him.

Michael Mulkay and G. Nigel Gilbert (1984), "Opening Pandora's Box: A Case for Developing a New Approach to the Analysis of Theory Choice in Science," *Knowledge and Society* 5:113–39.

Karin Knorr-Cetina and Michael Mulkay, eds. (1983), *Science Ob-*

served: Perspectives on the Social Studies of Science (Beverly Hills, Calif.: Sage); Karin Knorr-Cetina (1981), *The Manufacture of Knowledge* (New York: Pergamon). See the critical review by J. Agassi (1984), *Inquiry* 27:166–72.

Nils Roll-Hansen (1983), "The Death of Spontaneous Generation and the Birth of the Gene: Two Case Studies of Relativism," *Social Studies of Science* 13:481–519, discusses cases used to show the social relativism of science: W. Provine and D. MacKenzie and B. Barnes on the biometrician-Mendelian controversy [in (1979), *Natural Order,* ed. Barnes and Shapin (cited above), 191] and J. Farley and G. Geison on the Pasteur-Pouchet debate [*Bulletin of the History of Medicine* 48(1974):161]. The author argues that the relativists have misunderstood the scientific arguments at crucial points and that the cases really support a rationalist rather than a relativist view of science.

H. M. Collins and T. J. Pinch (1982), *Frames of Meaning: The Social Construction of Extraordinary Science* (London: Routledge and Kegan Paul), is on paranormal metal bending, Uri Geller, Kuhn's paradigms, and the philosophy of quantum mechanics.

Steve Woolgar (1981), "Interests and Explanations in the Social Study of Science," *Social Studies of Science* 11:365–94.

Augustine Brannigan (1981), *The Social Basis of Scientific Discoveries* (New York: Cambridge Univ. Press), includes a study of Mendel's theory (**4.3**).

Joseph Ben-David (1981), "Sociology of Scientific Knowledge," in *The State of Sociology,* ed. J. F. Short, Jr. (Beverly Hills, Calif.: Sage), 40–59, is a critical review of the Marx-Mannheim sociology of knowledge and its recent revival; states that P. Forman and others have failed to demonstrate any external influences on science.

Karin Knorr, Roger Krohn, and Richard Whitley, eds. (1980), *The Social Process of Scientific Investigation* (Boston: Reidel).

Larry Laudan (1981), "The Pseudo-Science of Science?" *Philosophy of the Social Sciences* 11:173–98, is a critique of the "strong programme"; see Bloor's reply, "The Strengths of the Strong Programme," ibid., 199–213. See also H. M. Collins, "What is TRASP?: The Radical Programme as a Methodological Imperative," ibid., 215–24, and other papers in this issue. Angus Gellatly (1980), "Logical Necessity and the Strong Programme for the Sociology of Knowledge," *Studies in History and Philosophy of Science* 11:325–39; Gad Freudenthal (1979), "How Strong is Dr. Bloor's 'Strong Programme'?" ibid., 10:67–83.

Everett Mendelsohn, Peter Weingart, and Richard Whitley, eds. (1977), *The Social Production of Scientific Knowledge* (Boston: Reidel), includes a general introduction by Mendelsohn, a detailed study of

David Bohm's and John von Neumann's work on hidden variables in quantum physics by Pinch (**11.2-*v***), a brief essay on the creation-evolution issue by Dorothy Nelkin, and several theoretical pieces.

Roy MacLeod (1977), "Changing Perspectives in the Social History of Science," in *Science, Technology and Society,* ed. I. Spiegel-Rösing and D. de S. Price (Beverly Hills, Calif.: Sage), 149–95.

Topic *iv:* Anthropology, Laboratory Life

Michael MacDonald (1983), "Anthropological Perspectives on the History of Science" in *Information Sources in the History of Science and Medicine,* ed. P. Corsi and P. Weindling (Boston: Butterworth Scientific), 61:80.

Mary Douglas, ed. (1982), *Essays in the Sociology of Perception* (Boston: Routledge and Kegan Paul); Douglas (1978), *Cultural Bias* (London: Royal Anthropological Institute).

Everett Mendelsohn and Yehuda Elkana, eds. (1981), *Sciences and Cultures* (Boston: Reidel), includes Yehuda Elkana "A Programmatic Approach at an Anthropology of Knowledge," 1–76; Kenneth L. Caneva, "What Should We Do with the Monster?: Electromagnetism and the Psychosociology of Knowledge," 101–31; Wolf Lepenies, "Anthropological Perspectives in the Sociology of Science," 245–61.

Bruno Latour and Steve Woolgar (1979), *Laboratory Life* (Beverly Hills, Calif.: Sage).

Topic *v:* Scientometrics

Michael H. MacRoberts and Barbara R. MacRoberts (1986), "Quantitative Measures of Communication in Science: A Study of the Formal Level," *Social Studies of Science* 16:151–72, is a critique of citation analysis on the grounds that very little "influence" appears as references in bibliographies.

Stephen Cole, J. R. Cole, and Gary A. Simon (1981), "Chance and Consensus in Peer Review," *Science* 214:881–86; (1981, 1982), letters to editor by R. C. Atkinson and others, *Science* 214:1292–94, 215:344–48.

Eugene Garfield (1979), *Citation Indexing* (New York: Wiley), especially Chap. 8, "Mapping the Structure of Science."

Yehuda Elkana, Joshua Lederberg, Robert K. Merton, Arnold Thackray, and Harriet Zuckerman, eds. (1978), *Toward a Metric of Science: The Advent of Science Indicators* (New York: Wiley/Interscience), includes Thackray, "Measurement in the Historiography of Science"; Gerald Holton, "Can Science Be Measured?"; Derek Price, "Toward a Model for Science Indicators"; Manfred Kochen, "Models of Scientific

Output"; Eugene Garfield, Morton V. Malin, and Henry Small, "Citation Data as Science Indicators," and other articles.

Jonathan R. Cole (1979), *Fair Science* (New York: Free Press); review by Karen Oppenheim Mason (1980), in *Science* 208:277–78.

D. Hywel White and Daniel Sullivan (1979), "Social Currents in Weak Interactions," *Physics Today* 32, no. 4:40–47, in which citation analysis is used to study the reward system in particle physics. White, Sullivan, and Edward J. Barboni (1979), "The Interdependence of Theory and Experiment in Revolutionary Science: The Case of Parity Violation," *Social Studies of Science* 9:303–27, use the Lakatos categories (**11.5**-*ii*) as a guide for citation analysis. Sullivan, White and Barboni (1977), "The State of a Science: Indicators of the Specialty of Weak Interactions," *Social Studies of Science* 7:167–200.

12

ASTRONOMY IN THE NINETEENTH CENTURY

12.1 WILLIAM HERSCHEL AND NEW DIRECTIONS IN ASTRONOMY CIRCA 1800

i. The solar spectrum; infrared radiation; Fraunhofer lines
ii. Discovery of new members of the solar system: Uranus, asteroids, Neptune; Bode's law; advance of Mercury's perihelion
iii. Attempts to find distances of stars; binary star systems; proper motion of stars and the Sun's motion through galaxy; structure of the universe
iv. Construction of large telescopes; role of instrumentation in astronomy; clock drive

READINGS

John Marks, *Science and the Making of the Modern World* (1983), 340–41 (*i*). A. Berry, *A Short History of Astronomy* (1898), 323–53. D. L. Anderson, *Discoveries in Physics* (1973), 5–30 (*ii*). G. E. Tauber, *Man's View of the Universe* (1979), 212–29. C. Whitney, *Discovery of Our Galaxy* (1971), 87–132.

Articles: Michael A. Hoskin, "William Herschel and the Making of Modern Astronomy," *Scientific American* 254, no. 2(1986):106–12; "Cosmology in the Eighteenth and Nineteenth Centuries," in *Human Implications of Scientific Advance,* ed. E. G. Forbes (1978), 544–52; "Apparatus and Ideas in Mid-Nineteenth-Century Cosmology," *Vistas in Astronomy* 9(1967):79–85. J. A. Bennett, "The Giant Reflector, 1770–1870," ibid. 553–58. C. L. Doolittle, "Some Advances Made in Astronomical Science during the Nineteenth Century," *Science* 14(1901):1–12.

SYNOPSIS

Modern astronomy may be said to begin with the work of William Herschel (1738–1822). His emphasis on the construction of large telescopes, his discovery of a new planet, his investigation of the stars, and his speculations about the large-scale structure of the universe, all opened up important new lines of research in astronomy.

Herschel's personal life and his family are worth some mention. He was born Friedrich Wilhelm Herschel in Hanover, Germany. His father played the oboe in the regimental band of the Hanoverian Foot-Guards, and William also learned the oboe. He went with his father's band to England in 1756 where he learned the English language. When the Guards were defeated by the French in 1757, he escaped to England where he managed to support himself by copying, teaching, performing, and composing music. He brought his sister Caroline Lucretia Herschel (1750–1848) from Germany to keep house and pursue her own career as a singer but eventually got her involved in his astronomical work. She became (somewhat reluctantly, it appears) the first major woman astronomer. William's son John Herschel became an influential scientist in Victorian England.

William Herschel's interest in the theory of music led him to read a book by Robert Smith on *Harmonics,* and then to a book on *Opticks* by the same author. The description of telescopes and the discoveries made with them, as recounted in Smith's book, aroused his interest in astronomy, and he soon became an expert at grinding mirrors for hugh reflecting telescopes. By about 1780 his instruments were the best in the world.

On 13 March 1781, while sweeping the heavens with his telescope in search of double stars, Herschel noticed an object he thought at first was a comet. Subsequently A. J. Lexell computed its orbit, which he found to be nearly circular with a diameter about twice that of Saturn, and concluded that it must be a planet. Herschel proposed the name *Georgium Sidus* in honor of King George III of England (who was also associated with Hanover). But the classical name Uranus, suggested by Johann Elbert Bode, was generally accepted. (Uranus was the father of Saturn in Roman mythology.)

As the first person in recorded history to discover a new planet, Herschel became famous overnight and was able to get substantial support for his research, especially for constructing larger telescopes. His initial goal in astronomy was to determine the actual distances of the stars from the Earth. Galileo had suggested that this could be done by the parallax effect if one could find a star located almost directly behind another star. It was for this reason that Herschel had been looking for double stars when he discovered Uranus. Though he didn't find a measurable parallax in this way, Herschel did find so many double stars that it became doubtful whether they could all be chance alignments of stars at different distances from the Earth. (That statement itself implies some kind of statistical reasoning and led to a debate among British scientists around 1850, just prior to Maxwell's introduction of statistical reasoning in kinetic theory—see **7.2**.) Instead, he found in 1802 that some of them are really binary star systems—two stars moving in orbit around their common center of mass. Analysis of such systems later provided the first definite evidence that Newton's law of gravity is valid for stars as well as within our solar system.

Herschel devoted much effort to making systematic surveys of stars in all parts of the sky to establish the overall structure of the universe (or, as we would now say, the "Milky Way" galaxy). He assumed that the number of stars in each direction is proportional to the distance to the edge of the galaxy. In this way he developed the now familiar picture of a lens-shaped galaxy, corresponding to the observation that many more stars can be seen in the equatorial plane than in directions perpendicular to that plane.

Herschel also studied the fuzzy objects known as "nebulae," which he tried to resolve into individual stars. He suggested that they may really be distant star systems comparable to our own galaxy. Later he recognized the existence of "true nebulosity," i.e., clouds of self-luminous matter, not resolvable into separate stars. He also proposed that such nebulae may evolve into star systems. This theory, published in 1811–1814, was apparently independent of the earlier hypothesis of Kant and Laplace, but helped to make it more plausible and gave it its name, "nebular hypothesis."

Herschel made two significant discoveries in connection with the sun. In 1783 he showed that existing data on the "proper motions" of the stars could be explained by assuming the sun, with the solar system, is moving toward a point in the constellation Hercules. Though originally based on rather crude data, this conclusion is now considered basically correct. Thus Herschel may be regarded as the successor to Copernicus; he showed the sun is not at rest and is not the center of the universe, but instead is moving through the universe.

The other solar discovery was made in 1800 when Herschel observed that rays beyond the visible part of the sun's spectrum produce the sensation of heat. He found that these "infrared rays" or "radiant heat" behave like light. This discovery, elaborated by other scientists, led to the idea that light and heat are essentially the same phenomenon, and as noted in **7.2** this idea resulted in the rejection of the caloric theory of heat after the wave theory of light was accepted in the 1820s.

In my course I use the above list of Herschel's discoveries as the starting point for a discussion of several lines of astronomical research in the nineteenth century. To begin with the sun: in 1802 W. H. Wollaston noticed some dark lines in the solar spectrum. The study of these dark lines was taken up more systematically by Joseph Fraunhofer (1787–1826). The quantitative study of spectra from various sources, as formulated by Kirchhoff and Bunsen, allowed a chemical analysis of astronomical objects (**9.3**) and provided the basis for atomic physics in the early twentieth century (**9.5**).

Fraunhofer was also important to the history of astronomy through his development of a clock drive for the telescope, which moves it at the correct rate to cancel the effects of the Earth's rotation. Thus one can keep a star at rest in the field of view, making possible much more precise measurements. Fraunhofer's "heliometer," designed for measuring the diameter of the sun or the distance between two stars, was used by Friedrich Wilhelm Bessel in his discovery of parallax (**12.2**).

To discuss the discovery of new members of the solar system it may be helpful to go back to 1776 when J. D. Titius proposed a formula for the distances of the planets from the sun. The formula was publicized by Johann Elbert Bode in 1772 and became known as the Titius-Bode law or, more frequently, "Bode's Law." It states that if you take the series n = 0, 1, 2, 4, 8, 16, 32, . . . , triple each number, add 4 and divide by 10, you get the radii of the planetary orbits measured in astronomical units (1 AU = average Earth-Sun distance).

i	n	3n	(3n + 4)/10		radius of orbit in AU
1	0	0	0.4	Mercury	0.39
2	1	3	0.7	Venus	0.72
3	2	6	1.0	Earth	1.00
4	4	12	1.6	Mars	1.53
5	8	24	2.8	?	
6	16	48	5.2	Jupiter	5.22
7	32	96	10.0	Saturn	9.6

The Titius-Bode law had to be taken seriously when the distance of Uranus was found to be 19.3 AU, very close to the "predicted" value of 19.6 for n = 64. Thus it was natural to look for something in the theoretical "gap" between Mars and Jupiter at n = 8.

On 1 January 1801 a small planet was discovered by Giuseppi

Piazzi. It was named Ceres, after the Roman goddess of agriculture, daughter of Saturn. Its distance from the sun is 2.77 AU, again very close to the theoretical value from the Titius-Bode law. Other minor planets or "asteroids" were found to exist in similar orbits forming a belt centered just inside the Titius-Bode orbit at 2.8 AU.

After Uranus had been observed for several years it was found that its orbit did not quite agree with Kepler's law, even when corrected for perturbations due to the other known planets. It was suspected by astronomers that another more distant planet might be involved. Attempts to locate the unknown planet by analysis of its effects on Uranus were made by two theorists: U. J. J. Le Verrier (1811–1877) in France, and J. C. Adams (1819–1892) in England. Both completed their calculations in 1846 and tried to get astronomers with high-powered telescopes to search a specified area of the sky where the hypothetical planet should be found.

Adams was unlucky because the best telescope in England for this task was controlled by James Challis at Cambridge. Challis, after considerable urging by Adams, did make two sweeps of the area but didn't bother to look at the results because he was too busy reducing his observations of comets. Le Verrier, in the meantime, had succeeded in getting the cooperation of J. G. Galle at the Berlin Observatory. Galle and a graduate student, H. L. d'Arrest, found the planet within a few hours after being notified of its predicted position by Le Verrier. Le Verrier thus became the official discoverer of the planet Neptune (named after the Roman god of water), much to the embarrassment of the British. Perhaps the British lacked faith in the power of theoretical calculation. In fact it was one of the most spectacular triumphs of Newtonian gravitational theory, although some luck was also required—it was later found that the orbit of Neptune does not for the most part agree with that computed by Adams and Le Verrier.

Although the Titius-Bode law was used to some extent as a guide in looking for the new planet, it turned out that the orbit of Neptune is much smaller than the law predicts: its distance is 30.2 AU instead of the theoretical value 38.8 for $n = 128$. But Pluto, discovered by C. W. Tombaugh in 1930, has a distance of 39.5 AU; one might guess that the law is still valid at those distances, but that Neptune doesn't really "belong" where it is. I think the case of the Titius-Bode law is useful in stimulating discussion about the nature of scientific theories: is a "mere" mathematical formula, lacking any derivation from physically plausible hypotheses, really a theory? Is such "numerology" of any value in science? Compare the case of Balmer's formula for the spectral lines of hydrogen (**9.3**), which had the same status as the Titius-Bode law in the nineteenth century but was later derived from an atomic model (**9.5**). Should the Titius-Bode law be taken as a clue to the way the solar system was formed?

Le Verrier continued his research on perturbations of planetary orbits. In 1859 he announced that the orbit of Mercury is undergoing a slow change that cannot be accounted for by the actions of the known planets: the position of the perihelion is advancing by 38 seconds of arc per century. Le Verrier proposed that this effect might be due to the presence of several small planets between Mercury and the sun, but extensive searches failed to discover them. The advance of the perihelion of Mercury was eventually explained by Albert Einstein's general theory of relativity in 1915, and is still regarded as one of the major pieces of evidence for that theory (**8.7**). Thus Le Verrier had, without realizing it, discovered a crack in the foundations of Newtonian physics. One should be cautious about identifying this anomaly as a *cause* of the replacement of Newtonian physics by general relativity, however.

BIBLIOGRAPHY

John Neu (1985), *ISIS Cumulative Bibliography 1966–1975,* vol. 2, *Subjects, Periods and Civilizations* (London: Mansell), 75, 84–90, 473–76.

Dieter B. Herrmann (1984), *The History of Astronomy from Herschel to Hertzsprung* (New York: Cambridge Univ. Press).

Kevin Krisciunas (1984), "The End of Pulkovo Observatory's Reign as the 'Astronomical Capital of the World,' " *Quarterly Journal of the Royal Astronomical Society* 25:301–5.

O. B. Sheynin (1984), "On the History of the Statistical Method in Astronomy," *Archive for History of Exact Sciences* 29:151–99, comments on applications to Titius-Bode law, distribution of minor planets, sunspots, double stars, proper motions of stars, etc.

David H. DeVorkin (1982), *The History of Modern Astronomy and Astrophysics: A Selected, Annotated Bibliography* (New York: Garland).

Topic *i:* Solar Spectrum

Frank A. J. L. James (1983), "The Debate on the Nature of the Absorption of Light, 1830–1835: A Core-Set Analysis," *History of Science* 21:235–68.

Brian Gee (1983), "Through a Hole in the Windowshutter," *Physics Education* 18:93–97, 140–43, discusses the experiments of Newton, Wollaston, and Fraunhofer.

Guenter D. Roth (1976), *Joseph von Fraunhofer: Handwerker,*

Forscher, Akademiemitgleid, 1787–1826 (Stuttgart: Wissenschaftliche Verlagsgesellschaft).

Alfred Leitner (1975), "The Life and Work of Joseph Fraunhofer (1787–1826)," *American Journal of Physics* 43:59–68.

Topic *ii:* New Members of the Solar System

Michael Martin Nieto (1985), "The Letters between Titius and Bonnet and the Titius-Bode Law of Planetary Distances," *American Journal of Physics* 53:22–25; (1972), *The Titius-Bode Law of Planetary Distances: Its History and Theory* (New York: Pergamon Press).

D. Rawlins (1984), "The Discovery of Neptune: Essential Revisions to the History," *Bulletin of the American Astronomical Society* 16:734.

Garry Hunt, ed., (1982), *Uranus and the Outer Planets* (New York: Cambridge Univ. Press), includes J. A. Bennett, "Herschel's Scientific Apprenticeship and the Discovery of Uranus," 35–53; Eric G. Forbes, "The Pre-Discovery Observations of Uranus," 67–79; Robert W. Smith, "The Impact on Astronomy of the Discovery of Uranus," 81–89.

Simon Schaffer (1981), "Uranus and the Establishment of Herschel's Astronomy," *Journal for the History of Astronomy* 12:11–26.

William G. Hoyt (1980), *Planets X and Pluto* (Tucson: Univ. of Arizona Press).

Morton Grosser (1962/1979), *The Discovery of Neptune* (New York: Dover).

D. Pascu (1978), "A History of the Discovery and Positional Observation of the Martian Satellites, 1877–1977," *Vistas in Astronomy* 22:141–8.

A. F. O'D. Alexander (1965), *The Planet Uranus: A History of Observation, Theory and Discovery* (London: Faber and Faber).

N. R. Hanson (1962), "Leverrier: the Zenith and Nadir of Newtonian Mechanics," *Isis* 53:359–78.

Topic *iii:* Stars, Structure of Universe

Michael Hoskin (1982), *Stellar Astronomy: Historical Studies* (Buckinghamshire, Eng.: Science History); (1982), "Herschel and the Construction of the Heavens," in *Uranus and the Outer Planets,* ed. G. Hunt (New York: Cambridge Univ. Press), 55–66; (1980), "Herschel's Determination of the Solar Apex," *Journal for the History of Astronomy* 11:153–63, concerns the motion of the sun in space; (1979), "William Herschel's Early Investigations of Nebulae: A Reassessment," ibid. 10:165–76; (1978), "Lambert and Herschel," ibid. 9:140–42.

Jacques Merleau-Ponty (1983), *La Science de l'Universe à l'âge du*

Positivisme: Étude sur les Origines de la Cosmologie Contemporaine (Paris: Vrin).

Simon Schaffer (1980), "Herschel in Bedlam: Natural History and Stellar Astronomy," *British Journal for the History of Science* 13:211–39.

S. L. Jaki (1978), "Lambert and the Watershed of Cosmology," *Scientia* 113:75–95.

Topic *iv:* Telescopes

Henry C. King (1955/1979) *The History of the Telescope* (New York: Dover).

A. J. Turner (1977), "Some Comments by Caroline Herschel on the Use of the 40-Ft Telescope," *Journal of the History of Astronomy* 8:196–98.

J. A. Bennett (1976), "On the Power of Penetrating into Space: The Telescopes of William Herschel," *Journal for the History of Astronomy* 7:75–108; (1976), "William Herschel's Large Twenty-Foot Telescope," *Quarterly Journal of the Royal Astronomical Society* 17:303–5.

12.2 QUANTITATIVE MEASUREMENTS ON STARS: BESSEL AND DOPPLER

i. The discovery of stellar parallax
ii. Implications of stellar distances for the age of the universe—refutation of "young-Earth creationism"
iii. Other discoveries by Bessel: unseen companions of Sirius and Procyon; the "personal equation" and the psychology of individual differences
iv. Doppler's principle and its application to stellar spectra

READINGS

Berry, *Short History of Astronomy,* 359–62, 391–2. Tauber, *Man's View of the Universe,* 234–39.

SYNOPSIS

The long-awaited discovery of stellar parallax was made almost simultaneously by three astronomers in the years 1833–1838. It should have been the "crucial experiment" to decide between the Copernican and Ptolemaic world systems, at least according to some views of scientific methodology. If the Earth does move around the sun, there should be a parallax effect; the Copernicans could explain away their failure to observe it only by postulating that the stars are all very far away. This might seem to be an *ad hoc* hypothesis invented to rescue the theory from refutation. Nevertheless, scientists accepted the heliocentric theory for other reasons, and by the time parallax was actually observed its value in confirming Copernicus was negligible. Instead, its significance was in providing for the first time reliable values for the distances of the closer stars, thus making astronomy three rather than two dimensional.

Of the three men who discovered parallax—Bessel, F. G. W. Struve, and Thomas Henderson—Bessel usually gets most of the credit not only because he published first but because his measurements were much more accurate. In view of some previous claims, which turned out to be the result of sloppy observations, the astronomical community had to take accuracy very seriously by this time.

Friedrich Wilhelm Bessel (1784–1846) was director of the observatory at Königsberg, at that time the capital of East Prussia and now part of the USSR under the name Kaliningrad. He made a survey of stars that had large "proper motion" and selected one, 61 Cygni, because it had very large proper motion (5.2 seconds per year). He considered this an indication that it is relatively close to the Earth—otherwise such a high angular velocity would imply an extremely large linear speed. After observing its position for several months with a Fraunhofer heliometer (**12.1**), he announced in 1838 that 61 Cygni has a parallax of 0.314 ± .02 seconds. The modern value is 0.292 ± .0045, within Bessel's error margin.

Recalling that the parallax of a star is defined as the angle subtended at the star by the radius of the Earth's orbit (1 AU), one can derive the standard formula for the distance of a star in AU,

$$d = 206{,}265/p$$

where p is the parallax angle measured in seconds. The *parsec* is the distance for which p = 1 second, i.e., 3.084×10^{16} meters = 3.26 light years. Thus 61 Cygni is a little more than 10 light years away.

F. G. W. Struve chose to study the star Vega because it is very bright and hence presumably fairly near, though its proper motion is small. His observations, announced shortly after those of Bessel, gave a parallax of 0.262 ± 03.7 seconds, which was later found to be considerably in error—the modern value is 0.121 ± 0.006 seconds.

The third discoverer of parallax was Thomas Henderson; he made his observations from the Cape of Good Hope (South Africa) in 1833–1834 but didn't get around to reducing them and publishing his results until several years later, so he lost his priority. He chose the star alpha

Centauri, visible from the southern hemisphere, which is bright and has a large proper motion; it is also a binary star that has a large orbit, another fact suggesting that it is fairly close to us. And indeed Henderson reported an extremely large parallax, 1.16 seconds; the modern value is 0.75 seconds, making this the closest star.

Herschel had assumed that all stars have the same intrinsic brightness so their different apparent brightnesses are due only to their being at different distances from us. As early as 1802 he had pointed out that from the known value of the speed of light one could infer that the most remote objects detected by his telescope must be nearly 2 million light years away, and hence must have already existed nearly 2 million years ago. Such an estimate was precarious as long as the only star whose distance and brightness were both known was the sun, but after the discovery of parallax a much more definite statement could be made. In 1843 Marcel de Serres pointed out that some nebulae must be at least 230,000 light years away; this conclusion does not depend on the assumption that all stars have the same intrinsic brightness, but only that *at least one star* in a nebula is as bright as 61 Cygni. The only way the argument could fail is if 61 Cygni just happened to be millions of times brighter (intrinsically) than *every other star,* a rather unlikely circumstance and one that was soon disposed of by measurements of the parallaxes and brightnesses of other stars.

Serres recognized the implications of this conclusion: it contradicted the older religious belief that the universe was created only about 6000 years ago. Now, astronomical research demonstrated that some stars in the sky must have existed much longer than 6000 years ago. Serres also pointed out the absurd consequence of the assumption that the stars were all created 6000 years ago: Adam would have seen no stars in the sky (other than the sun) for 10 years. After that they would have started to appear, one by one, as their light first reached the Earth. Throughout recorded history, according to this hypothesis, the number of stars seen in the sky would have increased every year until the present multitude is visible. Such a remarkable phenomenon, if it had really taken place, could hardly have gone unnoticed, especially by the early seafaring people who relied on the stars for navigation. A new star, rather than being such a rare event that it undermined the credibility of Aristotle's cosmology in 1572, would have been almost an everyday occurrence.

How did the theologians who believed in the Mosaic cosmogony, i.e., those religious fundamentalists now known as "creationists," reconcile their beliefs with the findings of astronomy? One way had already been suggested by François de Chauteaubriand (1802) and others who tried to defend the biblical account of creation against geological discoveries indicating immense periods of time (Krause 1980, has given a survey of these views). Chauteaubriand wrote: "God might have created, and doubtless did create, the world with all the marks of antiquity and completeness which it now exhibits." This is the "apparent age" argument, generally associated with P. H. Gosse's book *Omphalos* (1857). Adam, for example, had a navel (omphalos), as if he had been born of a woman, though "in reality" he was created a mature man. God could easily have created not only the fossils and geological formations that seem to indicate the antiquity of the Earth, but also the light waves entering our eyes that seem to have come from distant (hence very old) stars.

The "apparent age" argument was not taken seriously by scientists but has recently been revived by Henry Morris, Director of the Institute for Creation Research. In his book *The Remarkable Birth of Planet Earth* (1972) he suggested that God created the light in space as if coming from stars that did not yet exist. Thus he maintains that the universe is only a few thousands years old, and the creationists have been waging a campaign to force public school science classes to include this view. Yet many people who accept the argument that creationism should be given "equal time" with evolution in biology classes are reluctant to accept the implication that the same should be done in astronomy. This situation illustrates the need for teachers to stress the interconnectedness of the sciences. (See **4.7** for references on the creation-evolution issue.)

Bessel made two other discoveries whose significance did not become clear until much later. In 1844, as a result of his extensive study of proper motions, he found that two stars, Sirius and Procyon, have variable proper motions, which he attributed to their motion around an unseen massive companion star. The dark companion of Sirius was discovered in 1862 by A. G. Clark; the dark companion of Procyon was discovered in 1895 by J. M. Schaeberle. W. S. Adams found by spectroscopic methods that the first of these stars, Sirius B, is very hot but very small, and the name "white dwarf" was eventually used for this kind of star; it is now considered to be composed of very dense matter in a (quantum) degenerate state (**13.2**).

Bessel's other discovery was a contribution to psychology. In 1796 the British Astronomer Royal had fired his assistant because he always seemed to record the time of stellar transits about half a second too late. Since the method was thought to be accurate to a tenth of a second, the problem (as described in a published history of the Greenwich Observatory) posed a potentially serious threat to astronomical precision. Bessel investigated it by comparing transit observations of several pairs of astronomers including himself, and established that each observer has a characteristic "personal equation." Any two observers will disagree by a small but definite amount. Each observer's personal equation can be determined and used to correct his observations. Subsequently this discovery was seen as the beginning of the psychology of individual differences.

Johann Christian Doppler (1803–1853) suggested in 1842 that if a star is moving toward the observer, the wavelengths of the light it emits will appear to be shorter, whereas if it is moving away the wavelengths will seem longer. These two cases would correspond to a shift toward blue or red, respectively.

The Doppler effect was verified for sound in an experiment conducted by C. H. D. Buys-Ballot in 1845. He arranged for a railway locomotive to draw an open car with several people playing trumpets (trying to maintain a constant pitch) and found that the pitch perceived by a stationary listener does change as its source approaches or recedes.

Since the change in wavelength is proportional to the ratio of the speed of motion to the speed of the signal itself, the effect is extremely small for light emitted from macroscopic objects that can be studied in the laboratory, and thus the optical Doppler effect was not directly confirmed by terrestrial experiments until around 1901 (by A. A. Belopolsky).

In 1848 A.-H.-L. Fizeau pointed out that while the Doppler effect would shift *all* wavelengths of radiation emitted by a star and thus would have little visible effect on its color, one should be able to observe directly the shift in wavelength of a Fraunhofer dark line known to occur at a precise wavelength for the sun. Since this is the way the Doppler effect came to be applied in astronomy, it is sometimes known as the "Doppler-Fizeau effect."

William Huggins took up the study of the Doppler effect in stellar spectra. His work was stimulated by the demonstration of Kirchhoff and Bunsen in 1859–1860 that one could use the spectral lines emitted by a hot gas to determine its composition, and in particular that this method could be applied to the sun (**9.3**). Huggins found in 1868 that the star Sirius is moving away from the sun at a speed of 29 miles per second, a result later found to be rather inaccurate. The technique was substantially improved by H.C. Vogel, and his 1888 paper is sometimes considered the first reliable determination of Doppler shifts in stellar spectra.

The Doppler effect was extensively applied by J. E. Keeler in the 1890s to determine velocities of nebulae and to show that Saturn's rings are composed of small particles (confirming a theory of J. C. Maxwell).

Subsequently V. M. Slipher measured the velocities of several spiral nebulae and found that most of them are receding from us. This discovery, along with further measurements of the same kind and a new method for estimating the distances of distant stars, was the basis for E. P. Hubble's "expanding universe" (**13.1**).

BIBLIOGRAPHY

Jürgen Hamel (1984), *Friedrich Wilhelm Bessel* (Leipzig: Teubner).

Fritz Krafft (1982), *Das Selbstverständnis der Physik im Wandel der Zeit* (Weinheim: Physik-Verlag/Verlag Chemie), includes an essay on the development of astrophysics in nineteenth-century German science, emphasizing the role of spectroscopy in transforming traditional astronomy.

Walter Fricke (1970), "Bessel, Friedrich Wilhelm," *Dictionary of Scientific Biography* 1:97–102.

See the books by Herrmann and DeVorkin cited in **12.1** for further information.

Topic *i:* Discovery of Parallax

J. D. Fernie (1975), "The Historical Search for Stellar Parallax," *Journal of the Royal Astronomical Society of Canada* 69:153–61, 226–39.

Norriss S. Hetherington (1972), "The First Measurements of Stellar Parallax," *Annals of Science* 28:319–25.

Topic *ii:* Implications—Age of Universe

S. G. Brush (1982), "Finding the Age of the Earth: By Physics or by Faith?" *Journal of Geological Education* 30:34–58.

David J. Krause (1981), "Astronomical Distances, the Speed of Light, and the Age of the Universe," *Journal of the American Scientific Affiliation* (Dec.):235–39; (1980), "Apparent Age and Its Reception in the 19th Century," ibid. (Sept.):146–50.

Topic *iii:* Other Discoveries by Bessel

Norriss S. Hetherington (1980), "Sirius B and the Gravitational Redshift," *Quarterly Journal of the Royal Astronomical Society* 21:246–52.

D. B. Herrmann (1976), "Some Aspects of Positional Astronomy from Bradley to Bessel," *Vistas in Astronomy* 20:183–86.

Topic *iv:* Doppler Principle

Peter Brosche (1977), "Ein Vorläufer Christian Dopplers," *Physikalische Blätter* 33:24–29, discusses the work of Friedrich von Hahn in 1795.

D. B. Herrmann (1976), "Vogel, Hermann Carl," *Dictionary of Scientific Biography* 14:54–57.

Joachim Thiele (1971), "Zur Wirkungsgeschichte des Dopplerprinzips im neunzehnten Jahrhundert," *Annals of Science* 27:393–407.

12.3 THE RISE OF ASTRONOMY IN AMERICA

i. Nationalism in the historiography of science; comparison of the importance of discoveries in different countries, using astronomy as an example
ii. First phase of American astronomy (1750–1850); Benjamin Banneker, N. Bowditch, W. C. Bond, the Clark family of instrument-makers, J. W. Draper, L. M. Rutherfurd, Maria Mitchell
iii. Rise of American astronomy to world prominence after 1850; A. Hall, Simon Newcomb, G. W. Hill, Percival Lowell
iv. Financial support for big telescopes; G. E. Hale and "big science"
v. E. C. Pickering and the role of women in astronomy; W. P. Fleming, A. Maury, A. J. Cannon, H. S. Leavitt

READINGS

Articles: Michael Mendillo, David DeVorkin and Richard Berendzen, "History of American Astronomy," *Astronomy* 4, no. 7(1976):20–63, and following article by Trudy E. Bell, "History of Astrophotography," 66–79. D. H. Menzel, "The History of Astronomical Spectroscopy I. Qualitative Chemical Analysis and Radial Velocities," in *Education in and History of Modern Astronomy,* ed. R. Berendzen [*Annals of the New York Academy of Sciences* 198(1972)], 225–34. Vera Rubin, "Women's Work," *Science 86* 7, no. 6(1986):58–65.

SYNOPSIS

There are today a large and growing number of scholars who study "the history of American science." Almost all of them are located in the United States. Similarly there are scholars who study the history of Russian science, and most of them are located in Russia. In this section of my course I ask whether there is any justification for this fragmentation of the discipline of history of science into disjointed parts, and whether is it legitimate to study American science in isolation from European or world science. If not—if one decides that modern science is indeed an international enterprise, and that the significance of what is done in a country can be understood only in reference to what is done elsewhere—there still remains the question, does the *leadership* in the various sciences reside in a particular country at a particular time and perhaps pass from one country to another in a way that can be objectively studied? That is, would American and French historians of science agree on which country was dominant in astronomy in 1900? I argue that if one wants to discuss such questions as the social, economic, political, or cultural influences on science, or why science flourished in one time and place but died in another, one must first establish when and where science did flourish or die.

One reason for raising such questions is that some of my colleagues do not think these issues are relevant to the history of science. In particular, certain influential and brilliant scholars have argued that historians of "American science" need not worry about whether the contributions of American scientists were of any importance to science itself, as judged by scientists in other countries. The contribution of science and scientists to American civilization does not depend (they argue) on whether anyone outside the United States valued or even knew about the research done here. American science is to be studied because of its role in American society; the European reputation of an American scientist does him no good at home—unless he can convert it into the cash and prestige that go with a Nobel Prize.

Another group of scholars would object to the use of the male pronoun in the last sentence and would urge that the contributions of all female scientists be studied with equal seriousness, regardless of how the rest of the scientific community viewed their work. I would suggest, in the same spirit as above, that one should ask instead: how did it happen that science developed as a primarily masculine activity? When and how did women break into science? Which of them should be remembered as important scientists, not just as female scientists? What barriers to the full participation of women in science still remain? (See **6.7** for a more general discussion of these questions.)

I would argue that there is no such thing as American science (or female science). Science is the search for knowledge, and if the knowledge is valid it must be independent of the nationality, race, sex, etc., of the person who discovered it. Unless there is a distinctively American *method* for conducting scientific research (which is still a legitimate subject for study), American science in isolation from world science is an artificial category—or perhaps a subfield, not of the history of science, but of American history. One can study the population of American scientists—their institutions, journals, salaries, education, psychological characteristics, etc.—without worrying about whether their research had any recognized value in the world scientific community. This may indeed be a worthwhile scholarly activity but it is not the history of *science* unless one can establish the place of those Americans in the larger community; if they had any ambitions as *scientists,* they had to look to that community for problems worth solving and for recognition when they had solved them.

Historians of American science, as defined above, should presumably be concerned about how science is presented in textbooks and articles on American history. "Science" seems to mean only the atom bomb, social Darwinism, and various inventions that have changed our daily lives. Historians seem to assume that the great majority of Americans are like the majority of historians in having no interest in knowledge about nature for its own sake, but think of science only as it affects their safety, religious beliefs, or personal comfort. This brings me to the subject of this section: I think historians have greatly underestimated the interest of Americans in astronomy and have neglected to consider the question of how the United States managed to rise to world leadership in a relatively "pure" branch of science (practical applications such as navigation were not enough to account for this phenomenon) well before the immigration of European scientists in the 1930s, which is usually credited with producing our postwar superiority in fundamental research.

A glance at some of the most popular U.S. history textbooks published during the past two decades shows that while science is generally accorded a few pages, astronomy is barely mentioned at all; the reader is led to believe that there is no significant difference between the contributions of John Winthrop and Maria Mitchell, on the one hand, and those of George Ellery Hale and E. P. Hubble, on the other. Worse than that is the complete neglect of American astronomy in the late nineteenth century, which I would say is the most remarkable period in its history. But the general American historian can hardly be blamed for such oversights when there is no comprehensive survey of the history of American science *in its relation to world science.*

If we expect authors and lecturers who lack any detailed knowledge of astronomy to include that subject in their surveys of U.S. history, and indeed if we want to discuss the rise of American astronomy at all, we must be explicit about what were the most important astronomical discoveries in America, and who were the American astronomers most highly regarded in Europe. For this purpose I made a survey of the major events in astronomy and the major astronomers during the period 1800–1975, indicating which ones could be identified as American (see the article "Looking Up . . . " cited below). Having an idea of where the Americans stood on the international scene provides some of the perspective necessary to discuss what they were doing and makes it more meaningful to ask questions such as "why did American astronomy make such spectacular advances in the late nineteenth and early twentieth centuries?"

The survey shows that Germany enjoyed unquestioned leadership in astronomy during the first half of the nineteenth century, but suffered a precipitous decline thereafter, mitigated by the contributions of physicists (Kirchhoff, Helmholtz, and Einstein). German physics, on the other hand, was advancing rapidly after 1840. The British reached their astronomical peak in the 1860s and 1870s; their reputation was preserved in the twentieth century by Eddington and other theorists. The French lost their high standing in physical science after the death of Laplace (1827) and never recovered it despite the isolated triumphs of LeVerrier and Poincaré in celestial mechanics. Among the smaller countries Italy provided some important work in the nineteenth century but was replaced by Holland in the twentieth.

Probably no one will be surprised to hear that the United States had risen to world leadership in astronomy by the mid-twentieth century, but the rapidity of the ascent is remarkable. Starting from essentially zero at the beginning of the nineteenth century, the Americans had overtaken the Germans to jump into second place by the end of that century and were already challenging the British for the top spot. The discoveries announced by 1930 were sufficient to put the United States ahead of all other countries; the development of radio astronomy and the construction of the 200-inch telescope prepared the way for additional discoveries (not all of them American) after 1945.

The Europeans acknowledged American supremacy, sometimes by remarks on the public record and significantly by coming to the United States to do their research. Even before the great "intellectual migration" of the 1930s, astronomers were visiting American observatories and staying as long as possible, or, when they returned home, complaining about the relatively inadequate facilities in their own countries, while admiring the way the Americans obtained and used their instruments.

The early phase of American astronomy to about 1876 may be characterized as a period in which Americans brought themselves up to the level of European knowledge and began to show their aptitude in the use of new technology. I describe this period by mentioning six people. None of them can be called an astronomer of the first rank by world standards, but they show how things got started. The group consists of two Yankees—a glassmaker and a female librarian; a wealth New

Yorker; a chemist who immigrated from England; a white and a black from Maryland.

Benjamin Banneker (1731–1806), the black, attracted considerable attention because of the intellectual distance he had to travel from his own origins, rather than his ultimate achievements. He made a clock in 1753 without, apparently, ever having seen one, and starting in 1792, he published an almanac that involved astronomical calculations on a fairly high level. His example showed that a black could reach a level of scientific and mathematical competence far above that of the average white.

Alvan Clark (1804–1887) perfected the glassmaker's art of making optical instruments at his factory in Cambridgeport, Massachusetts; he became known as a maker of high-quality telescope lenses. His son, Alvan Graham Clark, continued the family business and in 1862 discovered the dark companion of Sirius predicted by Bessel (**12.2**-*iii*). Clark telescopes were responsible for much of the success of American astronomy in the latter part of the nineteenth century.

John William Draper (1811–1882), came from England to Virginia in 1832, became professor of chemistry and natural philosophy at New York University in 1839 and soon took up the new invention of photography. He took one of the first good pictures of the moon in 1840, and in 1844 he took the first photograph of a diffraction spectrum. He is also known for his 1874 book, *History of the Conflict between Religion and Science.*

John Draper's son, Henry Draper (1837–1882), was also known for work in astronomical photography. After his death his widow used the wealth inherited from her side of the family to support astronomical research at the Harvard College Observatory, especially the survey of stellar spectra organized by E. C. Pickering (see below).

Another pioneer of astronomical photography was Lewis Morris Rutherfurd (1816–1892), a wealthy New York amateur. In the 1860s he developed a practical method for photographing the spectra of the sun and stars so as to reveal a large number of Fraunhofer lines not visible with the ordinary spectroscope. He made diffraction gratings which he distributed to other astronomers; these were superseded in the 1880s by those made at Johns Hopkins University by Henry Rowland.

The net effect of the introduction of photographic methods was to make it possible for astronomers to take large numbers of observations in a fairly routine way at night, then analyze the results during the day, often with the help of assistants who were not professional astronomers. This is one way astronomy became "big science," involving team research; it is also one reason why opportunities were opened up for women in astronomy, even if at first their jobs were fairly routine and tedious.

The first woman in American astronomy was Maria Mitchell (1818–1889). She was librarian at the Atheneum in Nantucket, Massachusetts, when she discovered a comet with her telescope in 1847. For this she won a prize established in 1831 by the King of Denmark for the first comet to be discovered with a telescope. As a result of the publicity generated by this event, she gained some recognition from the scienctific community; after being employed for several years by the Nautical Almanac Office to compute ephemerides of Venus, she became Professor of Astronomy and Director of the Observatory at the new Vassar College for women.

Daniel Kirkwood (1814–1895) was the first American to make his reputation in theoretical astronomy. Born in Maryland, he eventually became professor of mathematics at Indiana University. In 1849 he proposed a formula relating the rotations and orbits of planets similar to Kepler's third ("harmonic") law for their revolution periods and distances; the apparent success of this formula in revealing a new regularity in the solar system earned him, for a brief time, the title of "the American Kepler." The "Kirkwood gap" in the asteroid belts, for which he is now best known, was originally thought to demonstrate yet another regularity in support of Laplace's nebular hypothesis (**2.1**).

For the most part, these six pioneers were self-educated in astronomy, worked alone, and had no institutional support until after they had established their reputations. It was still extremely difficult to become a professional astronomer in America (see Hetherington's article on Abbe, cited below). Yet Americans were quick to learn the new techniques that were revolutionizing astronomy—spectroscopy and photography—and they were beginning to have excellent telescopes available. In one small subfield Americans had already taken the lead: by 1876 they had discovered more asteroids than any other country.

Two other Americans became known as experts in theoretical astronomy. Simon Newcomb (1835–1909) used celestial mechanics to compute ever-more accurate orbits of all the planets so as to allow exact prediction of their positions at any time. In his later years he was the leader of the "old guard" of astronomers whose power had to be fought or circumvented by the younger generation that wanted to go in different directions. George William Hill (1838–1914) made substantial improvements in the theory of the motion of the moon, Jupiter, and Saturn; like Newcomb he was highly respected in the nineteenth century but his accomplishments now seem to be restricted to grinding out the next decimal place of accuracy from an established theory.

Insofar as the early development of American astronomy can be attributed to social or cultural factors, the factors seem to be the ones often associated with this period of American history: a democratic openness to bright, energetic people without professional training or certification by the establishment, and an emphasis on practical skills and technology. But the validity of these factors as general causes is undermined by the growing importance of Harvard and the "California-Wisconsin axis" in astronomical research and education, and the success of astronomers like Newcomb and Kirkwood in abstract theoretical investigations.

The year 1877 marked a turning point, when the opposition of Mars provided the opportunity for two sensational discoveries. Asaph Hall (1829–1907) used the 26-inch Clark refracting telescope at the Naval Observatory in Washington, D.C. (at that time the largest in the world) to discover the two moons of Mars. This confirmed a "prediction" made by Jonathan Swift in *Gulliver's Travels* (1727).

The other discovery of 1877 was made in Italy but exploited in America: Giovanni Schiaparelli's detailed examination of the surface of Mars revealed markings which he called "canali." The word really means "channels" in Italian, but the press seized on the idea that they are *canals* constructed by intelligent beings. Percival Lowell (1855–1916) pursued this idea and was responsible for creating much of the widespread fascination with "life on Mars" that continues to the present day. Indeed if the most pragmatic criterion of whether a hypothesis is taken seriously is willingness to spend billions of dollars to test it, then surely the question "is there life on Mars," which bolstered public support for the U.S. space program, ranks among the leading scientific problems of all time, and we have Percival Lowell to thank for making it so.

Lowell was a wealthy New Englander who could finance his own research and that of others. He was one of the first astronomers to act on the notion that observatories must be located someplace where atmospheric conditions are favorable for seeing, that is, away from urban centers. So he established the Lowell Observatory at Flagstaff, Arizona, in 1894. In addition to elaborating on Schiaparelli's "canali," Lowell described periodic darkenings of certain regions of Mars, which seemed to be correlated with the change in the size of the polar "ice cap." He interpreted the darkening as a seasonal growth of vegetation, stimulated by water melting from the icecaps and flowing through the canals, which appeared to be more prominent at those times.

Lowell also believed that there must be another planet beyond Neptune and organized a search for it at his observatory. Thanks to his persistence and his contagious enthusiasm for this project, another planet was in fact discovered after his death, by Clyde Tombaugh in 1930. Its name, Pluto, though of legitimate classical origin (Greek/Roman god of the underworld), was chosen to commemorate *P*ercival *L*owell by its first two letters.

Lowell's real achievement, not always appreciated by other astronomers, was to arouse popular interest in astronomy and thus gain support for research whose results were not so easy to understand. He was an excellent writer and speaker, a master of literary style as well as a competent astronomer and mathematician.

The years after 1877 saw the beginnings of major collaborative research programs and the construction of huge new telescopes in the United States. These projects were successful because of the enthusiasm and organizing talents of a handful of astronomers and the availability of money. Massive support for astronomy came from several Americans who had made their fortunes by methods that they or the public considered somewhat unethical; philanthropy allowed them to acquire social respectability and even immortality by giving them the means to have their names associated with a famous educational or scientific institution. Astronomy in particular appealed to those who were curious about the mysteries of the universe. Examples are James Lick, who made his fortune by land speculation in the California gold rush and endowed an observatory near San Francisco; Charles Tyson Yerkes, a streetcar tycoon who financed an observatory in Wisconsin for the University of Chicago; and Andrew Carnegie and John D. Hooker, who supported the Mt. Wilson Observatory in southern California.

George Ellery Hale (1868–1938) was the one person mainly responsible for the construction of the observatories and telescopes that have dominated twentieth-century astronomy, including the ones in Wisconsin and California mentioned above, and the Palomar Observatory (supported by the Rockefeller Foundation) whose 200-inch telescope was completed in 1948. Hale also helped to establish the California Institute of Technology as a major center of scientific research in the western United States.

Hale's own research was primarily concerned with the physical properties of the sun, and his major discovery of the magnetic properties of sunspots. More generally, he promoted the idea that the goal of astronomy should be to say *what* a heavenly body is, as well as *where* it is. Today this seems obvious, but it was not so in the mid-nineteenth century. August Comte, the founder of "positivism," (**11.1-*iv***) asserted in 1835 that reliable knowledge of the universe beyond our own solar system would be forever unattainable, so it is futile even to speculate about the nature of the stars. Here as in many other cases "positivism" turned out to be shortsighted "negativism"; astronomers were busily analyzing the chemistry of the stars only a few decades after Comte issued his dogmatic pronouncement. Not only could one identify the elements responsible for spectral lines, one could also learn something about the physical conditions of the stars (e.g., temperature, magnetic fields) by analyzing the fine structure of those lines. Thus spectrum analysis became an important part of astrophysics, a new science in which Hale was one of the leaders.

Another person who helped to convert astronomy to the "big science" style and the physical approach was Edward Charles Pickering (1846–1919). Like Hale, he is known not so much for his own discoveries as for his talent in organizing institutions in which other people could make discoveries. As Director of the Harvard College Observatory starting in 1877, he undertook a comprehensive survey of the brightness and later the spectra of all stars.

Just as Humphrey Davy is supposed to have said that his most important scientific discovery was Michael Faraday, one might say that Edward Pickering's most important scientific discovery was women. The Harvard Observatory had already initiated a policy of hiring women as computers in 1875, but they were not allowed to make observations in the unheated telescope room during the first major survey because "the fatigue and the exposure of the cold in winter are too great for a lady to undergo," as Pickering explained [Jones and Boyd, *The Harvard College Observatory* (1971), 188]. Later, the development of photographic methods, the improvement of physical facilities at the observatory, and a change in male attitudes allowed women to take a more prominent role in research there.

Williamina Paton Fleming (1857–1911), who had emigrated from Scotland as a cook in 1878, worked at the observatory starting in 1881, classifying stars; her classification system, as revised by Annie Jump Cannon, is the one now in use.

Pickering, pleased with the work done by the first woman he hired, published in 1882 a pamphlet on variable stars expressly designed to interest women in astronomy. His might be called the first "affirmative action" program in science although it was motivated in part by the stereotyped belief that women are best suited for tedious routine work requiring great attention to detail. In fact that is what astronomy needed at the time, since in studying the stars one had to deal with a large number of small pieces of information that must be classified before new theories could be established. In this respect, stellar astronomy around 1900, like biology in the eighteenth century, conformed fairly well to the Baconian conception of science as systematic observation and classification, though (like nineteenth-century biology) it was soon to break out of that mold with grand evolutionary schemes and bold conjectures.

In addition to Fleming, three other women made major contributions to stellar astronomy during the Pickering era at Harvard. The next, Antonia Maury (1866–1952), was not very happy with the role of diligent computer and refused to fit herself into the regular routine schedule of the observatory. She didn't classify as many stars as Fleming or Cannon, and her classification scheme was not as generally accepted by other astronomers as theirs was, but she did recognize one kind of distinction that turned out to be quite significant. She proposed a subdivision, called the "c" division, for stars with very narrow, sharply defined hydrogen and helium lines, and with very intense calcium lines. The Danish astronomer Ejnar Hertzsprung based his discovery of dwarf and giant stars on Maury's subdivision. He proposed that the "c" division contains giant stars of very great luminosity. This type plays an important role in the "Hertzsprung-Russell diagram" and in theories of stellar evolution (**13.2**).

Annie Jump Cannon (1863–1941) joined the observatory staff in 1896 and stayed there the rest of her life. She classified more than 200,000 stars for the *Henry Draper Catalogue,* including virtually all stars brighter than the ninth magnitude. She was the first woman to receive an honorary degree from Oxford University, in 1925; not until thirteen years after that did Harvard recognize her with a suitable title, "William Cranch Bond astronomer." She revised the Fleming classification system, establishing the present order of spectral types designated by letters in the sequence: O B A F G K M. (Students now memorize this sequence by means of the mnemonic, proposed by H. N. Russell, "Oh Be A Fine Girl, Kiss Me!") It was adopted by the International Solar Union in 1910 and is known as the "Harvard Spectral Classification."

Henrietta Swan Leavitt (1868–1921) was in charge of the project to establish stellar magnitudes by photography and specialized in variable stars, of which she discovered about 2400. On the basis of detailed study of one particular group known as the Cepheid variables, she proposed in 1908 the general rule that brighter variables have longer periods. Since in this particular group all of the stars were assumed to be at the same distance, this meant that the intrinsic brightness (luminosity) was directly correlated with the period of variation.

Hertzsprung was the first to recognize that this period-luminosity relation could be used to measure stellar distances; it is necessary only to have a good direct measurement of one of the Cepheid variables, and the Leavitt formula can then be used to find distances of other variables. By this method Hertzsprung was able to make the first reasonable estimate of the distance of the Small Magellanic Cloud in 1913: 30,000 light years. In 1922 Harlow Shapley and Donald Menzel at Harvard used a similar method to estimate the distance to the Large Magellanic Cloud; Shapley then used the Cepheid variables and the period-luminosity relation to construct a theory of the size and shape of our galaxy and the distance scale for other galaxies.

Meanwhile V. M. Slipher (1875–1969) was estimating the speeds of spiral nebulae by the Doppler effect. E. P. Hubble combined these results with distances estimated by the Leavitt-Hertzsprung-Shapley method to obtain his generalization that the other galaxies are receding from us at speeds proportional to their distances (**13.1**).

Some issues that might be discussed, using the rise of astronomy in America as a case study, are:

1. Is the concept of "professionalization" valid here, or—as Nathan Reingold (1976) has suggested—should one classify astronomers instead as "researchers," "practitioners," and "cultivators"? Surely one wants to distinguish between amateurs like L. M. Rutherfurd and P. Lowell, who became researchers able to compete with the professionals on more than equal terms, and those who remained cultivators, discovering an occasional comet or reporting on variable stars. Similarly, one wants to note that a professional like A. J. Cannon could evolve from a practitioner into a researcher when the scientific community wasn't looking.
2. What role did the organization of scientific societies play in promoting the growth of astronomy in the United States? Or did they retard that growth by enforcing methodological conformity and discouraging mavericks?
3. Did the removal of astronomy from school curricula around 1900 have any effect on recruitment into or support for the discipline in the twentieth century?
4. Did the international reputations of American astronomers give them any special power to influence the policies of their institutions or the attitudes of their society?
5. Was astronomy one of the few routes to achievement and intellectual gratification open to scientifically talented women in the late nineteenth century? (Did the rise of physics in the United States come too late to take advantage of this situation?)

6. Were the socioeconomic conditions that favored support for astronomy in late nineteenth-century America duplicated in other countries at other times with comparable results? How do these conditions differ from those favorable to sciences with more direct practical application?

7. Did American success in astronomy inspire research in other areas of science? Robert Seidel (1976) shows how this happened at Berkeley. Scientists elsewhere in the United States may have been encouraged by the feats of astronomers to believe that Americans could do first-rate work and overtake the Europeans if they tried hard enough. As pioneers of "big science," did the astronomers lay the foundations for future financial support of other sciences? They propagated the idea that basic research often requires very expensive equipment and the organization of large teams of people and convinced the public as well as wealthy donors that such research is worthwhile.

BIBLIOGRAPHY

Marc Rothenberg (1985), "History of Astronomy," *Osiris,* ser. 2, 1:117–31, surveys writings on the history of American astronomy.

Dieter B. Herrmann (1984), *The History of Astronomy from Herschel to Hertzsprung* (New York: Cambridge Univ. Press), has an appendix listing 92 events in astronomy from 1801 to 1931, with national breakdown (Table 7).

Clark A. Elliott (1979), *Biographical Dictionary of American Science: The Seventeenth through the Nineteenth Centuries* (Westport, Conn.: Greenwood Press), includes an index by fields of science.

S. G. Brush (1979), "Looking Up: The Rise of Astronomy in America," *American Studies,* 20, no. 2:41–67.

Topic *i:* Nationalism in Historiography of Science

David H. DeVorkin (1982), *The History of Modern Astronomy and Astrophysics: A Selected, Annotated Bibliography* (New York: Garland), gives numerous references on national and institutional histories, 49–106.

Nathan Reingold (1976), "Definitions and Speculations: The Professionalization of Science in America in the Nineteenth Century," in *The Pursuit of Knowledge in the Early American Republic,* ed. A. Olson and S. C. Brown (Baltimore: Johns Hopkins Univ. Press), 33–69.

Topic *ii:* First Phase 1750–1850

Emilia Pisani Belserene (1986), "Maria Mitchell: Nineteenth Century Astronomer," *Astronomy Quarterly* 5:133–50.

Norriss S. Hetherington (1983), "Mid-Nineteenth-Century American Astronomy: Science in a Developing Nation," *Annals of Science* 40:61–80.

John Lankford (1981), "Amateurs and Astrophysics: A Neglected Aspect in the Development of a Scientific Specialty,"*Social Studies of Science* 11:275–303.

I. B. Cohen, ed. (1980), *Aspects of Astronomy in America in the Nineteenth Century* (New York: Arno Press), is an anthology of original sources, mostly on the founding of observatories.

Deborah Jean Warner (1979), "Astronomy in Antebellum America," in *The Sciences in the American Context,* ed. N. Reingold (Washington, D.C.: Smithsonian Institution Press), 55–75.

Donald K. Yeomans (1977), "The Origins of North American Astronomy: Seventeenth Century," *Isis* 68:414–25.

Topic *iii:* Rise to Prominence after 1850

Richard C. Henry and Peter Beer, eds. (1986), "Henry Rowland and Astronomical Spectroscopy: Celebration of the 100th Anniversary of Henry Rowland's Introduction of the Concave Diffraction Grating," *Vistas in Astronomy* 29, part 2:119–235.

Donald E. Osterbrock (1985), *James E. Keeler, Pioneer American Astrophysicist and the Early Development of American Astrophysics* (New York: Cambridge Univ. Press); (1984), "The Rise and Fall of Edward S. Holden," *Journal for the History of Astronomy* 15:81–127, 151–76; (1979), "James E. Keeler, Pioneer Astrophysicist," *Physics Today* 32, no. 2(1979):40–47.

George E. Webb (1980), "The Planet Mars and Science in Victorian America," *Journal of American Culture* 3:573–80.

William Graves Hoyt (1980), *Planets X and Pluto* (Tucson: Univ. of Arizona Press); (1976), *Lowell and Mars* (Tucson: Univ. of Arizona Press); (1976), "W. H. Pickering's Planetary Predictions and the Discovery of Pluto," *Isis* 67:551–64.

Arthur L. Norberg (1978), "Simon Newcomb's Early Astronomical Career," *Isis* 69:209–25.

Howard Plotkin (1978), "Astronomers versus the Navy: The Revolt of American Astronomers over the Management of the United States Naval Observatory, 1877–1902," *Proceedings of the American Philosophical Society* 122:385–99.

Norriss S. Hetherington (1976), "Cleveland Abbe and a View of Science in Mid-Nineteenth Century America," *Annals of Science* 3:31–49.

Topic *iv:* Hale, Big Science

Donald E. Osterbrock (1985), "The Quest for More Photons: How Reflectors Supplanted Refractors as the Monster Telescopes of the Future at the End of the Last Century," *Astronomy Quarterly* 5:87–95; (1976), "The California-Wisconsin Axis in American Astronomy," *Sky and Telescope* 51:9–14, 91–97.

Richard Rhodes (1985), "Reflected Glory: How They Built Palomar," *American Heritage of Invention and Technology* 1, no. 1:12–21.

John Lankford (1983), "Photography and the Long-Focus Visual Refractor: Three American Case Studies, 1885–1914," *Journal of the History of Astronomy* 14:77–91, is on the Lick Observatory Experiment, G. E. Hale and the Yerkes Refractor, the evolution of the Thaw Refractor.

Topic *v:* Pickering, Role of Women

Peggy Kidwell (1986), "E. C. Pickering, Lydia Hinchman, Harlow Shapley, and the Beginning of Graduate Work at the Harvard College Observatory," *Astronomy Quarterly* 5:157–71.

Howard Plotkin (1982), "Henry Draper, Edward C. Pickering and the Birth of American Astrophysics," in *Symposium on the Orion Nebula to Honor Henry Draper,* ed. A. E. Glassgold et al. (New York: New York Academy of Sciences), 321–30; (1978), "Edward C. Pickering, the Henry Draper Memorial, and the Beginnings of Astrophysics in America," *Annals of Science* 35:365–77; (1978), "Edward C. Pickering and the Endowment of Scientific Research in America, 1877–1918," *Isis* 69:44–57.

David DeVorkin (1981), "Community and Spectral Classification in Astrophysics: The Acceptance of E. C. Pickering's System in 1910," *Isis* 72:29–49.

Margaret Walton Mayall (1981), "Vita: Annie Jump Connon, Ardent Astronomer: 1863–1941," *Harvard Magazine* 83, no. 4:34–35, reproduces a painting of her.

Deborah Jean Warner (1979), "Women Astronomers," *Natural History* 88, no. 5:12–26; (1978), "Science Education for Women in Antebellum America," *Isis* 69:58–67.

Helen Buss Mitchell (1976), "Henrietta Swan Leavitt and Cepheid Variables," *Physics Teacher* 14:162–67.

Robert Seidel (1976), "The Origins of Academic Physics Research in California," *Journal of College Science Teaching* 6:10–23.

Owen Gingerich (1974), "Maury, Antonia Caetana de Paiva Pereira," *Dictionary of Scientific Biography* 9:194–95; (1973), "Leavitt, Henrietta Swan," ibid. 8:105–6; (1972), "Fleming, Williamina Paton," ibid. 5:33–34; (1971), "Cannon, Annie Jump," ibid. 3:49–50.

12.4 HISTORY OF THE SOLAR SYSTEM

i. The Chamberlin-Moulton planetesimal/encounter theory; theories of James Jeans and Harold Jeffreys
ii. Refutation of the encounter theory by H. N. Russell (1935); lack of any plausible theory during next decade
iii. Revival of the nebular hypothesis: Hannes Alfven's magnetic braking mechanism; survival of planetesimals; reports of extrasolar planetary systems; Weizsaecker's vortex theory
iv. Age of the Earth

READINGS

S. G. Brush, "From Bump to Clump: Theories of the Origin of the Solar System 1900–1960," in *Space Science Comes of Age,* ed. P. A. Hanle and V. D. Chamberlain (1981), 78–100, 85–96. J. Singh, *Great Ideas and Theories of Modern Cosmology* (1966), 323–60. Marks, *Science,* 349–52 (on *vi*).

Articles: Thornton Page, "The Origin of the Earth," *Physics Today* 1, no. 6(1948):12–24. Andre Brahic, "Theories of the Origin of the Solar System: Some Historical Remarks," in *Formation of Planetary Systems,* ed. A. Brahic (Toulouse, France: Cepadues, 1982), 15–58. Stephen G. Brush, "Cooling Spheres and Accumulating Lead: The History of Attempts to Date the Earth's Formation," *Science Teacher* 54, no. 9 (1987):29–34.

SYNOPSIS

The nebular hypothesis (**2.1**) was overthrown at the beginning of the twentieth century by two Americans, the geologist T. C. Chamberlin and the astronomer F. R. Moulton. Chamberlin came to the subject originally as a result of his studies of glacial formations in North America; he examined contemporary attempts to explain the cause of the Ice Age, in particular the hypothesis that the Earth originally had an atmosphere rich in carbon dioxide. A drop in the CO_2 content supposedly reduced the absorption of solar heat and thus lowered global temperatures. But when Chamberlin learned of calculations based on the kinetic theory of gases showing that gases at high temperatures would have molecular velocities great enough to escape the Earth's gravitational field, he realized that the notion of a dense CO_2 atmosphere was inconsistent with the assumption that the Earth had once been a hot fluid ball; not only carbon dioxide but all the other gases in the atmosphere would have escaped, at temperatures high enough to melt rocks. Thus Chamberlin was led to challenge the prevailing doctrine that the Earth had been formed as a hot fluid ball, and to question Kelvin's estimate (based on this doctrine) that the Earth is only a few million years old. Even before Kelvin's estimate was refuted by research on radioactivity, Chamberlin had started to liberate geologists and evolutionary biologists from the tyranny of the truncated time scale (**2.4**).

Chamberlin started to look into the possibility that the Earth had been formed by the accretion of cold solid particles, at first only as an alternative working hypothesis. This was in the spirit of his well-known "method of multiple working hypotheses," a method intended to counteract the scientist's natural tendency to select only the evidence favorable to his own theory. (Cf., Einstein's "heuristic" particle theory of light, which was not intended to replace the wave theory, but merely to provide another way of dealing with certain phenomena; see **9.4**.) The first difficulty he encountered was theoretical: astronomers who had previously considered the idea, under the name "meteoric hypothesis," had concluded that planets formed in this way would have retrograde rotation. According to Kepler's third law, linear velocity decreases with distance from the sun, so the particle in the inner orbit would be moving faster than the one in the outer orbit just before they collided, and the combined body would have a net backwards rotation.

But the nineteenth-century astronomers who had rejected the meteoric hypothesis on the basis of Kepler's third law had forgotten to apply Kepler's other two laws. In general the particles would move in elliptical orbits (first law) and a given particle would move faster in the part of its orbit that is closer to the sun (second law). Chamberlin showed by analyzing several examples that unions of particles moving in intersecting elliptical orbits would be more likely to leave the resulting particle with direct rotation. This rather technical point (on which Chamberlin's conclusion has been confirmed by recent calculations) is worth mentioning because it illustrates the intellectual power that this geologist, in his fifties, was able to bring to bear on astronomy, a field in which he had no previous training. Equally important, he was able to put together an interdisciplinary team, beginning with the astronomer Moulton, to investigate a wide range of problems on the frontiers of geology, geophysics, and astronomy. (Funding from the Carnegie Institution of Washington made this possible.)

Chamberlin and Moulton put together a number of objections to the nebular hypothesis (many of them raised before but not worked out in such convincing detail) and made a good case for their own alternative theory. This had two components: (1) formation of the planets by accretion of cold solid particles—"planetesimals"; (2) formation of the earlier preplanetary nebula by encounter of two stars, the sun and an intruder that drew material out of the sun by tidal forces. Chamberlin originally identified this nebula with spiral nebulae found in telescopic observations but later abandoned the identification when the spiral nebulae were shown to be huge distant galaxies.

Henri Poincaré in 1911 proved a theorem that seemed to constitute another serious objection to the nebular hypothesis: if the present mass of the planets were spread out over the entire volume of the solar system, this material would be of such low density that it would dissipate into space before condensing under its own gravitational attraction. Thus the process postulated by Laplace could not have been initiated unless there were some external disturbance producing local regions of high density.

In 1916–1917 Harold Jeffreys and James Jeans in England independently adopted the second component of the Chamberlin-Moulton theory, but rejected the first. Jeffreys argued that high-speed collisions among the planetesimals would vaporize them so quickly that the material would remain gaseous until it collected into planets; thus he returned to the nineteenth-century assumption that the Earth was originally a hot fluid ball and has been cooling down. Jeans was more interested in developing idealized mathematical models to represent the initial ejection of material from the sun under the tidal influence of the other star. Thus Jeans concentrated on the astronomical side of the theory while Jeffreys developed it from a geophysical viewpoint.

The tidal theory, whether the Chamberlin-Moulton or Jeans-Jeffreys version, was generally accepted by astronomers until 1935, even though it was never worked out in sufficient detail to provide a convincing explanation of the quantitative properties of the solar system. Astronomers recognized that any theory requiring the encounter of two stars to form planets would entail an extremely small frequency of planetary systems in the universe. This was consistent with the failure (before 1940) to find any convincing evidence for extrasolar planetary systems. Jeans, in his popular book *The Mysterious Universe* (1930), seemed to take perverse pleasure in the idea that we are the result of a chance event that has only happened once in the universe and (because the stars are decaying and thinning out by expansion) will probably never happen again.

Henry Norris Russell raised two major objections to the assumption that material extracted from the sun by a passing star would condense into the planets of the present solar system. First, theories of stellar structure developed by A. S. Eddington and others in the 1920s indicated that gases from the interior of the sun would be at such a high temperature—on the order of a million degrees—that they would dissipate into space before they could condense into planets. (This process was later worked out quantitatively by L. Spitzer.) Second, a simple dynamical calculation showed that it would be impossible for the tidal encounter to leave enough material with the necessary angular momentum in orbits at distances from the Sun corresponding to the giant planets. These objections, presented in 1935, persuaded most astronomers that the encounter theory was untenable, but since supposedly fatal objections to the nebular hypothesis remained, the result was apparently that *no* theory of the origin of the solar system was satisfactory, even as a first approximation.

Incidentally, this seems to be a counterexample to the claim of Imre Lakatos (**11.5**) and others that scientists will not abandon a theory, no matter how defective it seems to be, unless a better one is available.

In the decade following H. N. Russell's refutation of the encounter theory no single theory was supported by more than a handful of astronomers. Nevertheless there were some significant developments in this decade that influenced later work:

1. revival of the planetesimal hypothesis;
2. the concept of "magnetic braking" of the sun's rotation;
3. suggestions that the sun had encountered an interstellar cloud and captured from it the material that later formed planets;
4. claims for discovery of extrasolar planetary systems;
5. research on cosmic abundances of the elements.

I will summarize these briefly.

1. Bertil Lindblad showed that partly inelastic collisions between particles initially moving with different speeds in eccentric orbits with different inclinations will tend to make all the particles move at similar velocities in circular orbits lying in a flat ring. Collisions between the particles would then occur with small relative velocities, thereby avoiding Jeffreys's argument that collisions would vaporize the particles. Lindblad suggested that a cold particle immersed in a hot gas would tend to grow by condensing the gas on its surface. Dirk ter Haar elaborated this idea by using the Becker-Doering kinetic theory of formation of drops in a saturated vapor, and reinforced Lindblad's proposal that solid particles could grow initially by nongravitational forces.

Jeffreys himself began to reconsider his objection to the planetesimal hypothesis and suggested that the vapor pressure of solids at very low temperatures might be below the pressure in the surrounding medium, so that condensation would outweigh the vaporizing effect of collisions. Alfred Parson published an estimate of the vapor pressure of iron that indicated condensation would be favored in interstellar space, and Jeffreys admitted in 1948 that his original objection to the planetesimal theory had thereby been answered.

2. Hannes Alfvén incorporated the concept of planetesimal accretion into his own theory but also added an important idea that removed a major objection to the nebular hypothesis. He showed that an ionized gas surrounding a rotating magnetized sphere will acquire rotation and thereby slow down the rotation of the sphere. Alfvén proposed that the early sun had a strong magnetic field and its radiation ionized a cloud of dust and gas, which then trapped the magnetic field lines and acquired most of the sun's original angular momentum. This mechanism of "magnetic braking" was later adopted by other theorists who rejected the rest of Alfvén's cosmogony.

If a star is rotating, the Doppler effect will cause a shift in the frequencies of spectral lines for radiation emitted by those parts momentarily moving toward or away from us, and by analysis of stellar spectra it is possible to estimate the speeds of rotation. Otto Struve and others found that stars in later stages of evolution generally rotate more slowly than those in earlier stages. There is apparently some fairly universal process by which a star loses most of its angular momentum at a particular stage of its evolution. Whether or not this process involves the formation of planets, at least one can no longer use the slow rotation of the sun as an argument against the nebular hypothesis.

3. Alfvén proposed that the Sun encountered a cloud of neutral gas that became more or less completely ionized at the distance of Jupiter; the magnetic field of the sun prevented it from moving any closer. Another cloud, consisting of dust particles, was postulated to account for the terrestrial planets. At about the same time (1944) Otto Schmidt in the USSR proposed that the sun had captured an interstellar cloud of meteorites; his theory was based on gravitational capture rather than electromagnetic effects. Like Alfvén he assumed that the Earth was formed by accretion of cold solid particles. This assumption provides a common basis for discussion of questions about the thermal history of the Earth, evolution of its core, etc., for scientists who may disagree on whether the sun itself was formed from this cloud or encountered it later.

4. In 1943 two reports of extrasolar planetary systems provided a new argument against all theories that treated the origin of the solar system as an extremely rare event. Both reports inferred the existence of an invisible third component, with mass much less than the smallest known stellar mass, from observations of a binary system. The validity of these

"discoveries" later became a matter of controversy; yet many scientists wanted to find life elsewhere in the universe and would not be happy with a theory that made it unlikely. In 1983 observations of Vega and other stars with the infrared astronomical satellite indicated that they are surrounded by rings of solid objects that may be forming into planetary systems. In 1984 ground-based photographs taken by Bradford Smith and Richard Terrile showed a disk of material around the star Beta Pictoris. These results seem to suggest that planetary systems develop around many stars.

5. Research on the cosmic abundances of chemical elements was also pursued through analysis of meteorites; much of this information was synthesized in a classic paper by V. Goldschmidt in 1937, and brought up to date in a review by Harrison Brown in 1949.

The postwar revival of the nebular hypothesis is mainly due to a paper by C. F. von Weizsaecker (1944). He postulated a gaseous envelope surrounding the sun and associated with its formation. Poincaré's objection that a low-density gas could not condense by its own gravitational attraction was evaded by assuming the mass of the original nebula is substantially greater than that of the present solar system. According to spectroscopic studies by Cecilia Payne in 1925 and H. N. Russell in 1929, the sun is mostly hydrogen, yet the terrestrial planets contain very little hydrogen. Weizsaecker assumed that the nebula originally had a solar composition, but the hydrogen that did not go into the sun was eventually dissipated into space. (Later it was suggested that the "T. Tauri solar wind" might provide a mechanism for blowing away the extra hydrogen.)

Whereas Laplace had assumed, rather implausibly, that the gaseous nebula would rotate like a rigid solid, Weizsaecker pointed out that there would be a tendency toward differential rotation with faster motion inside and slower outside, as is characteristic of the Keplerian orbits of particles governed by a gravitating center. But friction between adjacent streams would tend to equalize their speeds. This would create an instability, resulting in turbulent convection currents and eventually the formation of a pattern of vortex motions. Each vortex moves in a circular orbit around the sun, and there must be an integer number of vortices in a ring at a given distance from the sun. Weizsaecker assumed that the best place to accumulate particles into planets would be the regions where adjacent vortices come into contact producing violent turbulence.

Weizsaecker's theory was initially greeted with enthusiasm, especially in the United States. But subsequent work on turbulence theory indicated that the regular pattern of vortices originally postulated by Weizsaecker could not occur, but instead must be replaced by a range of eddy sizes.

The next major development was due to Gerard Kuiper, who envisaged planetary formation as a special case of a general process that ordinarily leads to binary stars. He rejected Weizsaecker's assumption that planets would be formed in the region between vortices, postulating instead that the nebula would first break up into giant protoplanets by gravitational action. The protoplanets destined to form terrestrial planets would lose their hydrogen before condensing. Kuiper adopted Alfvén's magnetic braking mechanism to explain the present slow rotation of the sun.

Since 1950 monistic theories have been popular although no single detailed model has been generally accepted. But one important fact has been established about the origin of the solar system: its date. The use of radiometric dating has made it possible to determine fairly accurately the age of the Earth and of the solar system.

In 1953 Clair Patterson estimated that the Earth was formed about 4.5 billion years ago by analyzing the proportions of lead and uranium isotopes in several rocks and in meteorites. He assumed that the proportions at the time of the Earth's formation could be inferred from the isotopic abundances of meteorites. Although initially his estimate seemed to be based on rather shaky grounds, it has been strongly confirmed by research since then, and in fact scientists now consider that the age of the Earth is known to an accuracy of better than 2 percent.

BIBLIOGRAPHY

D. ter Haar and A. G. W. Cameron (1963), "Historical Review of Theories of the Origin of the Solar System," in *Origin of the Solar System,* ed. R. Jastrow and A. G. W. Cameron (New York: Academic Press), 1–37.

Topic *i:* Chamberlin-Moulton Theory

S. G. Brush (1978), "A Geologist among Astronomers: The Rise and Fall of the Chamberlin-Moulton Cosmogony," *Journal for the History of Astronomy* 9:1–41, 77–104.

Topic *ii:* Refutation of the Encounter Theory

H. N. Russell (1935), *The Solar System and Its Origin* (New York: Macmillan).

Topic *iii:* Revival of the Nebular Hypothesis

D. ter Haar (1948), "Recent Theories About the Origin of the Solar System, *Science* 107:405–11.

Topic *iv:* Age of the Earth

Claude C. Albritton, Jr. (1984), "Geologic Time," *Journal of Geological Education* 32:29–37; (1980), *The Abyss of Time: Changing Conceptions of the Earth's Antiquity after the Sixteenth Century* (San Francisco: Freeman, Cooper).

S. G. Brush (1982), "Finding the Age of the Earth: By Physics or by Faith," *Journal of Geological Education 30:34–58.*

Henry Faul (1978), "A History of Geologic Time," *American Scientist* 66:159–65.

13

ASTRONOMY IN THE TWENTIETH CENTURY

13.1 NEBULAE AND THE EXPANDING UNIVERSE

i. Cosmology in 1900; models of the universe; nature of nebulae
ii. Distances and motions of nebulae: the period-luminosity relation for Cepheid variables; Doppler shifts; internal motions
iii. The "great debate" between Shapley and Curtis (1920)
iv. Hubble's law and the expanding universe
v. Research on galaxies after 1929; interstellar absorption; revision of the distance scale; spiral structure of the Milky Way

READINGS

John Marks, *Science and the Making of the Modern World* (1983), 341–43. A. E. E. McKenzie, *Major Achievements of Science* (1973), 330–34, 522–23. Mason, *History of the Sciences* (1962), 564–67. Timothy Ferriss *The Red Limit: The Search for the Edge of the Universe* (1977), 21–51. Joan Solomon, *The Structure of Space* (1974), 171–83. G. Tauber, *Man's View of the Universe* (1979), 251–63, 291–95. Otto Struve and Velta Zebergs, *Astronomy of the 20th Century* (1962), 60–74, 313–30, 408–73 (emphasis on observations rather than development of theories). Charles A. Whitney, *The Discovery of Our Galaxy* (1971), 191–278. Isaac Asimov, *The Universe* (1971), 56–95, 173–88, 201–8. A. Pannekoek, *A History of Astronomy* (1961), 467–90.

Articles: Richard Berendzen, "Geocentric to Heliocentric to Galactocentric to Acentric: The Continuing Assault to the Egocentric," *Vistas in Astronomy* 17(1975):65–81. Michael Mendillo, David DeVorkin, and Richard Berendzen, "The Universe Unfolds: 1900–1950," *Astronomy* 4, no. 7(1976):87–95. N. S. Hetherington, "Observational Cosmology in the 20th Century," in *Human Implications of Scientific Advance,* ed. E. G. Forbes (1978), 567–75. Richard Rhodes, "Reflected Glory: How They Built Palomar," *American Heritage of Invention and Technology* 1, no. 1(1985):12–21. Harlow Shapley, "On the Existence of External Galaxies," *Publications of the Astronomical Society of the Pacific* 31(1919):261–68 (a readable original source). E. Hubble, "The Problem of the Expanding Universe," *American Scientist* 30(1942):99–115. J. H. Oort, "The Development of Our Insight into the Structure of the Galaxy between 1920 and 1940," in *Education in and History of Modern Astronomy,* ed. R. Berendzen [*Annals of the New York Academy of Sciences* 198 (1972)], 255–66. Allan Sandage, "Inventing the Beginning," *Science 84* 5, no. 9(1984):111–13.

SYNOPSIS

During the nineteenth century there were two opposing views of the nature of nebulae: the island universe (IU) theory—that they are huge stellar systems outside of and comparable to our own "Milky Way" galaxy (the latter being often called the Galaxy), and the gaseous cloud (GC) theory—that they are relatively small patches of diffuse matter within our galaxy that might be in the process of developing into stars and planetary systems. The latter view was closely tied to the "nebular hypothesis" for the origin of the solar system (**12.4**). Rosse's success in resolving some nebulae into stars in the 1840s suggested that *all* nebulae are systems of stars and thus supported the IU theory. But any observations of change or motion in nebulae tended to undermine this view.

The GC theory was dominant in the late nineteenth century because of several observations that seemed incompatible with the IU theory. In 1864 Huggins examined the spectrum of a planetary nebula in Draco and found only a single bright line, which indicated to him that it was gaseous. In 1885 a nova in Andromeda was observed to attain a brightness equal to one-tenth that of the entire nebula; it would seem impossible for a single star to emit one-tenth of the energy of a galaxy containing millions of stars. In 1888 Isaac Roberts exhibited photographs of the Andromeda nebula that made it look like a stage of the nebular hypothesis.

Another argument against the IU theory was that nebulae are not observed near the plane of our galaxy, whereas if they are external and unrelated to our galaxy they should be randomly distributed throughout the sky ("zone-of-avoidance" argument).

The astronomical basis for the modern "expanding universe" theory was developed in the early twentieth century. Vesto Melvin Slipher (1875–1969) at the Lowell Observatory in Arizona began an extensive program of measuring velocities of nebulae in 1912, using the Doppler effect (**12.2**). By 1925 he had studied forty nebulae, most of which were found to be moving away from us. Several astronomers suggested a linear relation between redshifts and distance, as predicted from Willem de Sitter's relativistic model (**13.3**), though in this model the redshift is due to time dilation rather than recessional velocity.

Before 1920 it was difficult to determine whether the nebulae are really outside our galaxy, because there was no reliable method for determining their distances. It was necessary to supplement the parallax method with some other method of estimating the distances of those stars too far away to have a measurable parallax. This was done by the Danish astronomer Ejnar Hertzsprung (1873–1967) and the American astronomer Harlow Shapley (1885–1972) on the basis of a discovery made by Henrietta Swan Leavitt at the Harvard Observatory (**12.3-*v***). She found that the period of a Cepheid variable is determined by its intrinsic brightness (luminosity) and thus can be used to estimate its distance from its apparent brightness.

The IU theory was revived after 1900 when several observations suggested that nebulae are not part of our galaxy. For example, Slipher's first measurement of the Doppler shift of Andromeda (1913) indicated that it is moving toward us at 300 km/sec, and soon afterwards he found two other spiral nebulae receding from us at 1100 km/sec. Such high speeds were unknown for stars in our galaxy, and some astronomers concluded that the spiral nebulae were thus outside our galaxy.

Heber D. Curtis at Lick Observatory was a strong advocate of the IU theory. He refuted the zone of avoidance objection by noting that several spiral nebulae that are seen edge-on have "dark lanes," which he attributed to obscuring material; it is reasonable to suppose that our galaxy also has such a dark lane, and that it would prevent us from seeing nebulae in the galactic plane.

A related problem was the size of our galaxy and our location within it. Statistical studies of the distribution of stars, by J. C. Kapteyn and others, led to a lens-shaped model of the Galaxy with the sun near the center and a radius of about 30,000 light years. But in 1918 Harlow Shapley announced that the Galaxy has a diameter of 300,000 light years (LY), with the sun located several tens of thousands of LY away from the center. His model of the Galaxy was based partly on his use of the period-luminosity relation to estimate the distance of globular star clusters. Shapley's "big galaxy" theory led him initially to reject the IU theory because he thought it unlikely that the spiral nebulae could be as large as our galaxy.

In 1916 Adrian van Maanen reported measurements of internal

motions in the spiral nebula M101, and later reported similar motions in other spirals. These results appeared to refute the IU theory since if the spirals are really external galaxies, any such motions would be impossible to observe. A major problem for astronomers during the next few years was whether to believe van Maanen's observations, which seemed to conflict sharply with other information about nebulae.

On 26 April 1920 Curtis and Shapley presented their views at the National Academy of Sciences in Washington, D.C. This has been called astronomy's "great debate," although there seems to be some confusion about the issues and some discrepancy between what the speakers said and the later published versions of their lectures. There seem to have been two points of contention: the IU vs. the GC theory of nebulae, and the size and structure of our galaxy. Curtis, the advocate of the IU theory, attacked the use of the period-luminosity relation to estimate distances of globular clusters and argued that the Galaxy is only about 30,000 LY in diameter (one-tenth Shapley's value). This would be consistent with the IU theory if one assumes that nebulae are external galaxies comparable in size to ours; whereas if they were as big as Shapley's big galaxy, they would have to be extremely distant and the novae we see in them would be impossibly bright. Shapley defended his big galaxy theory but didn't give much attention to the nebulae.

In 1923 E. P. Hubble found a Cepheid variable in the Andromeda nebula, and in 1924 several more there and in another spiral nebula; using the period-luminosity relation he estimated their distances to be about 800,000 LY. In retrospect this discovery is considered to have established the IU theory once and for all, though acceptance of the result was somewhat delayed because the Cepheid yardstick was not generally adopted by astronomers, and van Maanen's results could still be cited on the other side. (In fact it was Shapley, rather than supporters of the IU theory, who placed confidence in the Cepheid method for measuring distances; but since he apparently didn't think that spirals contained stars, he didn't look for Cepheids there.) Hubble still used the term "extragalactic nebulae" rather than "galaxies"—he accepted Jeans's theory that the central parts of nebulae are gas-dust clouds, not collections of stars. It also appeared that the nebulae are much smaller than our galaxy, if Shapley's estimate is correct for the latter.

Thus while nebulae were shown to be outside our galaxy, it was not yet clear that they are really galaxies themselves; if they are, then our galaxy would seem to be exceptionally large compared to all others. This problem was partly resolved when R. J. Trumpler in 1930 established the existence of interstellar absorption. This effect made the clusters appear farther away then they really are. Thus our galaxy is smaller (by a factor of 3) than Shapley thought, though other features of his big galaxy model were subsequently confirmed—the eccentric position of the sun and the role of globular clusters in outlining the Galaxy. The rest of the discrepancy was removed when Walter Baade revised the distance scale in 1952, making the nebulae farther away and hence larger than previously thought, and closer in size to our galaxy.

The outcome of the "great debate" was a victory for each astronomer on the issue he considered most important: Curtis's IU theory was adopted along with Shapley's model of our galaxy.

In the meantime astronomers, inspired in part by W. de Sitter's 1917 solution of Einstein's general-relativistic field equations (**13.3**), had been attempting to find a relation between Doppler shifts of nebulae and their distance. E. P. Hubble proposed such a relation in 1929,

$$v = Hd$$

with H, the "Hubble constant," equal to the inverse of the time for the expansion. (Note, however, that the interpretation of a measured redshift as a recession velocity is correct only for sufficiently small distances, i.e., such that the curvature of space can be neglected.) Hubble's original estimate for 1/H was about a billion years. Since radiometric estimates of the age of the Earth were in the range 2 to 6 billion years at this time, there seemed to be a conflict between astronomy and geology; it was eventually resolved in favor of geology when the astronomers revised their estimates of distances and ages upwards (**13.3**).

BIBLIOGRAPHY

Norriss S. Hetherington (1983), "Just How Objective Is Science," *Nature* 306:727–30, discusses cases in modern astronomy where personal bias affected reports of observations.

David H. DeVorkin (1982), *The History of Modern Astronomy and Astrophysics: A Selected, Annotated Bibliography* (New York: Garland).

Robert W. Smith (1982), *The Expanding Universe: Astronomy's "Great Debate" 1900–1931* (New York: Cambridge Univ. Press).

Kenneth R. Lang and Owen Gingerich, eds. (1979), *Source Book in Astronomy and Astrophysics, 1900–1975* (Cambridge: Harvard Univ. Press), unlike many "source books," gives long enough selections and commentary to provide a comprehensive reference work.

Richard Berendzen, Richard Hart, and Daniel Seeley (1976), *Man Discovers the Galaxies* (New York: Scientific History), Sections I–III, gives a comprehensive account based on published and archival material.

Elizabeth L. Scott (1976), "Statistics in Astronomy in the United States," in *On the History of Statistics and Probability,* ed. D. B. Owen (New York: Dekker), 319–31, concerns the determination of distances and spatial distributions of stars, work of C. D. Shane.

Richard Berendzen and David DeVorkin (1973), "Resource Letter EMAA-1: Educational Materials in Astronomy and Astrophysics," *American Journal of Physics* 41:783–808.

Harlow Shapley (1959), "On the Astronomical High Lights of a Half Century," *Irish Astronomical Journal* 5:190–94, lists the dozen most important discoveries, compiled during the preparation of his (1960) *Source Book in Astronomy, 1900–1950* (Cambridge: Harvard Univ. Press); see similar article (1962) with expanded list, in *Popular Astronomy* 56:28–30.

Topic *i:* Cosmology in 1900

E. Robert Paul (1986), "J. C. Kapteyn and the Early Twentieth-Century Universe," *Journal for the History of Astronomy* 17:155–82.

Richard F. Hirsh (1979), "The Riddle of the Gaseous Nebulae," *Isis* 70:197–212.

Topic *ii:* Distances and Motions of Nebulae

M. A. Hoskin (1976), "Ritchie, Curtis and the Discovery of Novae in Spiral Nebulae," *Journal for the History of Astronomy* 7:47–53.

Topic *iii:* Great Debate (Curtis vs. Shapley)

Harlow Shapley and Heber D. Curtis (1921), "The Scale of the Universe," *Bulletin of the National Research Council* 2, no. 11; reprinted (1972) in *The Realm of Science,* ed. Stanley B. Brown (Louisville, Ky.: Touchstone), 145–75.

M. A. Hoskin (1976), "The 'Great Debate': What Really Happened," *Journal for the History of Astronomy* 7:169–82.

Harlow Shapley (1969), *Through Rugged Ways to the Stars* (New York: Scribner), reminiscences about his galactic research and interactions with van Maanen, Hubble, and Russell.

Topic *iv:* Hubble's Law

Jeremy Bernstein and Gerald Feinberg, eds. (1986), *Cosmological Constants: Papers in Modern Cosmology* (New York: Columbia Univ. Press), 77–91.

E. P. Hubble (1936), *The Realm of the Nebulae* (New Haven: Yale Univ. Press).

Norriss S. Hetherington (1986), "Edwin Hubble: Legal Eagle," *Na-*

ture 319:189–90; (1983), "Philosophical Values and Observation in Edwin Hubble's Choice of a Model of the Universe," *Historical Studies in the Physical Sciences* 13:41–68.

Robert W. Smith (1982), *The Expanding Universe* (New York: Cambridge Univ. Press); (1979), "The Origins of the Velocity-Distance Relation," *Journal for the History of Astronomy* 10:133–65.

Topic *v*: Galaxies after 1929

P. W. Hodge (1984), "The Cosmic Distance Scale," *American Scientist* 72:474–82; (1981), "The Extragalactic Distance Scale," *Annual Review of Astronomy and Astrophysics* 19:357–72, reviews research on the expansion time ("age of the universe").

D. Seeley and R. Berendzen (1972), "The Development of Research in Interstellar Absorption, c. 1900–1930," *Journal of the History of Astronomy* 3:52–64.

13.2 STELLAR EVOLUTION

i. Theories of stellar evolution before 1900: Kelvin-Helmholtz contraction; Norman Lockyer

ii. Classification of stellar spectra; the H-R diagram and Russell's theory of stellar evolution

iii. Theories of stellar structure based on the new atomic theories; speculations about the source of stellar energy in 1920s; degenerate gases and white dwarfs; Chandrasekhar's limiting mass; gravitational collapse; speculations about neutron stars in the 1930s

iv. Use of spectroscopy and statistical mechanics to determine composition and temperatures of stellar atmospheres; the predominance of hydrogen in the sun and stars (Russell, 1929; Bengt Stroemgren) and the reversal of earlier ideas about motion along the main sequence

v. Bethe's theory of energy generation and nucleosynthesis in stars; further developments of the theory; late stages of evolution

iv. Discovery of pulsars and their identification with neutron stars

vii. The black hole

READINGS

Marks, *Science*, 347–49. Struve and Zebergs, *Astronomy of the 20th Century,* 186–208 (on *ii*), 209–38 (on *iv*), 239–57 (on *iii*), 258–83. Pannekoek, *History of Astronomy,* 429–66, 491–96. Isaac Asimov, *The Universe,* 128–72, 276–80. J. Singh, *Great Ideas* (1966), 33–74, 81–91,

305–22. Tauber, *Man's View,* 246–50, 277–90. H. Gursky and R. Ruffini, "Introduction" to *Neutron Stars . . .* (1975), 1–12. R. Berendzen, ed., *Education in and History of Modern Astronomy* (1972), article by D. H. Menzel on "The History of Astronomical Spectroscopy II. Quantitative Chemical Analysis and the Structure of the Solar Atmosphere," 235–44, and B. Stroemgren, "The Rise of Astrophysics," 245–54. Stroemgren, "On the Development of Astrophysics during the last Half Century," in *Astrophysics,* ed. J. A. Hynek (1951), 1–11. Rudolf Kippenhahn, *100 Billion Suns. The Birth, Life, and Death of the Stars* (1983) (popular book with illustrations).

Articles: David H. DeVorkin (1978), "Steps toward the Hertzsprung-Russell Diagram," *Physics Today* 31, no. 3:32–39. Hans A. Bethe, "Energy Production in Stars," *American Scientist* 30(1942):243–64.

The Center for History of Physics has produced a packet of audio tapes, slides, illustrated scripts, and teacher's guide on "A Pulsar Discovery" in its *Moments of Discovery* series. The tape includes a recording made during the discovery of the first known optical pulsar.

SYNOPSIS

George Ellery Hale, one of the founders of modern astronomy in the United States (**12.3**), pointed out in 1908 that Darwin's *Origin of Species* had exerted an enormous influence on the thinking of scientists in many fields. Astronomers in particular had adopted the evolutionary viewpoint and often used it to arrange and interpret their data. The study of the life history of the universe as a whole would in fact require the efforts of several sciences; he wrote,

> We are now in a position to regard the study of evolution as that of a single giant problem, beginning with the origin of the stars in the nebulae and culminating in those difficult and complex sciences that endeavor to account, not merely for the phenomena of life, but for the laws which control a society composed of human beings. [*The Study of Stellar Evolution,* 3]

But the term "evolution" in astronomy is not limited to the specifically Darwinian process of natural selection; it simply means a well-defined temporal process governed by natural laws, extended to social as well as biological phenomena in the manner of Robert Chambers and Herbert Spencer (**1.1**, **3.3**).

Stellar evolution became an active field of research in the early decades of the twentieth century, just at the time when cultural evolution was falling into disfavor under the attacks of Franz Boas and other (**3.4**). Just as nineteenth-century anthropologists had placed the various races of man at different points on an evolutionary scale with the help of quantitative measures such as facial angle and cephalic index, twentieth-century astronomers found it convenient to characterize every star by two parameters: luminosity and spectral type. The latter could be translated directly into surface temperature, and graphs showing the distribution of stars according to these two parameters—the "Hertzsprung-Russell diagram"—irresistibly suggested an evolutionary progression of the stars through a continuous sequence of physical states. This interpretation broke down by the 1930s, to be replaced later by a very much different and more complicated evolutionary theory encompassing not only the stars but the chemical elements and associated in a manner still not clearly understood with the evolution of the entire universe.

As mentioned in **12.3**, Harvard astronomers developed a spectral classification system in which each star is assigned a letter in the sequence O B A F G K M. Ejnar Hertzsprung in 1911, and independently Henry Norris Russell in 1913, constructed plots of the magnitudes and spectral classes of large numbers of stars using the Harvard data. Such a plot is now called an H-R diagram. They found that most stars fall roughly along a line in the magnitude-spectral class plane, known as the "main sequence," while a few fell in another line known as the "giant sequence."

While many astronomers suspected that there is some connection between spectral class and temperature, the exact nature of the connection was not understood until after the development of Niels Bohr's quantum theory of the atom in 1913 (**9.5**). In 1920 the Indian physicist M. N. Saha applied Bohr's theory to explain the effects of temperature and pressure on the ionization of different kinds of atoms and showed that the differences in spectral class should not be attributed to differences in chemical composition but primarily to temperature. Most stellar atmospheres probably contain a mixture of various elements, but we will observe their characteristic frequencies of radiation only when the temperature is in the appropriate range to cause electrons in those atoms to jump back and forth among their possible orbits as determined by Bohr's theory.

According to Saha's theory the Hertzsprung-Russell diagram has a very simple physical meaning: it is a plot of the absolute magnitude or luminosity of each star against its surface temperature. Moreover, according to the Stefan-Boltzmann law (**9.3**) the rate of energy radiation from a hot body is (under certain conditions that are fairly well satisfied for stars) proportional to the fourth power of the absolute temperature.

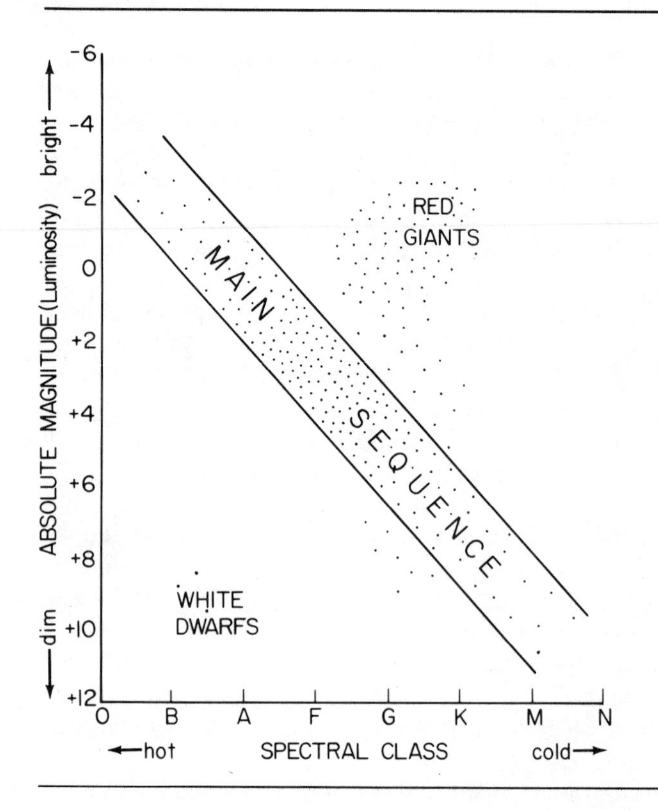

Fig. 13.1. Hertzsprung-Russell (H-R) diagram. The number of dots in each region is proportional to the number of stars observed to have the corresponding luminosity and spectral class.

Since the rate of energy radiation (energy flux) is equal to the luminosity divided by the surface area, this means that the luminosity and temperature together determine the radius of the star: the radius is proportional to the square root of the luminosity, divided by the square of the temperature. This is why stars with high luminosity and low temperature must have large radius and are called "giants." Conversely, when W. S. Adams found in 1914 that the dark companion of Sirius (**12.2-*iii***) belongs to a spectral class corresponding to a high temperature, it could be inferred that its radius is very small; this kind of star is therefore known as a "white dwarf."

The fundamental problem of stellar evolution is to translate a collection of data representing the state of stars at some particular time into a description of how an individual star passes through those states during its own life span. In an abstract sense this astronomical problem is like a biological one, as H. N. Russell pointed out in 1927:

> Following an analogy suggested by [William] Herschel, suppose that an intelligent observer, who had never before seen a tree were permitted an hour's walk in a forest. During the space he would not see a single leaf unfold; yet he could find sprouting seeds, small saplings, young, full-grown, and decrepit trees, and fallen trunks moldering back into earth, and in that brief hour he might form a correct idea of the life history of a tree.
>
> In the same manner our task is to take the various types and classes of stars with whose properties we have become familiar, and arrange them in some rational scheme of evolution,—some orderly sequence of development. [Russell, Dugan, and Stewart, *Astronomy,* 908]

The evolutionary interpretation of the H-R diagram will obviously depend on what theory one adopts about the structure and dynamic properties of stars. Russell's own first theory, proposed in 1914, was strongly influenced by the nineteenth-century scheme of cosmic evolution based on Laplace's nebular hypothesis (**2.1**). This scheme postulated that all celestial bodies, including the Earth and sun, are gradually cooling down, and that the sun derived its energy from the gravitational contraction of an extended gaseous cloud. One can trace this idea back to 1692 when Isaac Newton (in a letter to his friend Richard Bentley) suggested that the sun and stars could have been formed by gravitational forces acting to condense particles of matter that were originally scattered throughout infinite space. In the middle of the nineteenth century, Hermann von Helmholtz and William Thomson (later known as Lord Kelvin) suggested that as matter falls into the primitive sun, gravitational energy will be converted into kinetic energy and thence into heat, raising its temperature sufficiently to make it radiate energy outwards.

In 1869 the American scientist J. H. Lane showed theoretically that a contracting gaseous sphere will gain more heat by contraction than it loses by radiation, so its temperature will increase during this process. But this contraction would presumably stop or slow down markedly when the pressure and density become so high that the gas changes to a liquid or solid. After that it will radiate away the heat it has acquired and gradually cool down. The British scientist Norman Lockyer thus proposed in 1887 that stars would go through a sequence of rising and then falling temperatures, passing through the same spectral class twice in their life span.

Russell revived Lockyer's theory and used it to interpret the much more extensive data that had become available in the meantime. He suggested that the normal path followed by a star would begin with the red giants, move to the left in the H-R diagram as the star grew hotter, then move *down* the "main sequence" as it cooled off. (The "white dwarfs" were not yet included in the scheme he proposed in 1914.)

As more precise information about stellar luminosities, temperatures, and masses became available, it became clear that the Helmholtz-Kelvin contraction process could not provide enough energy to keep a star shining very long. But two discoveries in physics at the beginning of the twentieth century suggested another possible source of energy. One was radioactivity, which according to Rutherford involved the actual transmutation of one chemical element to another (**9.2**). The other was Einstein's relativity theory, which included the prediction that mass can be changed into energy by the famous equation $E = mc^2$ (**8.6**).

By 1920 several scientists had suggested that stars obtain most of their energy by destroying mass. But there was considerable disagreement about just how this is done. The most obvious candidate was radioactive decay of heavy elements like uranium and radium, since such decays had already been observed to produce energetic radiation in terrestrial laboratories. But spectral analysis of the sun and other stars did not indicate that they contain large amounts of these elements.

A more plausible way of getting energy was by nuclear reactions in which hydrogen atoms combined to produce helium and heavier elements. While there was not yet any direct evidence that such "fusion" reactions could occur, they would fit in nicely with the theory of the atom being developed at that time (**9.2**). Rutherford had proposed in 1911 that the atom consists of an extremely small positively charged nucleus that had almost all of the mass of the atom, with the remaining space being empty except for light negatively charged electrons. H. G. J. Moseley showed that the amount of positive charge on the nucleus is simply proportional to the atomic number of the element. The atomic weights of the elements were known to be fairly close to integer multiples of that of hydrogen, though it was clear from chemical experiments that they were not exactly whole numbers (**7.1**). Thus one could assume that the nucleus is composed of protons and electrons; the difference between the actual mass of the nucleus and the sum of its constituents could then be identified with the "binding energy" of the forces holding them together, using Einstein's formula. (See **10.1** for discussion of the neutron and the proton-neutron model of the nucleus.)

A simple calculation shows that if hydrogen atoms could be fused to form helium, a substantial amount of energy could be released. Of course it would first be necessary to give the hydrogen atoms a rather high speed so that they could overcome the electrical repulsive forces between them; presumably these speeds would occur at the high temperatures expected inside stars.

Not only would nuclear fusion provide a possible source of energy for the stars, it might also explain how the various chemical elements had been formed, starting with hydrogen. Thus one would no longer have to accept the doctrine that matter is composed of qualitatively different elements—92 were known by the beginning of the twentieth century—but one would have only nuclear compounds of hydrogen. In this way the evolutionary viewpoint would be extended to include the creation of the elements themselves; each element is like a "species" of atom that has evolved from a simpler species by a definite physical process, rather than originating by a separate "special creation."

Eddington surveyed the situation in a speech to the British Association for the Advancement of Science in 1920. He noted that recent precise measurements of atomic masses by F. W. Aston indicated that the mass of a helium atom is less than that of 4 hydrogen atoms by about 1 part in 120; from this he concluded that if 5 percent of a star's mass consists of hydrogen atoms that are gradually combined to form helium and other elements, the problem of an energy source would be solved. Looking ahead a little farther, he said:

> I think the suspicion has been generally entertained that the stars are the crucibles in which the lighter atoms which abound in nebulae are compounded into more complex elements. If, indeed, the subatomic energy in the stars is being freely used to maintain their great furnaces, it seems to bring a little nearer to fulfillment our dreams of controlling this latent power for the well being of the human race—or for its suicide.

In addition to fission and fusion, there was a third way in which Einstein's mass-energy conversion formula could be applied, at least in the speculations of astrophysicists of the 1920s: to the complete annihilation of particles of matter. It was thought that a proton and an electron could come together and destroy each other, leaving nothing but electromagnetic radiation. Two influential British scientists, Jeans and Eddington, argued that such a process was the most likely source of stellar energy. They also believed that an empirical correlation between mass and luminosity governed the evolution of stars: as a star loses mass it becomes dimmer. Thus the stars shine by turning their substance into radiation and gradually slide down the main sequence in the H-R diagram as they cool off.

The annihilation theory of stellar evolution suggested an even more pessimistic view of cosmic history than the nineteenth-century "heat death" theory (**7.3**). Not only are all hot bodies cooling off, but, as Jeans wrote in 1928, we have a picture of the universe "slowly but inexorably dissolving into radiation." There will not even be enough time for new planetary systems to be born before this happens; the "most eventful part" of our universe's life is already over.

In the early 1920s it was not yet clear that fusion was the most likely source of stellar energy because the sun and other stars were believed to consist mainly of heavier elements. Saha's ionization theory, as modified to take account of the excited and ionized states of atoms in a manner consistent with quantum theory by R. H. Fowler and E. A. Milne, could be used to estimate the abundances of elements from spectra data. This was first attempted by Cecilia Payne in 1925; her results indicated that hydrogen and helium were the dominant constituents of stellar atmospheres. But this went against the generally accepted view that the sun's atmosphere has the same chemical composition as the Earth, since the Earth was formed from gases drawn out of the sun's atmosphere by a passing star according to the "encounter theory" accepted at that time (**12.4**). Payne therefore doubted her own result, and it was not until 1929 that H. N. Russell, on the basis of a very comprehensive analysis of all available data for the sun's atmosphere, concluded that the sun is in fact mostly hydrogen.

Another result of quantum theory was found to be directly applicable to the theory of stellar structure. As early as 1917 Eddington had postulated that atoms in the interior of stars are completely ionized, and that the resulting mixture of nuclei and electrons might behave in some way like an ideal gas even though its density was much greater than that of ordinary solids. With the introduction of Pauli's exclusion principle (**9.5**) and the Fermi-Dirac statistics (**9.7**), it was possible to treat the electrons in such a gas as if they simply filled up a series of quantum states. Even at fairly high temperatures the electrons will tend to occupy the lowest possible states, and only those in the highest states will be able to move to unfilled states above them. Most of the electrons cannot absorb small amounts of energy because the states into which they might go are already filled. Thus the system cannot respond like an ordinary gas to small changes in energy. Conversely, even when the Fermi-Dirac gas of electrons is cooled to absolute zero temperature it has a finite energy and exerts a finite pressure that depends only on its density, hardly at all on its temperature. This behavior was called "degeneracy" by physicists.

In 1926 R. H. Fowler showed that the peculiar features of white dwarf stars might be explained by the quantum phenomenon of gas degeneracy. A star would not stop contracting when it reaches the density of an ordinary liquid or solid but could make use of the empty space inside the atom and pack the nuclei and electrons much more closely together. If classical statistical mechanics were applicable, it would be difficult to understand why this would not be the fate of matter in general, since the collapsed configuration would appear to have much lower energy than the normal one, even if there were no external pressure. But with quantum (Fermi-Dirac) statistics, the zero-point energy is dominant at low temperature and high pressures; if the stellar material were not confined by the pressure of the material sitting on top of it, it would expand back to the normal state.

In the early 1930s S. Chandrasekhar extended the theory of Fermi-Dirac gases to include relativistic effects and showed that the star can be stable only if its mass is less than a critical amount, which is about 1.5 times that of the sun. (The exact value of the "Chandrasekhar limit" depends on the chemical composition; it also depends on the speed of rotation.) It was not clear at that time what would happen to a more massive star if it continued to contract under its own gravitational attraction: it might get rid of the excess mass by an explosion (producing a "nova" or "supernova"), or it might continue to contract to a density so high that the nuclei would crumble into their constituent parts. Following the adoption of Heisenberg's proposal that nuclei consist of protons and electrons and Fermi's theory of beta decay, Baade and Zwicky proposed that the result of this crumbling would be a "neutron star" (**10.1**). This possibility was investigated theoretically by L. D. Landau in the USSR and by J. Robert Oppenheimer and his students in the United States.

These theoretical studies in the 1930s cast some light on a further stage of stellar evolution. Oppenheimer and G. M. Volkoff reached the conclusion that the collapse of stars would probably *not* lead to stable neutron stars and therefore that these hypothetical objects would not play a part in stellar evolution. Although this conclusion later proved to be incorrect, it led Oppenheimer to consider, together with H. Snyder, the consequences of gravitational collapse based on the equations of Einstein's general theory of relativity. These equations had recently been formulated for spherical masses of fluid by R. C. Tolman. Oppenheimer and Snyder in 1939 found that a dense neutron star would continue to contract without ever reaching a stable equilibrium state; but when its size approached the "Schwarzschild radius" (**8.7**) peculiar things would begin to happen. As the density increased, the red shift due to the intense gravitational field of the star would become so large that eventually no radiation could escape at all. (Another way of describing this phenomenon is to say that space-time becomes so strongly curved that it closes off the star from the rest of the universe.) Thirty years later this kind of object was baptized the "black hole" by John Wheeler.

Speculations about the beginning as well as the end of the process of stellar evolution were also becoming more refined during this period. In 1932–1933 Bengt Stroemgren concluded that stars on the main sequence of the H-R diagram consist mostly of hydrogen. "Perhaps," he suggested, "the simplest hypothesis that can be made is that the stars start as pure hydrogen stars in which, in the course of time, hydrogen is transformed to complex elements, the energy radiated away in the successive equilibrium configurations being equal to the energy set free by the transformation." As the hydrogen content decreases, Stroemgren argued, the star would expand but its mass would remain roughly the same. It would eventually leave the main sequence, going upwards and to the right in the H-R diagram, and becoming a "red giant."

Although Stroemgren was not yet able to present a complete theory of stellar evolution, he did accomplish a radical break with previous ideas: he insisted that the main sequence is *not* a path of evolution for an individual star but only the locus of equilibrium states for stars of various masses. A star would spend a relatively long time close to one position in the main sequence, that position being determined primarily by its mass. The reason the sequence looks like a continuous curve is that stars can exist with a continuous range of masses; the more massive stars would be higher on the sequence (higher luminosity, higher temperature).

Stroemgren's assumption that there would be no appreciable loss of mass by stars that might cause them to move *down* the main sequence was supported by the new ideas about nuclear reactions developed after 1932. The possibility of complete annihilation of particles, favored by Jeans and Eddington as an energy source, was ruled out; the hydrogen fusion process and other possible reactions investigated by Bethe and others (**10.1**) involved changing less than 1 percent of the mass of a nucleus into energy. Indeed, rather than moving *down* the sequence, it appeared that a young star would more likely move upwards in the initial stage of contraction just before it reached its equilibrium position, and its subsequent track in leaving the sequence would also be in the upward direction.

In 1942 M. Shonberg and S. Chandrasekhar suggested that the reason why a star leaves the main sequence is that after a significant amount of hydrogen has been converted to helium, the pressure generated by reactions in the core is no longer sufficient to support the overlying layers. The collapse of these layers toward the center liberates so much new gravitational energy that the temperature rises and the outer layers rapidly expand. The result is that the outside of the star becomes much cooler; since it is the temperature of the outside that determines the color and spectral type, this means that the star moves to the right on the H-R diagram. Because of its greatly increased surface area it is much more luminous, hence it also moves *up* in the diagram. This is how the star gets to be a "red giant." Gamow, in 1945, suggested that the red giant develops a core of degenerate matter, which ultimately becomes a white dwarf star when the outer layers of the giant are ejected.

After World War II, scientists returned to the problem of stellar evolution armed with new data on nuclear reactions and electronic computers that helped them to develop theoretical models of stars. The problem of how elements heavier than helium can be synthesized was not solved until the 1950s. E. E. Salpeter argued that two helium nuclei could stick together as a beryllium nucleus which, though unstable, would last long enough to capture another helium nucleus and form a carbon nucleus. Salpeter's "tri-alpha" reaction ($3\ {}_2He^4 \rightarrow {}_6C^{12}$) was shown to be feasible by laboratory experiments performed at Caltech by C. Cook, W. A. Fowler, C. C. Lauritsen and T. Lauritsen in 1957. While this theory did not account for the evolution of all the elements, it gave astrophysicists considerable confidence that reasonable models of the next stage of stellar evolution could be developed and that there was some hope of finding a more complete theory to explain the synthesis of other elements as well.

Salpeter's tri-alpha reaction was used in a detailed theoretical study of evolution toward the red giant region by Fred Hoyle and Martin Schwarzschild in 1955. They found that as a star with mass slightly greater than the sun evolves off the main sequence along the path MN in Figure 13.2, the proportion of helium reaches about 50 percent before the internal temperature becomes high enough (about 100 million degrees) to start the helium fusion reaction. (This fraction is correct for a star whose mass is about the same as the sun's.) Even though the temperature is high, the density is so great that the electrons in the core form a degenerate quantum gas. This means that when energy generation from helium fusion does begin, the gas cannot respond in the normal way by increasing the pressure, which would force the star to expand and thereby cool itself. Instead there is an unstable "runaway" process in which the star gets even hotter, which in turn boosts the rate of energy production. There is a sudden increase of luminosity called the "helium flash" as the star jumps from N up to O on the H-R diagram (see Fig. 13.2). The flash is terminated only when the temperature gets

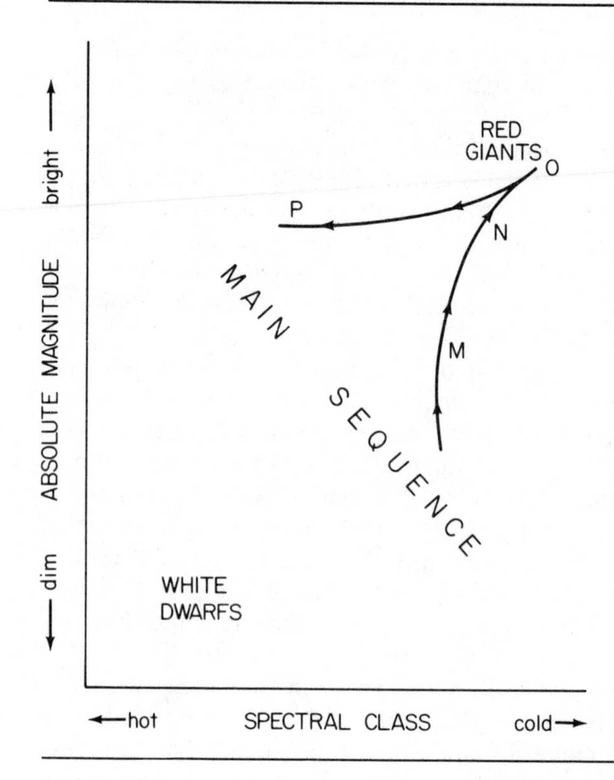

Fig. 13.2. Theoretical evolutionary track of a solar-mass star after it uses up most of its hydrogen according to Hoyle and Schwarzschild (1955).

so high that the electrons are kicked up to high energy levels, removing the degeneracy and converting the interior of the star momentarily back to a normal (nondegenerate) state. The temperature and energy generation then drop and the star moves back to N. It then continues to move to the left along NP back toward the main sequence.

Around 1957, as a result of these calculations and theoretical ideas, astronomers believed that the complete evolutionary track for a star whose mass is about the same as the sun's would look like Figure 13.3. The life history of a star would thus consist of 4 basic stages:

1. Gravitational contraction, ending when the internal temperature is high enough to start hydrogen fusion.

2. The star arrives at a point on the main sequence determined by its mass, and remains there (climbing a little way up the sequence) in an

equilibrium state as the hydrogen fusion reaction gradually proceeds. This stage may last 100 to 1000 times as long as the first stage, which is why a relatively large number of stars are observed to be somewhere on the main sequence.

3. As the hydrogen fuel is exhausted the star moves off to the right; it become a red giant with low surface temperature, getting hotter on the inside until the helium in the core begins to fuse to carbon. After a brief rise in temperature and luminosity due to the helium flash, it turns around and moves to the left.

4. During its final stage the star builds heavier elements, e.g., two

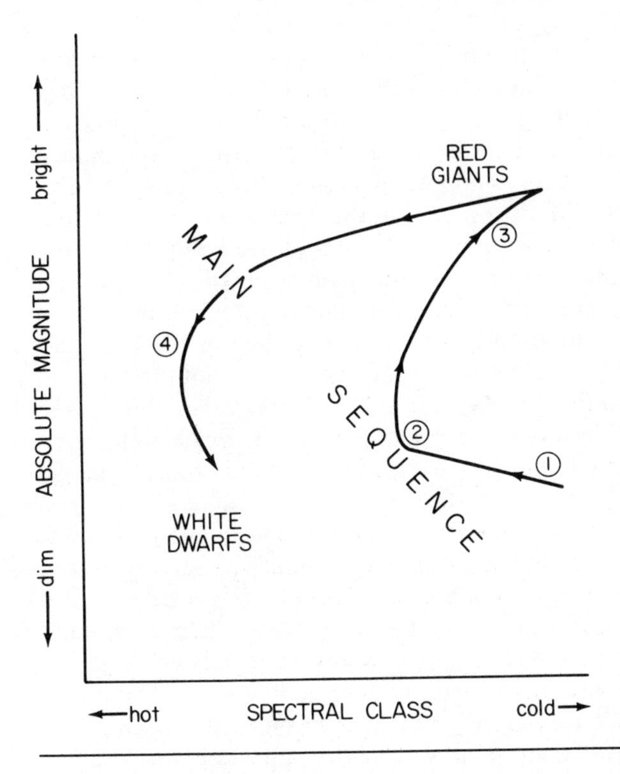

Fig. 13.3. Evolutionary track of a solar-mass star according to astrophysical theories in the late 1950s. A star will stay very nearly at the same place in the main sequence ②, determined by its mass, until it has used up most of its hydrogen. Much more massive stars can't stop at the white dwarf stage but collapse to neutron stars or black holes.

carbon nuclei may fuse to form magnesium. If it has too much mass (more than about 1.3 times the sun's mass) it cannot reach a stable state and will have to undergo one or more explosions to get rid of excess mass. Finally it becomes a white dwarf, a burned-out star consisting mainly of degenerate matter. Possibly a substantial amount of the nuclei will have been fused together to form iron, the nucleus that represents the lowest (most stable) point on the binding-energy curve.

Since 1957 further calculations have changed the details of this theoretical life-history of a star in the middle stages (2) and (3); more dramatic changes have occurred in ideas about the last stage.

Chushiro Hayashi and his colleagues R. Hoshi and D. Sugimoto put together a comprehensive account of all stages of stellar evolution in 1962, based on detailed calculations with models for each stage. They pointed out that the last stage, in which a red giant burns up its nuclear fuel and collapses into a degenerate white dwarf, is strangely reminiscent of the evolutionary theory adopted by Russell, Jeans, and others at the beginning of the twentieth century. In that theory a star begins as a gaseous red giant, collapses to a hot white star at the upper left of the main sequence, and then slides down the sequence as it changes to a liquid star, losing mass and cooling down. In the new theory one still has a movement to the left along a nearly horizontal line in the H-R diagram, but the subsequent slide down a diagonal path is not on the main sequence but displaced to a parallel track. The phase change from gas to liquid, which was responsible for stopping the contraction and changing the direction of the path in the old theory, has been replaced by a change from "normal" to "degenerate" matter.

There is a further similarity between the old and the new theories. Before the development of the quantum theory of atomic structure, one would have expected that a cooling star would ultimately freeze; so a solid would be the final state of stellar evolution. Then Jeans and Eddington pointed out that at the high pressures that must exist in stellar interiors, the atoms would be ionized and behave like a gas. The nuclei would have positive electric charges, hence they would repel each other. According to older ideas about states of matter, a solid is formed as a result of short-range attractive forces between particles, and it would seem that particles that repel each other at all distances could not solidify. However, in 1960 D. A. Kirzhnits proposed that the charged nuclei embedded in a degenerate electron gas in a white dwarf star would condense into a solid- or liquidlike structure. A few months later A. A. Abrikosov and E. E. Salpeter independently published similar suggestions, arguing that a regular solid lattice would be the most likely arrangement until the density becomes so high that quantum effects become important for the nuclei. (The nuclei themselves would eventually behave like a degenerate gas.) The phase transition was quantitatively confirmed, for a classical statistical-mechanical model of charged particles, with the help of a high-speed electronic computer (S. G. Brush, H. L. Sahlin, and E. Teller 1966).

In 1967 a new discovery by radio astronomers led to a radical change in ideas about the next stage of stellar evolution beyond white dwarfs. A group led by Anthony Hewish began a survey of galaxies that emitted electromagnetic radiation at frequencies around 100 megahertz (near the FM radio band). Their original purpose was to study a "scintillation" effect produced when a very narrow beam of radiation passes through irregularities in the ionized gases (plasma) in the solar system; the effect is analogous to the "twinkling" of stars produced by irregularities in the Earth's atmosphere. Jocelyn Bell, a graduate student, noticed that one of the radio sources was producing scintillations even though it was on the opposite side from the sun and its radiation would not be expected to pass through much of the interplanetary plasma. Further analysis showed that the source was emitting *pulses* of radiation at very precisely fixed intervals of about one second. Bell searched the records for other sources and found three more with similar behavior; they were soon given the name *pulsars*.

The announcement of the discovery of pulsars by Hewish, Bell, J. D. H. Pilkington, P. F. Scott, and R. A. Collins stimulated a worldwide search for more of these peculiar objects by astronomers and various attempts by theorists to explain how they could produce such regular signals. After some initial speculation that they might be generated as messages by some form of intelligent life, Thomas Gold proposed a theory that was generally accepted. He showed that a rapidly rotating neutron star emitting a beam of radiation along its magnetic axis (not the same as its axis of rotation) could reproduce such pulses. Moreover, the expected slowing down of its rotation produces a very slow but measurable increase in the period of the pulses, in agreement with observations. Thus a pulsar is like a lighthouse whose beam sweeps across us as it rotates, though the exact mechanism by which such directional radiation could be produced is still not very well understood.

The role of neutron stars in stellar evolution was further confirmed when David Staelin and Edward Reifenstein, using the 300-foot radio telescope at Green Bank, West Virginia, discovered a pulsar at the center of the Crab Nebula in 1968. It had been suggested earlier by several astronomers that this nebula is the remnant of a supernova explosion observed by Chinese astronomers in 1054 A.D. Using the present observed rate of expansion of this nebula one could estimate that it had originated in a point explosion about 900 years ago. While Zwicky had been arguing that neutron stars could be produced by such supernova explosions as early as 1939, there was no convincing evidence for their existence. The identification of the "Crab pulsar" at a location where there was already some reason to expect a supernova remnant has made this object one of the most frequently studied in modern astronomy.

The study of the final stage of stellar evolution was taken up in the late 1950s by John Wheeler and his colleagues at Princeton. They revived the suggestion of Oppenheimer and his students that a neutron star might not be the final state for a very massive star. What happens next may be determined primarily by the relation between mass and the curvature of space, as described in Einstein's general theory of relativity, and in particular in Schwarzschild's 1916 solution of the Einstein-Hilbert equations (**8.7**). As Oppenheimer and Snyder pointed out in 1939, the curvature of space in a region of high gravitational field causes a strong red shift of all radiation emitted and ultimately prevents it from getting out at all. On the other hand, radiation and matter can still be sucked into this region, once it has been created by the collapse of a star beyond the point where the repulsive forces between neutrons can resist the attractive force of gravity. (The possibility of an impenetrable "hard core" of a neutron is ruled out by relativity—impulses through such a hard core would travel instantaneously, violating the speed limit c).

The object produced by such a gravitational collapse was given the name "black hole" by Wheeler in 1967. It was originally thought that a black hole by its nature would be undetectable unless one were close enough to notice a small region of space (diameter 10 kilometers or less) that absorbed all incoming radiation but was surrounded by a strong gravitational field. But further theoretical analysis suggested that material being sucked into the hole would still be able to emit some radiation, especially if the hole is rotating. In addition to electromagnetic radiation, probably in the very high frequency (X ray) part of the spectrum, there is some possibility that gravitational waves, whose existence was predicted by Einstein in 1916, may be emitted by black holes.

The black hole quickly became the fashionable explanation for all new or mysterious phenomena. The first definite evidence for its existence came from observations of X-ray sources by the Uhuru satellite designed by R. Giaconni and his colleagues. The sources designated Cygnus X-1 and X-3 were found to produce radiation that has been interpreted as black hole emissions. (Only Cygnus X-1 is known to be a stellar-mass black hole; its mass exceeds the limiting value for a black hole.)

BIBLIOGRAPHY

David H. DeVorkin (1982), *The History of Modern Astronomy and Astrophysics: A Selected, Annotated Bibliography* (New York: Garland), 316–25.

Thornton Page and Lou Williams Page eds. (1968), *The Evolution of Stars* (New York: Macmillan), is an anthology of short selections from *Sky and Telescope.*

Bengt Stroemgren (1951), "The Growth of Our Knowledge of the Physics of the Stars," in *Astrophysics: A Topical Symposium,* ed. J. A. Hynek (New York: McGraw-Hill), 172–258.

Topic *i:* Theories before 1900

Frank A. J. L. James (1982), "Thermodynamics and Sources of Solar Heat, 1846–1862," *British Journal for the History of Science* 15:155–81.

Peggy Aldrich Kidwell (1981), "Prelude to Solar Energy: Pouillet, Herschel, Forbes and the Solar Constant," *Annals of Science* 38:457–76.

A. J. Meadows (1972), *Science and Controversy: A Biography of Sir Norman Lockyer* (Cambridge: MIT Press), Chaps. 6 and 7; (1970), *Early Solar Physics* (New York: Pergamon), reviews ideas about the sun in the late nineteenth century and reprints a number of papers from this period.

Topic *ii:* H-R Diagram

David H. DeVorkin (1981), "Community and Spectral Classification in Astrophysics: The Acceptance of E. C. Pickering's System in 1910," *Isis* 72:29–49.

A. G. Davis Philip and Louis C. Green (1978), "Henry Norris Russell and the H-R Diagram," *Sky and Telescope* 55:306–10.

A. G. Davis Philip and David H. DeVorkin, eds. (1977), *In Memory of Henry Norris Russell*, includes articles on the history of the H-R diagram.

Topic *iii:* Stellar Structure: Theory of Collapsed Matter

S. Chandrasekhar (1984), "On Stars, Their Evolution and Their Stability," *Reviews of Modern Physics* 56:137–47, is his Nobel Prize lecture reviewing his work on this subject; (1964), "The Case of Astronomy," *Proceedings of the American Philosophical Society* 108:1–6.

Kameshwar C. Wali (1982), "Chandrasekhar vs. Eddington: An Unanticipated Confrontation," *Physics Today* 35, no. 10:33–40, describes the dispute about relativistic degeneracy. See the comments by

Herbert Jehle (1983), "Chandrasekhar and Eddington," ibid. 36, no. 3:15, 101, and reply by Wali, ibid.

M. Schwarzschild (1978), "The Study of Stellar Structure," in *Theoretical Principles in Astrophysics and Relativity,* ed. N. R. Lebowitz et al. (Chicago: Univ. of Chicago Press), 1–14, reviews the contributions of Chandrasekhar and others.

W. J. Luyten (1960), "White Dwarfs and Stellar Evolution," *American Scientist* 48:30–39, includes personal accounts of his observations beginning in 1921.

Topic *iv:* Hydrogen's Predominance; Reinterpretation of H-R

J. B. Hearnshaw (1986), *The Analysis of Starlight: One Hundred and Fifty Years of Astronomical Spectroscopy* (New York: Cambridge Univ. Press).

Peggy A. Kidwell (1984), "Cecilia Payne-Gaposchkin: The Making of an Astrophysicist," in *Making Contributions,* ed. B. Lotze (College Park, Md.: American Association of Physics Teachers).

Katherine Haramundanis, ed. (1984), *Cecilia Payne-Gaposchkin: An Autobiography and Other Recollections* (New York: Cambridge Univ. Press), includes essays by J. L. Greenstein and P. A. Kidwell. See the review by John Lankford (1985), "Explicating an Autobiography," *Isis* 76:80–83, suggesting a feminist interpretation of Payne-Gaposchkin's career in astrophysics.

David H. DeVorkin and Ralph Kenat (1983), "Quantum Physics and the Stars (I): The Establishment of a Stellar Temperature Scale" and ". . . (II) Henry Norris Russell and the Abundances of the Elements in the Atmospheres of the Sun and Stars," *Journal for the History of Astronomy* 14:102–32, 180–222.

Owen Gingerich (1982), "Henry Draper's Scientific Legacy," in *Symposium on the Orion Nebula to Honor Henry Draper,* ed. A. E. Glassgold et al. (New York: New York Academy of Sciences), 308–20, includes the "smoking gun" letter from Russell that persuaded Cecilia Payne to discount her results on the abundance of hydrogen.

Elske V. P. Smith (1980), "Cecilia Payne-Gaposchkin," *Physics Today* 33, no. 6:64–65.

Topic *v:* Nucleosynthesis; Stages of Evolution

Andre Maeder and Alvio Renzini, eds. (1984), *Observational Tests of the Stellar Evolution Theory* (Boston: Reidel).

A. V. Sweigart (1976), "Evolution of Red-Giant Stars," *Physics Today* 29, no. 1:256–32.

Topic *vi:* Pulsars; Neutron Stars

M. M. Waldrop (1986), "Is Cygnus X-3 a Quark Star?" *Science* 231:336–38, reports work of Francis Hazen suggesting that "neutron stars" like Cygnus X-3 may actually be made of degenerate quark matter rather than neutrons.

Michael N. McMorris (1982), "Pulsar Research as Normal Science," *Physis* 24:265–79; (1981), "The Ancestry of Pulsars," ibid. 23:473–84.

Herbert Gursky (1976), "Neutron Stars, Black Holes, and Supernovae," in *Frontiers of Astrophysics,* ed. E. H. Avrett (Cambridge: Harvard Univ. Press), 147–202.

A. Hewish (1975), "Pulsars and High Density Physics" *Science* 188:1079–83, is a Nobel Prize lecture.

Arthur J. Meadows and J. G. O'Connor (1971), "Bibliographical Statistics as a Guide to Growth Points in Science," *Science Studies* 1:95–99, analyzes papers on pulsars.

F. G. Smith et al. (1968), *Pulsating Stars* (New York: Plenum), is a selection of 50 articles that appeared in *Nature* in 1968, with an introduction by F. G. Smith and A. Hewish.

Topic *vii:* Black Holes

Walter J. Wild (1987), "Black Holes: A Historical Perspective," *Astronomy Quarterly* 5:207–26.

Virginia Trimble and Lodewijk Woltjer (1986), "Quasars at 25," *Science* 234:155–61, describes the evidence for the present view that the energy source for a quasar is accretion onto a massive black hole.

J. B. Hutchings (1985), "Observational Evidence for Black Holes," *American Scientist* 73:52–59.

J. Eisenstaedt (1982), "Histoire et Singularités de la Solution de Schwarzschild (1915–1923)," *Archive for History of Exact Sciences* 27:157–98.

James W. Montgomery, Jr. (1982), "Nathanial Bowditch's Classical 'Black Hole' Calculation," *Journal for the History of Astronomy* 13:54–55.

John A. Wheeler (1981), "The Lesson of the Black Hole," *Proceedings of the American Philosophical Society* 125:25–37.

Steven Detweiler (1981), "Resource Letter BH-1: Black Holes," *American Journal of Physics* 49:394–400.

Simon Schaffer (1979), "John Michell and Black Holes," *Journal for the History of Astronomy* 10:42–43.

E. Broda (1971), "Origin of the Black Hole," *Physics Today* 24, no. 12:11.

13.3 COSMOLOGY

i. The expanding universe in general relativity: models of W. de Sitter, A. Friedmann, G. Lemaitre, E. A. Milne
ii. The Big Bang theory of G. Gamow et al.
iii. The continuous-creation theory (F. Hoyle, Bondi, T. Gold)
iv. New evidence for the Big Bang: radio source counts, primeval fireball radiation, revision of distance scale
v. Modern cosmology: the inflationary universe; relation to elementary particle physics; will the universe expand forever?

READINGS

Marks, *Science,* 343–7. Solomon, *Structure of Space,* 184–212. F. P. Dickson, *The Bowl of Night* (1968), 126–91. Tauber, *Man's View,* 296–311, 264–81. Ferris, *Red Light,* 55–245. I. Asimov, *The Universe* (1971), 188–229. Frank Durham and Robert D. Purrington, *Frame of the Universe* (1983), 175–215. *Cosmology + 1* (1977, readings from *Scientific American*), articles by G. Gamow, "The Evolutionary Universe" (Sept. 1956), 12–19; J. J. Callahan, "The Curvature of Space in a Finite Universe," (Aug. 1976), 20–30; D. Sciama, "Cosmology before and after Quasars" (Sept. 1967), 31–33; A. Webster, "The Cosmic Background Radiation" (August 1974), 34–41.

Articles: Lloyd Motz, "Cosmology since 1850," *Dictionary of the History of Ideas,* ed. P. P. Wiener 1(1973):554–70. G. J. Whitrow, "Theoretical Cosmology in the Twentieth Century," in *Human Implications of Scientific Advance,* ed. E. G. Forbes (1978), 576–93. E. P. Hubble, "The Problem of the Expanding Universe," *American Scientist* 30(1942):95–115. G. Gamow, "The Origin and Evolution of the Universe," *American Scientist* 39(1951):393–407. H. Bondi and T. Gold, "The Steady State Theory of the Expanding Universe," *Monthly Notices of the Royal Astronomical Society* 108(1948):252–70 (an original source, most of which can be understood by students; 260–61 could be omitted). Martin Ryle, "Radio Astronomy and Cosmology," *American Scientist* 50(1962):92–98. M. M. Waldrop, "The New Inflationary Universe," *Science* 219(1983):375–77; "The Large-Scale Structure of the Universe," *Science* 219(1983):1050–52. D. H. Smith, "The Inflationary Universe Lives?" *Sky and Telescope* 65(1983):207–10. Jeremy Bernstein, "The Birth of Modern Cosmology," *American Scholar* 55, no. 1(1985–1986):7–18. George Field, "Astronomy of the Twentieth Century," *American Scientist* 74(1986):173–81. Paul Davies, "New Physics and the New Big Bang," *Sky and Telescope* 70(1985):406–10. Steven Weinberg, "Origins," *Science* 230(1985):15–18.

The concept of the "expanding universe" emerged from the mathematical calculations of Aleksandr Friedman, Willem de Sitter, and Georges Lemaitre, based on Einstein's general relativity theory (**8.7**) and the analysis of astronomical evidence on the distances and velocities of nebulae Edwin P. Hubble (**13.1**). This section covers cosmology from 1930 to the present, leaving the topics of stellar evolution and the origin of the elements to **13.2**, and the origin of the solar system to **12.4**.

SYNOPSIS

Lemaitre in 1927 proposed that the universe originated as a single "primeval atom." He arrived at this conception by considering how the second law of thermodynamics should be expressed in terms of quantum theory. Natural processes that correspond to increasing entropy should be reducible to processes in which the number of quanta increases; a high-energy quantum, by interacting with matter, is transformed into two or more quanta of lower energy. Moreover, because of Einstein's relation between mass and energy, such a process is equivalent to the division of a single large mass into two or more smaller masses. Conversely, if we trace the history of the universe backwards we should arrive at an initial state in which the total mass-energy is concentrated in a single superquantum or superatom.

Acceptance of such a theory would require us to abandon the assumption, implicit in Newtonian physics and articulated by Laplace, that the entire history of the world is determined by its state at any time. As Lemaitre put it:

> Clearly the initial quantum could not conceal in itself the whole course of evolution; but according to the principle of indeterminacy, this is not necessary. Our world is now understood to be a world where something really happens; the whole story of the world need not have been written down in the first quantum like a song on the disc of a phonograph. The whole matter of the world must have been present at the beginning, but the story it has to tell may be written step by step.

I interpret this as a bold assertion that *randomness* supplies novelty in the world; something new and unpredictable is always coming into existence as the primeval atom subdivides into smaller and smaller portions of mass-energy.

Lemaitre assumed that the entire mass of the universe, estimated at about 10^{54} grams, started as a single atomic nucleus that occupied a sphere of radius 10^{13} centimeters, i.e., about 1 astronomical unit (Earth-sun distance) when the distance between particles was the same as that between protons in an ordinary nucleus, about 10^{-13} centimeters. This sphere is also identical with the entire space of the universe. It then begins to expand, the rate of expansion being governed by the force of "cosmic repulsion" (Einstein's "cosmological constant") and gravitational attraction. There will be three major phases in the expansion:

1. fragmentation of the primeval atom, giving high-energy radiation and high-velocity particles;
2. a period of deceleration when gravitational attraction and cosmic repulsion are roughly balanced; in this period there will be condensations, forming galaxies and clusters of stars;
3. renewed expansion as the repulsion dominates. [According to Petrosian (1974) Lemaitre's justification for the expansion—that condensation would reduce the pressure, thereby reducing the gravitational attraction and allowing the repulsion to overcome it—is incorrect because it ignores the increase in radiation pressure caused by condensation. So it is doubtful whether Lemaitre's model would actually expand or collapse after the second phase.]

The present radius of the universe, according to Lemaitre, is about 10^{10} LY; so far, telescopes had seen out to a distance of about 5×10^{8} LY, a not insignificant part of the entire universe.

Lemaitre himself gave a memorable summary of his cosmogony:

> The evolution of the world can be compared to a display of fireworks that has just ended: some few red wisps, ashes and smoke. Standing on a well-chilled cinder, we see the slow fading of the suns, and we try to recall the vanished brilliance of the origin of worlds. [1931, *The Primeval Atom,* 78]

Among the many cosmological theories in the 1930s, one that attracted considerable attention was E. A. Milne's "kinematic relativity." Milne objected to Lemaitre's theory because, first, it did not even attempt to explain why space is expanding rather than contracting; and, second, the curvature of space was introduced merely as an ad hoc hypothesis to account for the data.

Milne proposed a "cosmological principle," which he attributed to Einstein: *all places in the universe are equivalent.* (It had previously been stated by Einstein in a somewhat more restricted form in 1931, as applying primarily to the average density of matter.) Milne interpreted this to mean that any observer, measuring the positions and velocities of the stars and galaxies, ought to find the same relation between position and velocity.[1] This principle seems to have been intended to dispel the idea that "expansion" of the universe implies a unique *center* from which things are moving away. Milne proved mathematically that his cosmological principle implies that each observer will find a law of the form $v = Hd$, the Hubble velocity-distance law.

Milne also pointed out that expansion is practically inevitable if one makes a very general assumption about the state of the universe at some time in the past. Suppose one has a finite number of "particles" (stars or galaxies) moving with various speeds within a finite region of space. If each particle keeps moving forever with approximately the same velocity and interactions between the particles have negligible effect, sooner or later all the particles will leave the region and go off in all directions. An observer looking out from inside that region will discover that the particles farthest away are moving fastest, and Milne showed that after a long enough time Hubble's velocity-distance law will be obeyed. Thus the fact that the most distant galaxies are receding from us at the greatest speeds does not necessarily mean that there is a "cosmic repulsion"; the explanation may be that they are farthest away *because* they started with the highest speeds.

In Milne's theory every observer may consider himself to be at the center of the universe since he sees the other galaxies receding in the same fashion in every direction. The number of galaxies is actually infinite—for otherwise some observers would have to be at the edge and would see nothing on one side. But to each observer all the galaxies seem to be squeezed into a finite volume, a sphere of radius ct where $c =$ speed of light and $t =$ time since the beginning of the expansion ($t = 1/H$ in Hubble's law). The density of galaxies seems to go to infinity near the boundaries of the sphere, but this is only an optical illusion caused by the relativistic distortion of time and space measurements; you can never actually travel to the edge of the universe to check what is going on there, for it is receding from you at the speed of light. On the other hand if you stay at home you will eventually see nothing in the sky except your own galaxy; the others will disappear into the dim background at the edge of the universe, where galaxies recede at the speed of light and their light never gets back to us.

Milne's cosmological theory, though widely discussed in the 1930s,

[1]Obviously this is unlikely to be exactly true for all observers in the real universe; Milne simply wanted to construct an ideal universe that could be used as a first approximation. He defined a basic set of fundamental observers for which his principle is valid; the observer at any particular place is assumed to share the motion of the "substratum" that may participate in the overall expansion of the universe.

was eventually discarded because it involved too many implausible assumptions and dubious arguments. But it provided part of the inspiration for the Bondi-Gold-Hoyle "continuous creation" cosmology, in which the "cosmological principle" of the equivalence of all observers was extended to include observers at all *times* as well as all *places* (see below).

The so-called "Big Bang" theory of the origin of the universe was proposed by George Gamow (formerly a student of A. A. Friedmann, in Russia) in a series of articles beginning in 1942. Gamow started from the problem of explaining the abundance of the chemical elements. In his earlier work with Edward Teller he had found that some features of the abundance curve (as a function of atomic number Z) can be explained by nuclear reactions now going on in stars, for example the destruction of lithium, beryllium, and boron by reaction with protons to form helium and carbon. But attempts to explain the building up of heavy elements in stars (e.g., by C. v. Weizsaecker in 1938) were unsatisfactory. As an alternative one could assume that these elements were created at an early stage of the universe, when according to the expanding-universe theory there was a state of extremely high density and temperature. S. Chandrasekhar and L. R. Henrich (1942) had shown that by assuming that the elements had been formed in thermodynamic equilibrium at a density of 10^7 grams per cubic centimeter and a temperature of 8×10^9 °C one could explain the abundances of the first 20 elements. However, this theory failed to explain why the abundances of all the elements in the second half of the period table are roughly the same; one would expect from the thermodynamic equilibrium model that they would fall off exponentially with atomic number Z.

In his first paper on the subject (1942) Gamow postulated that in the initial superdense state of the universe there would be a large number of transuranium elements; these would fission into two or more roughly equal parts. The free neutrons produced by fission will decay to protons (beta decay), thus giving a large amount of hydrogen. It is also possible that superheavy nuclei can fission into three or more parts, which would thus yield nuclei with atomic numbers less than 50. In later papers Gamow abandoned this idea, working instead from the assumption that all the elements are built up from hydrogen by a process of successive absorption of single neutrons followed by beta decay.

To construct such a theory it is essential to know the probability that each nucleus, when bombarded by neutrons, will absorb a neutron and turn into the next heavier nucleus. That information was also needed for a large number of nuclei to design atomic bombs during

World War II and nuclear reactors afterwards. Early in 1946 Donald J. Hughes of Brookhaven National Laboratory announced measurements of these probabilities, known as "neutron cross sections," for a large number of nuclei; from his data it appeared that there was an inverse correlation of the element's cross section with its abundance in the universe. This suggested to Gamow that it would be fruitful to develop his "nonequilibrium" theory of the formation of elements, based on successive neutron capture, as distinct from the "equilibrium" theory in which the abundances would be determined primarily by binding energies.

Gamow's theory was worked out in detail by Ralph Alpher and Robert Herman, beginning in 1947. (Alpher was a graduate student at George Washington University, where Gamow was a professor of physics, and also worked at the Johns Hopkins Applied Physics Laboratory where Herman was employed.) The theory became known as the "alpha-beta-gamma theory" because of a paper that appeared in 1948 under the names R. A. Alpher, H. Bethe, and G. Gamow. Bethe was brought in mainly to complete the alphabetical sequence. Gamow tried unsuccessfully to persuade Herman to change his name to Delter (cf., Gamow, 1949, 369). J. S. Smart, Enrico Fermi, and Anthony Turkevich also contributed to the theory.

According to the alpha-beta-gamma theory the universe began as a very hot dense gas of neutrons, which immediately started to expand and decay into protons and electrons. This primordial substance was called *ylem,* from a Greek work meaning the matter from which the elements were formed. (English words derived from this root usually begin with *hyl-*; thus "hylozois," the theory that matter is alive; "hylotheism," the doctrine that God and matter are identical.) The temperature would have to be at least 10^9 °K to start the nuclear fusion reactions, e.g., formation of deuterium from a neutron and a proton. Because of the rapid expansion and cooling of the ylem, most of the elements would have to be formed within the first half hour while the neutrons are still available at sufficiently high density.

(Alternatively one may assume that the primordial matter consists of protons and electrons. Because of the extremely high temperature during the first few seconds, a temporary thermodynamic equilibrium would result in the formation of a substantial number of neutrons, and the outcome would be about the same as if one had started with neutrons.)

The theory at first appeared to give a good qualitative explanation for the overall shape of the abundance curve. It also predicted that a substantial amount of helium would be formed, amounting to between 22 and 28 percent (by mass) of the total, and this result was in striking agreement with the observed fact that their abundance in most parts of the universe is around 25 percent. As Gamow had already pointed out in his 1942 paper, the abnormally low abundances of lithium, beryllium, and boron could be easily explained by taking account of the reactions with protons that would be formed by beta decay of the neutrons in about fifteen minutes.

A major stumbling block of the alpha-beta-gamma theory was the failure to explain how an element of mass-number 5 or 8 could be built up by adding particles to nuclei of mass-numbers 4 or 7, respectively. The helium nucleus (2 protons, 2 neutrons) is very stable, while the nucleus formed by adding another proton is quite unstable. The barrier could apparently be overcome only by adding two or three particles at once, and the probability of such a reaction seemed extremely small under the conditions postulated. The only way out seemed to be to assume that heavier elements are formed in stars at a later stage of the evolution of the universe (**13.2**).

A distinctive feature of the alpha-beta-gamma theory is that an enormous amount of radiation is present in the early stages of the universe. This is simply a consequence of the very high temperature assumed and the Stefan-Boltzmann law (**9.3-*iv***), which states that the radiation energy is proportional to the fourth power of the temperature. In fact, in this "primeval fireball" as it is now called, most of the energy of the universe is in the form of radiation. As the system cools off and expands, the proportion of matter increases while the proportion of radiation decreases. There will be a crossover point at about 10 million years, when the amounts are roughly the same. Only then would it be possible for matter to condense into galaxies by the "gravitational instability" effect discovered by James Jeans in 1901—random fluctuations in density can be self-reinforcing.

If the alpha-beta-gamma theory is correct, the universe even today should be filled with a uniform background of black-body radiation with a frequency distribution given by Planck's law (**9.4**). Alpher and Herman (1948) estimated that the present background radiation should correspond to a temperature of about 5 °K. For this temperature most of the radiation is at extremely long wavelengths and thus could not be detected directly with the equipment available at the time.

Another objection to the alpha-beta-gamma theory, or to any expanding-universe theory, was that the rate of expansion determined by Hubble's velocity-distance law (**13.1**) indicated that all the material in the universe was collected in one place about 1.8 billion years ago. Yet the Earth itself, according to radiometric dating methods, was more than twice as old as that. Theories of stellar evolution also indicated that some stars must have ages of several billion years. Gamow in 1949 admitted that this is "the first serious disagreement between the conclusions of relativistic cosmology and the observed facts." He attempted to evade it by adding the cosmological constant to the Einstein-Hilbert equations, as Einstein had originally done (**8.7**). This would mean postulating a long-range repulsive force between galaxies, so that the rate of expansion is greater now than in the past, and thus the age of the universe could be made greater than 1.8 billion years. In this way Gamow could force the theory into agreement with the estimated age of the universe—about 3 billion years in 1949—yet the value of the cosmological constant would be so small that it would have no detectable effect on the motion of planets in the solar system.

Another way to avoid the time-scale difficulty was to go back to the second phase in Lemaitre's model (see above). This is a period when there is no expansion, and its duration is somewhat indefinite. The "age" deduced from Hubble's law refers only to the third (expanding) phase. So it is possible that the stars and our solar system were formed earlier in the second phase before the present phase of expansion got started. This is not a very satisfactory answer since one would expect the theory to state just how long the second phase lasts, rather than use this period to evade observational disproofs.

The major alternative to the alpha-beta-gamma or "Big Bang" theory was proposed in 1948 by Hermann Bondi, Thomas Gold, and Fred Hoyle. Their theory became known as the *continuous-creation* or *steady-state* cosmology.

Bondi and Gold started from the axiom that any experimental result in physical science should be independent of both the place and the time at which the experiment is performed. By "experimental result" they did not mean measurements of local or temporary properties such as temperature and pressure, but laws and constants supposed to have universal validity. They argued that this axiom would be violated if the large-scale structure of the universe were changing with time. In particular, *Mach's Principle* states that the distribution of all matter in the universe determines the local frame of reference with respect to which an object may be rotating. We cannot postulate a laboratory completely isolated from all outside influences; the structure of space-time inside the laboratory must be affected, according to general relativity theory, by what is outside it.

The axiom that an observer (i.e., a "fundamental" observer defined as in Milne's theory, see footnote 1) must find the same laws and physical constants at any time or place was called the *perfect cosmological principle*. It might seem that this axiom requires the overall structure of the universe to be static as well as uniform and would thus contradict the idea that the universe is expanding, since in an expanding universe the density of matter is continually decreasing. Bondi and Gold however postulated that *matter is continually being created* at a rate sufficient to keep the average density constant. The rate would be too small to be directly observable: one new atom of hydrogen would appear in each cubic meter of space every 300,000 years. The actual time and place of its creation would be random.

According to the continuous-creation theory the universe did not begin at some particular time—rather, it has always existed. Moreover, it would look the same to an observer despite the expansion; the galaxies that pass out of the visible portion of the universe are replaced by new ones, so that the number of stars visible in the sky at any given time (either by naked eye or with a telescope of given power) is the same.

The idea that matter simply comes into existence at various times and places seems implausible to many people. On the other hand, as Hoyle pointed out, it is really no less plausible than the idea that *all* the matter in the universe was suddenly created at *one* time and place; in fact it is easy to describe continuous creation by a simple mathematical formula of universal validity, whereas the Big Bang requires an arbitrary selection of the instant of creation. If you believe that matter was created at all, why should it be more likely to have been created at one time rather than another? The Big Bang theory describes what happens to matter *after* it is created but says nothing about creation itself; whereas the continuous-creation theory brings the creation of matter within the domain of scientific theory.

Another difference between the two cosmologies is what they predict for the future. According to the Big Bang theory, almost all of the stars in other galaxies will have disappeared from our sky 10 billion years from now; they will have moved so far away and their red shifts will be so great that we could not see them even with the most powerful telescopes imaginable. The continuous-creation theory predicts that while the stars we now see will have disappeared in this way, an equal number of new ones will have taken their place.

Aside from its philosophical merits, the only evidence favoring the continuous-creation theory in 1948 was the discrepancy between the time scale for the conventional expanding universe (Hubble's 1.8 billion years) and the age of the Earth (at least 3 billion years). The continuous-creation theory could eliminate this discrepancy since it pictured the universe as being essentially the same several billion years ago as now, rather than concentrated in a small space; it did not place any limits on the time available for the formation of stars and planets.

The "short" time scale of 1.8 billion years was based on the galactic distances determined by the use of Henrietta Leavitt's period-luminosity relation for the Cepheid variables (**12.3**). During the 1940s, evidence was accumulated (e.g., by the Swedish astronomer K. Lundmark) that these distances were too small, because they led to estimated absolute magnitudes disagreeing with those obtained by other methods. In 1951 Alfred Behr reanalyzed Hubble's data and corrected the distance scale by a factor 2.2, making the time since the start of the expansion 3.8 billion years. Finally in 1952 Walter Baade showed that the Andromeda galaxy is at least twice as far away as the accepted Cepheid scale indicated; he concluded that there are two different kinds of Cepheid variables with different period-luminosity relations. Further research by Allan Sandage and others led to more accurate distance determinations and increased the Hubble expansion time to more than 12 billion years by 1972. This would be the age of the universe if it had expanded at constant speed beginning from a point; in most versions of the Big Bang theory one has to multiply this time by a "deceleration factor" to take account of the change in speed due to gravitational attraction and/or cosmic repulsion.

In any case the time-scale difficulty was definitely removed after 1952, since more recent estimates of the age of the Earth have not kept up with the increasing values of the estimated age of the universe. Clair Patterson's 1953 value from uranium-lead dating, 4.5 ± 0.3 billion years, has been repeatedly confirmed during the past thirty years, even though there have been major changes in our theories about the Earth's past. Big Bang advocates could emphasize that the time-scale difficulty was the main observational reason for adopting a steady-state cosmology, and so one could now forget about that cosmology. The steady-state theorists regarded this point as irrelevant and wanted their theory to be judged on the basis of its own specific predictions. In fact Bondi several times declared himself a follower of Karl Popper's philosophy of falsificationism (**11.5**); he argued that the steady-state theory is a "good scientific hypothesis" in Popper's sense precisely because it makes very definite statements that should be vulnerable to disproof. He described it as being very "inflexible" compared to evolutionary models (including Big Bang) that use adjustable parameters (e.g., the cosmological constant). On the other hand its predictions differed from those of evolutionary models just at the limits of observation where errors are large, and thus the first results may contradict the steady-state theory even if it is really correct.

A classic example of a premature refutation, frequently cited by steady-state theorists, was the "Stebbins-Whitford effect." J. Stebbins and A. G. Whitford announced in 1948 that they had found a reddening in the spectra of distant elliptical galaxies (apart from the red shift of individual lines). This would imply some kind of evolutionary effect—a difference between properties of new and old galaxies—contrary to the postulates of the steady-state theory. A few years later it was found that the Stebbins-Whitford effect does not exist after all; the original observations were simply incorrect. Logically the steady-state theory should have been unaffected by this mistake once it was exposed, but psychologically it seems to have helped a little. (Modern observations have shown that evolutionary effects are observed for giant elliptical galaxies. This affects the determination of the deceleration parameter.)

The steady-state theory at first seemed less comprehensive than the Big Bang theory because it did not explain the formation of the chemical elements. Gamow, Alpher, and Herman had shown that helium and other elements could be built up from hydrogen under conditions of extremely high pressure and temperature such as would exist a few minutes after the Big Bang, although they had some difficulty in finding a plausible theoretical scheme for getting past mass-numbers 4 and 7. This problem was partly solved in 1957 when Margaret and Geoffrey Burbidge, William Fowler, and Fred Hoyle showed how heavy elements could be synthesized by neutron reactions in stars, especially supernovae. Thus a Big Bang was not needed to account for the formation of the elements; nevertheless the Big Bang theory is still considered to give a more satisfactory explanation for the fact that 25 to 30 percent of the mass of the universe is in the form of helium.

One would expect that cosmological theories can best be tested by looking at the most distant objects whose radiation was emitted several billion years ago. According to the Big Bang theory, the first galaxies and stars were being formed at that time, so we might well expect to see things look different from what we find in the closer (i.e., more recent) part of the universe.

The rapid growth of radio astronomy in the 1950s promised to allow just this kind of cosmological test to be made. Martin Ryle and his colleagues at Cambridge University counted the number of radio sources of various intensities and compared the results with those predicted by the steady-state and other theories. The first results (1955) were said to provide conclusive evidence against the steady-state theory but were later criticized as inaccurate. Subsequent results, though subject to some disagreement, were still unfavorable. According to a summary by Ryle:

> As we proceed outwards from the most intense—and presumably nearest sources, we find a great excess of fainter ones. Now this suggests that in the past either the power, or the space density of the sources was greater than it is now. Whichever way it is, the universe must have changed radically within the time-span accessible to our radio telescopes.
>
> But at still smaller intensities we find a sudden reversal of this trend—a dramatic reduction in the number of the faintest sources. This convergence is so abrupt that we must suppose that before a certain epoch in the past, there were no radio sources. Both these observations, therefore, seem to indicate that we are living in an evolving universe—which has not always looked the same. . . .
>
> The abrupt reduction in the number of very faint sources corresponds to an age about 1/10th of the present age. This is probably associated with the actual formation of galaxies from the primeval gas. . . . So the picture presented by the radio source observations supports the idea of an expanding universe which evolves with time, from an initial state of very high temperature and high density. [M. Ryle in John, 1973, 41]

The radio-source evidence was somewhat confused by the recognition in the 1960s of a separate category of peculiar sources, the quasi-stellar objects or *quasars*. All quasars have very large red shifts, and if one converts the red shifts to distance by the usual Doppler formula one must assume that they are very far away and generate exceptionally large amounts of energy. If that is the correct interpretation of their red shifts, then quasars are objects that exist only long ago and far away and thus contradict the steady-state theory. Some theorists have argued, however, that the red shifts of quasars are due to some other cause and that they are really much closer and less energetic objects. Some such explanation is needed to save the steady-state theory; if quasars are nearby rather than distant objects, their existence would be compatible with the perfect cosmological principle. Moreover, by subtracting them from Ryle's radio-source counts of distant objects, one would get results that do not disagree with the predictions of the steady-state theory.

It now appears that the steady-state theory has been definitely refuted by the discovery of the cosmic background radiation predicted by the alpha-beta-gamma theory. The discovery was made by Arno Penzias and Robert Wilson at the Bell Telephone Laboratory in New Jersey in 1965, as part of an effort to build sensitive microwave receiving systems for satellite communication. Their results, together with several other measurements at various wavelengths, established that space in the vicinity of the Earth is filled with black-body radiation—i.e., a frequency distribution given by the Planck law (**9.4**)—corresponding to a temperature of 2.7 °K. Considering the degree of accuracy of most cosmological calculations, this is astonishingly close to the value of 5 °K estimated by Alpher and Herman in 1948. The only known explanation for the existence of this radiation is that given by the Big Bang theory—the universe has evolved from a state of very high temperature and density several billion years ago.

The radiation is very nearly the same in all directions, which implies that the Earth is nearly at rest relative to the rest of the universe. If the Earth were moving through the radiation, the Doppler effect would make it appear hotter on one side than the other. Experiments done since 1970 indicate that the earth is indeed moving at a speed of 250 to 300 km/sec through the radiation field; this is consistent with the fact that the Earth moves around the sun at about 30 km/sec, the sun goes around the center of the galaxy at 230 km/sec, and the galaxy moves relative to other galaxies in the "local group" or "supergalaxy" at a speed of about 100 km/sec. The results also indicate that the universe as a whole is not rotating. The radiation field may be able to perform some of the functions of the old "ether" by establishing a preferred frame of reference, and by giving some substance to "Mach's principle."

The cosmic background radiation does create a serious theoretical problem because it is *too* uniform. After the first fraction of a second, the different parts of the universe could no longer be causally connected with each other; this is because light signals could not pass from one to another in time to have any influence. Thus the fact that we observe radiation to have the same temperature in different directions becomes a new mystery.

One proposed cosmology that may deal with this problem is the "New Inflationary Universe" of Alan Guth. With the help of the "grand unified theory" of elementary particles, one may imagine a phase transition (breaking the symmetry between the fundamental forces) that suddenly releases enough energy to produce a rapid expansion (exponential with time). Regions small enough to have been causally connected before the phase transition now become so inflated in size that they may encompass all of the universe which we now observe after the inflationary phase is over.

The major question that still remains, if we accept the Big Bang cosmology in some version, is whether the universe will continue to expand indefinitely or eventually start contracting. The answer depends primarily on the density of matter in the universe; if it is great enough, the gravitational attraction can slow down the expansion and reverse it, just as the Earth's gravity forces an object thrown upward to fall down again. We need about one atom per million cubic centimeters to brake the expansion; the amount of material we can see in galaxies is only about one-thirtieth of this. Hence the evidence seems to suggest indefinite expansion. However, it is always possible that there is enough interstellar dust, black holes, or neutrinos (if they are found to have finite mass) to make up the difference.

If the universe does contract at some time in the future, it would presumably collapse into a black hole. It might then be reborn in a new Big Bang. John Wheeler has suggested that every time this happens, the various dimensionless physical constants such as the proton-electron mass ratio may acquire new values. So far, theoretical physicists have not been able to explain why those constants have the values they do. But Wheeler and other cosmologists have shown that if these constants were very much different from their present values, the formation of planetary systems and the evolution of higher forms of life might be impossible. This observation has given rise to the so-called "anthropic principle" suggested by Brandon Carter: the reason why the physical constants have the values they do is that if they didn't, we would not be here to measure them (**11.2**). There are two interpretations of this principle: first (what one might call the "weak" anthropic principle), many universes have been and will be created with different values of the physical constants, but life can exist in only a few of them. Second ("strong" anthropic principle), there is a coupling between the creation of the universe and future evolution of life in that universe, because the universe cannot have a real existence unless it is observed by a conscious intelligence.

BIBLIOGRAPHY

M. K. Munitz, ed. (1957), *Theories of the Universe* (New York: Free Press), 302–439, contains extracts from writings of W. de Sitter, A. S. Eddington, G. Lemaitre, E. A. Milne, H. P. Robertson, G. Gamow, H. Bondi, D. W. Sciama, F. Hoyle.

L. C. Shepley and A. A. Strassenburg, eds. (1979), *Cosmology: Selected Reprints* (Stony Brook, N.Y.: American Association of Physics Teachers), includes M. P. Ryan and L. C. Shepley, "Resource Letter RC-1: Cosmology," reprinted from *American Journal of Physics* 44(1976):223–30; reprints of papers by C. W. Misner, A. Einstein and E. G. Straus, H. Reeves, R. W. Wilson, etc.

Edward Harrison (1985), *Masks of the Universe* (New York: Macmillan).

John Lankford (1985), "Discovery in Modern Astronomy," *4S Review* 3, no. 1:16–21, is an essay review of recent books by R. F. Hirsh and M. Harwit.

Owen Gingerich, ed. (1984), *General History of Astronomy,* vol. 4, Part A, *Astrophysics and Twentieth-Century Astronomy to 1950* (New York: Cambridge Univ. Press).

David H. DeVorkin (1982), *The History of Modern Astronomy and Astrophysics: A Selected, Annotated Bibliography* (New York: Garland), 339–46.

C. M. Copp (1982), "Relativistic Cosmology II: Social Structure, Skepticism, and Cynicism," *Astronomy Quarterly* 4, no. 16:179–88.

Bernard Lovell (1981), *Emerging Cosmology* (New York: Columbia Univ. Press).

R. Berendzen, Richard Hart, and Daniel Seeley (1976), *Man Discovers the Galaxies* (New York: Science History), Sect. 4: "The Birth of Modern Cosmology."

Jacques Merleau-Ponty and Bruno Morando (1976), *The Rebirth of Cosmology* (New York: Knopf).

Topic *i*: Expanding Universe in General Relativity

Georges Lemaitre (1950), *The Primeval Atom* (New York: Van Nostrand Reinhold), is a translation of *L'Hypothèse de l'Atome Primitif.*

H. Bondi (1948), "Review of Cosmology," *Monthly Notices of the Royal Astronomical Society* 108:104–20, is a critique that laid the foundations for the introduction of the continuous creation theory.

Helge Kragh (1987), "The Beginning of the World: Georges Lemaitre and the Expanding Universe," *Centaurus* 30:114–39.

O. Godart and M. Heller (1986), *Cosmology of Lemaitre* (Tucson, Ariz.: Pachart); M. Heller and O. Godart (1985), *The Expanding Universe: Lemaitre's Unknown Mansucript* (Tucson, Ariz.: Pachart); O. Godart and J. Turek (1982), "Le Développment de l'Hypothèse de l'Atome Primitif," *Revue des Questions Scientifiques* 153:145–71; M. Heller and O. Godart (1981), "Origins of Relativistic Cosmology," *Astronomy Quarterly* 4, no. 13:27–33; O. Godart and M. Heller (1979), "Einstein-Lemaitre: Rencontre d'idées," *Revue des Questions Scientifiques* 150:23–43.

Norriss S. Hetherington (1983), "Philosophical Values and Observation in Edwin Hubble's Choice of a Model of the Universe," *Historical Studies in the Physical Sciences* 13:41–68.

S. Chandrasekhar (1980), "The 1979 Milne Lecture: Edward Arthur Milne: His Part in the Development of Modern Astrophysics," *Quarterly Journal of the Royal Astronomical Society* 21:93–107.

V. Petrosian (1974), "Confrontation of Lemaitre Models and the Cosmological Constant with Observation," in *Confrontation of Cosmological Theories with Observational Data,* ed. M. S. Longair (Boston: Reidel), 31–46.

C. W. Kilmister (1973), *General Theory of Relativity* (New York: Pergamon Press), includes reprints of original sources.

Topic *ii:* Big Bang

George Gamow (1970), *My World Line: An Informal Autobiography* (New York: Viking), includes recollections of A. A. Friedmann; (1949), "On Relativistic Cosmogony," *Reviews of Modern Physics* 21:367–73.

Frederick Reines, ed. (1972), *Cosmology, Fusion and Other Matters: George Gamow Memorial Volume* (Boulder: Colorado Associated Universities Press), includes recollections by Alpher and Herman.

Topic *iii:* Continuous Creation Theory

H. Bondi et al. (1960), *Rival Theories of Cosmology* (London: Oxford Univ. Press), has short presentations by Bondi, W. B. Bonnor, and R. A. Lyttleton, followed by a discussion chaired by G. J. Whitrow.

Fred Hoyle (1982), "Steady-State Cosmology Revisited,"in *Cosmology and Astrophysics,* ed. Y. Terzian and E. M. Bilson (Ithaca: Cornell Univ. Press), 17–57, with comments by H. Bondi and T. Gold, 58–61, 62–65; (1982), "The Universe: Past and Present Reflections," *Annual Review of Astronomy and Astrophysics* 20:1–35.

B. R. Martin (1978), "Radio Astronomy Revisited: A Reassessment of the Role of Competition and Conflict in the Development of Radio Astronomy," *Sociological Review* 26:27–56, concerns the effect of the steady-state theory on funding for radio astronomy.

R. Schlegel (1958), "Steady-State Theory at Chicago," *American Journal of Physics* 26:601–4, discusses MacMillan's papers of 1918 and 1925.

Topic *iv:* New Evidence for Big Bang

Bernard Lovell (1985), "Martin Ryle," *Quarterly Journal of the Royal Astronomical Society* 26:358–68, is an obituary notice.

Jeremy Bernstein (1984), "Three Degrees above Zero," *New Yorker* 60, no. 27:42–70, is a profile of R. W. Wilson and A. A. Penzias.

R. W. Wilson (1979), "The Cosmic Microwave Background Radiation," *Science* 205:866–74.

Steven Weinberg (1977), *The First Three Minutes* (New York: Basic Books), Chap. 6, relates the history of the discovery (and nondiscovery) of the cosmic background radiation.

G. L. Trigg (1975), *Landmark Experiments in Twentieth Century Physics* (New York: Crane, Russak), Chap. 16, is on the Penzias-Wilson observation.

Laurie John, ed. (1973), *Cosmology Now* (London: British Broadcasting Corp.), includes M. Ryle, "Looking with New Eyes," and other views on the steady-state/Big Bang controversy.

Topic *v:* Modern Cosmology

Marcia Bartusiak (1987), "The Genesis of the Inflationary Universe Hypothesis," *Mercury* 16:34–45.

John D. Barrow and Frank J. Tipler (1986), *The Anthropic Cosmological Principle* (New York: Oxford Univ. Press); J. D. Barrow (1981), "The Lore of Large Numbers: Some Historical Background to the Anthropic Principle," *Quarterly Journal of the Royal Astronomical Society* 22:388–420.

Jeremy Bernstein and Gerald Feinberg, eds. (1986), *Cosmological Constants: Papers in Modern Cosmology* (New York: Columbia Univ. Press), 92–238.

C. M. Copp (1985), "Professional Specialization, Perceived Anomalies, and Rival Cosmologies," *Knowledge: Creation, Diffusion, Utilization* 7:63–95; (1982, 1983), "Relativistic Cosmology I. Paradigm Commitment and Rationality" and ". . . II. Social Structure, Skepticism, and Cynicism," *Astronomy Quarterly* 4:103–16, 179–88.

Paul Hodge (1984), "The Cosmic Distance Scale," *American Scientist* 72:474–82, focuses on two approaches that give different values for Hubble's constant.

G. F. R. Ellis (1984), "Alternatives to the Big Bang," *Annual Review of Astronomy and Astrophysics* 22:157–84.

David N. Schramm (1983), "The Early Universe and High-Energy Physics," *Physics Today* 36, no. 4:27–33.

Sidney van der Bergh (1981), "Size and Age of the Universe," *Science* 213:825–30.

R. J. Tayler (1980), "Cosmology, Astrophysics and Elementary Particle Physics," *Reports on Progress in Physics* 43:253–99.

Harry Woolf, ed. (1980), *Some Strangeness in the Proportion* (Reading, Mass: Addison-Wesley), contains papers by M. F. Rees, J. A.

Wheeler, D. W. Sciama, C. W. Misner, G. de Vaucouleurs, and Y. Ne'eman.

James E. Gunn and Beatrice M. Tinsley (1975), "Will the Universe Expand Forever?" *Nature* 257:454–57; another article with same title (1976) by J. R. Gott, J. E. Gunn, D. N. Schramm, and B. M. Tinsley, *Scientific American* 234, no. 3:62–72.

George B. Field, Halton Arp, and J. N. Bahcall (1973), *The Redshift Controversy* (Reading, Mass.: Benjamin), proceedings of a 1972 symposium, with about 180 pages of reprinted earlier papers starting with Hubble (1929).

BOOK LIST

THIS LIST includes all books containing suggested student readings. Boldface numbers at the end of the entry indicate sections in which the book is cited. Dates in parentheses immediately after the title indicate the original publication, when different from that of the current edition or publisher. Information on publishers and prices is taken from *Books in Print* (1985–1986) or from recent publishers' announcements. "O/P" means that this information could not be found, so the book is probably out of print.

Alexander, Franz G., and Sheldon T. Selesnick. 1966. *The History of Psychiatry: An Evaluation of Psychiatric Thought and Practice from Prehistoric Times to the Present.* New York: Harper and Row. O/P. **5.1, 5.2**

Allen, Garland E. 1978. *Life Science in the Twentieth Century.* New York: Cambridge Univ. Press. $34.50, pb. $11.95. **4.3–6**

American Institute of Physics, *see* Center for History of Physics.

Anderson, David. 1973. *Discoveries in Physics.* New York: Holt, Rinehart and Winston. O/P. **9.2, 10.1–2, 10.4, 12.1**

Andrade, E. N. daC. N.d. *Rutherford and the Nature of the Atom* (1964). Reprinted. Magnolia, Mass.: Peter Smith. $11.25. **10.1**

Appleman, Philip. 1979. *Darwin, A Norton Critical Edition.* 2d ed. New York: Norton. $24.95, pb. $8.95. **2.2–5**

Aris, Rutherford, H. Ted Davis, and Roger H. Stuewer, eds. 1983. *Springs of Scientific Creativity: Essays on Founders of Modern Science.* Minneapolis: Univ. of Minnesota Press. $32.50. **8.6, 9.6**

Asimov, Isaac. 1980. *The Universe: From Flat Earth to Quasar* (1971). New York: Avon-Discus (1976), pb $3.95. Rev. ed., New York: Walker, $15.95. **13.1, 13.3**

Asquith, Peter, and Thomas Nickles, eds. 1982, 1983. *PSA 1982: Proceedings of the 1982 Biennial Meeting of the Philosophy of Science Association.* East Lansing, Mich.: Philosophy of Science Association. Vol. 1, $21.00, pb $19.00. Vol. 2, $25. **8.6**

Bannister, Robert C. 1979. *Social Darwinism: Science and Myth in American Social Thought.* Philadelphia: Temple Univ. Press. $34.95. **3.3**

Barber, Bernard, and Walter Hirsch, eds. 1978. *The Sociology of Science* (1962). Reprinted. Westport, Conn.: Greenwood Press. $55.00. **11.6**

Barzun, Jacques. 1965. *Race: A Study in Superstition.* Rev. ed. New York: AMS Press. $28.50. **3.1**

Ben-David, Joseph. 1984. *The Scientist's Role in Society: A Comparative Study.* 2d ed. Chicago: Univ. of Chicago Press. $20.00, pb. $8.95. **1.4, 11.6**

Bernal, J. D. 1980. *The Social Function of Science* (1939). Philadelphia: Richard West. $40.00. **11.6**

Berry, Arthur. 1961. *A Short History of Astronomy: From Earliest Times through the Nineteenth Century* (1898). Reprinted. New York: Dover. Pb $8.95. **12.1, 12.2**

Bloor, David. 1976. *Knowledge and Social Imagery.* Boston: Routledge and Kegan Paul. $16.95. **11.6**

Boller, Paul F. 1981. *American Thought in Transition: The Impact of Evolutionary Naturalism 1865–1900* (1969). Reprinted. Lanham, Md.: Univ. Press of America. $23.50, pb $12.00. **3.3**

Boorse, H. A., and L. Motz, eds. 1966. *The World of the Atom.* 2 vols. New York: Basic Books. O/P. **10.4**

Bordeau, Sanford P. 1982. *Volts to Hertz: The Rise of Electricity.* Minneapolis: Burgess. $18.95. **8.1, 8.2**

Bowler, Peter. 1984. *Evolution: The History of an Idea.* Berkeley: Univ. of California Press. $29.95, pb $10.95. **2.1–5, 3.1, 3.3, 4.3, 4.5–7**

Boyer, Carl B. 1985. *A History of Mathematics* (1968). Princeton, N.J.: Princeton Univ. Press. Pb. $12.50. **1.3, 7.5, 11.3**

Brahic, A., ed. 1982. *Formation of Planetary Systems: Proceedings of Summer School, Centre National d'Études Spatiales, August 1980.* Toulouse, France: Cepadues. Fr. 295. **12.4**

Bronowski, J. 1974. *The Ascent of Man.* Boston: Little, Brown. **1.2**

Brush, Stephen G. 1983. *Statistical Physics and the Atomic Theory of Matter: From Boyle and Newton to Landau and Onsager.* Princeton, N.J.: Princeton Univ. Press. $45.00, pb $14.50. **9.7**

———. 1978. *The Temperature of History: Phases of Science and Culture in the 19th Century.* New York: Burt Franklin. $18.95. **3.3**

Burke, John G., ed. 1987. *Science & Culture in the Western Tradition: Sources and Interpretations.* Scottsdale, Ariz.: Gorsuch Scarisbrick. Pb $14.00. **1.2**

Bynum, W. F., et al., eds. 1982. *Dictionary of the History of Science.* Princeton, N.J.: Princeton Univ. Press. $50.00, pb $12.95. **1.3**

Campbell, Jeremy. 1982. *Grammatical Man: Information, Entropy, Language and Life.* New York: Simon and Schuster. $16.95, pb $8.95. **11.4**

Casper, Barry, and Richard J. Noer. 1972. *Revolutions in Physics.* New York: Norton. $20.95. **8.5, 8.6**

Center for History of Physics, American Institute of Physics. 1984. *Moments of Discovery* (Set of Audio Tapes, Slides, Illustrated Scripts and Teachers Guides). New York: American Institute of Physics. $85.00. **10.2, 13.2**

Chant, Colin, and John Fauvel, eds. 1981. *Darwin to Einstein: Historical Studies on Science and Belief* (1980). White Plains, N.Y.: Longman. $27.00. **8.6, 11.4**

Chodorow, M., H. E. Rorschach, and A. L. Schawlow, eds. 1980. *Felix Bloch and Twentieth-Century Physics.* Houston, Tex.: Rice University. $25.00, pb $15.00. **9.7**

Cline, Barbara L. 1987. *Men Who Made a New Physics: Physicists and the Quantum Theory* (1965). Originally published as *The Questioners*. Reprinted. Chicago: Univ. of Chicago Press. **9.1, 9.4–6, 11.2**

Cohen, I. Bernard. 1985. *Birth of a New Physics*. New York: Norton. $17.95. **1.4**

Coleman, William. 1978. *Biology in the Nineteenth Century: Problems of Form, Function, and Transformation*. New York: Cambridge Univ. Press. Pb $11.95. **2.1–3**

Coley, Noel G., and M. D. Hall, eds. 1981. *Darwin to Einstein: Primary Sources on Science and Belief*. New York: Longman. O/P. **2.1, 2.4, 4.3, 4.5, 7.3, 8.6, 11.2, 11.4**

Corsi, Pietro, and Paul Weindling, eds. 1983. *Information Sources in the History of Science and Medicine*. Boston: Butterworth Scientific. $75.00. **11.5–6**

Corsini, R. J., ed. 1984. *Encyclopedia of Psychology*, 4 vols. New York: Wiley. $249.95. **6.3**

Cosmology + 1. 1977. Readings from *Scientific American*, with Introduction by Owen Gingerich. San Francisco: W. H. Freeman. O/P. **13.3**

Cravens, Hamilton. 1978. *The Triumph of Evolution: American Scientists and the Heredity-Environment Controversy, 1900–1941*. Philadelphia: Univ. of Pennsylvania Press. $26.00. **3.3**

Cronin, Vincent. 1981. *The View from Planet Earth*. New York: Morrow. $15. Reprinted (1983). New York: Quill. Pb. $6.95. **2.1**

Curti, Merle. 1980. *Human Nature in American Thought: A History*. Madison, Wisc.: Univ. of Wisconsin Press. $35.00. **5.1–2, 6.1–4, 6.8**

Dampier, W. C. 1965. *A History of Science and Its Relations with Philosophy and Religion* (1929). 4th ed. Reprinted with a Postscript by I. Bernard Cohen. New York: Cambridge Univ. Press. Pb $21.95. **1.4, 2.2, 3.4, 4.3**

Darwin, Charles. 1958. *The Autobiography of Charles Darwin and Selected Letters*, ed. Francis Darwin (1892). Reprinted. New York: Dover. Pb $5.95. Also: Magnolia, Mass.: Peter Smith. $14.00. **2.3**

———. 1975. *The Origin of Species* (1859). Facsimile of the 1st ed. Cambridge, Mass.: Harvard Univ. Press. Pb $8.95. **2.3**

———. 1979. *The Voyage of the Beagle*. Totowa, N.J.: Biblio Distribution Center. $10.95, pb $6.95. Also: 1962. New York: Natural History Press. Pb $6.95. **2.3**

Davis, Philip J., and Reuben Hirsh. 1981. *The Mathematical Experience*. Cambridge: Birkhauser. $27.95. **11.3**

De Beer, Gavin. N.d. *Charles Darwin* (1964). Reprinted. Westport, Conn.: Greenwood Press. $27.50. **2.3**

Delmont, Sara, and Lorne Duffin, eds. 1978. *The Nineteenth-Century Woman: Her Cultural and Physical World*. Totowa, N.J.: Barnes and Noble Books-Imports. $27.50. **4.1**

Dennis, Jack, ed. 1984. *The Nuclear Almanac: Confronting the Atom in War and Peace*. Reading, Mass.: Addison-Wesley. Pb $20.00. **10.2, 10.3**

De Waal Malefijt, Annemarie. 1974. *Images of Man: A History of Anthropological Thought*. New York: Knopf. O/P. **3.1, 3.4, 4.1–2**

Dickson, F. P. 1968. *The Bowl of Night: The Physical Universe and Scientific Thought*. Cambridge, Mass: MIT Press. O/P. **8.5–6, 13.3**

Douglas, Mary, ed. 1982. *Essays in the Sociology of Perception*. Boston: Routledge and Kegan Paul. Pb $17.95. **11.6**

Dubbey, J. M. 1975. *Development of Modern Mathematics* (1972). Reprinted. New York: Crane, Russak. Pb $8.75. **7.5**

Durant, John, ed. 1985. *Darwinism and Divinity: Essays on Evolution and Religious Belief*. New York: Blackwell. $24.95. **2.4, 4.7**

Durbin, Paul, ed. 1980. *A Guide to the Culture of Science, Technology, and Medicine*. New York: Free Press. $65.00. Reprinted with additions, 1984. New York: Free Press. Pb $19.95. **1.3**

Durham, F., and R. D. Purrington. 1983. *The Frame of the Universe: A History of Physical Cosmology*. New York: Columbia Univ. Press. $25.00, pb $12.50. **8.7, 13.3**

Eastwood, Bruce. 1979. *Directory of Audio-Visual Sources: History of Science, Medicine, and Technology*. New York: Watson. $20.00. **1.3**

Einstein, Albert. N.d. *Relativity: The Special and the General Theory* (1917). Reprint of English translation. Magnolia, Mass.: Peter Smith. $13.25. Also: New York: Crown. Pb $3.95. **8.6–7**

Einstein, Albert, and Leopold Infeld. 1967. *The Evolution of Physics* (1938). Reprinted. New York: Simon and Schuster. Pb $9.95. **1.4, 7.2, 8.1, 8.3, 8.5–7, 9.4–6, 11.2**

Eiseley, Loren. 1958. *Darwin's Century*. New York: Doubleday. Pb $6.50. **2.1–5**

Elkana, Yehuda, et al., eds. 1978. *Towards a Metric of Science*. New York: Wiley-Interscience. $38.50. **11.6**

Eves, Howard. 1983. *An Introduction to the History of Mathematics*. 5th ed. Philadelphia: Saunders College Publishing. $39.95. **7.5**

Fancher, Raymond. 1979. *Pioneers of Psychology*. New York: Norton. Pb $9.95. **1.4, 5.1, 6.1–4**

———. 1985. *The Intelligence Men: Makers of the IQ Controversy*. New York: Norton. $17.95. **6.4–6, 6.8**

Ferris, Timothy. 1983. *The Red Limit: The Search for the Edge of the Universe*. Rev. ed. New York: Morrow. Pb $9.70. **13.1, 13.3**

Feyerabend, P. K. 1981. *Philosophical Papers*. 2 vols. New York: Cambridge Univ. Press. $57.50, Pb $14.95 + $44.50, pb $13.95. **11.5**

Forbes, E. G., ed. 1978. *Human Implications of Scientific Advance: Proceedings of the XV International Congress of the History of Science, Edinburgh, 10–15 August 1977*. Edinburgh: Edinburgh Univ. Press. £15. **12.1, 13.1, 13.3**

Frank, Phillipp G. 1975. *Modern Science and Its Philosophy* (1941). Reprinted. Salem, N.H.: Ayer. $27.00. **11.1, 11.5**

Freud, Sigmund. 1985. *On the History of the Psychoanalytic Movement* (1917). New York: Macmillan. Pb $4.95. **5.3**

Fyrth, H. J., and Maurice Goldsmith. 1969. *Science, History and Technology*. Book 1, *A.D. 800 to the 1840's*. Book 2, Part I, *The Age of Confidence: The 1840's to the 1880's*. Book 2, Part II, *The Age of Uncertainty: The 1880's to the 1940's*. Book 2, Part III, *The Age of Choice: The 1940's to the 1960's*. London: Cassell. O/P. **2.2, 4.3, 5.3, 7.2, 8.2, 9.1–2, 10.3**

Garber, Edward D., ed. 1985. *Genetic Perspectives in Biology and Medicine*. Chicago: Univ. of Chicago Press. Pb $12.00. **4.5**

Gilgen, Albert R. 1982. *American Psychology since World War II*. Westport, Conn.: Greenwood Press. $29.95. **5.3, 6.3**

Gillispie, Charles Coulston. 1960. *The Edge of Objectivity*. Princeton, N. J.: Princeton Univ. Press. Pb $12.95. **1.3, 2.2, 2.4**

Gjertsen, Derek. 1984. *The Classics of Science: A Study of Twelve Enduring Scientific Works*. New York: Lilian Barber Press. $24.95. **1.2**

Goldberg, Stanley. 1984. *Understanding Relativity: Origins and Impact of a Scientific Revolution*. Cambridge: Birkhauser. $24.95. **8.5–6**

Gould, Stephen Jay. 1981. *The Mismeasure of Man*. New York: Norton. $14.95, pb $5.95. **3.1–3, 4.1, 6.4–6, 6.8**

Gowing, Margaret, and Lorna Arnold. 1979. *The Atomic Bomb*. Science in a Social Context Series. London: Butterworths. Pb £1.95. **10.2**

Graetzer, H. G., and D. L. Anderson. 1981. *The Discovery of Nuclear Fission* (1971). Reprinted. Salem, N.H.: Ayer. $12.00. **10.2**

Graham, Loren R. 1981. *Between Science and Values*. New York: Columbia Univ. Press. $24.00, pb $12.50. **1.2, 11.4**

Greene, John C. 1959. *The Death of Adam: Evolution and Its Impact on Western Thought*. Ames: Iowa State Univ. Press. Pb $9.95. **2.1–3, 2.5**

———. 1981. *Science, Ideology, and World View*. Berkeley: Univ. of California Press. $19.50, pb $5.95. **1.4, 2.2–4**

Guillemin, Victor. 1968. *The Story of Quantum Mechanics*. New York: Charles Scribner's. O/P. **9.1, 9.5–6, 10.4, 11.2**

Gursky, H., and R. Ruffini, eds. 1975. *Neutron Stars, Black Holes and Binary X-Ray Sources*. Hingham, Mass.: Kluwer Academic. $71.00, pb $34.00. **13.2**

Hacking, Ian, ed. 1981. *Scientific Revolutions*. New York: Oxford Univ. Press. Pb $7.95. **11.5**

Hall, A. R., and M. B. Hall. 1988. *A Brief History of Science* (1964). Reprinted. Ames: Iowa State Univ. Press. Pb $10.95. **1.2**

Hankins, Thomas. 1985. *Science and the Enlightenment*. New York: Cambridge Univ. Press. $29.95, pb $9.95. **1.4**

Hanle, Paul A., and Von Del Chamberlain, eds. 1981. *Space Science Comes of Age: Perspectives in the History of the Space Sciences*. Washington, D.C.: Smithsonian Institution Press. $22.50, pb $12.50. **12.4**

Hanson, Robert W., ed. 1986. *Science and Creation: Geological, Theological and Educational Perspectives*. New York: Macmillan. $24.95. **4.7**

Hays, H. R. 1979. *From Ape to Angel: An Informal History of Social Anthropology* (1958). Reprinted. Westport, Conn.: Greenwood Press. $32.50. **3.1, 3.3–4, 4.1–2**

Henaham, John F., ed. 1975. *The Ascent of Man: Sources and Interpretations*. Boston: Little, Brown. O/P. **1.2**

Hilgard, Ernest R. 1987. *Psychology in America: A Historical Survey*. New York: Harcourt Brace Jovanovich. $34.95. **1.2**

History of Science Society. 1986. *ISIS Guide to the History of Science*. Philadelphia: History of Science Society. **1.3**

Hoffman[n], Banesh. 1983. *Relativity and Its Roots*. New York: Scientific American Books/W. H. Freeman. $16.95, pb $9.95. **8.1, 8.5–7**

———. 1959. *The Strange Story of the Quantum*. 2d. ed. New York: Dover. Pb $4.95. Also: Magnolia, Mass.: Peter Smith. $14.50. **9.5, 9.6**

Holton, Gerald. 1978. *The Scientific Imagination: Case Studies*. New York: Cambridge Univ. Press. $47.50, pb $13.95. **1.3**

———. 1986. *The Advancement of Science, and Its Burdens*. New York: Cambridge Univ. Press. $39.50, pb $11.95. **11.4**

Holton, Gerald, and Stephen G. Brush. 1985. *Introduction to Concepts and*

Theories in Physical Science. 2d ed. (1973). Reprinted Princeton, N.J.: Princeton Univ. Press. Pb $19.95. **8.1, 8.3–7, 9.2–6, 11.2, 11.5**

Holton, Gerald, and Yehuda Elkana, eds. 1982. *Albert Einstein: Historical and Cultural Perspectives.* Princeton, N.J.: Princeton Univ. Press. $40.00, pb $12.50. **11.4**

Hothersall, David. 1984. *History of Psychology.* Philadelphia: Temple Univ. Press. $34.95. **6.1, 6.4–5**

Howson, Colin, ed. 1976. *Method and Appraisal in the Physical Sciences.* New York: Cambridge Univ. Press. $52.50. **1.3, 11.5**

Hynek, J. A., ed. 1951. *Astrophysics.* New York: McGraw-Hill. O/P. **13.2**

Irvine, William. 1983. *Apes, Angels, and Victorians: The Story of Darwin, Huxley, and Evolution* (1955). Reprinted. Lanham, Md.: University Press of America. Pb $18.75. **2.3–4**

James Clerk Maxwell: A Commemoration Volume. 1931. Cambridge: Cambridge Univ. Press, 1931. O/P. **8.3**

Jayawardene, S. A. 1982. *Reference Books for the Historian of Science: A Handlist.* London: Science Musem Library. Pb £2.50. **1.3**

Johanson, Donald C., and M. A. Edey. 1981. *Lucy: The Beginnings of Humankind.* New York: Warner Books. Pb $9.95. **2.5**

Jones, Howard Mumford. 1959. *One Great Society: Humane Learning in the United States.* New York: Harcourt, Brace. O/P. **11.4**

Joravsky, David. 1970. *The Lysenko Affair.* Cambridge, Mass.: Harvard Univ. Press. $20.00. O/P. **2.4**

Kardiner, Abram, and Edward Preble. 1961. *They Studied Man.* Cleveland: World. O/P. **2.5, 3.1, 3.4, 4.2**

Keller, Alex. 1983. *The Infancy of Atomic Physics.* New York: Oxford Univ. Press. $19.95. **9.1–2, 9.4–5**

Kevles, Daniel. 1985. *In the Name of Eugenics: Genetics and the Uses of Human Heredity.* New York: Knopf. $22.95. **3.3**

———. 1978. *The Physicists: The History of a Scientific Community in Modern America.* New York: Knopf. $15.95. Also: New York: Random House, pb $10.95. **10.2–3**

Kippenhahn, Rudolf. 1983. *100 Billion Suns: The Birth, Life and Death of Stars.* New York: Basic Books. $25.00. **13.2**

Klein, Viola. 1972. *The Feminine Character: The History of an Ideology.* Urbana: Univ. of Illinois Press. Pb $5.95. **5.4, 6.7**

Klemke, E. D., R. Hollinger, and A. D. Kline, eds. 1980. *Introductory Readings in the Philosophy of Science.* Buffalo: Prometheus. Pb $14.95. **11.5**

Kline, Morris. 1964. *Mathematics in Western Culture* (1953). Reprinted. New York: Oxford Univ. Press. Pb $12.95. **7.5, 8.4–6**

———. 1980. *Mathematics: The Loss of Certainty.* New York: Oxford Univ. Press. $25.00, pb $9.95. **11.3**

Kneebone, G. T. 1963. *Mathematical Logic and the Foundations of Mathematics: An Introductory Survey.* New York: D. Van Nostrand. O/P. **11.3**

Knight, David M. 1975. *Sources for the History of Science 1660–1914.* Ithaca, N.Y.: Cornell Univ. Press. O/P. **1.3**

Koestler, Arthur. 1972. *The Case of the Midwife Toad.* New York: Random House. Pb $3.50. **2.4**

Königsson, L. K., ed. 1980. *Current Argument on Early Man.* New York: Pergamon Press. $81.00. **2.5**

Kooser, Robert G., and R. Lance Factor. 1983. *Cubes, Eights and Dots: A Student's Guide to the Octet Rule and Its History*. Galesburg, Ill.: Knox College. Not available commercially; for information write to R. Lance Factor, Department of Chemistry, Box 6, Knox College, Galesburg, Ill. 61401. **9.7**

Kramer, Edna E. 1982. *The Nature of Growth of Modern Mathematics* (1970). Reprinted. Princeton, N.J.: Princeton Univ. Press. $48.00, pb $12.50. **7.5**

Kuhn, Thomas S. 1970. *The Structure of Scientific Revolutions* (1962). 2d ed. Chicago: Univ. of Chicago Press. $17.50, pb $5.95. **11.5**

———. 1977. *The Essential Tension: Selected Studies in Scientific Tradition and Change*. Chicago: Univ. of Chicago Press. $25.00, pb $12.00. **1.3**

Lanczos, C. 1970. *Space through the Ages: The Evolution of Geometrical Ideas from Pythagoras to Hilbert and Einstein*. New York: Academic Press. O/P. **8.5**

Laudan, Larry. 1978. *Progress and Its Problems*. Berkeley: Univ. of California Press. $23.00, pb $7.95. **1.2**

Leahey, Thomas H. 1979. *A History of Psychology*. Englewood Cliffs, N.J.: Prentice Hall. $28.95. **5.1, 6.1–3**

Levine, George, and Owen Thomas. 1963. *The Scientist versus the Humanist*. New York: Norton. Pb $3.95. **11.4**

Lewin, Miriam, ed. 1984. *In the Shadow of the Past: Psychology Portrays the Sexes*. New York: Columbia Univ. Press. $32.00. **4.1, 5.4, 6.7**

Lindberg, David C., and Ronald L. Numbers, eds. 1986. *God and Nature: Historical Essays on the Encounter between Christianity and Science*. Berkeley: Univ. of California Press. $50.00, pb $17.95. **1.3**

Lindsay, R. B. 1979. *Early Concepts of Energy in Atomic Physics*. New York: Academic Press. O/P. **9.1, 9.4–5**

Losee, John. 1980. *A Historical Introduction to the Philosophy of Science*. 2d ed. New York: Oxford Univ. Press. Pb. $7.95. **11.1, 11.5**

McKenzie, A. E. E. 1988. *The Major Achievements of Science* (1973). Reprinted. Ames: Iowa State Univ. Press. Pb. $14.95. **1.4, 2.1, 2.3–4, 4.3, 4.5–6, 7.1–4, 8.1, 8.3–7, 9.1–2, 9.4–6, 11.1–2, 13.1**

Magner, Lois. 1979. *A History of the Life Sciences*. New York: Dekker. $34.50. **1.4, 2.1–3, 4.3, 4.5**

Marks, John. 1983. *Science and the Making of the Modern World*. Exeter, N.H.: Heinemann. Pb $20.00. **1.4, 2.1, 2.3–4, 4.3, 4.5, 5.2, 6.1, 7.1–3, 7.5, 8.1–3, 8.5–7, 9.5–7, 10.3, 11.1–2, 11.5,11.6, 12.1, 12.4, 13.1–3**

Mason, Stephen F. 1962. *A History of the Sciences*. New York: Macmillan. Pb $7.95. **1.4, 2.1–4, 3.1, 4.3, 4.5, 7.1–4, 9.1, 13.1–3**

Mayr, Ernst. 1982. *The Growth of Biological Thought*. Cambridge, Mass: Harvard Univ. Press. $30.00, pb $12.95. **4.6**

Mayr, Ernst, and William B. Provine, eds. 1980. *The Evolutionary Synthesis: Perspectives on the Unification of Biology*. Cambridge, Mass.: Harvard Univ. Press. $25.00. **4.6**

Medvedev, Zhores. N.d. *The Rise and Fall of T. D. Lysenko* (1969). Reprinted. Ann Arbor, Mich.: Books on Demand. Pb $76.00. **2.4**

Menard, Henry. 1971. *Science: Growth and Change*. Cambridge, Mass.: Harvard Univ. Press. $12.50. **11.6**

Mendelssohn, K. 1977. *The Quest for Absolute Zero: The Meaning of Low Temperature Physics*. New ed. London: Taylor and Francis. Pb £5.00. **9.7**

Merton, Robert K. 1968. *Social Theory and Social Structure.* New York: Free Press. $24.95. **11.6**

———. 1973. *The Sociology of Science.* Chicago: Univ. of Chicago Press. $30.00. **11.6**

Miller, Jonathan. 1982. *Darwin for Beginners.* New York: Pantheon (Random House). Pb $4.95. **2.3**

Nelkowski, H., A. Hermann, H. Poser, R. Schrader, and R. Seiler, eds. *Einstein Symposion Berlin aus Anlass der 100. Wiederkehr seines Geburtstages 25. bis März 1979* (Lecture Notes in Physics, 100). New York: Springer-Verlag. Pb $29.20. **8.7**

Oldroyd, David. 1983. *Darwinian Impacts.* 2d. ed. Atlantic Highlands, N.J.: Humanities Press International. Pb $14.00. **2.2–4, 3.1, 3.3, 4.3**

Olson, Richard, ed. 1971. *Science as Metaphor: The Historical Role of Scientific Theories in Forming Western Culture.* Belmont, Calif.: Wadsworth. O/P. **2.4, 5.2, 5.3, 6.3, 8.7, 11.4**

Open University. 1976. *History of Mathematics* (course AM 289). Milton Keynes, England: Open Univ. Press. O/P. **7.5**

———. 1974. *Science and Belief: From Copernicus to Darwin* (course AMST 183). Milton Keynes, England: Open Univ. Press. O/P. **1.4, 2.1–4**

———. 1986. *Science and Belief: From Darwin to Einstein* (course A381) (1981). Philadelphia: Taylor and Francis. $130. Seven booklets sold separately, contact publisher for price. **1.4, 2.1, 2.4, 4.3, 4.7, 7.2–3, 8.5, 9.6, 11.2, 11.4**

———. 1973. *Science and the Rise of Technology since 1800* (course AST 281). Milton Keynes, England: Open Univ. Press. **7.1, 8.1–2, 8.4, 9.1**

Pagels, Heinz. 1982. *The Cosmic Code: Quantum Physics as the Law of Nature.* New York: Simon and Schuster. $16.95. **8.5–7, 9.5–6, 10.4, 11.2**

Pannekoek, A. 1961. *A History of Astronomy.* New York: Rowman. $19.50. O/P. **13.1, 13.2**

Poliakov, Leon. 1974. *The Aryan Myth: A History of Racist and Nationalist Ideas in Europe.* New York: Basic Books. O/P. **3.1**

Popper, Karl. 1968. *Conjectures and Refutations: The Growth of Scientific Knowledge* (1962). Reprinted. New York: Harper and Row. Pb $8.95. **5.5**

Portugal, Franklin H., and Jack S. Cohen. 1977. *A Century of DNA.* Cambridge: MIT Press. $35.00, pb $7.95. **4.3, 4.5**

Price, Derek J. de Solla. 1963. *Little Science, Big Science.* New York: Columbia Univ. Press. O/P. **1.3, 11.6**

Price, W. C., S. S. Chissick, and T. Ravensdale, eds. 1973. *Wave Mechanics: The First Fifty Years.* New York: Wiley. O/P. **9.7**

Project Physics Course. 1975. New York: Holt, Rinehart and Winston. O/P. **10.1**

Purcell, Edward A., Jr. 1973. *The Crisis of Democratic Theory: Scientific Naturalism and the Problem of Value.* Lexington: Univ. Press of Kentucky. Pb $10.00. **11.4**

Quen, Jacques M., and Eric T. Carlson, eds. 1978. *American Psychoanalysis: Origins and Development.* New York: Brunner/Mazel. O/P. **5.3**

Reingold, Nathan, ed. 1979. *The Sciences in the American Context: New Perspectives.* Washington, D.C.: Smithsonian Institution Press. Pb $9.95. **1.3, 4.5**

Robinson, Paul. 1976. *The Modernization of Sex: Havelock Ellis, Alfred Kinsey, William Masters and Virginia Johnson.* New York: Harper and Row. O/P. **4.1**

Rosenberg, Charles. 1976. *No Other Gods: On Science and American Social Thought.* Baltimore: Johns Hopkins Univ. Press. Pb $6.95. **3.3, 4.1, 4.5**

Ross, Dale. 1972. *Studies in Biology.* London: Hulton Educational Publications. O/P. **2.2**

Rucker, Rudy. 1982. *Infinity and the Mind: The Science and Philosophy of the Infinite.* Cambridge, Mass.: Birkhauser. $19.95. **11.3**

Ruse, Michael. 1979. *The Darwinian Revolution: Science Red in Tooth and Claw.* Chicago: Univ. of Chicago Press. $20.00, pb $10.95. **2.1–4**

Schneer, Cecil J. 1984. *The Evolution of Physical Science: Major Ideas from Earliest Times to the Present* (1960). Lanham, Md.: University Press of America. Pb $12.75. **1.4, 2.1, 7.1–5, 8.1, 8.3, 8.5–7, 9.1**

———. 1988. *Mind and Matter: Man's Changing Concepts of the Material World* (1969). Reprinted. Ames: Iowa State Univ. Press. Pb. $9.95. **7.1–2**

Schrödinger, Erwin. 1951. *Science and Humanism: Physics in Our Time.* New York: Cambridge Univ. Press, 1951. O/P. **11.4**

Schultz, Duane. 1981. *A History of Modern Psychology.* 3d ed. New York: Academic Press. $21.75. **1.2**

Segre, Emilio. 1980. *From X-Rays to Quarks.* San Francisco: Freeman. Pb $12.95. **9.1–6, 10.1–2, 10.4**

———. 1984. *From Falling Bodies to Radio Waves.* New York: Freeman. Pb $13.95. **7.2–3, 8.1, 8.3–5**

Sharlin, Harold I. 1966. *The Convergent Century: The Unification of Science in the Nineteenth Century.* New York: Abelard-Schumann. O/P. **7.1–2, 8.3, 9.1**

———. 1963. *The Making of the Electrical Age.* New York: Abelard-Schuman. O/P. **8.1–4**

Singer, Charles. 1959. *A History of Biology to about the Year 1900.* 3d ed. New York: Abelard-Schuman. O/P. **2.2**

Singh, J. 1970. *Great Ideas and Theories of Modern Cosmology.* New York: Dover. Pb $6.75. **8.7, 12.4, 13.2**

Skinner, B. F. 1972. *Cumulative Record: A Selection of Papers.* 3d ed. New York: Appleton-Century-Crofts. O/P. **6.3**

———. 1965. *Science and Human Behavior.* New York: Free Press. Pb $9.95. **6.3**

Snow, C. P. 1969. *The Two Cultures and the Scientific Revolution* (1960). Reprinted with additional material as *Two Cultures and a Second Look.* New York: Cambridge Univ. Press. $19.95, pb $5.95. **11.4**

Solomon, Joan. 1974. *The Structure of Matter: The Growth of Man's Ideas on the Nature of Matter.* New York: Wiley/Halsted. O/P. **9.1–2, 9.4, 10.2–3**

———. 1974. *The Structure of Space: The Growth of Man's Ideas on the Nature of Forces, Fields, and Waves.* New York: Wiley/Halsted. O/P. **7.2, 8.1, 8.3–7, 13.1, 13.3**

Steele, David, ed. 1970. *The History of Scientific Ideas: A Teachers' Guide.* London: Hutchinson Educational. O/P. **1.3**

Stocking, George W., Jr. 1982. *Race, Culture, and Evolution* (1971). Chicago: Univ. of Chicago Press. Pb $12.00. **3.4**

Stone, Irving. 1980. *The Origin: A Biographical Novel of Charles Darwin.* Garden City, N.Y.: Doubleday. $17.95. Also: New York: New American Library/Signet Books. Pb $4.95. **2.3**

Struve, Otto and Velta Zebergs. 1962. *Astronomy of the 20th Century.* New York: Macmillan. O/P. **13.1–2**

Suppe, Frederick, ed. 1977. *The Structure of Scientific Theories.* Urbana: Univ. of Illinois Press. $35.00, pb $14.50. **1.3**

Swenson, Lloyd S. 1979. *Genesis of Relativity: Einstein in Context.* New York: Burt Franklin. $29.00. **8.3–7.**

Tauber, G. E. 1979. *Man's View of the Universe.* New York: Crown. $19.95. **12.1–3, 13.1–3**

Thomson, Robert. 1968. *The Pelican History of Psychology.* Baltimore: Penguin Books. O/P. **5.1–4, 6.1–5**

Toulmin, Stephen, ed. 1970. *Physical Reality.* New York: Harper and Row. O/P. **11.2**

Toulmin, Stephen, and June Goodfield. 1982. *The Architecture of Matter* (1962). Reprinted. Chicago: Univ. of Chicago Press. Pb $10.95. **1.4, 2.2, 4.3, 7.1–3**

———. N.d. *The Discovery of Time* (1965). Reprinted. New York: Octagon Books. $23.00. Also: Chicago: Univ. of Chicago Press. Pb $8.95. **1.4, 2.1–4**

———. 1961. *The Fabric of the Heavens.* New York: Harper and Row. Pb $7.95. **1.4**

Trenn, Thaddeus J. 1981. *Transmutation: Natural and Artificial.* New York: Wiley. $29.95, pb $19.95. **9.2, 10.1–2**

Trigg, George L. 1975. *Crucial Experiments in Modern Physics.* New York: Crane, Russak. Pb $7.95. **9.2, 9.4, 9.6**

———. 1975. *Landmark Experiments in Twentieth Century Physics.* New York: Crane, Russak. Pb $12.50. **10.2, 10.4**

Turner, F. M. 1974. *Between Science and Religion: The Reaction to Scientific Naturalism in Late Victorian England.* New Haven: Yale Univ. Press. O/P. **2.4**

Vicinus, Martha, ed. 1972. *Suffer and Be Still: Women in the Victorian Age.* Bloomington: Indiana Univ. Press. $18.50, pb $9.95. **4.1**

Watson, Robert. 1978. *The Great Psychologists.* 4th ed. New York: Harper and Row. Pb $19.95. **5.2–3, 6.1–2, 6.4**

Weiner, Charles, ed. 1977. *History of Twentieth Century Physics: Proceedings of the International School of Physics "Enrico Fermi," Course LVII, Varenna, 1972.* New York: Academic Press. O/P. **9.4–5**

Wendt, Herbert. 1973. *In Search of Adam: The Story of Man's Quest for the Truth about His Earliest Ancestors* (1956). Reprinted. Westport, Conn.: Greenwood Press. $39.75. **2.2–5**

Wertheimer, Michael. 1979. *A Brief History of Psychology.* 2d ed. New York: Holt, Rinehart and Winston. Pb $16.95. **1.2**

White, G. K. 1980. *Changing Views of the Physical World 1954–1979.* Canberra: Australian Academy of Science. O/P. **11.3**

Whiteside, Haven. 1971. *Elementary Particles.* New York: Holt, Rinehart and Winston. O/P. **10.4**

Whitney, Charles. 1988. *The Discovery of Our Galaxy* (1971). Reprinted. Ames: Iowa State Univ. Press. Pb. 9.95. **2.1, 13.1**

Wiener, Philip P., ed. 1980. *Dictionary of the History of Ideas* (1973). 5 vols. Reprinted. New York: Scribner. Pb $75.00. **11.5, 13.3**

Williams, L. Pearce. 1980. *The Origins of Field Theory* (1966). Reprinted. Lanham, Md.: Univ. Press of America. $19.00, pb $8.25. **7.2, 8.1, 8.3**

Williams, L. Pearce, ed. 1968. *Relativity Theory: Its Origins and Impact on Modern Thought.* New York: Wiley. O/P. **11.4**

Williams, L. Pearce, and Henry John Steffens. 1978. *A History of Science in*

Western Civilization. Vol. 3, *Modern Science 1700–1900.* Lanham, Md.: Univ. Press of America. Pb $15.50. **2.2, 7.1, 7.2**

Wilson, David B., ed. 1983. *Did the Devil Make Darwin Do It? Modern Perspectives on the Creation-Evolution Controversy.* Ames: Iowa State Univ. Press. $25.00, pb $12.95. **2.4**

Woodward, W. R., and M. G. Ash, eds. 1982. *The Problematic Science: Psychology in Nineteenth Century Thought.* New York: Praeger. $41.95. **5.1**

Zetterberg, J. Peter, ed. 1983. *Evolution versus Creationism: The Public Education Controversy.* Phoenix, Ariz.: Oryx Press. $37.50. **4.7**

SUBJECT INDEX

This index contains only entries that are major topics for the sections indicated.